Y0-AAF-622

Rings That Are Nearly Associative

This is a volume in
PURE AND APPLIED MATHEMATICS

A Series of Monographs and Textbooks

Editors: SAMUEL EILENBERG AND HYMAN BASS

A list of recent titles in this series appears at the end of this volume.

RINGS THAT ARE NEARLY ASSOCIATIVE

K. A. ZHEVLAKOV
A. M. SLIN'KO
I. P. SHESTAKOV
A. I. SHIRSHOV

Institute of Mathematics
Academy of Sciences of the U. S. S. R.,
Siberian Branch
Novosibirsk, U. S. S. R.

TRANSLATED BY

Harry F. Smith

Department of Mathematics
Iowa State University
Ames, Iowa

1982

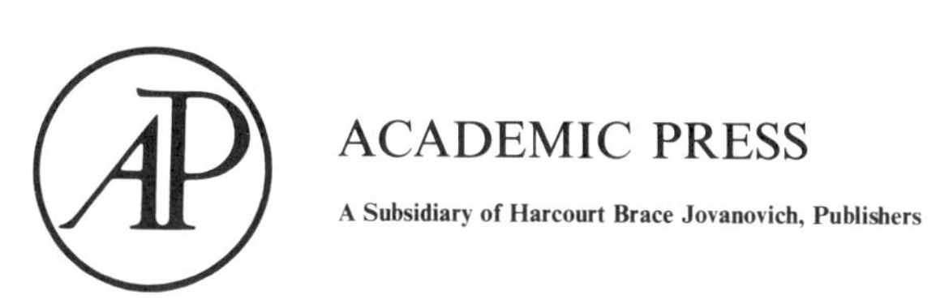

New York London
Paris San Diego San Francisco São Paulo Sydney Tokyo Toronto

ACADEMIC PRESS, INC.
111 Fifth Avenue, New York, New York 10003

United Kingdom Edition published by
ACADEMIC PRESS, INC. (LONDON) LTD.
24/28 Oval Road, London NW1 7DX

Library of Congress Cataloging in Publication Data

Main entry under title:

Rings that are nearly associative.

(Pure and applied mathematics)
Translation of: Kolʹt͡sa blizkie k assot͡siativnym.
Bibliography: p.
Includes index.
1. Associative rings. I. Zhevlakov, Konstantin Aleksandrovich. II. Title. III. Series.
QA3.P8 [QA251.5] 510s [512'.4] 82-4065
ISBN 0-12-779850-1 AACR2

RINGS THAT ARE NEARLY ASSOCIATIVE
Translated from the original Russian edition entitled

КОЛЬЦА, БЛИЗКИЕ К АССОЦИАТИВНЫМ

Published by Principal Editorship of Physical-Mathematical Literature of the Publisher "Nauka" Moscow, 1978.

PRINTED IN THE UNITED STATES OF AMERICA

82 83 84 85 9 8 7 6 5 4 3 2 1

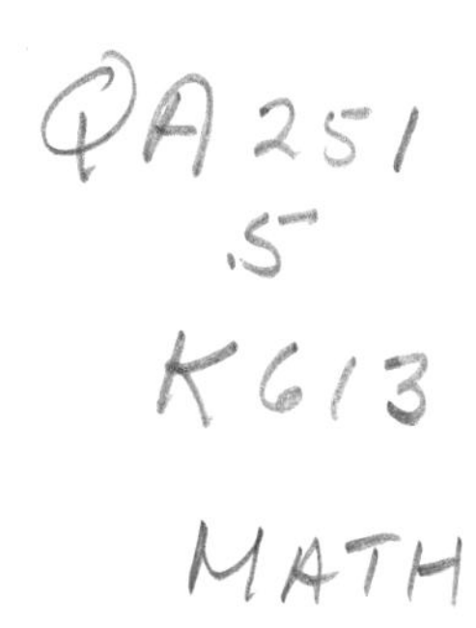

Contents

Preface

Until thirty years into this century the theory of rings developed mainly as the theory of associative rings. However, even in the middle of the last century there arose mathematical systems that satisfied all the axioms of a ring except associativity. For example, the algebra of Cayley numbers, which was constructed in 1845 by the English mathematician Arthur Cayley, is such a system. An identity found by Degen in 1818, which represents the product of two sums of eight squares again in the form of a sum of eight squares, was rediscovered by means of this algebra. The algebra of Cayley numbers is an eight-dimensional division algebra over the field of real numbers, which satisfies the following weakened identities of associativity: $(aa)b = a(ab)$, $(ab)b = a(bb)$.

Algebras satisfying these two identities were subsequently named alternative. Such a name stems from the fact that in every algebra satisfying the two indicated identities, the associator $(x, y, z) = (xy)z - x(yz)$ is an alternating (skew-symmetric) function of its arguments. The theory of alternative algebras attracted the serious attention of mathematicians after the discovery of its deep connection with the theory of projective planes, which was actively developed at the beginning of this century. In this regard, it was discovered that alternative algebras are "sufficiently near" to associative ones. The essence of this nearness is exhibited by the theorem of Artin, which asserts that in every alternative algebra the subalgebra generated by any two elements is associative.

Another important class of nonassociative algebras is Jordan algebras, which are defined by the identities $xy = yx$, $(x^2y)x = x^2(yx)$. They arose in the work of the German physicist Jordan, that dealt with the axiomatization of the foundations of quantum mechanics. Jordan algebras are connected with associative algebras in the following manner.

Let A be an associative algebra over a field of characteristic $\neq 2$. We define on the vector space of the algebra A a new operation of multiplication, $a \odot b = \frac{1}{2}(ab + ba)$, and denote the algebra obtained by $A^{(+)}$. This algebra is Jordan. If a subspace J of the algebra A is closed with respect to the operation $a \odot b$, then J together with this operation forms a subalgebra of the algebra $A^{(+)}$, and consequently is a Jordan algebra.

A similar connection exists between associative algebras and Lie algebras, which are defined by the identities $x^2 = 0$, $(xy)z + (zx)y + (yz)x = 0$. If there is defined on an associative algebra A a new multiplication given by the formula $[a, b] = ab - ba$, then the obtained algebra $A^{(-)}$ is a Lie algebra. For Lie algebras over fields there is also a connection in the reverse direction. By the Poincaré–Birkhoff–Witt theorem, every Lie algebra over an arbitrary field is a subalgebra of an algebra $A^{(-)}$ for some suitable associative algebra A.

The situation is different for the theory of Jordan algebras. Over any field there exist Jordan algebras that are not subalgebras of algebras $A^{(+)}$ for any associative algebra A. If Jordan algebra J is a subalgebra of an algebra $A^{(+)}$, where A is an associative algebra, then J is called a special Jordan algebra.

A connection also exists in the reverse direction for Jordan algebras, although it is not as strong as in the case of Lie algebras. The basic result in this direction, namely, that any Jordan algebra with two generators is special, is due to one of the authors.

The theories of alternative, Lie, and Jordan algebras, which are closely interlaced among themselves, have extensive contacts with different areas of mathematics, and together with the highly developed theory of associative rings comprise the basic nucleus of the modern theory of rings. Algebras from these three classes are combined under the general title "algebras that are nearly associative." This term was introduced in a survey article by A. I. Shirshov [279], where an extensive program for further investigations was outlined.

Naturally, we could not include the whole theory of rings in our book. The title of the book itself indicates that the theory of associative rings is excluded. However, for reading of the basic text, only a knowledge of such classical material as the structure of Artinian rings or the theory of the Jacobson radical is required of the reader, and this can be read in any book on the theory of associative rings. In addition, our book completely avoids Lie algebras, to which several books in the Russian language are already devoted.

We present in the most detail the theory of alternative algebras, with which the reader previously could become acquainted only through journals. There exist some omissions in our presentation of the theory of Jordan algebras. Many of the results not included are contained in the monograph of Jacobson [47], with which we have tried not to intersect. We have made our accent on the achievements of the last decade.

The first chapter of the book contains reference material. In it are found collections of basic facts on varieties of algebras and linearizations of identities. It is recommended that the reader turn to it as needed. It is also not necessary to read the remaining chapters in succession. We recommend two courses to the reader:

"alternative algebras"—2, 3.1, 3.2, 5–13;

"Jordan algebras"—2–4, 5.1–5.4, 8.1, 8.2, 14, 15;

(indicating only the numbers of chapters and sections). The last chapter, which is devoted to right alternative algebras, contains a basic survey.

Almost every section of the book is provided with exercises, in which it is often suggested that the reader prove a theorem of entirely independent value. In these cases detailed hints are given. The exercises are not used anywhere in the basic text.

In each chapter the numbering of theorems, lemmas, and formulas is its own. In references to theorems (formulas) of another chapter, the number of the appropriate chapter is written before the number of the theorem (formula).

A list of literature, which is not claimed to be complete, is found at the end of the book. The surveys and monographs cited in that list are starred.

Throughout the book there is material used in special courses, which was repeatedly presented by A. I. Shirshov at MGU and NGU,† and also used in special courses presented in various years at NGU by the other authors.

The authors thank Yu. A. Medvedev and V. G. Skosyrskiy for their great assistance in putting together the bibliography.

† MGU and NGU are the transliterated initials for Moscow State University and Novosibirsk State University, respectively (Translator).

CHAPTER **1**

Varieties of Algebras

1. FREE ALGEBRAS

Let Φ be an associative and commutative ring with identity element 1. A set A is called an *algebra* over the ring Φ if there is defined on A the structure of a unital Φ-module and an operation of multiplication connected with the module operations by the relations

$$(a + b)c = ac + bc, \qquad a(b + c) = ab + ac,$$
$$\alpha(ab) = (\alpha a)b = a(\alpha b),$$

for $a, b, c \in A$ and $\alpha \in \Phi$. We shall call algebras over a ring Φ simply *Φ-algebras*. Sometimes they are also called *Φ-operator rings*. The ring Φ is called a *ring of scalars* or *ring of operators*. If Φ is a field, then every customary algebra over the field Φ is obviously a Φ-algebra. Every ring is an algebra over the ring of integers $\boldsymbol{Z}$. Thus the concept of an algebra over a ring combines the concepts of a ring and an algebra over a field.

Let A and B be two Φ-algebras. A mapping $\phi: A \to B$ which is a homomorphism of the Φ-modules of these algebras with

$$\phi(ab) = \phi(a)\phi(b) \qquad \text{for} \quad a, b \in A$$

is called a *homomorphism* of the Φ-algebra A into the Φ-algebra B. Homomorphisms of Φ-algebras correspond to Φ-admissible ideals, i.e., ideals invariant under multiplication by elements from Φ.

We fix an arbitrary set $X = \{x_\alpha\}$ and add to it two more symbols, left and right parentheses, to obtain a set $X^* = X \cup \{(,)\}$. Consider all possible finite sequences of elements from the set X^*. Two sequences $a_1 a_2 \ldots a_m$ and $b_1 b_2 \ldots b_n$, where $a_i, b_j \in X^*$, are considered equal if $m = n$ and $a_i = b_i$ for $i = 1, 2, \ldots, m$. We define inductively a set $V[X]$ of those sequences of elements from the set X^* which we shall call *nonassociative words* from elements of the set X. First, all elements of the set X belong to the set $V[X]$. Second, if $x_1, x_2 \in X$ and $u, v \in V[X] \backslash X$, then the sequences $x_1 x_2$, $x_1(u)$, $(v)x_2$, and $(u)(v)$ also belong to the set $V[X]$. No other sequences are contained in $V[X]$. For example, the sequence $(x_1(x_2 x_3))x_4$ is a nonassociative word from elements of the set $X = \{x_1, x_2, x_3, x_4\}$, but the sequence $(x_1(x_2 x_3)x_4)$ is not. The number of elements from the set X which appear in a word v is called the *length of the nonassociative word* v, and it is denoted by $d(v)$.

PROPOSITION 1. Let v be a nonassociative word from elements of some set. Then

(1) the number of left parentheses in v equals the number of right parentheses;

(2) in any initial subsequence of the word v, the number of left parentheses is not less than the number of right parentheses.

The proof is carried out by an obvious induction on the length of the word v.

We define on the set $V[X]$ a binary operation, denoted by $\cdot$, according to the following rules. Let $x_1, x_2 \in X$ and $u, v \in V[X] \backslash X$. We set

$$x_1 \cdot x_2 = x_1 x_2, \qquad x_1 \cdot u = x_1(u),$$
$$v \cdot x_2 = (v)x_2, \qquad u \cdot v = (u)(v).$$

PROPOSITION 2. Every nonassociative word v with $d(v) \geq 2$ has a unique representation in the form of a product of two nonassociative words of lesser length.

PROOF. The existence of such a representation follows directly from the definition. Suppose that the word v has two different representations, $v = u \cdot w = u' \cdot w'$. Consider the case when the words u, w, u', w' all have length greater than 1. Then

$$v = (u)(w) = (u')(w'),$$

and also $d(u) + d(w) = d(u') + d(w')$. If $d(u) = d(u')$, then it is obvious $u = u'$, $w = w'$. If $d(u) > d(u')$, then the subsequence $u')$ is an initial subsequence of the word u, which by the first part of Proposition 1 has more right parentheses than left. But this contradicts the second part of that same proposition. The analysis of the remaining cases is even simpler. This proves the proposition.

We now consider the free unital Φ-module $\Phi[X]$ from the set of generators $V[X]$, and we extend the operation of multiplication defined on $V[X]$ to $\Phi[X]$ by the rule

$$\left(\sum_i \alpha_i u_i\right) \cdot \left(\sum_j \beta_j v_j\right) = \sum_{i,j} \alpha_i \beta_j (u_i \cdot v_j),$$

where $\alpha_i, \beta_j \in \Phi$ and $u_i, v_j \in V[X]$. We have obtained a Φ-algebra $\Phi[X]$ which is called the *free algebra* over the ring Φ from set of generators X. Free algebras possess the following universal property.

THEOREM 1. Let A be an arbitrary Φ-algebra and θ be some mapping of X into A. Then θ extends uniquely to a homomorphism of the algebra $\Phi[X]$ into A.

PROOF. We first extend θ to $V[X]$. Suppose that θ is already defined on the set of all nonassociative words of length less than n, and let the length of the word w equal n. By Proposition 2, w is uniquely representable in the form of a product of two words of lesser length, $w = u \cdot v$. The elements $\theta(u)$ and $\theta(v)$ are already defined, so therefore we set $\theta(w) = \theta(u) \cdot \theta(v)$. This mapping is well defined since u and v are uniquely determined by the word w. We now extend θ to $\Phi[X]$ by setting

$$\theta\left(\sum_i \alpha_i u_i\right) = \sum_i \alpha_i \theta(u_i)$$

for all $\alpha_i \in \Phi$ and $u_i \in V[X]$. This can be done since $\Phi[X]$ is a free Φ-module. It is not difficult now to see that the indicated mapping θ is a homomorphism of $\Phi[X]$ into A. Its uniqueness is clear. This proves the theorem.

The elements of the algebra $\Phi[X]$ are called *nonassociative polynomials* from elements of the set X. A nonassociative polynomial of the form αv, where $\alpha \in \Phi$ and $v \in V[X]$, is called a *nonassociative monomial.* The length of the word v is called the *degree of the monomial.* The maximum of the degrees of the monomials whose sum forms a polynomial is called the *degree of the polynomial.*

2. VARIETIES. ALGEBRAS FREE IN A VARIETY

We fix a countable set of symbols $X = \{x_1, x_2, \ldots\}$. Let f be an arbitrary element from $\Phi[X]$. There appear in its list only a finite number of elements from X, for example $x_1, x_2, \ldots, x_n$. In this case we shall write $f = f(x_1, x_2, \ldots, x_n)$.

Let A be some Φ-algebra and $a_1, a_2, \ldots, a_n \in A$. By Theorem 1 there exists a unique homomorphism $\theta: \Phi[X] \rightarrow A$ which sends x_i to a_i for $i = 1,$

$2, \ldots, n$, and sends the other elements of the set X to zero. We shall denote the image of the element f under the homomorphism θ by $f(a_1, a_2, \ldots, a_n)$, and say that the element $f(a_1, a_2, \ldots, a_n)$ is obtained by substitution of the elements $a_1, a_2, \ldots, a_n$ in the nonassociative polynomial $f(x_1, x_2, \ldots, x_n)$.

A nonassociative polynomial $f = f(x_1, x_2, \ldots, x_n) \in \Phi[X]$ is called an *identity* of the algebra A if $f(a_1, a_2, \ldots, a_n) = 0$ for any $a_1, a_2, \ldots, a_n \in A$. We also say that A satisfies the identity f or that the identity f is valid in A. The collection of all identities for a given algebra is an ideal of the algebra $\Phi[X]$, which is called the *identities ideal* (*T-ideal*) of the algebra A and is denoted $T(A)$. The collection of all identities which are satisfied by each algebra from some class of algebras $\mathfrak{M}$ is also an ideal in $\Phi[X]$. It is called the *identities ideal* (*T-ideal*) of the class of algebras $\mathfrak{M}$ and is denoted $T(\mathfrak{M})$.

Let I be a subset in $\Phi[X]$. The class $\mathfrak{M}$ of all Φ-algebras satisfying each identity from I is called the *variety* of Φ-algebras defined by the set of identities I. I is called a set of *defining identities* for the variety $\mathfrak{M}$. For example, all associative Φ-algebras form a variety with the one defining identity $f = (x_1x_2)x_3 - x_1(x_2x_3)$. The class of all Lie Φ-algebras is also a variety, whose defining identities are $f_1 = x_1^2$, $f_2 = (x_1x_2)x_3 + (x_2x_3)x_1 + (x_3x_1)x_2$. The identities $f_1 = x_1^2x_2 - x_1(x_1x_2)$ and $f_2 = x_1x_2^2 - (x_1x_2)x_2$ define the variety of alternative algebras. Individually these identities define the varieties of left alternative and right alternative algebras, respectively. The variety of Jordan algebras is defined by the identities $f_1 = x_1x_2 - x_2x_1$, $f_2 = (x_1^2x_2)x_1 - x_1^2(x_2x_1)$.

The variety consisting of only the zero algebra is called *trivial*.

Let $\mathfrak{M}$ be a nontrivial variety of Φ-algebras and F be an algebra from $\mathfrak{M}$ with set of generators Y. The algebra F is called *free in the variety* $\mathfrak{M}$ (*or* $\mathfrak{M}$-*free*) *with set of free generators* Y if every mapping of the set Y into an arbitrary algebra A from $\mathfrak{M}$ can be extended uniquely to a homomorphism of the algebra F into A.

Let I be some set of polynomials from $\Phi[X]$. We shall denote by $I(A)$ the ideal of the Φ-algebra A generated by all possible elements of the form $f(a_1, a_2, \ldots, a_n)$, where $f \in I$ and $a_1, a_2, \ldots, a_n \in A$.

THEOREM 2. Let $\mathfrak{M}$ be a nontrivial variety of Φ-algebras with system of defining identities I. Then for any set Y the restriction to Y of the canonical homomorphism $\sigma: \Phi[Y] \to \Phi[Y]/I(\Phi[Y])$ is injective and the algebra $\Phi_{\mathfrak{M}}[Y^\sigma] = \Phi[Y]/I(\Phi[Y])$ is free in the variety $\mathfrak{M}$ with set of free generators Y^σ. Any two free algebras in $\mathfrak{M}$ with equivalent sets of free generators are isomorphic.

PROOF. The algebra $\Phi_{\mathfrak{M}}[Y^\sigma]$ is generated by the set Y^σ and belongs to $\mathfrak{M}$ since it satisfies all identities from I. We shall show that it is free in $\mathfrak{M}$ with set of free generators Y^σ. Let $A \in \mathfrak{M}$ and $\tau: Y^\sigma \to A$ be some mapping of

Y^σ into A. Then by Theorem 1 the mapping $\sigma \circ \tau: Y \to A$ extends to a homomorphism $\eta: \Phi[Y] \to A$).[†] But the ideal $I(\Phi[Y])$ is contained in the kernel of any homomorphism of the algebra $\Phi[Y]$ into an algebra of the variety $\mathfrak{M}$, and so in particular it is in the kernel of the homomorphism η. Therefore there exists a homomorphism $\phi: \Phi[Y]/I(\Phi[Y]) \to A$ which makes the following diagram commutative:

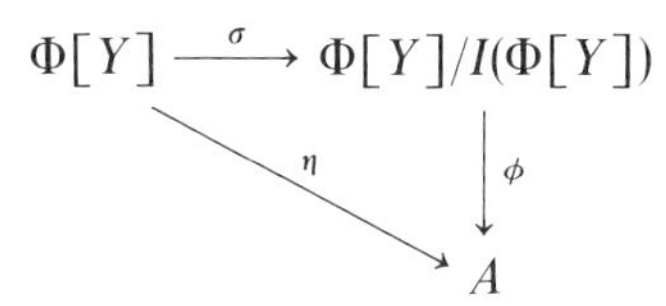

that is, such that $\sigma \circ \phi = \eta$. If $y \in Y$, then $\phi(y^\sigma) = y^{\sigma \circ \phi} = y^\eta = y^{\sigma \circ \tau} = \tau(y^\sigma)$, and this means ϕ extends the mapping τ. Consequently ϕ is the desired homomorphism. Its uniqueness is clear.

Thus $\Phi_{\mathfrak{M}}[Y^\sigma]$ is a free algebra in the variety $\mathfrak{M}$, and Y^σ is the set of its free generators. We shall prove that the restriction of σ to Y is injective. Let $y_1 \neq y_2$ but $y_1^\sigma = y_2^\sigma$. We consider a nonzero algebra A from $\mathfrak{M}$ and a nonzero element a from A. There exists a homomorphism $\tau: \Phi[Y] \to A$ such that $y_1^\tau = a$, $y_2^\tau = 0$. Since $I(\Phi[Y])$ is contained in the kernel of τ, there exists a homomorphism $\phi: \Phi_{\mathfrak{M}}[Y^\sigma] \to A$ such that $\tau = \sigma \circ \phi$. But now, since $y_1^\sigma = y_2^\sigma$, we have $a = y_1^\tau = (y_1^\sigma)^\phi = (y_2^\sigma)^\phi = y_2^\tau = 0$. This contradiction proves the injectivity of the restriction of σ to Y.

We have proved that Y^σ is a set of free generators for the algebra $\Phi_{\mathfrak{M}}[Y^\sigma]$ which is equivalent to Y. By means of a standard procedure we can now construct a free algebra $\Phi_{\mathfrak{M}}[Y]$ with set of free generators Y.

Now suppose the sets Y and Y' are equivalent. We shall prove that the algebras $\Phi_{\mathfrak{M}}[Y]$ and $\Phi_{\mathfrak{M}}[Y']$ are isomorphic. Consider a bijection $\sigma: Y \to Y'$. By what has already been proved, there exists a unique homomorphism $\tau: \Phi_{\mathfrak{M}}[Y] \to \Phi_{\mathfrak{M}}[Y']$ extending σ. Consider also the bijection $\sigma^{-1}: Y' \to Y$ and the homomorphism $\tau': \Phi_{\mathfrak{M}}[Y'] \to \Phi_{\mathfrak{M}}[Y]$ extending σ^{-1}. Then $\tau \circ \tau'$ homomorphically maps the algebra $\Phi_{\mathfrak{M}}[Y]$ to itself, and the restriction of $\tau \circ \tau'$ to Y is the identity mapping. By what has been proved, however, there is only one homomorphism of the algebra $\Phi_{\mathfrak{M}}[Y]$ into $\Phi_{\mathfrak{M}}[Y]$ that is the identity on Y. That homomorphism is the identity mapping, which means $\tau \circ \tau' = \mathrm{id}$. Analogously we have $\tau' \circ \tau = \mathrm{id}$. Consequently τ is an isomorphism.

This proves the theorem.

As is easy to note, the varieties of associative, alternative, right alternative, and Jordan algebras are nontrivial (e.g., they all contain the algebras with zero multiplication). Therefore they possess free algebras from any set of generators Y, which we shall denote by Ass$[Y]$, Alt$[Y]$, RA$[Y]$, and J$[Y]$, respectively.

[†] Here and subsequently $(\phi \circ \psi)(x) = \psi(\phi(x))$.

COROLLARY. If I is a system of defining identities for the variety $\mathfrak{M}$, then $T(\mathfrak{M}) = I(\Phi[X])$.

PROOF. The ideal $T(\mathfrak{M})$ is contained in the kernel of any homomorphism of the algebra $\Phi[X]$ onto an algebra from $\mathfrak{M}$. In particular, it is in the kernel of the canonical homomorphism $\sigma:\Phi[X] \to \Phi_{\mathfrak{M}}[X]$, which equals $I(\Phi[X])$. Since the reverse containment is clear, this proves the corollary.

REMARK 1. In the study of algebras from a specific variety $\mathfrak{M}$ we shall often call an element $f \in \Phi_{\mathfrak{M}}[X]$ an identity of an algebra A from $\mathfrak{M}$, meaning by this that some inverse image (and consequently all inverse images) in $\Phi[X]$ of the element f under the canonical homomorphism of $\Phi[X]$ onto $\Phi_{\mathfrak{M}}[X]$ is an identity of the algebra A.

3. HOMOGENEOUS IDENTITIES AND HOMOGENEOUS VARIETIES

Every nonassociative polynomial $f \in \Phi[X]$ can be uniquely decomposed into an irreducible sum of monomials. We shall say that a monomial αv, where $\alpha \in \Phi$ and $v \in V[X]$, has *type* $[n_1, n_2, \ldots, n_k]$ if the nonassociative word v contains x_i exactly n_i times, and also $n_k \neq 0$ but $n_j = 0$ for $j > k$. For example, the monomial $((x_1x_2)x_2)(x_1x_4)$ has type $[2, 2, 0, 1]$. We shall call the number n_i the *degree of the monomial* αv *in* x_i. If all the monomials in an irreducible listing of the polynomial f have one and the same degree n_i in x_i, then we call the polynomial f *homogeneous in* x_i of degree n_i. Monomials having one and the same type are said to be of the *same-type*. A nonassociative polynomial (identity of an algebra) is called *homogeneous* if all its monomials in an irreducible listing are of the same-type. In other words, a polynomial f is homogeneous if it is homogeneous in each variable. For example, the polynomial $(x_1^2x_2)x_4 + ((x_1x_4)x_2)x_1 - x_1^2(x_2x_4) - (x_1x_4)(x_2x_1)$ is homogeneous since each of its monomials has type $[2, 1, 0, 1]$. The polynomial $x_1^2x_2 - (x_1x_3)x_1 + x_1^2$ is homogeneous in x_1 of degree 2; however, it is not homogeneous. A homogeneous polynomial of type $[n_1, n_2, \ldots, n_k]$, where all $n_i \leq 1$, is called *multilinear*.

Since the free algebra $\Phi[X]$ is a free Φ-module with set $V[X]$ of free generators, it has a decomposition into a direct sum of submodules $\Phi^{[n_1, \ldots, n_k]}[X]$ consisting of the homogeneous polynomials of type $[n_1, \ldots, n_k]$

$$\Phi[X] = \bigoplus \Phi^{[n_1, \ldots, n_k]}[X].$$

The *degree of a nonassociative polynomial in* x_i is defined as the maximum degree in x_i of its monomials.

Let $f \in \Phi[X]$ be an arbitrary nonassociative polynomial. We group together the same type monomials of the polynomial f. Then f is represented in the form of a sum of homogeneous polynomials. These homogeneous polynomials are called the *homogeneous components* of the polynomial f.

THEOREM 3. Let $f \in \Phi[X]$ be an identity of a Φ-algebra A of degree k_i in x_i. Set $k = \max k_i$. Suppose that in Φ there are $k+1$ elements $\alpha_1, \alpha_2, \ldots, \alpha_{k+1}$ such that for the Vandermonde determinant

$$d = \begin{vmatrix} 1 & \alpha_1 & \alpha_1^2 & \cdots & \alpha_1^k \\ 1 & \alpha_2 & \alpha_2^2 & \cdots & \alpha_2^k \\ \vdots & \vdots & \vdots & \vdots & \vdots \\ 1 & \alpha_{k+1} & \alpha_{k+1}^2 & \cdots & \alpha_{k+1}^k \end{vmatrix} = \prod_{1 \le i < j \le k+1} (\alpha_j - \alpha_i)$$

the equation $d \cdot a = 0$ implies $a = 0$ for any $a \in A$. Then every homogeneous component of the identity f is an identity of the algebra A.

PROOF. First we represent f in the form $f = f_0 + f_1 + \cdots + f_{k_1}$, where f_i is the sum of all monomials of the polynomial f having degree i in x_1. Let $f = f(x_1, x_2, \ldots, x_n)$ and $a_1, a_2, \ldots, a_n$ be arbitrary elements of the algebra A. For brevity we shall denote $f_i(a_1, a_2, \ldots, a_n)$ by $f_i(a)$. For $i = 1, 2, \ldots, k_1 + 1$ we have the relation

$$f_0(a) + \alpha_i f_1(a) + \alpha_i^2 f_2(a) + \cdots + \alpha_i^{k_1} f_{k_1}(a) = f(\alpha_i a_1, a_2, \ldots, a_n) = 0,$$

whence it follows (see S. Lang, "Algebra," p. 376, Mir, 1968) that $d_1 f_j(a) = 0$, $j = 0, 1, \ldots, k_1$, where

$$d_1 = \begin{vmatrix} 1 & \alpha_1 & \alpha_1^2 & \cdots & \alpha_1^{k_1} \\ 1 & \alpha_2 & \alpha_2^2 & \cdots & \alpha_2^{k_1} \\ \vdots & \vdots & \vdots & \vdots & \vdots \\ 1 & \alpha_{k_1+1} & \alpha_{k_1+1}^2 & \cdots & \alpha_{k_1+1}^{k_1} \end{vmatrix}.$$

Since d_1 is a divisor of d, from $d_1 f_j(a) = 0$ it follows that $f_j(a) = 0$. Since $a_1, a_2, \ldots, a_n$ were arbitrary, this means that $f_0, f_1, \ldots, f_{k_1}$ are identities of the algebra A. Performing the same operation with the polynomials $f_j, j = 0, 1, 2, \ldots, k_1$, and also with the variable x_2 etc., we finally prove the theorem.

The conditions of the theorem are automatically satisfied in each of the following cases:

(1) there exist $\alpha_1, \alpha_2, \ldots, \alpha_{k+1} \in \Phi$ such that the Vandermonde determinant $d = \prod_{i<j}(\alpha_j - \alpha_i)$ is invertible in Φ;

(2) the algebra A is a free Φ-module and there exist elements $\alpha_1, \alpha_2, \ldots, \alpha_{k+1} \in \Phi$ such that $d = \prod_{i<j}(\alpha_j - \alpha_i) \neq 0$;

(3) Φ is a field containing more than k elements.

COROLLARY 1. Let f be an identity of the Φ-algebra A of degree k_i in x_i and $k = \max k_i$. Then in any of the cases (1)–(3) every homogeneous component of f is an identity of the algebra A.

A variety $\mathfrak{M}$ is called *homogeneous* if for every $f \in T(\mathfrak{M})$ all homogeneous components of f also belong to $T(\mathfrak{M})$.

COROLLARY 2. Every variety of algebras over an infinite field is homogeneous.

We note that for a homogeneous variety $\mathfrak{M}$

$$T(\mathfrak{M}) = \bigoplus_{[n_1, \ldots, n_k]} (\Phi^{[n_1, \ldots, n_k]}[X] \cap T(\mathfrak{M})).$$

Therefore the decomposition of the algebra $\Phi[X]$ into a direct sum of the subspaces $\Phi^{[n_1, \ldots, n_k]}[X]$ also induces an analogous decomposition for the free algebra $\Phi_{\mathfrak{M}}[X]$ of a variety $\mathfrak{M}$, which is isomorphic to the quotient algebra $\Phi[X]/T(\mathfrak{M})$:

$$\Phi_{\mathfrak{M}}[X] = \bigoplus_{[n_1, \ldots, n_k]} \Phi_{\mathfrak{M}}^{[n_1, \ldots, n_k]}[X],$$

where

$$\Phi_{\mathfrak{M}}^{[n_1, \ldots, n_k]}[X] = \Phi^{[n_1, \ldots, n_k]}[X]/\Phi^{[n_1, \ldots, n_k]}[X] \cap T(\mathfrak{M}).$$

Therefore for elements of the free algebra $\Phi_{\mathfrak{M}}[X]$ of a variety $\mathfrak{M}$ the concepts of homogeneity, degree, and degree in each of the variables can be reasonably defined.

Later on we shall need the following criterion for homogeneity of a variety.

LEMMA 1. For the homogeneity of a variety $\mathfrak{M}$ it is necessary and sufficient that $T(\mathfrak{M})$ as an ideal of the algebra $\Phi[X]$ has a system of homogeneous generating elements.

PROOF. If $\mathfrak{M}$ is homogeneous, then the collection of all homogeneous components of elements of $T(\mathfrak{M})$ is contained in $T(\mathfrak{M})$ and generates this ideal even as a Φ-module. If $T(\mathfrak{M})$ is generated as an ideal by a set of homogeneous polynomials $\{f_\alpha\}$, then $T(\mathfrak{M})$ is generated as a Φ-module by the homogeneous elements of the form

$$v(u_1, \ldots, u_k, f_\alpha, u_{k+1}, \ldots, u_m),$$

where $v, u_1, \ldots, u_m$ are monomials from $\Phi[X]$. The presence in $T(\mathfrak{M})$ of a system of homogeneous generators as a Φ-module implies the homogeneity of the variety $\mathfrak{M}$. This proves the lemma.

4. PARTIAL LINEARIZATIONS OF IDENTITIES

We fix some element $y \in \Phi[X]$ and define a family $\Delta(y) = \{\Delta_i^k(y) \mid i = 1, 2, \ldots; k = 0, 1, 2, \ldots\}$ of linear mappings of the algebra $\Phi[X]$ into itself, letting

(1) $\Delta_i^0(y) = \mathrm{id}$—the identity mapping;
(2) $x_s \Delta_i^k(y) = 0$ if $k > 1$ or $k = 1$, $i \neq s$;
(3) $x_i \Delta_i^1(y) = y$;
(4) $(u \cdot v)\Delta_i^k(y) = \sum_{r+s=k}(u\Delta_i^r(y)) \cdot (v\Delta_i^s(y))$;

where $x_j \in X$ and u, v are arbitrary monomials from $\Phi[X]$. As is easy to see, conditions (1)–(4) together with the condition of linearity uniquely define the action of the mappings $\Delta_i^k(y)$ on the algebra $\Phi[X]$.

LEMMA 2. Let $v = v(x_1, \ldots, x_n)$ be a multilinear monomial. Then

$$v(f_1, \ldots, f_n)\Delta_i^k(y) = \sum_{k_1 + \cdots + k_n = k} v(f_1\Delta_i^{k_1}(y), \ldots, f_n\Delta_i^{k_n}(y))$$

for any $f_1, f_2, \ldots, f_n \in \Phi[X]$.

PROOF. For brevity we shall denote $\Delta_i^k(y)$ simply by Δ_i^k and $v(f_1, \ldots, f_n)$ by $v(f)$. We carry out an induction on n. For $n = 1$ everything is clear. Now let $n > 1$, so that $v = v_1 \cdot v_2$ where $d(v_1) = m < n$, $d(v_2) = l < n$. From the definition of the operator Δ_i^k and the induction assumption, we obtain

$$\begin{aligned} v(f)\Delta_i^k &= (v_1(f) \cdot v_2(f))\Delta_i^k = \sum_{r+s=k}(v_1(f)\Delta_i^r)(v_2(f)\Delta_i^s) \\ &= \sum_{r+s=k}\left(\sum_{r_1+\cdots+r_m=r} v_1(f_1\Delta_i^{r_1}, \ldots, f_m\Delta_i^{r_m})\right) \\ &\quad \times \left(\sum_{s_1+\cdots+s_l} v_2(f_{m+1}\Delta_i^{s_1}, \ldots, f_n\Delta_i^{s_l})\right) \\ &= \sum_{k_1+\cdots+k_n=k} v_1(f_1\Delta_i^{k_1}, \ldots, f_m\Delta_i^{k_m})v_2(f_{m+1}\Delta_i^{k_{m+1}}, \ldots, f_n\Delta_i^{k_n}) \\ &= \sum_{k_1+\cdots+k_n=k} v(f_1\Delta_i^{k_1}, \ldots, f_n\Delta_i^{k_n}). \end{aligned}$$

This proves the lemma.

COROLLARY. Let $u = u(x_1, \ldots, x_n)$ be an arbitrary monomial from $\Phi[X]$ of degree m in x_i. Then for $k \leq m$

$$u\Delta_i^k(y) = u_1 + u_2 + \cdots + u_{\binom{m}{k}},$$

where

$$u_1, u_2, \ldots, u_{\binom{m}{k}},$$

are all the possible monomials obtainable from u by substituting y for k occurrences of x_i, and for $k > m$

$$u\,\Delta_i^k(y) = 0.$$

For the proof, it suffices to consider the multilinear monomial

$$v(x_{11}, \ldots, x_{1k_1}, \ldots, x_{n1}, \ldots, x_{nk_n}),$$

for which $u(x_1, \ldots, x_n) = v(x_1, \ldots, x_1, \ldots, x_n, \ldots, x_n)$, and to use Lemma 2 as well as condition (2) in the definition of the operator $\Delta_i^k(y)$.

We list some examples:

$$\begin{aligned}
&[(x_1^2x_2)x_1]\,\Delta_1^1(y) = (x_1^2x_2)y + [(x_1y)x_2]x_1 + [(yx_1)x_2]x_1,\\
&[(x_1^2x_2)x_1]\,\Delta_1^2(y) = [(x_1y)x_2]y + [(yx_1)x_2]y + (y^2x_2)x_1,\\
&[(x_1^2x_2)x_1]\,\Delta_1^3(y) = (y^2x_2)y,\\
&[(x_1^2x_2)x_1]\,\Delta_1^4(y) = 0.
\end{aligned}$$

Note that the symbol i in the notation $\Delta_i^k(y)$ is needed only as an indication of the generator x_i on which the operator $\Delta_i^k(y)$ acts in a nontrivial fashion. If the elements of the set X are not equipped with indices, we shall write the generating element itself in place of the index. For example,

$$[(x^2z)t]\,\Delta_x^1(y) = [(xy)z]t + [(yx)z]t.$$

LEMMA 3. Let $f = f(x_1, \ldots, x_n)$ be a polynomial from $\Phi[X]$ of degree m in x_i. Then

$$f(x_1, \ldots, x_i + y, \ldots, x_n) = \sum_{k=0}^{m} f\,\Delta_i^k(y).$$

PROOF. In view of the corollary to Lemma 2 and the linearity of the operator $\Delta_i^k(y)$, it suffices to consider the case when f is a monomial. We carry out an induction on the degree of f. If $d(f) = 1$, then everything is clear. Now let $d(f) > 1$, so that $f = f_1 \cdot f_2$ where $d(f_i) < d(f)$, and let f_1 have degree m_1 in x_i, f_2 have degree m_2 in x_i. By the induction assumption and by the definition of the operator $\Delta_i^k(y)$, we obtain

$$\begin{aligned}
f(x_1, \ldots, x_i + y, \ldots, x_n) &= f_1(x_1, \ldots, x_i + y, \ldots, x_n)f_2(x_1, \ldots, x_i + y, \ldots, x_n)\\
&= \left[\sum_{k=0}^{m_1} f_1\,\Delta_i^k(y)\right]\left[\sum_{r=0}^{m_2} f_2\,\Delta_i^r(y)\right]\\
&= \sum_{j=0}^{m}\left[\sum_{k+r=j} (f_1\,\Delta_i^k(y))(f_2\,\Delta_i^r(y))\right] = \sum_{j=0}^{m} f\Delta_i^j(y).
\end{aligned}$$

This proves the lemma.

Consider some nonassociative polynomial $f = f(x_1, \ldots, x_n) \in \Phi[X]$, and let $x_j \in X\backslash\{x_1, \ldots, x_n\}$. We shall call the polynomials $f_i^{(k)}(x_1, \ldots, x_n; x_j) = f\Delta_i^k(x_j)$ the *partial linearizations of the polynomial f in x_i of degree k*. Let the polynomial f have degree s in x_i. In view of the corollary to Lemma 2, the polynomial $f_i^{(k)}(x_1, \ldots, x_n; x_j)$ is homogeneous in x_j of degree k for $k \leq s$ and equals zero for $k > s$. If f is a homogeneous polynomial of type $[k_1, \ldots, k_i, \ldots, k_n]$, then the polynomial $f_i^{(k)}(x_1, \ldots, x_n; x_{n+1})$ is also homogeneous of type $[k_1, \ldots, k_i - k, \ldots, k_n, k]$ (for $k \leq k_i$). We note here that for any $y \in \Phi[X]$ the element $f\Delta_i^k(y)$ is obtained by substituting the element y in place of the variable x_j in the polynomial $f_i^{(k)}(x_1, \ldots, x_n; x_j)$.

PROPOSITION 3. If a variety $\mathfrak{M}$ is homogeneous, then for any $f \in T(\mathfrak{M})$ and for any $y \in \Phi[X]$ the following containment is valid:

$$f\Delta(y) = \{f\Delta_i^k(y) \mid i = 1, 2, \ldots; k = 0, 1, 2, \ldots\} \subseteq T(\mathfrak{M}).$$

PROOF. We note first of all that if $f(x_1, \ldots, x_n) \in T(\mathfrak{M})$, then also $f(y_1, \ldots, y_n) \in T(\mathfrak{M})$ for any elements $y_1, \ldots, y_n \in \Phi[X]$. Therefore for the proof of the containment $f\Delta_i^k(y) \in T(\mathfrak{M})$, it suffices to prove that $f_i^{(k)}(x_1, \ldots, x_n; x_j) \in T(\mathfrak{M})$, where $x_j \in X\backslash\{x_1, \ldots, x_n\}$. By Lemma 3 we have $T(\mathfrak{M}) \ni f(x_1, \ldots, x_i + x_j, \ldots, x_n) \sum_{k=0}^{n_i} f\Delta_i^k(x_j)$, where n_i is the degree of f in x_i. As was already noted, each of the polynomials $f\Delta_i^k(x_j)$ is homogeneous in x_j of degree k. In view of the homogeneity of $\mathfrak{M}$, it thus follows that $f\Delta_i^k(x_j) = f_i^{(k)}(x_1, \ldots, x_n; x_j) \in T(\mathfrak{M})$, and hence also $f\Delta_i^k(y) = f_i^{(k)}(x_1, \ldots, x_n; y) \in T(\mathfrak{M})$. This proves the proposition.

It is easy to see that the converse assertion is also valid: if $f\Delta(y) \subseteq T(\mathfrak{M})$ for any $f \in T(\mathfrak{M})$ and $y \in \Phi[X]$, then the variety $\mathfrak{M}$ is homogeneous. We show now that, in fact, for the homogeneity of a variety $\mathfrak{M}$, the realization of a much weaker condition is sufficient.

Let f be a nonassociative polynomial. We denote by $f\Delta$ the set of all polynomials from $\Phi[X]$ which can be obtained from f by means of consecutive partial linearizations:

$$f\Delta = \{g \in \Phi[X] \mid g = f\Delta_{i_1}^{j_1}(x_{k_1}) \cdots \Delta_{i_s}^{j_s}(x_{k_s})\}.$$

THEOREM 4. Let a variety $\mathfrak{M}$ have a system I of defining identities such that $f\Delta \subseteq T(\mathfrak{M})$ for all $f \in I$. Then $\mathfrak{M}$ is homogeneous.

PROOF. We note first of all that we can assume without loss of generality the system I consists of homogeneous polynomials. In fact, let $f = f(x_1, \ldots, x_n) \in I$ and $f = f_1 + f_2$ where f_1 is homogeneous in x_1 of degree m and f_2 has degree less than m in x_1. Further, let x_j be a generator not appearing in the listing of f. Then we have $f\Delta_1^m(x_j) = f_1\Delta_1^m(x_j) + f_2\Delta_1^m(x_j) = f_1(x_j, x_2, \ldots, x_n) \in f\Delta$, and furthermore $f_1(x_j, x_2, \ldots, x_n)\Delta_j^m(x_1) = f_1(x_1, x_2, \ldots, x_n) = f_1 \in f\Delta \subseteq T(\mathfrak{M})$, i.e., $f_1, f_2 \in T(\mathfrak{M})$. We show that $f_1\Delta \cup f_2\Delta \subseteq T(\mathfrak{M})$. Since

$f_1 \in f\Delta$, it is clear that $f_1\Delta \subseteq f\Delta \subseteq T(\mathfrak{M})$. Now let $g \in f_2\Delta$. Then $g = f_2\Delta_1 \cdots \Delta_k$, where $\Delta_i = \Delta_{t_i}^{r_i}(x_{s_i})$. We have $g = (f - f_1)\,\Delta_1 \cdots \Delta_k = f\Delta_1 \cdots \Delta_k - f_1\,\Delta_1 \cdots \Delta_k \in T(\mathfrak{M})$. Thus we can replace the polynomial f in the system I by the polynomials f_1 and f_2. Continuing this process, we arrive at a homogeneous system I' which satisfies the condition of the theorem. According to Lemma 1 the homogeneity of the variety $\mathfrak{M}$ is equivalent to the ideal $T(\mathfrak{M})$ having a system of homogeneous generators. By the corollary to Theorem 2 this ideal is generated by elements of the form $f(g_1, g_2, \ldots, g_n)$, where $f \in I$ and $g_1, g_2, \ldots, g_n \in \Phi[X]$. We decompose the polynomials $g_1, \ldots, g_n$ into monomials, $g_i = \sum_{k=1}^{s_i} g_{ik}$. Then by Lemma 3 the element $h = f(g_1, \ldots, g_n)$ is representable in the form

$$h = \sum_m f_m(g_{11}, \ldots, g_{1s_1}, \ldots, g_{n1}, \ldots, g_{ns_n}),$$

where $f_m = f_m(x_{11}, \ldots, x_{1s_1}, \ldots, x_{n1}, \ldots, x_{ns_n}) \in f\Delta$. By the homogeneity of f, all of the elements f_m are homogeneous. In addition, by assumption they are in $T(\mathfrak{M})$. But then all the elements

$$f_m(g_{11}, \ldots, g_{1s_1}, \ldots, g_{n1}, \ldots, g_{ns_n})$$

are also homogeneous and belong to $T(\mathfrak{M})$. Consequently, the collection of these elements is the desired system of homogeneous generators for the ideal $T(\mathfrak{M})$.

This proves the theorem.

We now clarify under what sort of conditions the partial linearizations of some identity are themselves identities.

THEOREM 5. Let $f \in \Phi[X]$ be an identity of a Φ-algebra A, which is homogeneous in x_i of degree k. Suppose there are $k - 1$ elements $\alpha_1, \alpha_2, \ldots, \alpha_{k-1}$ in Φ such that for the determinant

$$d = \begin{vmatrix} \alpha_1 & \alpha_1^2 & \cdots & \alpha_1^{k-1} \\ \alpha_2 & \alpha_2^2 & \cdots & \alpha_2^{k-1} \\ \vdots & \vdots & & \vdots \\ \alpha_{k-1} & \alpha_{k-1}^2 & \cdots & \alpha_{k-1}^{k-1} \end{vmatrix} = \left(\prod_{s=1}^{k-1} \alpha_s\right)\left(\prod_{i<j} (\alpha_j - \alpha_i)\right)$$

the equation $d \cdot a = 0$ implies $a = 0$ for any $a \in A$. Then all partial linearizations in x_i of the polynomial f are identities of the algebra A.

PROOF. Let $f = f(x_1, \ldots, x_n)$ and $a_1, a_2, \ldots, a_n, a_j$ be arbitrary elements of the algebra A. As in the proof of Theorem 3, for brevity we denote the elements $f_i^{(m)}(a_1, \ldots, a_n; a_j)$ by $f_i^{(m)}(a)$. By Lemma 3 we have

$$0 = f(a_1, \ldots, a_{i-1}, a_i + a_j, a_{i+1}, \ldots, a_n) = \sum_{s=0}^{k} f_i^{(s)}(a) = \sum_{s=1}^{k-1} f_i^{(s)}(a),$$

since

$$f_i^{(0)}(a) = f(a_1, \ldots, a_{i-1}, a_i, a_{i+1}, \ldots, a_n) = 0$$

and

$$f_i^{(k)}(a) = f(a_1, \ldots, a_{i-1}, a_j, a_{i+1}, \ldots, a_n) = 0.$$

Substituting successively into the obtained equation the elements $\alpha_1 a_j$, $\alpha_2 a_j, \ldots, \alpha_{k-1} a_j$ in place of a_j, we obtain

$$\begin{gathered}
\alpha_1 f_i^{(1)}(a) + \alpha_1^2 f_i^{(2)}(a) + \cdots + \alpha_1^{k-1} f_i^{(k-1)}(a) = 0, \\
\alpha_2 f_i^{(1)}(a) + \alpha_2^2 f_i^{(2)}(a) + \cdots + \alpha_2^{k-1} f_i^{(k-1)}(a) = 0, \\
\vdots \\
\alpha_{k-1} f_i^{(1)}(a) + \alpha_{k-1}^2 f_i^{(2)}(a) + \cdots + \alpha_{k-1}^{k-1} f_i^{(k-1)}(a) = 0,
\end{gathered}$$

whence it follows that $d \cdot f_i^{(s)}(a) = 0$ for $1 \leq s \leq k-1$. This means $f_i^{(s)}(a) = 0$, which, since the elements $a_1, \ldots, a_n, a_j \in A$ are arbitrary, proves the validity of the identity $f_i^{(s)}(x_1, x_2, \ldots, x_n; x_j)$ in A. This proves the theorem.

COROLLARY. If a variety $\mathfrak{M}$ of Φ-algebras is defined by a system I of homogeneous identities of degree $\leq k$ in each of the variables and there are $k-1$ elements $\alpha_1, \alpha_2, \ldots, \alpha_{k-1}$ in Φ such that for $d = (\prod_{s=1}^{k-1} \alpha_s)(\prod_{i<j}(\alpha_j - \alpha_i))$, any algebra $A \in \mathfrak{M}$, and $a \in A$ the equation $d \cdot a = 0$ implies $a = 0$, then the variety $\mathfrak{M}$ is homogeneous. In particular, every variety defined by homogeneous identities of degree ≤ 2 in each of the variables is homogeneous.

PROOF. If there can be found nonzero elements $\alpha_1, \alpha_2, \ldots, \alpha_{k-1}$ in Φ, then by Theorem 5 we shall have $f\,\Delta \subseteq T(\mathfrak{M})$ for each $f \in I$. Hence by Theorem 4 the variety $\mathfrak{M}$ is homogeneous. If all the elements in I have degree not greater than 2 in each of the variables, then we can take the identity element 1 as α_1 and apply the first part of the corollary.

The varieties of associative, alternative, and right alternative algebras are defined by homogeneous identities of degree not greater than 2, and therefore they are homogeneous. The defining identities for the variety of Jordan algebras are also homogeneous, but one of them has degree 3 in x_1. This is the reason that for some rings of operators Φ, for example the field $\boldsymbol{Z}_2$ of two elements, the variety of Jordan algebras is not homogeneous. If the equation $2x = 1$ is solvable in Φ, however, then the situation changes radically. In fact, if we set $\alpha_1 = 1$ and as α_2 take the solution of this equation (we denote it by $\frac{1}{2}$), then for $d = 1 \cdot \frac{1}{2} \cdot (1 - \frac{1}{2}) = (\frac{1}{2})^2$ the implication $da = 0 \Rightarrow a = 0$ holds, since $4da = a$. By the corollary to Theorem 5, in this case we conclude that the variety of Jordan algebras is homogeneous.

THEOREM 6. Let A be an algebra over a ring Φ which belongs to a homogeneous variety $\mathfrak{M}$ and let B be an associative–commutative Φ-algebra. Then the Φ-algebra $C = B \otimes_\Phi A$ also belongs to the variety $\mathfrak{M}$.

PROOF. It suffices to show that C satisfies all homogeneous identities from $T(\mathfrak{M})$. Let $f(x_1, x_2, \ldots, x_n)$ be such an identity. Then for any $c_1, c_2, \ldots, c_n \in C$ the element $f(c_1, c_2, \ldots, c_n)$ is representable in the form of a sum of elements of the form $g(b_1 \otimes a_1, \ldots, b_s \otimes a_s)$, $a_i \in A$, $b_j \in B$, where g is a homogeneous polynomial from $f\Delta \subseteq T(\mathfrak{M})$. If the polynomial $g(x_1, x_2, \ldots, x_s)$ has type $[i_1, i_2, \ldots, i_s]$, then

$$g(b_1 \otimes a_1, \ldots, b_s \otimes a_s) = b_1^{i_1} b_2^{i_2} \cdots b_s^{i_s} \otimes g(a_1, a_2, \ldots, a_s) = 0.$$

But this means that also $f(c_1, c_2, \ldots, c_n) = 0$. This proves the theorem.

5. MULTILINEAR IDENTITIES. COMPLETE LINEARIZATION OF IDENTITIES

In this section we shall prove the existence of some identity in an algebra A also implies the existence of some multilinear identity in the algebra A.

We introduce the following notation: a roof $\hat{}$ over x_i in a sum $x_1 + \cdots + \hat{x}_i + \cdots + x_n$ denotes that x_i appears in the sum with zero coefficient, that is,

$$x_1 + \cdots + \hat{x}_i + \cdots + x_n = \sum_{k=1}^{n} x_k - x_i.$$

For example, $x_1 + \hat{x}_2 + x_3 + \hat{x}_4 + x_5 = x_1 + x_3 + x_5$.

Let $f = f(x_1, x_2, \ldots, x_n)$ be some nonassociative polynomial from $\Phi[X]$ and $y_1, y_2, \ldots, y_k \in X \setminus \{x_1, x_2, \ldots, x_n\}$. For each $i = 1, 2, \ldots, n$ we define a nonassociative polynomial $f\boldsymbol{L}_i^k$ by the formula

$$\begin{aligned} f\boldsymbol{L}_i^k&(x_1, \ldots, x_{i-1}, y_1, y_2, \ldots, y_k, x_{i+1}, \ldots, x_n) \\ &= f(x_1, \ldots, x_{i-1}, y_1 + y_2 + \cdots + y_k, x_{i+1}, \ldots, x_n) \\ &\quad - \sum_{q=1}^{k} f(x_1, \ldots, x_{i-1}, y_1 + \cdots + \hat{y}_q + \cdots + y_k, x_{i+1}, \ldots, x_n) \\ &\quad + \sum_{1 \le q_1 \le q_2 \le k} f(x_1, \ldots, x_{i-1}, y_1 + \cdots + \hat{y}_{q_1} + \cdots + \hat{y}_{q_2} + \cdots \\ &\quad + y_k, x_{i+1}, \ldots, x_n) \\ &\quad - \cdots + (-1)^{k-1} \sum_{q=1}^{k} f(x_1, \ldots, x_{i-1}, y_q, x_{i+1}, \ldots, x_n). \end{aligned}$$

The following is obvious.

PROPOSITION 4. Let P be a subsemigroup and Q be a subgroup of the additive group of the Φ-algebra A. Then if for a nonassociative polynomial $f = f(x_1, x_2, \ldots, x_n)$ and any elements $a_1, a_2, \ldots, a_n \in P$ we have $f(a_1, a_2, \ldots, a_n) \in Q$, then for any $a_1, \ldots, a_{i-1}, b_1, b_2, \ldots, b_k, a_{i+1}, \ldots, a_n \in P$ we shall also have $f\boldsymbol{L}_i^k(a_1, \ldots, a_{i-1}, b_1, b_2, \ldots, b_k, a_{i+1}, \ldots, a_n) \in Q$. In particular, if $f \in T(A)$, then also $f\boldsymbol{L}_i^k \in T(A)$.

Thus the operator $\boldsymbol{L}_i^k$ transforms an identity of the algebra A into an identity again of the algebra A. In order to elucidate the mechanics of its operation, we first prove a lemma.

LEMMA 4. Let $g: A \times A \times \cdots \times A \to A$ be a function on n variables, which is defined for a Φ-algebra A and is linear in each argument. Then for any $a_1, a_2, \ldots, a_k \in A$, where $k \geq n$,

$$
\begin{aligned}
&g(a_1 + a_2 + \cdots + a_k, \ldots, a_1 + a_2 + \cdots + a_k) \\
&\quad - \sum_{q=1}^{k} g(a_1 + \cdots + \hat{a}_q + \cdots + a_k, \ldots, a_1 + \cdots + \hat{a}_q + \cdots + a_k) \\
&\quad + \sum_{1 \leq q_1 < q_2 \leq k} g(a_1 + \cdots + \hat{a}_{q_1} + \cdots + \hat{a}_{q_2} + \cdots + a_k, \ldots, a_1 + \cdots \\
&\quad + \hat{a}_{q_1} + \cdots + \hat{a}_{q_2} + \cdots + a_k) \\
&\quad + \cdots + (-1)^{k-1} \sum_{q=1}^{k} g(a_q, a_q, \ldots, a_q) \\
&= \begin{cases} \displaystyle\sum_{(i_1 \cdots i_n) \in S_n} g(a_{i_1}, a_{i_2}, \ldots, a_{i_n}), & \text{if } k = n, \\ 0, & \text{if } k > n. \end{cases}
\end{aligned}
$$

PROOF. Let $k \geq n$. Using the linearity of the function g, we remove all $+$ signs from within the arguments of this function. Then the left side of the equality to be proved is itself represented as a linear combination of elements of the form $g(a_{j_1}, a_{j_2}, \ldots, a_{j_n})$ with integer coefficients. If there are s different indices among $j_1, j_2, \ldots, j_n$ and $s < k$, then the coefficient for $g(a_{j_1}, a_{j_2}, \ldots, a_{j_n})$ equals the alternating sum

$$1 - \binom{k-s}{1} + \binom{k-s}{2} - \cdots + (-1)^{k-s}\binom{k-s}{k-s},$$

which equals zero. We note that s does not exceed n, and k by assumption is not less than n. Therefore $s \geq k$ can occur only for $s = k = n$. In this case the coefficient for $g(a_{j_1}, a_{j_2}, \ldots, a_{j_n})$ clearly equals one. This proves our formula, and consequently also the lemma.

PROPOSITION 5. For any $f, f' \in \Phi[X]$ on which the operator L_i^k is defined, $(f + f')L_i^k = fL_i^k + f'L_i^k$. If $f = f(x_1, x_2, \ldots, x_m)$ is a monomial of degree n in x_i and $g = g(x_1, \ldots, x_{i-1}, y_1, y_2, \ldots, y_n, x_{i+1}, \ldots, x_m)$ is a monomial linear in $y_1, y_2, \ldots, y_n \in X \backslash \{x_1, \ldots, x_m\}$ such that $f(x_1, x_2, \ldots, x_m) = g(x_1, \ldots, x_{i-1}, x_i, \ldots, x_i, x_{i+1}, \ldots, x_m)$, then

$$fL_i^k(x_1, \ldots, x_{i-1}, z_1, \ldots, z_k, x_{i+1}, \ldots, x_m)$$
$$= \begin{cases} \sum_{(i_1 \cdots i_n) \in S_n} g(x_1, \ldots, x_{i-1}, z_{i_1}, \ldots, z_{i_n}, x_{i+1}, \ldots, x_m), & \text{if } k = n, \\ 0, & \text{if } k > n. \end{cases}$$

PROOF. The linearity of the operator L_i^k follows directly from the definition. If f and g are the monomials indicated in the assumptions of the proposition, then fixing $x_1, \ldots, x_{i-1}, x_{i+1}, \ldots, x_n$ we can consider the monomial g as a function of the n variables $y_1, y_2, \ldots, y_n$. Applying Lemma 4 we obtain the formula for fL_i^k. This proves the proposition.

We illustrate this proposition by examples:

$$[x_1^2(x_2x_1)]L_1^3 = (y_1y_2)(x_2y_3) + (y_2y_1)(x_2y_3) + (y_1y_3)(x_2y_2) + (y_3y_1)(x_2y_2)$$
$$+ (y_2y_3)(x_2y_1) + (y_3y_2)(x_2y_1),$$
$$[(x_1^2x_2)x_1^2]L_1^5 = 0,$$
$$[x_1^2x_2^2]L_1^2L_2^2 = (y_1y_2)(z_1z_2) + (y_2y_1)(z_1z_2) + (y_1y_2)(z_2z_1) + (y_2y_1)(z_2z_1).$$

Let $f = f(x_1, x_2, \ldots, x_n)$ be a homogeneous polynomial of type $[k_1, k_2, \ldots, k_n]$. Then we shall call the multilinear polynomial $fL_1^{k_1}L_2^{k_2} \cdots L_n^{k_n}$ the *complete linearization of the polynomial* f.

THEOREM 7. Let a nonassociative polynomial $f = f(x_1, x_2, \ldots, x_n)$ vanish for any substitution of elements from a subgroup P of the additive group of the algebra A. Then the complete linearization of any of its homogeneous components with maximal degree also vanishes on P.

PROOF. Let $f = f_1 + f_2 + \cdots + f_s$ be a decomposition of the polynomial f into a sum of homogeneous components, and also let the degree of f_1 be maximal. Suppose that some variable x_j does not appear in f_1 but does appear in other homogeneous components. Then the polynomial

$$f'(x_1, x_2, \ldots, x_n) = f(x_1, \ldots, x_{j-1}, 0, x_{j+1}, \ldots, x_n)$$

likewise vanishes on P, has f_1 as a homogeneous component of maximal degree, but f' does not contain x_j. Furthermore, if x_j appears in f_1 but does not appear in some other homogeneous component of the polynomial f', then we can consider the new polynomial

$$f''(x_1, x_2, \ldots, x_n) = f'(x_1, x_2, \ldots, x_n) - f'(x_1, \ldots, x_{j-1}, 0, x_{j+1}, \ldots, x_n).$$

This polynomial also vanishes on P, but x_j now appears in all homogeneous components of the polynomial f'', and, as before, f_1 is a homogeneous component of maximal degree in f''.

Thus, renumbering the variables if necessary, we can assume at the outset that the homogeneous component f_1 of the polynomial f has type $[k_1, k_2, \ldots, k_n]$, where $k_i \neq 0$ for $i = 1, 2, \ldots, n$, and the remaining homogeneous components (f_2 for example) have type $[m_1, m_2, \ldots, m_n]$, where $m_i \neq 0$ for $i = 1, 2, \ldots, n$, and where the strict inequality $m_j < k_j$ holds for some j. Since $m_j < k_j$, then $f_2 \boldsymbol{L}_1^{k_1} \boldsymbol{L}_2^{k_2} \cdots \boldsymbol{L}_n^{k_n} = 0$ by Proposition 5. Consequently,

$$f_1 \boldsymbol{L}_1^{k_1} \boldsymbol{L}_2^{k_2} \cdots \boldsymbol{L}_n^{k_n} = f \boldsymbol{L}_1^{k_1} \boldsymbol{L}_2^{k_2} \cdots \boldsymbol{L}_n^{k_n}.$$

By Proposition 4 wc now conclude that the complete linearization

$$f_1 \boldsymbol{L}_1^{k_1} \boldsymbol{L}_2^{k_2} \cdots \boldsymbol{L}_n^{k_n}$$

of the component f_1 vanishes on P. This proves the theorem.

COROLLARY. If an algebra A satisfies some identity, then it also satisfies some multilinear identity.

6. ADJOINING AN IDENTITY ELEMENT

Let A be some algebra over the ring Φ. We consider the ring Φ as a module over itself. The identity element 1 of the ring Φ is a generating element for this Φ-module, $\Phi = \Phi \cdot 1$. We consider the direct sum $A^{\#} = A \oplus \Phi \cdot 1$ of the two Φ-modules A and $\Phi \cdot 1$, and we define a multiplication on it in the following manner:

$$(a + \alpha \cdot 1)(b + \beta \cdot 1) = (ab + \alpha b + \beta a) + \alpha\beta \cdot 1,$$

where $\alpha, \beta \in \Phi$ and $a, b \in A$. We shall call the algebra $A^{\#}$ the *algebra obtained by the formal adjoining of an identity element to the algebra A*. It is easy to see that 1 is an identity element for the algebra $A^{\#}$ and A is a subalgebra in $A^{\#}$. We call a variety $\mathfrak{M}$ *unitally closed* if, for any algebra A from $\mathfrak{M}$, the algebra $A^{\#}$ also belongs to $\mathfrak{M}$.

We shall subsequently call the algebra $\Phi[X]^{\#}$, obtained by adjoining an identity element to the free algebra $\Phi[X]$, the *free algebra with identity element*.

PROPOSITION 6. A variety $\mathfrak{M}$ defined by set of identities I is unitally closed if and only if $I(\Phi[X]^{\#}) \subseteq T(\mathfrak{M})$.

PROOF. We consider the $\mathfrak{M}$-free algebra $\Phi_{\mathfrak{M}}[X]$ and the $\mathfrak{M}$-free algebra with identity element $\Phi_{\mathfrak{M}}[X]^{\#}$. As is easy to see, $\mathfrak{M}$ is unitally closed if and only if $\Phi_{\mathfrak{M}}[X]^{\#} \in \mathfrak{M}$. The latter condition is equivalent to $I(\Phi_{\mathfrak{M}}[X]^{\#}) = (0)$. We note now that the algebra $\Phi_{\mathfrak{M}}[X]^{\#}$ is isomorphic to the quotient algebra $\Phi[X]^{\#}/T(\mathfrak{M})$, and therefore the equality $I(\Phi_{\mathfrak{M}}[X]^{\#}) = (0)$ is equivalent to the containment $I(\Phi[X]^{\#}) \subseteq T(\mathfrak{M})$. This proves the proposition.

We denote by Δ_i^k the mapping of the algebra $\Phi[X]$ into the algebra $\Phi[X]^{\#}$, which pairs the polynomial

$$f = f(x_1, \ldots, x_n)$$

from $\Phi[X]$ with the element

$$f\,\Delta_i^k(1) = f_i^{(k)}(x_1, \ldots, x_n; 1)$$

from the algebra $\Phi[X]^{\#}$. The mapping Δ_i^k is also defined linearly on the whole algebra $\Phi[X]$. The action of the operator Δ_i^k is analogous to "partial differentiation" with respect to the variable x_i. For example,

$$\begin{aligned}
[(x_1^2x_2)x_1]\,\Delta_1^1 &= x_1^2x_2 + 2(x_1x_2)x_1,\\
[(x_1^2x_2)x_1]\,\Delta_1^2 &= x_2x_1 + 2x_1x_2,\\
[(x_1^2x_2)x_1]\,\Delta_1^3 &= x_2,\\
[(x_1^2x_2)x_1]\,\Delta_1^4 &= 0.
\end{aligned}$$

If the elements of the set X are not equipped with indices, then we shall replace the index i by the notation for the corresponding generator. For example,

$$[(x^2y)x]\,\Delta_x^3 = y.$$

We shall denote by $f\Delta'$ the collection of all elements which can be obtained from the polynomial f by applying the operators Δ_i^k.

THEOREM 8. Let a variety $\mathfrak{M}$ have a system I of defining identities such that $f\Delta' \subseteq T(\mathfrak{M})$ for all $f \in I$. Then $\mathfrak{M}$ is unitally closed.

PROOF. In view of Proposition 6, we need to prove that $I(\Phi[X]^{\#}) \subseteq T(\mathfrak{M})$. Since $T(\mathfrak{M})$ is an ideal in $\Phi[X]^{\#}$, it suffices to prove that $f(g_1, \ldots, g_n) \in T(\mathfrak{M})$ for any $f = f(x_1, \ldots, x_n) \in I$ and any $g_1, \ldots, g_n \in \Phi[X]^{\#}$. We have $g_i = y_i + \alpha_i \cdot 1$, where $y_i \in \Phi[X]$, $\alpha_i \in \Phi$, $i = 1, 2, \ldots, n$. Now by Lemma 3

$$\begin{aligned}
f(g_1, \ldots, g_n) &= f(y_1 + \alpha_1 \cdot 1, \ldots, y_n + \alpha_n \cdot 1)\\
&= \sum_{i_1, \ldots, i_n} \alpha_1^{i_1} \cdots \alpha_n^{i_n} f\,\Delta_1^{i_1} \cdots \Delta_n^{i_n}(y_1, \ldots, y_n).
\end{aligned}$$

Since, by assumption, $f\Delta_1^{i_1}\cdots\Delta_n^{i_n} = f\Delta_1^{i_1}\cdots\Delta_n^{i_n}(x_1,\ldots,x_n)\in T(\mathfrak{M})$, then also $f(g_1,\ldots,g_n)\in T(\mathfrak{M})$. This proves the theorem.

We call a variety $\mathfrak{M}$ *strongly homogeneous* if $\mathfrak{M}$ is homogeneous and $f\Delta' \subseteq T(\mathfrak{M})$ for any $f\in T(\mathfrak{M})$.

PROPOSITION 7. A variety $\mathfrak{M}$ is strongly homogeneous if and only if $\mathfrak{M}$ is homogeneous and unitally closed.

PROOF. If $\mathfrak{M}$ is strongly homogeneous, then the assertion follows from Theorem 8. Conversely, let $\mathfrak{M}$ be homogeneous and unitally closed. We consider $f = f(x_1,\ldots,x_n)\in T(\mathfrak{M})$. By Proposition 3 we have $f\Delta_i^k(x_j) = f_i^{(k)}(x_1,\ldots,x_n;x_j)\in T(\mathfrak{M})$, where $x_j\in X\backslash\{x_1,\ldots,x_n\}$. Since the set of identities $T(\mathfrak{M})$ defines the variety $\mathfrak{M}$, then by Proposition 6 we obtain $f\Delta_i^k = f_i^{(k)}(x_1,\ldots,x_n;1)\in T(\mathfrak{M})$. This proves the proposition.

From Theorems 4 and 8 we now obtain

COROLLARY 1. Let the variety $\mathfrak{M}$ have a system I of defining identities such that $f\Delta \cup f\Delta' \subseteq T(\mathfrak{M})$ for all $f\in I$. Then $\mathfrak{M}$ is strongly homogeneous.

In particular, we have

COROLLARY 2. The varieties of associative, alternative, and right alternative algebras are strongly homogeneous. If there is an element $\frac{1}{2}$ in Φ, then the variety of Jordan Φ-algebras is also strongly homogeneous.

PROOF. As we have already seen in Section 4, all of these varieties are homogeneous. We show that the variety Jord of Jordan algebras is strongly homogeneous. This variety is defined by the identities $f_1 = x_1x_2 - x_2x_1$ and $f_2 = (x_1^2x_2)x_1 - x_1^2(x_2x_1)$. It is easy to see that $f_i\Delta_l^k = 0$ in all cases except when $i = 2$, $k = l = 1$. In that case

$$\begin{aligned} f_2\Delta_1^1 &= 2(x_1x_2)x_1 + x_1^2x_2 - 2x_1(x_2x_1) - x_1^2x_2 \\ &= 2(x_1x_2 - x_2x_1)x_1 + 2(x_2x_1)x_1 - 2x_1(x_2x_1) \\ &= 2f_1(x_1,x_2)x_1 + 2f_1(x_2x_1,x_1)\in T(\text{Jord}). \end{aligned}$$

Thus $f_1\Delta' \cup f_2\Delta' \subseteq T(\text{Jord})$, and the variety Jord is strongly homogeneous. The proofs for the other varieties are analogous. This proves the corollary.

We now prove a property for the free algebras of strongly homogeneous varieties.

PROPOSITION 8. Let $\mathfrak{M}$ be a strongly homogeneous variety. Then for any $f\in\Phi_{\mathfrak{M}}[X]$ and for any $x\in X$ each of the equalities $f\cdot x = 0$, $x\cdot f = 0$ implies $f = 0$.

PROOF. We denote by $d_x(f)$ the degree of the polynomial f in the variable x, and by R_z the operator of right multiplication by the element z, $yR_z = y \cdot z$. We shall show by induction on $d_x(f)$ that for any natural number m the equality $fR_x^m = 0$ implies $f = 0$. Because of the strong homogeneity of the variety, for $d_x(f) = 0$ we have $0 = (fR_x^m)\Delta_x^m = f$, which gives us the basis for the induction. Let $d_x(f) = n > 0$. Using again the strong homogeneity of the variety $\mathfrak{M}$ and Lemma 2, we obtain

$$0 = (fR_x^m)\Delta_x^m = \sum_{i=0}^{m} \binom{m}{i}(f\Delta_x^i)R_x^i = f + \sum_{i=1}^{m} \binom{m}{i}(f\Delta_x^i)R_x^i,$$

whence $f = f_1R_x$, where $d_x(f_1) < n$. We now have

$$0 = fR_x^m = f_1R_x^{m+1}.$$

The induction assumption gives us $f_1 = 0$ and $f = 0$.

In particular, we have proved that if $f \cdot x = 0$, then $f = 0$. It is proved analogously that the equality $x \cdot f = 0$ implies $f = 0$.

This proves the proposition.

An element a of an algebra A is called a *right* (*left*) *annihilator* of A if $A \cdot a = (0)$ ($a \cdot A = (0)$, respectively).

COROLLARY. A free algebra of a strongly homogeneous variety does not contain nonzero right and left annihilators.

In particular, free alternative, right alternative, and Jordan algebras do not contain nonzero annihilators.

In conclusion we prove a proposition which sometimes facilitates the proof of identities in strongly homogeneous varieties. Its assertion is often called the *Koecher principle.*

PROPOSITION 9. Let $\mathfrak{M}$ be a strongly homogeneous variety and $f = f(x_1, \ldots, x_n)$ be some homogeneous nonassociative polynomial. Suppose that for any algebra A from $\mathfrak{M}$ which is finitely generated as a Φ-module, the polynomial f vanishes for any substitution of elements of the form $1 + x$, where the ideal generated in A by the element x is nilpotent. Then f is an identity of the variety $\mathfrak{M}$.

PROOF. It suffices to establish that $f(x_1, \ldots, x_n) = 0$ in the $\mathfrak{M}$-free algebra $F = \Phi_{\mathfrak{M}}[x_1, \ldots, x_n]$. Let the degree of the polynomial f equal m. We consider the quotient algebra $G = F/F^{m+1}$ and the elements $y_1, \ldots, y_n$ in G that are images of the elements $x_1, \ldots, x_n$ under the canonical homomorphism of F onto G. Adjoining an identity element to this algebra, we obtain the algebra $G^{\#} \in \mathfrak{M}$. The elements $y_1, \ldots, y_n$ generate nilpotent ideals in this algebra. Therefore

$$f(1 + y_1, \ldots, 1 + y_n) = 0.$$

But then, by strong homogeneity, in the algebra $F^{\#}$ we have the relation

$$f(1 + x_1, \ldots, 1 + x_n) = 0.$$

The polynomial $f(x_1, \ldots, x_n)$ is a homogeneous component of the polynomial $f(1 + x_1, \ldots, 1 + x_n)$, and therefore equals zero. This proves the proposition.

In all the classical varieties elements of the form indicated in the conditions of the proposition are invertible, so that in these varieties it suffices to prove identities only for invertible elements.

CHAPTER 2

Composition Algebras

One of the most important properties of the complex numbers is expressed by the identity

$$|zz'| = |z| \cdot |z'|.$$

If we denote $z = a_1 + a_2 i$, $z' = b_1 + b_2 i$, then this identity is rewritten in the form

$$(a_1^2 + a_2^2)(b_1^2 + b_2^2) - (a_1 b_1 - a_2 b_2)^2 + (a_1 b_2 + a_2 b_1)^2.$$

Put somewhat loosely, this formula can be read as the product of a sum of two squares times the sum of two squares is again the sum of two squares.

The question that naturally arises is: Do there exist analogous identities for a greater number of squares? How are all such identities described?

Examples of identities for the sum of four squares were already known to Euler and Lagrange. In 1818 Degen[†] found the first example of an identity for the sum of eight squares, which, unfortunately, then went unnoticed. In 1843 Hamilton noted that the existence of an identity for the sum of n squares is equivalent to the existence of a division algebra of a definite form with dimension n over the field of real numbers. For example, the algebra of

[†] Degen C. F., *Mem. Acad. Sci. St. Petersbourg* **8**, (1818), 207–219.

complex numbers is such an algebra in the case $n = 2$. Hamilton constructed a four-dimensional division algebra which gave Euler's identity for the sum of four squares. In 1845 Cayley constructed an eight-dimensional division algebra giving an identity for the sum of eight squares. An analogous algebra was constructed independently by Graves in 1844. It was not clarified until much later whether or not analogous identities exist for other values of n. Only in 1898 did Hurwitz prove the identities of interest to us are possible only for $n = 1, 2, 4, 8$.

The algebras constructed by Hamilton and Cayley, now known, respectively, as the algebra of quaternions and algebra of Cayley numbers, also proved to be very interesting from many other points of views. Thus in 1878 Frobenius proved that the algebra of quaternions and the field of complex numbers are the only finite-dimensional associative division algebras with dimension greater than 1 over the field of real numbers. Later the algebra of Cayley numbers found an important place in the theory of alternative and Jordan algebras.

In connection with the Hurwitz theorem arises a more general question. What happens if, instead of the sum of squares, some nondegenerate quadratic form over an arbitrary field is considered? For which quadratic forms do results analogous to the Hurwitz theorem hold? This question, known as the Hurwitz problem, was solved comparatively recently in works of Albert, Kaplansky, and Jacobson. It turns out that in this case the problem also reduces to the description of a certain class of algebras—the so-called "composition algebras." Composition algebras are a natural generalization of the algebras of complex numbers, quaternions, and Cayley numbers, and they play an important role in the theory of alternative and Jordan algebras. This chapter is devoted to the study of these algebras.

1. DEFINITION AND SIMPLE PROPERTIES OF COMPOSITION ALGEBRAS

Let F be an arbitrary field and A be a vector space of nonzero dimension over F. A mapping $f: A \times A \to F$ is called a *bilinear form* if for any $x, x', y, y' \in A$ and $\alpha \in F$

(1) $f(x + x', y) = f(x, y) + f(x', y)$;
(2) $f(x, y + y') = f(x, y) + f(x, y')$;
(3) $f(\alpha x, y) = f(x, \alpha y) = \alpha f(x, y)$.

A bilinear form f is called *symmetric* if $f(x, y) = f(y, x)$ for all $x, y \in A$. A symmetric bilinear form f is called *nondegenerate* if, from $f(a, x) = 0$ for all $x \in A$, it follows that $a = 0$.

A mapping $n: A \to F$ is called a *quadratic form* if

(1) $n(\lambda x) = \lambda^2 n(x)$, where $x \in A$, $\lambda \in F$;
(2) the function $f(x, y) = n(x + y) - n(x) - n(y)$ is a bilinear form on A.

A quadratic form $n(x)$ is called *strictly nondegenerate* if the symmetric bilinear form $f(x, y)$ which corresponds to it is nondegenerate, and it is called *nondegenerate* if from $n(a) = f(a, x) = 0$ for all $x \in A$ it follows that $a = 0$. Every strictly nondegenerate quadratic form is nondegenerate, and the converse is true if the characteristic of the field F is different from 2. A form $n(x)$ is said to *admit composition* if there exists a bilinear binary operation (composition) xy in A such that

$$n(x)n(y) = n(xy). \tag{1}$$

The *Hurwitz problem* consists of the determination of nondegenerate quadratic forms admitting composition. If $A = E^n$ is n-dimensional Euclidean space and $n(x) = (x, x)$ is the scalar square of the vector x, then we obtain the classical formulation of the Hurwitz problem. In this case, equality (1) written in an orthogonal basis gives the desired identity for the sum of squares.

As an example of a quadratic form admitting composition and different from the sum of squares, the following form from two variables can be used:

$$n(x) = n(x_1, x_2) = x_1^2 + kx_1x_2 + tx_2^2, \qquad k, t \in F.$$

Actually, it is easy to verify that

$$n(x_1, x_2) \cdot n(y_1, y_2) = n(x_1y_1 - tx_2y_2, x_1y_2 + x_2y_1 + kx_2y_2).$$

If the form $n(x)$ admits composition, then the vector space A becomes an algebra with respect to the operations of addition, scalar multiplication, and the multiplication xy. In addition, the multiplication is connected with the quadratic form $n(x)$ by equality (1). Conversely, if a quadratic form $n(x)$ connected with the multiplication by equality (1) is defined on some algebra A, then the form $n(x)$ admits composition. Consequently, the Hurwitz problem is equivalent to the problem of describing the algebras on which there is defined a nondegenerate quadratic form $n(x)$ satisfying (1).

As we shall see later, if the form $n(x)$ is strictly nondegenerate and admits composition, then as in the Hurwitz theorem the dimension of the corresponding algebra A can only equal 1, 2, 4, 8. At the same time, over each imperfect field of characteristic 2 there exists an infinite number of nondegenerate quadratic forms which are not strictly nondegenerate and do admit composition. In addition, the dimension of the corresponding algebra A can be an infinite number of different values and can even be infinite. In this connection, in order to protect ourselves from "pathological" cases in characteristic 2 and at the same time not impose a restriction on the characteristic, we shall subsequently consider only strictly nondegenerate quadratic forms. Nondegenerate quadratic forms admitting composition are considered in the exercises for Section 2.

An algebra A over a field F with quadratic form $n(x)$ is called a *composition algebra* if

(1) $n(xy) = n(x)n(y)$;
(2) the form $n(x)$ is strictly nondegenerate;
(3) there is an identity element 1 in A.

PROPOSITION 1. Let $n(x)$ be a strictly nondegenerate quadratic form on a finite-dimensional vector space A. Then if $n(x)$ admits some kind of composition xy on A, $n(x)$ admits a composition $x \cdot y$ with respect to which A is a composition algebra.

PROOF. By assumption, the algebra A with multiplication xy and the form $n(x)$ satisfy conditions (1) and (2). By condition (2) there exists an element $a \in A$ for which $n(a) \neq 0$. Let $u = a^2/n(a)$. Then $n(u) = 1$, and consequently $n(xu) = n(ux) = n(x)$. Hence it follows that $f(xu, yu) = f(ux, uy) = f(x, y)$. By condition (2) this means that the linear mappings $R_u: x \to xu$ and $L_u: x \to ux$ are nonsingular. In view of the finite dimensionality of A, the mappings R_u and L_u are invertible. We now define a new multiplication on A by setting $x \cdot y = (xR_u^{-1})(yL_u^{-1})$. Then $u^2 \cdot x = (u^2R_u^{-1})(xL_u^{-1}) = u(xL_u^{-1}) = x$. Analogously, $x \cdot u^2 = x$. Consequently, u^2 is an identity element with respect to the new multiplication. Furthermore,

$$n(x \cdot y) = n((xR_u^{-1})(yL_u^{-1})) = n(xR_u^{-1})n(yL_u^{-1}) = n(x)n(y).$$

Thus we can replace the composition xy by $x \cdot y$ and obtain a composition algebra. This proves the proposition.

Thus, for strictly nondegenerate quadratic forms defined on finite-dimensional spaces, the Hurwitz problem is equivalent to the problem of describing finite-dimensional composition algebras.

We shall describe all composition algebras without any sort of restriction on the dimension.

We shall need the following definition.

An algebra R is called *alternative* if $x^2y = x(xy)$, $yx^2 = (yx)x$ for all x, y from R.

LEMMA 1. Let A be a composition algebra. Then A is alternative and each element from the algebra A satisfies a quadratic equation with coefficients from F (i.e., the algebra A is quadratic over F).

PROOF. We shall identify the field F with the subalgebra $F \cdot 1$ of the algebra A. Substituting $y + w$ in place of y in (1), we obtain $n(x)n(y + w) = n(xy + xw)$. Subtracting identity (1) from this equality, and also the identity obtained from (1) by replacing y by w, we obtain

$$n(x)f(y, w) = f(xy, xw). \tag{2}$$

Performing the same procedure with x, we obtain

$$f(x, z)f(y, w) = f(xy, zw) + f(xw, zy). \tag{3}$$

We now substitute $z = 1$, $y = xu$ in (3)[†]:

$$f(x, 1)f(xu, w) = f(x \cdot xu,\ w) + f(xw, xu). \tag{4}$$

Since by (2) $f(xw, xu) = n(x)f(w, u)$, then (4) can be rewritten in the form

$$f(x \cdot xu,\ w) + n(x)f(w, u) - f(x, 1)f(xu, w) = 0,$$

which by the bilinearity and symmetry of the form f is equivalent to the identity

$$f(x \cdot xu + n(x)y - f(x, 1)xu, w) = 0.$$

Since the form $f(x, y)$ is nondegenerate and w is arbitrary, it follows from this that

$$x \cdot xu + n(x)u - f(x, 1)xu = 0 \tag{5}$$

for any x, y from A. Now putting $u = 1$ in (5), we obtain

$$x^2 - f(x, 1)x + n(x) = 0, \tag{6}$$

which proves the second half of the lemma. It just remains to prove the algebra A is alternative.

Multiplying (6) on the right by u and comparing with (5), we obtain $x^2u = x(xu)$. It is proved analogously that $ux^2 = (ux)x$. Consequently the algebra A is alternative, which proves the lemma.

An endomorphism ρ of a vector space A is called an *involution* of the algebra A if $\rho(\rho(a)) = a$ and $\rho(ab) = \rho(b)\rho(a)$ for all $a, b \in A$.

LEMMA 2. The mapping $a \to \bar{a} = f(1, a) - a$ is an involution of the algebra A which leaves elements of the field F fixed. Also, the elements $t(a) = a + \bar{a}$ and $n(a) = a\bar{a}$ lie in F for all a from A. Moreover, a satisfies the equality

$$a^2 - t(a)a + n(a) = 0.$$

[†] Here and subsequently, in order to avoid a profusion of parentheses we shall sometimes use dots in place of some of them; for example, $xy \cdot zt = (xy)(zt)$.

PROOF. We shall verify, in turn, each of the properties for an involution:

(a) $\overline{a+b} = \bar{a} + \bar{b}$. We have $\overline{a+b} = f(1, a+b) - (a+b) = f(1,a) - a + f(1,b) - b = \bar{a} + \bar{b}$.

(b) Let $\lambda \in F$. Then $\overline{\lambda a} = f(1, \lambda a) - \lambda a = \lambda(f(1,a) - a) = \lambda\bar{a}$.

(c) $\bar{\bar{a}} = a$. We have $\bar{\bar{a}} = f(1,\bar{a}) - \bar{a} = f(1, f(1,a) - a) - f(1,a) + a = f(1,1)f(1,a) - 2f(1,a) + a$. Furthermore, as is easy to see, $n(1) = 1$. Therefore, $f(1,1) = n(2) - n(1) - n(1) = 2$. Consequently, $\bar{\bar{a}} = a$.

(d) $\overline{ab} = \bar{b}\bar{a}$. From identity (6), substituting in turn $a + b$, a, and b in place of x, and then subtracting from the first identity obtained the other two, we have

$$ab + ba - f(1,a)b - f(1,b)a + f(a,b) = 0. \tag{7}$$

Furthermore, from (3) we obtain for $x = y = 1$, $z = a$, and $w = b$,

$$f(1,a)f(1,b) = f(1,ab) + f(a,b). \tag{8}$$

Substituting this in (7), we obtain

$$ab + ba - f(1,a)b - f(1,b)a + f(1,a)f(1,b) - f(1,ab) = 0, \tag{9}$$

whence

$$\begin{aligned}(f(1,a) - a)(f(1,b) - b) &= ab - f(1,a)b - f(1,b)a + f(1,a)f(1,b) \\ &= f(1,ab) - ba.\end{aligned}$$

From (8) it follows that $f(1,ab) = f(1,ba)$. Consequently $\bar{a}\bar{b} = \overline{ba}$, which is what was to be proved.

Furthermore, if $\lambda \in F$, then $\bar{\lambda} = f(1,\lambda) - \lambda = \lambda f(1,1) - \lambda = \lambda$. Finally, $a + \bar{a} = f(1,a) \in F$, and by (6) $a\bar{a} = f(1,a)a - a^2 = n(a) \in F$. Since $t(a) = f(1,a)$, the last assertion of the lemma also follows from (6). This proves the lemma.

LEMMA 3. Let A be an algebra over a field F with identity element 1 and with involution $x \to \bar{x}$, where for every $x \in A$ the elements $x + \bar{x}$ and $x\bar{x}$ belong to F. Then if the quadratic form $n(x) = x\bar{x}$ is strictly nondegenerate on A, the algebra A is either simple or isomorphic to the direct sum $F \oplus F$ with involution $\overline{(a,b)} = (b,a)$.

PROOF. We prove first that A is simple as an algebra with involution. Let $I \neq A$ be an ideal of the algebra A and $\bar{I} = I$. Since $I \cap F = (0)$, we have $a + \bar{a} = 0$, $a\bar{a} = 0$ for any $a \in I$. Furthermore, for any elements $a \in I$, $x \in A$ we have $a\bar{x} \in I$, whence by the preceding $f(a,x) = a\bar{x} + x\bar{a} = a\bar{x} + \overline{(a\bar{x})} = 0$. In view of the nondegeneracy of the form f, it follows from this that $I = (0)$.

Now let T be an arbitrary ideal of the algebra A, $T \neq (0)$ and $T \neq A$. We set $I = T + \bar{T}$. Then $\bar{I} = I$ and $I \neq (0)$, whence by what was proved above

$I = A$. Furthermore, we set $J = T \cap \bar{T}$. Then $\bar{J} = J$ and $J \neq A$, and again by what was proved above $J = (0)$. Consequently, the algebra A is the direct sum of the ideals T and $\bar{T}$. It remains to prove that the ideal T is one-dimensional over the field F. Let $t, s \in T$ with $t \neq 0$, and $\lambda = t + \bar{t}$, $\mu = s + \bar{s}$ so that $\lambda, \mu \in F$. We note that $\lambda \neq 0$, since otherwise $t = -\bar{t} \in T \cap \bar{T} = (0)$. Furthermore, we have $T \cdot \bar{T} + \bar{T} \subseteq T \cap \bar{T} = (0)$, whence $\lambda s = (t + \bar{t})s = ts$, $\mu t = t(s + \bar{s}) = ts$. Consequently, $\lambda s = \mu t$ and $s = (\lambda^{-1}\mu)t$, that is, the one-dimensionality of the ideal T is proved.

Thus, if the algebra A is not simple, then it is the direct sum of two copies of the field F. This proves the lemma.

We see that the problem of describing composition algebras brings us in a natural manner to the study of simple alternative algebras. In Section 3 we show that, in turn, with the exception of some algebras of characteristic 2, each simple quadratic alternative algebra is a composition algebra. In Chapter 7 we shall obtain a definitive description of simple alternative algebras that are not associative.

2. THE CAYLEY–DICKSON PROCESS. GENERALIZED THEOREM OF HURWITZ

Let A be an algebra over a field F with identity element 1 and with an involution $a \to \bar{a}$, where $a + \bar{a}, a\bar{a} \in F$ for all $a \in A$. By means of the so-called *Cayley–Dickson process* we shall construct a new algebra with involution which contains A as a subalgebra. In addition, if the dimension of the algebra A equals m, then the dimension of the new algebra will equal $2m$.

We fix $0 \neq \alpha \in F$, and we denote by (A, α) the collection of all ordered pairs (a_1, a_2), $a_i \in A$, with operations of componentwise addition and scalar multiplication and the multiplication

$$(a_1, a_2)(a_3, a_4) = (a_1a_3 + \alpha a_4\bar{a}_2, \bar{a}_1a_4 + a_3a_2).$$

As is easy to see, (A, α) is an algebra over F. The element $(1,0)$ is an identity element of the algebra (A, α). The set $A' = \{(a,0) | a \in A\}$ is a subalgebra of the algebra (A, α) which is isomorphic to the algebra A. Let $v = (0,1)$. Then $v^2 = \alpha(1,0)$, and (A, α) is a direct sum of the vector spaces A' and vA'. If we identify A' with A, the elements of the algebra (A, α) are represented in the form $x = a_1 + va_2$, where the $a_i \in A$ are uniquely determined by the element x and the multiplication in (A, α) is given by

$$(a_1 + va_2)(a_3 + va_4) = (a_1a_3 + \alpha a_4\bar{a}_2) + v(\bar{a}_1a_4 + a_3a_2).$$

For an arbitrary element $x = a_1 + va_2 \in (A, \alpha)$, we set $\bar{x} = \bar{a}_1 - va_2$.

LEMMA 4. The mapping $x \to \bar{x}$ is an involution of the algebra (A, α). In addition, $x + \bar{x}, x\bar{x} \in F$ for all $x \in (A, \alpha)$. If the quadratic form $n(a) = a\bar{a}$ is strictly nondegenerate on A, then the quadratic form $n(x) = x\bar{x}$ is strictly nondegenerate on (A, α).

PROOF. It is clear that the mapping $x \to \bar{x}$ is linear and that $\bar{\bar{x}} = x$ for any $x \in (A, \alpha)$. Let $x = a_1 + va_2, y = a_3 + va_4$, where $a_i \in A$. Then we have

$$\bar{y}\bar{x} = (\bar{a}_3 - va_4)(\bar{a}_1 - va_2) = (\bar{a}_3\bar{a}_1 + \alpha a_2\bar{a}_4) - v(a_3a_2 + \bar{a}_1a_4).$$

On the other hand,

$$\overline{xy} = (\bar{a}_3\bar{a}_1 + \alpha a_2\bar{a}_4) - v(\bar{a}_1a_4 + a_3a_2) = \bar{y}\bar{x}.$$

Consequently the mapping $x \to \bar{x}$ is an involution of the algebra (A, α). Furthermore, $x + \bar{x} = a_1 + \bar{a}_1 \in F$ and $x\bar{x} = a_1\bar{a}_1 - a_2\bar{a}_2 \in F$.

Now let the quadratic form $n(a)$ be strictly nondegenerate on A. The bilinear form $f(x, y) = x\bar{y} + y\bar{x} = (a_1\bar{a}_3 + a_3\bar{a}_1) - \alpha(a_2\bar{a}_4 + a_4\bar{a}_2)$, where $x = a_1 + va_2$, $y = a_3 + va_4$, and $a_i \in A$, corresponds to the quadratic form $n(x) = x\bar{x}$. Let $f(x, y) = 0$ for all $y \in (A, \alpha)$. Setting $y = a_3 \in A$, we obtain $a_1\bar{a}_3 + a_3\bar{a}_1 = 0$ for all $a_3 \in A$. Hence, in view of the strict nondegeneracy of the form $n(a) = a\bar{a}$, it follows $a_1 = 0$. Since $\alpha \neq 0$, analogously we obtain $a_2 = 0$, whence $x = 0$. This means the quadratic form $n(x) = x\bar{x}$ is strictly nondegenerate, which proves the lemma.

Now let A be a composition algebra with quadratic form $n(a)$. By Lemma 2 there exists an involution $a \to \bar{a}$ on A such that $n(a) = a\bar{a}$ and $t(a) = a + \bar{a} \in F$ for all $a \in A$. This means we can apply the Cayley–Dickson process to A. We next find the conditions under which the obtained algebra (A, α) with quadratic form $n(x)$ is also a composition algebra.

LEMMA 5. If A is a composition algebra, then (A, α) is a composition algebra if and only if the algebra A is associative.

PROOF. It was shown above that the algebra (A, α) has an identity element. In addition, by Lemma 4 the quadratic form $n(x) = x\bar{x}$ is strictly nondegenerate on (A, α). This means (A, α) is a composition algebra if and only if the form $n(x)$ admits composition. We note that if $x = a_1 + va_2$, then $n(x) = n(a_1) - \alpha n(a_2)$. Let $y = a_3 + va_4$. We then have

$$\begin{aligned} n(xy) - n(x)n(y) &= n(a_1a_3 + \alpha a_4\bar{a}_2) - \alpha n(\bar{a}_1a_4 + a_3a_2) \\ &\quad - [n(a_1) - \alpha n(a_2)][n(a_3) - \alpha n(a_4)] \\ &= n(a_1a_3) + \alpha^2 n(a_4\bar{a}_2) + \alpha f(a_1a_3, a_4\bar{a}_2) \\ &\quad - \alpha n(\bar{a}_1a_4) - \alpha n(a_3a_2) - \alpha f(a_1\bar{a}_4, a_3a_2) \\ &\quad - n(a_1)n(a_3) + \alpha n(a_1)n(a_4) + \alpha n(a_2)n(a_3) \\ &\quad - \alpha^2 n(a_2)n(a_4). \end{aligned}$$

Since $n(ab) = n(a)n(b)$ and $n(\bar{a}) = n(a)$ for any $a, b \in A$, we obtain

$$n(xy) - n(x)n(y) = \alpha f(a_1 a_3, a_4 \bar{a}_2) - \alpha f(\bar{a}_1 a_4, a_3 a_2).$$

In view of (3) we have

$$\begin{aligned} f(a_1 a_3, a_4 \bar{a}_2) &= f(a_1 a_3 \cdot 1, a_4 \bar{a}_2) = -f(a_1 a_3 \cdot \bar{a}_2, a_4) + f(a_1 a_3, a_4) f(1, \bar{a}_2) \\ &= f(a_1 a_3 \cdot (f(1, \bar{a}_2) - \bar{a}_2), a_4) = f(a_1 a_3 \cdot a_2, a_4), \end{aligned}$$

and analogously

$$f(\bar{a}_1 a_4, a_3 a_2) = f(a_4, a_1 \cdot a_3 a_2).$$

Consequently

$$n(xy) - n(x)n(y) = \alpha f(a_1 a_3 \cdot a_2 - a_1 \cdot a_3 a_2, a_4).$$

Since $\alpha \neq 0$ and the bilinear form f is nondegenerate, it follows from this that the quadratic form $n(x)$ admits composition if and only if the algebra A is associative. This proves the lemma.

We can now cite the following examples of composition algebras.

I. F is a field of characteristic $\neq 2$, and $n(\alpha) = \alpha^2$ for all $\alpha \in F$. We note that in the case of characteristic 2, the field F is not a composition algebra. In fact, in that case the bilinear form $f(x, y) = (x + y)^2 - x^2 - y^2 \equiv 0$ on F, that is, the quadratic form $n(x) = x^2$ is not strictly nondegenerate. It is possible to change the definition of a composition algebra so that a field of characteristic 2 is a composition algebra, but there then appears an entire supplement of composition algebras of characteristic 2, which can even be infinite-dimensional (see Exercise 6).

II. $\boldsymbol{K}(\mu) = F + Fv$, where $v_1^2 = v_1 + \mu$ and $4\mu + 1 \neq 0$, the field F is arbitrary, the involution is $\overline{\alpha + \beta v_1} = (\alpha + \beta) - \beta v_1$, and the quadratic form is $n(a) = a\bar{a}$. If the polynomial $x^2 - x - \mu$ is irreducible in $F[x]$, then the algebra $\boldsymbol{K}(\mu)$ is a field (a separable quadratic extension of the field F); otherwise $\boldsymbol{K}(\mu) = F \oplus F$. If the characteristic of the field F does not equal 2, then the element $v = v_1 - \frac{1}{2}$ satisfies the equality $v^2 = \alpha$ where $\alpha = \frac{1}{4}(4\mu + 1) \neq 0$. In this case $\boldsymbol{K}(\mu) = (F, \alpha)$. Conversely, if F is a field of characteristic $\neq 2$, then the algebra (F, α) is an algebra of type II. In fact, setting $v_1 = v + \frac{1}{2}$ we obtain $v_1^2 = v_1 + (\alpha - \frac{1}{4})$, and moreover $4(\alpha - \frac{1}{4}) + 1 = 4\alpha \neq 0$. In addition, $\overline{\gamma + \beta v_1} = \gamma + \beta\overline{(v + \frac{1}{2})} = \gamma - \beta v + \frac{1}{2}\beta = (\gamma + \beta) - \beta v_1$. We note here that if F is a field of characteristic 2, then the condition $4\mu + 1 \neq 0$ is automatically satisfied.

III. $\boldsymbol{Q}(\mu, \beta) = (\boldsymbol{K}(\mu), \beta)$ with $\beta \neq 0$ is the *algebra of generalized quaternions*. As is easy to see, the algebra $\boldsymbol{Q}(\mu, \beta)$ is associative but not commutative.

IV. $\boldsymbol{C}(\mu, \beta, \gamma) = (\boldsymbol{Q}(\mu, \beta), \gamma)$ with $\gamma \neq 0$ is the *Cayley–Dickson algebra.* It is easy to verify (see Exercise 2) that the Cayley–Dickson algebra $\boldsymbol{C}(\mu, \beta, \gamma)$ is not associative, and therefore by Lemma 5 it is impossible to continue further our inductive process for the construction of composition algebras.

We shall now prove that all the composition algebras are accounted for by our indicated examples.

Let A be a composition algebra and $n(x)$ be the corresponding quadratic form defined on A. By Lemma 2 there is an involution $a \to \bar{a}$ on the algebra A such that $n(a) = a\bar{a}$ and $a + \bar{a} \in F$ for all $a \in A$. In addition, $\bar{\alpha} = \alpha$ for any $\alpha \in F$. Let $f(x, y)$ be the bilinear form associated with the quadratic form $n(x)$. Then $f(x, y) = x\bar{y} + y\bar{x}$. If B is some subspace of the space A, we shall denote by $B^{\perp}$ the orthogonal complement of the subspace B with respect to the form $f(x, y)$, $B^{\perp} = \{a \in A \mid f(a, B) = 0\}$.

We shall need the following:

LEMMA 6. Let B be a subalgebra of a composition algebra A which contains the identity element 1 of the algebra A. Then $B^{\perp}B + BB^{\perp} \subseteq B^{\perp}$, and for every $a, b \in B$ and $v \in B^{\perp}$ the following relations are valid:

$$\bar{v} = -v, \qquad av = v\bar{a}, \tag{10}$$

$$a(vb) = v(\bar{a}b), \qquad (vb)a = v(ab), \tag{11}$$

$$(va)(vb) = -n(v)b\bar{a}. \tag{12}$$

PROOF. From the equalities $f(v, 1) = 0, f(a, v) = 0$ we obtain $v + \bar{v} = 0$, $a\bar{v} + v\bar{a} = 0$, whence follows (10). Furthermore, by (3) we have

$$f(a, vb) = f(a \quad 1, vb) = -f(ab, v) + f(a, v)f(1, b) = 0,$$

and analogously

$$f(a, bv) = 0.$$

Since the elements $a, b \in B$ and $v \in B^{\perp}$ are arbitrary, this means that $BB^{\perp} + B^{\perp}B \subseteq B^{\perp}$. Furthermore, by Lemma 1 the identity $x(xy) = x^2y$ is satisfied in the algebra A. Hence it follows easily that A also satisfies the relation

$$x(\bar{x}y) = (x\bar{x})y = n(x)y. \tag{13}$$

Linearizing this relation in x (see Chapter 1), we obtain

$$x(\bar{z}y) + z(\bar{x}y) = f(x, z)y. \tag{14}$$

If we set here $x = a$, $y = b$, $z = v$, we obtain

$$a(\bar{v}b) = -v(\bar{a}b),$$

from which follows the first of equalities (11). Acting on both sides of this equality with the involution, we obtain the second of equalities (11). Finally, by identity (14) and in view of (2), (13), and (11), we have

$$\begin{aligned}(va)(vb) &= -\bar{v}(\overline{va} \cdot b) + f(va, \bar{v})b = v(\bar{a}\bar{v} \cdot b) - f(va, v)b \\ &= -v(\bar{a}v \cdot b) - n(v)f(a, 1)b = -v(va \cdot b) - n(v)f(a, 1)b \\ &= v(\bar{v} \cdot ba) - n(v)f(a, 1)b = n(v)(ba - f(a, 1)b) = -n(v)b\bar{a}.\end{aligned}$$

This proves the lemma.

We can now prove

THEOREM 1. Let A be a composition algebra. Then A is isomorphic to one of the above-mentioned algebras of types I–IV.

PROOF. By the term "subalgebra" we shall mean a subalgebra containing the identity element 1. Since $a + \bar{a} \in F$, every such subalgebra is invariant under the involution.

Let B be a finite-dimensional subalgebra of the algebra A on which the restriction of the form $f(x, y)$ is nondegenerate. As is well known (e.g., see A. I. Mal'cev, "Foundations of Linear Algebra," p. 284, Nauka, 1975.), A then decomposes into a direct sum of subspaces $A = B + B^{\perp}$. Moreover, the restriction of the form $f(x, y)$ to $B^{\perp}$ is also nondegenerate. We assume that $B \neq A$. Then we can find a $v \in B^{\perp}$ such that $n(v) = -\alpha \neq 0$. From (2) we obtain $f(va, vb) = n(v)f(a, b) = -\alpha f(a, b)$. Since f is nondegenerate on B, it follows from this that the mapping $x \to vx$ of the subspace B onto vB is one-to-one. Consequently, B and vB have the same dimension. In addition, the relation obtained shows that the subspace vB is nondegenerate with respect to $f(x, y)$. This means the subspace $B_1 = B + vB$, which by Lemma 6 is the orthogonal sum of two nondegenerate subspaces, is also nondegenerate. Relations (11) and (12) show that $B_1 = B + vB$ is the subalgebra of the algebra A obtained from B by means of the Cayley–Dickson process, $B_1 = (B, \alpha)$. Since $\bar{v} = -v$ and $\overline{a + vb} = \bar{a} - \bar{b}v = \bar{a} - vb$, the involution induced on B_1 by the involution on A coincides with the involution obtained in the Cayley–Dickson process. Finally, the subalgebra B_1 is nondegenerate with respect to $f(x, y)$ and satisfies the same conditions as B. Therefore we can repeat the same process with the algebra B_1.

We now return to the algebra A and consider separately two cases.

(1) F is a field of characteristic $\neq 2$. In this case the subalgebra F is nondegenerate with respect to $f(x, y)$, and therefore we can set $B = F$. If $F \neq A$, then there is a subalgebra B_1 of type II contained in A. If $B_1 \neq A$, then there is a subalgebra B_2 of type III contained in A. Finally, if $B_2 \neq A$, then A contains a subalgebra B_3 of type IV. With this the process must

terminate, since otherwise by Lemma 5 there would be in the algebra A a subalgebra B_4 which is not a composition algebra, and that is impossible. Consequently $A = B_3$.

(2) F is a field of characteristic 2. As we have seen, in this case $A \neq F$. We shall show there exists $a \in A \backslash F$ for which $t(a) \neq 0$. In fact, let $t(x) = f(1, x) = 0$ for every $x \in A \backslash F$. Since $t(\alpha) = 2\alpha = 0$ for every $\alpha \in F$, then in general $t(x) = f(1, x) = 0$ for all $x \in A$, which contradicts the nondegeneracy of the form f. Let $t(a) \neq 0$ with $a \notin F$, and set $s = a/t(a)$. Then $t(s) = 1$, that is, $s^2 = s + \mu$ where $\mu \in F$. As is easy to see, the subalgebra $F + Fs$ is nondegenerate with respect to $f(x, y)$. Furthermore, $\overline{\alpha + \beta s} = \alpha + \beta - \beta s$, so that the subalgebra $F + Fs$ is an algebra of type II. Setting $B = F + Fs$, we again obtain that the algebra A is one of the algebras of type II, III, or IV.

This proves the theorem.

COROLLARY. A strictly nondegenerate quadratic form $n(x)$ defined on a vector space V over a field F admits composition if and only if $\dim_F V = 1, 2, 4, 8$, and with respect to some basis for the space V the form $n(x)$ corresponds to one of the forms

(1) $n(x) = x^2$ (char $F \neq 2$);
(2) $n(x) = x_1^2 + x_1x_2 - \mu x_2^2$;
(3) $n(x) = x_1^2 + x_1x_2 - \mu x_2^2 - \beta x_3^2 - \beta x_3x_4 + \beta\mu x_4^2$;
(4) $n(x) = x_1^2 + x_1x_2 - \mu x_2^2 - \beta x_3^2 - \beta x_3x_4 + \beta\mu x_4^2 - \gamma x_5^2 - \gamma x_5x_6 + \gamma\mu x_6^2 + \beta\gamma x_7^2 + \beta\gamma x_7x_8 - \beta\gamma\mu x_8^2$,

where $\mu, \beta, \gamma \in F$, $(4\mu + 1)\beta\gamma \neq 0$. If the characteristic of the field F does not equal 2, then it is possible to choose a basis for V in which the form $n(x)$ has one of the forms

(1) $n(x) = x^2$;
(2) $n(x) = x_1^2 - \alpha x_2^2$;
(3) $n(x) = x_1^2 - \alpha x_2^2 - \beta x_3^2 + \alpha\beta x_4^2$;
(4) $n(x) = x_1^2 - \alpha x_2^2 - \beta x_3^2 + \alpha\beta x_4^2 - \gamma x_5^2 + \alpha\gamma x_6^2 + \beta\gamma x_7^2 - \alpha\beta\gamma x_8^2$,

where $\alpha, \beta, \gamma \in$ F, $\alpha\beta\gamma \neq 0$.

Exercises

1. Let $A = F + Fs$, where $s^2 = s + \alpha$, $\alpha \in F$, $4\alpha + 1 \neq 0$. Prove there exists a unique nontrivial involution on A, and also A is a composition algebra with respect to this involution (the algebra $\boldsymbol{K}(\mu)$ of type II).

2. Let A be an algebra over a field F with identity element 1 and with involution $a \to \bar{a}$ such that $a + \bar{a}$, $\bar{a}a \in F$ for every $a \in A$. Also, let (A, α) be the algebra obtained from A by means of the Cayley–Dickson process.

Prove that

(a) the algebra (A, α) is associative if and only if A is associative and commutative;

(b) the algebra (A, α) is alternative if and only if the algebra A is associative.

3. Prove that in a Cayley–Dickson algebra $\boldsymbol{C}(\mu, \beta, \gamma)$ over a field F of characteristic $\neq 2$ it is possible to choose a basis $e_0 = 1, e_1, e_2, \ldots, e_7$ with the following multiplication table ($e_0 = 1$ is the identity element of the algebra $\boldsymbol{C}(\mu, \beta, \gamma)$):

	e_1	e_2	e_3	e_4	e_5	e_6	e_7
e_1	α	e_3	αe_2	e_5	αe_4	$-e_7$	$-\alpha e_6$
e_2	$-e_3$	β	$-\beta e_1$	e_6	e_7	βe_4	βe_5
e_3	$-\alpha e_2$	βe_1	$-\alpha\beta$	e_7	αe_6	$-\beta e_5$	$-\alpha\beta e_4$
e_4	$-e_5$	$-e_6$	$-e_7$	γ	$-\gamma e_1$	$-\gamma e_2$	$-\gamma e_3$
e_5	$-\alpha e_4$	$-e_7$	$-\alpha e_6$	γe_1	$-\alpha\gamma$	γe_3	$\alpha\gamma e_2$
e_6	e_7	$-\beta e_4$	βe_5	γe_2	$-\gamma e_3$	$-\beta\gamma$	$-\beta\gamma e_1$
e_7	αe_6	$-\beta e_6$	$\alpha\beta e_4$	γe_3	$-\alpha\gamma e_2$	$\beta\gamma e_1$	$\alpha\beta\gamma$

where $\alpha = \frac{1}{4}(4\mu + 1) \neq 0$. If $\alpha = \beta = \gamma = -1$ and F is the field of real numbers, then we obtain the classical *algebra of Cayley numbers*, which resolves the Hurwitz problem for the sum of eight squares.

4. Prove that over the field of real numbers the quadratic form $n(x) = x_1^2 + x_2^2 + x_3^2 - x_4^2$ does not admit composition.

5. Determine up to equivalence all nondegenerate quadratic forms over the field of real numbers which admit composition. (There exist seven such forms in all.)

6. Prove that if in the definition of a composition algebra the requirement of strict nondegeneracy of the form $n(x)$ is changed to nondegeneracy, then besides the algebras of types I–IV there appear the following composition algebras: a field F of characteristic 2 and arbitrary purely inseparable quadratic extensions of the field F for which the dimension either equals 2^t or is infinite. *Hint.* If $f(x, y) \equiv 0$, then the mapping $x \to n(x)$ is a ring homomorphism with zero kernel.

3. SIMPLE QUADRATIC ALTERNATIVE ALGEBRAS

In this section we shall prove that with the exception of some algebras of characteristic 2 every simple quadratic alternative algebra is a composition algebra.

We need some simple results on alternative algebras.

Let A be some algebra over an arbitrary ring of operators Φ and $x, y, z \in A$. We denote by $(x, y, z) = (xy)z - x(yz)$ the *associator* of the elements x, y, z; by $[x, y] = xy - yx$ the *commutator* of elements x, y; and by $x \circ y = xy + yx$ the *Jordan product* of the elements x, y. With this notation, the identities defining the variety of alternative algebras are written in the form

$$(x, x, y) = 0, \qquad (x, y, y) = 0.$$

The first of these identities is called the *left alternative identity*, and the second is the *right alternative identity*.

Linearizing the left and right alternative identities, we obtain the identities

$$(x, z, y) + (z, x, y) = 0, \qquad (x, y, z) + (x, z, y) = 0,$$

from which it follows that in an alternative algebra, the associator is a skew-symmetric function of its arguments. In particular, every alternative algebra satisfies the identity

$$(x, y, x) = 0.$$

Algebras satisfying this identity are called *flexible*, and this identity is itself called the flexible identity. For example, every commutative or anticommutative algebra is flexible. In view of flexibility, in an alternative algebra we can (and shall) subsequently write the product xyx without indicating an arrangement of parentheses.

LEMMA 7. In every alternative algebra A the following identities are valid:

$$x(yzy) = [(xy)z]y, \qquad \text{the right Moufang identity,}$$
$$(yzy)x = y[z(yx)], \qquad \text{the left Moufang identity,}$$
$$(xy)(zx) = x(yz)x, \qquad \text{the middle Moufang identity.}$$

PROOF. We shall first prove that the algebra A satisfies the identity

$$(x^2, y, x) = 0. \tag{15}$$

By the flexible and left alternative identities we have

$$(x^2y)x = [x(xy)x] = x(xyx) = x^2(yx),$$

that is, (15) is proved. Now by (15)

$$xy^3 = (xy^2)y - (x, y^2, y) = (xy^2)y = [(xy)y]y.$$

In view of the corollary to Theorem 1.5, the variety of alternative algebras is homogeneous, and therefore by Proposition 1.3 the algebra A satisfies the identity

$$\begin{aligned}0 &= \{xy^3 - [(xy)y]y\}\,\Delta_y^1(z)\\ &= x(yzy) + x(y^2 \circ z) - [(xy)z]y - (xy^2)z - (xz)y^2\\ &= x(yzy) - [(xy)z]y - (x, y^2, z) - (x, z, y^2)\\ &= x(yzy) - [(xy)z]y.\end{aligned}$$

The left Moufang identity is proved analogously. Finally,

$$\begin{aligned}(xy)(zx) - x(yz)x &= -(xy, z, x) + (x, y, z)x\\ &= (z, xy, x) + (z, x, y)x\\ &= [z(xy)]x - z(xyx) + [(zx)y]x - [z(xy)]x = 0.\end{aligned}$$

This proves the lemma.

COROLLARY. Every alternative algebra satisfies the identities

$$(x, xy, z) = (x, y, z)x, \tag{16}$$

$$(x, yx, z) = x(x, y, z), \tag{16'}$$

$$(x^2, y, z) = (x, x \circ y, z), \tag{17}$$

$$(x^2, y, z) = x \circ (x, y, z). \tag{17'}$$

PROOF. Identities (16) and (16′) are equivalent to the right and left Moufang identities, respectively. In fact,

$$\begin{aligned}(z, xy, x) + (z, x, y)x &= -z(xyx) + [(zx)y]x,\\ (x, yx, z) + x(y, x, z) &= (xyx)z - x[y(xz)].\end{aligned}$$

Furthermore, from (15) we obtain

$$0 = (x^2, y, x)\Delta_x^1(z) = (x \circ z, y, x) + (x^2, y, z),$$

whence follows (17). Finally, (17′) follows from (17), (16), and (16′). This proves the corollary.

We shall subsequently also call identities (16)–(17) the *Moufang identities.*

THEOREM 2 (*Artin*). In an alternative algebra A any two elements generate an associative subalgebra.

PROOF. It clearly suffices to show that if $u_i = u_i(x_1, x_2)$ for $i = 1, 2, 3$, are arbitrary nonassociative words from the generators x_1, x_2, then for any $a, b \in A$ the equality $(\bar{u}_1, \bar{u}_2, \bar{u}_3) = 0$ is valid, where $\bar{u}_i = u_i(a, b)$. We shall prove this assertion by induction on the number $n = d(u_1) + d(u_2) + d(u_3)$, where $d(u_i)$ is the length of the word u_i. If $n = 3$, then everything follows from the left and right alternative and flexible identities. Now let $n > 3$ and for any nonassociative words $v_1(x_1, x_2)$, $v_2(x_1, x_2)$, $v_3(x_1, x_2)$ such that $d(v_1) + d(v_2) + d(v_3) < n$ let the equality $(\bar{v}_1, \bar{v}_2, \bar{v}_3) = 0$ hold. Then by the induction assumption the associator $(\bar{u}_1, \bar{u}_2, \bar{u}_3)$ does not depend on the way the parentheses are arranged in the words u_i, and therefore it is possible to assume that each u_i has parentheses arranged from the right, that is,

$$u_i = \{\cdots [(x_1^{(i)} x_2^{(i)}) x_3^{(i)}] \cdots \} x_{k_i}^{(i)},$$

where $x_j^{(i)} \in \{x_1, x_2\}$ and $k_i = d(u_i)$. In this case some two of the words u_1, u_2, u_3 terminate in the same generator. For example, let $u_1 = (v_1)x_1$, $u_2 = (v_2)x_1$. It is easy to see that if v_1 or v_2 is missing, then by the induction assumption and the Moufang identities $(\bar{u}_1, \bar{u}_2, \bar{u}_3) = 0$. Consequently, it is possible to assume that $d(v_1) \geq 1$ and $d(v_2) \geq 1$. Then by the linearized Moufang identity we have

$$\begin{aligned}(\bar{u}_1, \bar{u}_2, \bar{u}_3) &= (\bar{v}_1 a, \bar{v}_2 a, \bar{u}_3) \\ &= -(\bar{v}_1 \bar{u}_3, \bar{v}_2 a, a) + a(\bar{v}_1, \bar{v}_2 a, \bar{u}_3) + \bar{u}_3(\bar{v}_1, \bar{v}_2 a, a) \\ &= -(\bar{v}_1 \bar{u}_3, \bar{v}_2 a, a) = -a(\bar{v}_1 \bar{u}_3, \bar{v}_2, a) = 0.\end{aligned}$$

This proves the theorem.

An algebra A is called a *power-associative algebra* if each element of the algebra A generates an associative subalgebra.

COROLLARY. Every alternative algebra is power associative.

We note that it also follows from Artin's theorem that an alternative algebra can be defined as an algebra in which any two elements generate an associative subalgebra.

We recall that an algebra A over a field F with identity element 1 is called *quadratic* over F if each element x from A satisfies an equality of the form

$$x^2 - t(x)x + n(x) = 0; \qquad t(x), n(x) \in F \tag{18}$$

(in addition, we identify the field F with the subalgebra $F \cdot 1$ of the algebra A). We shall call the elements $t(x)$ and $n(x)$ the *trace* and *norm*, respectively, of the element x. By definition we set $t(\alpha) = 2\alpha$, $n(\alpha) = \alpha^2$ for $\alpha \in F$. Then the trace $t(x)$ and norm $n(x)$ are uniquely defined for each $x \in A$.

THEOREM 3. Let A be a quadratic algebra over a field F which contains at least three elements. Then the trace $t(x)$ is a linear function, and the norm $n(x)$ is a quadratic form on A. If, in addition, the algebra A is alternative, then the quadratic form $n(x)$ satisfies relation (1).

PROOF. Let $\alpha, \beta \in F$ with $\alpha \neq \beta$, $\alpha\beta \neq 0$. From (18) for any $x, y \in A$ we have

$$0 = (\alpha x + \beta y)^2 - t(\alpha x + \beta y)(\alpha x + \beta y) + n(\alpha x + \beta y) - \alpha\beta[(x + y)^2 - t(x + y)(x + y) + n(x + y)],$$

and furthermore,

$$0 = [\alpha^2 t(x) - \alpha t(\alpha x + \beta y) - \alpha\beta t(x) + \alpha\beta t(x + y)]x + [\beta^2 t(y) - \beta t(\alpha x + \beta y) - \alpha\beta t(y) + \alpha\beta t(x + y)]y + [-\alpha^2 n(x) - \beta^2 n(y) + n(\alpha x + \beta y) + \alpha\beta n(x) + \alpha\beta n(y) - \alpha\beta n(x + y)].$$

Now if $1, x, y$ are linearly independent over F, then we have

$$\alpha[\alpha t(x) - t(\alpha x + \beta y) - \beta t(x) + \beta t(x + y)] = 0,$$
$$\beta[\beta t(y) - t(\alpha x + \beta y) - \alpha t(y) + \alpha t(x + y)] = 0,$$

whence

$$(\alpha - \beta)[t(x) + t(y) - t(x + y)] = 0$$

and

$$t(x + y) = t(x) + t(y). \tag{19}$$

Now let the elements $1, x, y$ be linearly dependent over F and, for example, $y = \alpha x + \beta$. By direct calculation we obtain from (18) $t(y) = \alpha t(x) + 2\beta$ and $t(x + y) = t((1 + \alpha)x + \beta) = (1 + \alpha)t(x) + 2\beta = t(x) + t(y)$. Thus equality (19) is proved for all x and y.

Furthermore, for any $\lambda \in F$ we have

$$\lambda^2 x^2 - \lambda t(x)\lambda x + \lambda^2 n(x) = 0,$$

whence $t(\lambda x) = \lambda t(x)$ and $n(\lambda x) = \lambda^2 n(x)$. By the same token the linearity of the trace $t(x)$ is established. For the proof that the norm $n(x)$ is a quadratic form on A, it remains for us to show that the function $f(x, y) = n(x + y) - n(x) - n(y)$ is a bilinear form on A. From the equality

$$(x + y)^2 - t(x + y)(x + y) + n(x + y) = 0,$$

in view of (19) we obtain

$$f(x, y) = t(x)y + t(y)x - x \circ y. \tag{20}$$

Hence the bilinearity of the function $f(x, y)$ follows from the linearity of the trace. At the same time the first part of Theorem 3 is proved.

Now let A be an alternative algebra. We shall prove equality (1). By Artin's theorem the subalgebra generated by elements x and y is associative.

Multiplying equality (20) on the right by xy, we obtain

$$f(x, y)xy = t(x)yxy + t(y)x^2y - (xy)^2 - yx^2y.$$

By (18) we have

$$\begin{aligned} yx^2y &= y(t(x)x - n(x))y \\ &= t(x)yxy - n(x)t(y)y + n(x)n(y), \\ t(y)x^2y &= t(y)t(x)xy - t(y)n(x)y. \end{aligned}$$

Substituting these two equalities into the previous one, we obtain

$$(xy)^2 - (-f(x, y) + t(y)t(x))xy + n(x)n(y) = 0,$$

whence follows (1) if $xy \notin F$. Now let $xy = \alpha \in F$. Since $n(\lambda x) = \lambda^2 n(x) = n(\lambda)n(x)$ for any $\lambda \in F$, it suffices for us to consider the case when $x \notin F$, $y \notin F$. Multiplying equality (18) on the right by y^2, we obtain

$$\alpha^2 - \alpha t(x)y + n(x)y^2 = 0.$$

Let $n(x) = 0$. If in this case $\alpha \neq 0$, then we have $\alpha = t(x)y$, whence $y \in F$, which is not so. This means $\alpha = 0$ and $n(xy) = 0 = n(x)n(y)$. If $n(x) \neq 0$, then we obtain

$$\frac{\alpha^2}{n(x)} - \frac{\alpha t(x)}{n(x)} y + y^2 = 0,$$

whence in view of the fact that $y \notin F$ it follows $n(y) = \alpha^2/n(x)$, and $n(x)n(y) = \alpha^2 = n(xy)$. Equality (1) is established, and this proves the theorem.

THEOREM 4. Let A be a simple quadratic alternative algebra over a field F containing at least three elements. Then either A is a composition algebra or A is some field of characteristic 2.

PROOF. By Theorem 3 there is defined on the algebra A a quadratic form $n(x)$ connected with the multiplication in the algebra A by equality (1). Let $f(x, y)$ be the bilinear form which corresponds to the quadratic form $n(x)$. We consider the set $W = \{a \in A \mid f(a, A) = 0\}$, the kernel of the form $f(x, y)$, and we show that W is an ideal in A. It is clear that W is a subspace of the vector space A. As in the proof of Lemma 1, linearizing relation (1) we obtain that relation (3) is valid for the algebra A. Let $a \in W$ and $x, y \in A$. Then from (3) we obtain

$$f(ax, y) = f(ax, y \cdot 1) = -f(a, yx) + f(a, y)f(x, 1) = 0,$$

that is, W is a right ideal of the algebra A. It is proved analogously that W is a left ideal in A. In view of the simplicity of the algebra A, either $W = (0)$

or $W = A$. In the first case the form $n(x)$ is strictly nondegenerate, and A is a composition algebra. Now let $W = A$. In this case $f(x, y) \equiv 0$ on A, so that $n(x + y) = n(x) + n(y)$ for any $x, y \in A$. Consequently the mapping $x \to n(x)$ is a homomorphism of the ring A into the ring F. As is easy to see, the kernel of this homomorphism is an ideal of the algebra A. Since the algebra A is simple and $n(1) = 1 \neq 0$, this kernel equals zero, that is, the ring A is isomorphic to some subring of the ring F. Furthermore, from the equality $0 = f(1, 1) = 2$ it follows that in this case the field F has characteristic 2. It is also clear that the ring A is a subfield of the field F.

This proves the theorem.

Exercises

1. (Generalized theorem of Artin) Let A be an alternative algebra and $a, b, c \in A$. Prove that if $(a, b, c) = 0$, then the subalgebra generated by the elements a, b, c is associative.

In Exercises 2–4 D is a finite-dimensional alternative algebra without divisors of zero over the field of real numbers $\boldsymbol{R}$.

2. Prove that D is a division algebra, that is, for any $a, b \in D$ with $a \neq 0$ there are solutions to the equations

$$ax = b, \qquad ya = b.$$

3. Prove that there is an identity element 1 in D.

4. Prove that the algebra D is quadratic over $\boldsymbol{R}$. *Hint.* Each element of the algebra D generates an associative subalgebra. Therefore the concept of a minimal polynomial for an element of a finite-dimensional algebra makes sense in D.

5. (Generalized theorem of Frobenius) Let D be a finite-dimensional alternative algebra without divisors of zero over the field of real number $\boldsymbol{R}$. Then D is isomorphic to one of the following algebras over $\boldsymbol{R}$: (1) $\boldsymbol{R}$; (2) the field of complex numbers $\boldsymbol{K}$; (3) the division ring of quaternions $\boldsymbol{Q}$; (4) the algebra of Cayley numbers $\boldsymbol{C}$.

6. Let A be an algebra with identity element 1 over a field F, and let K be some extension of the field F. Prove that if the algebra $A_K = K \otimes_F A$ is quadratic over K, then the algebra A is quadratic over F.

4. FURTHER PROPERTIES OF COMPOSITION ALGEBRAS

In this section we shall prove some properties of composition algebras which we need later. We recall that in every composition algebra A, in addition to the quadratic form $n(x)$, there are also defined the bilinear form $f(x, y)$ connected with $n(x)$ and the linear trace form $t(x) = f(1, x)$. Furthermore, each element of the algebra A satisfies equality (18). If B is some subspace of the composition algebra A, then, as before, we denote by $B^\perp$ the orthogonal complement of the space B with respect to the form $f(x, y)$.

Composition algebras different from a field F can be represented in the form

$$\begin{aligned} \boldsymbol{K}(\mu) &= F + Fv_1, \\ v_1^2 &= v_1 + \mu, \qquad 4\mu + 1 \neq 0, \\ \boldsymbol{Q}(\mu, \beta) &= (\boldsymbol{K}(\mu), \beta) = \boldsymbol{K}(\mu) + v_2\boldsymbol{K}(\mu), \\ v_2 &\in \boldsymbol{K}(\mu)^\perp, \qquad v_2^2 = \beta \neq 0, \\ \boldsymbol{C}(\mu, \beta, \gamma) &= (\boldsymbol{Q}(\mu, \beta), \gamma) = \boldsymbol{Q}(\mu, \beta) + v_3\boldsymbol{Q}(\mu, \beta), \\ v_3 &\in \boldsymbol{Q}(\mu, \beta)^\perp, \qquad v_3^2 = \gamma \neq 0. \end{aligned}$$

We note that since by (10) $v_2 k = \bar{k} v_2$, $v_3 q = \bar{q} v_3$ for any $k \in \boldsymbol{K}(\mu)$, $q \in \boldsymbol{Q}(\mu, \beta)$, the algebras $\boldsymbol{Q}(\mu, \beta)$ and $\boldsymbol{C}(\mu, \beta, \gamma)$ can also be represented in the form

$$\boldsymbol{Q}(\mu, \beta) = \boldsymbol{K}(\mu) + \boldsymbol{K}(\mu)v_2, \qquad \boldsymbol{C}(\mu, \beta, \gamma) = \boldsymbol{Q}(\mu, \beta) + \boldsymbol{Q}(\mu, \beta)v_3.$$

Hence it follows that in the algebras $\boldsymbol{K}(\mu)$, $\boldsymbol{Q}(\mu, \beta)$, and $\boldsymbol{C}(\mu, \beta, \gamma)$, respectively, it is possible to choose the bases

$$\{1, v_1\}, \quad \{1, v_1, v_2, v_1v_2\}, \quad \{1, v_1, v_2, v_3, v_1v_2, v_1v_3, v_2v_3, (v_1v_2)v_3\}.$$

Also, $v_1^2 = v_1 + \mu$, $v_2^2 = \beta$, $v_3^2 = \gamma$; $(4\mu + 1)\beta\gamma \neq 0$; $\bar{v}_1 = 1 - v_1$, $\bar{v}_2 = -v_2$, $\bar{v}_3 = -v_3$; $v_iv_j = v_j\bar{v}_i$ for $j > i$.

If F is a field of characteristic $\neq 2$, then each composition algebra over F that is different from F is representable in the form $F + F^\perp$. In addition, in any case the algebras $\boldsymbol{Q}(\mu, \beta)$ and $\boldsymbol{C}(\mu, \beta, \gamma)$ are representable in the form $\boldsymbol{K}(\mu) + \boldsymbol{K}(\mu)^\perp$, and the algebra $\boldsymbol{C}(\mu, \beta, \gamma)$ in the form $\boldsymbol{Q}(\mu, \beta) + \boldsymbol{Q}(\mu, \beta)^\perp$.

Finally, we note that by Lemma 3 the algebras $\boldsymbol{Q}(\mu, \beta)$ and $\boldsymbol{C}(\mu, \beta, \gamma)$ are always simple algebras.

Now let A be some algebra. We denote by $N(A)$ the collection of all elements n from A such that $(n, a, b) = (a, n, b) = (a, b, n) = 0$ for any elements $a, b \in A$. The set $N(A)$ is called the *associative center* of the algebra A. The collection of all elements $z \in N(A)$ such that $[z, a] = 0$ for all $a \in A$ is called the *center* of the algebra A and is denoted by $Z(A)$.

THEOREM 5. Let A be a composition algebra over a field F. If $\dim_F A \geq 4$, then $Z(A) = F$. If $A = \boldsymbol{C}(\mu, \beta, \gamma)$ is a Cayley–Dickson algebra, then also $N(A) = F$.

PROOF. It is clear that $F \subseteq Z(A) \subseteq N(A)$. For the proof of the containment $Z(A) \subseteq F$, it suffices to show that if $[a, x] = 0$ for all x from A, then $a \in F$. As was noted above, $A = \boldsymbol{K}(\mu) + \boldsymbol{K}(\mu)^\perp$. Let $a = k + a'$ where $k \in \boldsymbol{K}(\mu)$,

$a' \in \boldsymbol{K}(\mu)^{\perp}$. We have $0 = [a, v_1] = [a', v_1]$. But in view of (10), $v_1 a' = a'\bar{v}_1 = a' - a'v_1$, and therefore we obtain $a' = 2a'v_1 = 4a'v_1^2 = 4a'(v_1 + \mu) = 2a' + 4a'\mu = (2 + 4\mu)a'$. Since $4\mu + 1 \neq 0$, it follows from this that $a' = 0$, that is, $a = k \in \boldsymbol{K}(\mu)$. Let $a = \alpha + \beta v_1$, where $\alpha, \beta \in F$. Then we have $0 = [a, v_2] = \beta[v_1, v_2] = \beta(v_1 v_2 - \bar{v}_1 v_2) = \beta(2v_1 v_2 - v_2)$. Hence $\beta = 0$ and $a \in F$. By the same token the equality $Z(A) = F$ is proved.

Now let $A = \boldsymbol{C}(\mu, \beta, \gamma)$ and $a \in N(A)$. We again can write $a = q + a'$ where $q \in \boldsymbol{Q}(\mu, \beta)$, $a' \in \boldsymbol{Q}(\mu, \beta)^{\perp}$. By assumption and by (11) we have $0 = (a, v_1, v_2) = (a', v_1, v_2) = (a'v_1)v_2 - a'(v_1 v_2) = a'[v_2, v_1] = a'(v_2 - 2v_1 v_2)$. Hence $a'v_2 = 2a'(v_1 v_2)$, and again by (11) $a'v_2 = 2(a'v_2)v_1 = 4(a'v_2)v_1^2 = (2 + 4\mu)a'v_2$. Since $1 + 4\mu \neq 0$, we obtain $a'v_2 = 0$. Hence $0 = (a'v_2)v_2 = a'v_2^2 = \beta a'$, and $a' = 0$. Thus $a = q \in \boldsymbol{Q}(\mu, \beta)$. Now let b be an arbitrary element from $\boldsymbol{Q}(\mu, \beta)$. Then we have $0 = (v_3, a, b) = (v_3 a)b - v_3(ab)$, whence by (11) we obtain $0 = v_3[a, b]$. Furthermore, $0 = v_3(v_3[a, b]) = \gamma[a, b]$, that is, $[a, b] = 0$. We have obtained that $a \in Z(\boldsymbol{Q}(\mu, \beta))$, whence by what has been proved $a \in F$.

This proves the theorem.

LEMMA 8. Let A be a composition algebra. Then $x^2, x \circ y \in F$ for any $x, y \in A$ such that $t(x) = t(y) = 0$. In particular, for any $x, y, z, r, s, t \in A$ the following containments hold:

$$[x, y]^2, \quad [x, y] \circ [z, r], \quad (x, y, z)^2, \quad (x, y, z) \circ [r, s],$$
$$(x, y, z) \circ (r, s, t) \in F.$$

PROOF. Identity (9) is valid in the algebra A and can be rewritten in the form

$$x \circ y = t(x)y + t(y)x - t(x)t(y) + t(xy). \tag{21}$$

Hence follows the first assertion of the lemma. For the proof of the second assertion it suffices to note that

$$t([x, y]) = t((x, y, z)) = 0. \tag{22}$$

Since $x + \bar{x} = t(x) \in F$, we then have

$$[\bar{x}, y] = -[x, y], \qquad (\bar{x}, y, z) = (x, \bar{y}, z) = (x, y, \bar{z}) = -(x, y, z).$$

Now

$$t([x, y]) = [x, y] + \overline{[x, y]} = [x, y] - [\bar{x}, \bar{y}] = 0,$$
$$t((x, y, z)) = (x, y, z) + \overline{(x, y, z)} = (x, y, z) - (\bar{z}, \bar{y}, \bar{x})$$
$$= (x, y, z) + (z, y, x) = 0,$$

that is (22) is established, and at the same time this proves the lemma.

COROLLARY. In every composition algebra the following identities are valid:

$$
\begin{aligned}
[[x, y]^2, t] &= [[x, y] \circ [z, t], r] = 0, \\
[(x, y, z)^2, t] &= [(x, y, z) \circ [r, s], t] \\
&= [(x, y, z) \circ (r, s, t), u] = 0, \\
([x, y]^2, s, t) &= ([x, y] \circ [z, r], s, t) = 0, \\
((x, y, z)^2, s, t) &= ((x, y, z) \circ [r, s], u, t) \\
&= ((x, y, z) \circ (r, s, t), u, v) = 0.
\end{aligned}
$$

LEMMA 9. For a composition algebra A the following conditions are equivalent:

(a) $n(x) = 0$ for some $x \neq 0$ from A;
(b) there are zero divisors in A.

PROOF. If $x \neq 0$ and $n(x) = x\bar{x} = 0$, then it is clear that x is a zero divisor. Conversely, let $xy = 0$ where $x, y \in A$ and $x \neq 0$, $y \neq 0$. Then we have $0 = n(xy) = n(x)n(y)$, whence either $n(x) = 0$ or $n(y) = 0$. This proves the lemma.

If one of the conditions (a), (b) of Lemma 9 is satisfied in a composition algebra A, then A is called *split*.

LEMMA 10. A composition algebra A is split if and only if A contains an idempotent $e \neq 0, 1$.

PROOF. If there is an idempotent $e \neq 0, 1$ in A, then we have $e^2 = e$, whence $n(e) = 0$ and the algebra A is split. Conversely, let the algebra A be split. If there can be found an element x in A such that $n(x) = 0$, $t(x) = \alpha \neq 0$, then the element $\alpha^{-1}x$ will be the desired idempotent. Now suppose such elements do not exist, that is, for every $x \in A$ the equality $n(x) = 0$ implies $t(x) = 0$. Since A is split, there exists $0 \neq a \in A$ for which $n(a) = 0$. For every $x \in A$ we have $n(ax) = n(a)n(x) = 0$, so by our assumption also $t(ax) = 0$. Now by (8) we obtain $f(a, x) = t(a)t(x) - t(ax) = 0$. In view of the nondegeneracy of the form $f(x, y)$, it follows from this that $a = 0$. We have obtained a contradiction which proves the lemma.

THEOREM 6. Any two split composition algebras of the same dimension over a field F are isomorphic.

PROOF. Let A be a split composition algebra and B be a nondegenerate subalgebra of the algebra A which contains the identity element 1 and an idempotent $e \neq 0, 1$. We shall prove that if $B \neq A$ then there exists $v \in B^\perp$ such that $v^2 = 1$. As in the proof of Theorem 1 we obtain that there exists $u \in B^\perp$ for which $u^2 = -n(u) = \alpha \neq 0$, and also $uB \subseteq B^\perp$. We now set $v = u + \alpha^{-1}(1-\alpha)ue$. Then $v \in B^\perp$. In addition, by the middle Moufang identity we have $(ue)^2 = (ue)(\bar{e}u) = u(e\bar{e})u = 0$, whence

$$\begin{aligned} v^2 &= (u + \alpha^{-1}(1-\alpha)ue)^2 = u^2 + \alpha^{-1}(1-\alpha)(u^2e + ueu) \\ &= \alpha + (\alpha^{-1} - 1)(\alpha e + u(u\bar{e})) = \alpha + (1-\alpha)(e + \bar{e}) = 1. \end{aligned}$$

Repeating now the arguments carried out for the proof of Theorem 1, by Lemma 10 we obtain that the algebra A is isomorphic to either the algebra $\boldsymbol{K}(0)$, or the algebra $\boldsymbol{Q}(0,1)$, or the algebra $\boldsymbol{C}(0,1,1)$. Hence follows the assertion of the theorem.

COROLLARY. Over an algebraically closed field F there exist, in all, 4 nonisomorphic composition algebras.

PROOF. It suffices to show that every composition algebra A over the field F of dimension $n > 1$ is split. Let $a \in A$, $a \notin F$, and let $\alpha \in F$ be a root of the equation $x^2 - t(a)x + n(a) = 0$. Then we have

$$a^2 - \alpha^2 = t(a)(a - \alpha),$$

whence

$$(a - \alpha)(a + \alpha - t(a)) = 0,$$

that is, there are zero divisors in A. This proves the corollary.

We consider in greater detail the structure of the split composition algebras.

First of all it is clear that $\boldsymbol{K}(0) \cong F \oplus F$, the corresponding isomorphism established by the mapping

$$\alpha + \beta v_1 \rightarrow (\alpha + \beta, \alpha).$$

In addition, the involution $\overline{(\alpha, \beta)} = (\beta, \alpha)$ in the algebra $F \oplus F$ corresponds to the involution of the algebra $\boldsymbol{K}(0)$.

Furthermore, we consider the algebra F_2 of matrices of order 2×2 with elements from F. As is easy to see F_2 is a composition algebra with respect to the quadratic form $n(x) = \det x$. Also, in view of the presence of zero divisors, the algebra F_2 is split. By Theorem 6 every four-dimensional split composition algebra over F is isomorphic to F_2. Consequently $\boldsymbol{Q}(0,1) \cong F_2$. This isomorphism can also be established directly by considering the correspondence

$$\alpha + \beta v_1 + \gamma v_2 + \delta v_1 v_2 \rightarrow \begin{pmatrix} \alpha + \beta & \gamma + \delta \\ \gamma & \alpha \end{pmatrix}.$$

As we already noted, the corresponding quadratic form on the algebra F_2 is the determinant of the matrix. We note, too, that the linear trace form $t(x)$ on the algebra F_2 is the usual trace of a matrix. In addition, by the Hamilton–Cayley theorem equality (18) is valid in F_2. Finally, the involution $a \to \bar{a} = t(a) - a$ on the algebra F_2 appears in the manner

$$\overline{\begin{pmatrix} \alpha & \beta \\ \gamma & \delta \end{pmatrix}} = \begin{pmatrix} \delta & -\beta \\ -\gamma & \alpha \end{pmatrix}. \tag{23}$$

We now turn to consideration of the Cayley–Dickson algebra $\boldsymbol{C}(0,1,1)$. By what was already proved, we have $\boldsymbol{C}(0,1,1) \cong (F_2, 1)$, that is, $\boldsymbol{C}(0,1,1) = F_2 + vF_2$, where $v \in F_2^{\perp}$, $v^2 = 1$. Let $e_{11}, e_{12}, e_{21}, e_{22}$ be the matrix units of the algebra F_2. We set

$$\begin{aligned} e_{12}^{(1)} &= e_{12}, & e_{12}^{(2)} &= ve_{22}, & e_{12}^{(3)} &= ve_{21}; \\ e_{21}^{(1)} &= e_{21}, & e_{21}^{(2)} &= ve_{11}, & e_{21}^{(3)} &= -ve_{12}. \end{aligned}$$

Then the elements $e_{11}, e_{22}, e_{12}^{(k)}, e_{21}^{(k)}$ for $k = 1, 2, 3$ form a basis of the algebra $\boldsymbol{C}(0,1,1)$.

LEMMA 11. The elements e_{ii}, $e_{ij}^{(k)}$ $(i, j = 1,2;\ k = 1,2,3)$ satisfy the relations

$$\begin{aligned} &\text{(a)} && e_{ii}^2 = e_{ii}, \quad e_{ii}e_{jj} = 0; \\ &\text{(b)} && e_{ii}e_{ij}^{(k)} = e_{ij}^{(k)}e_{jj} = e_{ij}^{(k)}, \quad e_{jj}e_{ij}^{(k)} = e_{ij}^{(k)}e_{ii} = 0; \\ &\text{(c)} && (e_{ij}^{(k)})^2 = e_{ij}^{(k)} \circ e_{ij}^{(l)} = 0 \quad (k \neq l); \\ &\text{(d)} && e_{ij}^{(k)}e_{ji}^{(k)} = e_{ii}, \quad e_{ij}^{(k)}e_{ji}^{(l)} = 0 \quad (k \neq l); \\ &\text{(e)} && e_{ij}^{(k)}e_{ij}^{(k+1)} = (j - i)e_{ji}^{(k+2)} \quad (k \bmod 3). \end{aligned} \tag{24}$$

PROOF. Relations (a) are satisfied in the algebra F_2. The validity of relations (b–e) is verified directly from the definition of multiplication in the algebra $(F_2, 1)$. For example, we shall prove the validity of relation (d). We have

$$e_{ij}^{(1)}e_{ji}^{(1)} = e_{ij}e_{ji} = e_{ii}.$$

Furthermore, it follows from (23) that $\bar{e}_{ii} = e_{jj}$, $\bar{e}_{ij} = -e_{ij}$. Therefore

$$\begin{aligned} e_{ij}^{(2)}e_{ji}^{(2)} &= (ve_{jj})(ve_{ii}) = e_{ii}\bar{e}_{jj} = e_{ii}^2 = e_{ii}, \\ e_{ij}^{(3)}e_{ji}^{(3)} &= -(ve_{ji})(ve_{ij}) = -e_{ij}\bar{e}_{ji} = e_{ij}e_{ji} = e_{ii}. \end{aligned}$$

On the other hand, we have

$$\begin{aligned} e_{ij}^{(1)}e_{ji}^{(2)} &= e_{ij}(ve_{ii}) = v(\bar{e}_{ij}e_{ii}) = -v(e_{ij}e_{ii}) = 0, \\ e_{ij}^{(1)}e_{ji}^{(3)} &= \pm e_{ij}(ve_{ij}) = \pm v(\bar{e}_{ij}e_{ij}) = 0, \\ e_{ij}^{(2)}e_{ji}^{(3)} &= \pm(ve_{jj})(ve_{ij}) = \pm e_{ij}\bar{e}_{jj} = \pm e_{ij}e_{ii} = 0. \end{aligned}$$

We have proved that $e_{ij}^{(k)}e_{ji}^{(l)} = 0$ for $k < l$. Since $t(e_{ij}^{(k)}) = 0$, then by (21) we have $e_{ij}^{(k)} \circ e_{ji}^{(l)} = 0$ for $k \neq l$, hence also $e_{ij}^{(k)}e_{ji}^{(l)} = 0$ for $k > l$. By the same token this proves (d). The remaining relations are proved analogously. This proves the lemma.

The eight-dimensional algebra over a field F with basis e_{ii}, $e_{ij}^{(k)}$ $(i, j = 1, 2;\ k = 1, 2, 3)$ and multiplication table (24) is called the *Cayley–Dickson matrix algebra* over F. We shall denote this algebra by $\boldsymbol{C}(F)$. We shall call the elements e_{ii}, $e_{ij}^{(k)}$ the *Cayley–Dickson matrix units*. From what was proved above it follows that every split Cayley–Dickson algebra over a field F is isomorphic to the Cayley–Dickson matrix algebra $\boldsymbol{C}(F)$.

Thus is valid:

THEOREM 7. Every split composition algebra over a field F is isomorphic to either the algebra $F \oplus F$, or the algebra of matrices F_2, or the Cayley–Dickson matrix algebra $\boldsymbol{C}(F)$.

The Cayley–Dickson matrix algebra can be considered not only over a field but also over an arbitrary associative–commutative ring Φ. It is clear that the algebra $\boldsymbol{C}(\Phi)$ will also be alternative.

The name "matrix algebra" stems from the fact that elements from $\boldsymbol{C}(F)$ can be represented by matrices in the following manner. Let $a \in \boldsymbol{C}(F)$ with

$$a = \alpha_{11}e_{11} + \alpha_{22}e_{22} + \sum_{k=1}^{3} (\alpha_{12}^{(k)}e_{12}^{(k)} + \alpha_{21}^{(k)}e_{21}^{(k)}).$$

We associate the element a with the matrix

$$\begin{pmatrix} \alpha_{11} & a_{12} \\ a_{21} & \alpha_{22} \end{pmatrix}, \qquad \text{where} \quad a_{ij} = (\alpha_{ij}^{(1)}, \alpha_{ij}^{(2)}, \alpha_{ij}^{(3)}) \in F^3. \tag{25}$$

Addition and scalar multiplication of elements of the algebra $\boldsymbol{C}(F)$ will then correspond to the usual addition and scalar multiplication of matrices of the form (25). However, multiplication of elements of the algebra $\boldsymbol{C}(F)$ will correspond to the following multiplication of matrices of form (25):

$$\begin{pmatrix} \alpha_{11} & a_{12} \\ a_{21} & \alpha_{22} \end{pmatrix}\begin{pmatrix} \beta_{11} & b_{12} \\ b_{21} & \beta_{22} \end{pmatrix}$$
$$= \begin{pmatrix} \alpha_{11}\beta_{11} + (a_{12}, b_{21}) & \alpha_{11}b_{12} + \beta_{22}a_{12} - a_{21} \times b_{21} \\ \beta_{11}a_{21} + \alpha_{22}b_{21} + a_{12} \times b_{12} & \alpha_{22}\beta_{22} + (a_{21}, b_{12}) \end{pmatrix},$$

where for vectors $x = (x_1, x_2, x_3)$, $y = (y_1, y_2, y_3) \in F^3$ by $(x, y) = x_1y_1 + x_2y_2 + x_3y_3$ is denoted their scalar product, and by $x \times y = (x_2y_3 - x_3y_2, x_3y_1 - x_1y_3, x_1y_2 - x_2y_1)$ their "vector product." We note, too, that for this representation the trace of the element a is the usual trace of the matrix,

$t(a) = \alpha_{11} + \alpha_{22}$. However, the corresponding quadratic form (or norm) is the following analog of the determinant for matrices of form (25): $n(a) = \alpha_{11}\alpha_{22} - (a_{12}, a_{21})$. Also, the action of the involution $a \to \bar{a}$ on the algebra $\boldsymbol{C}(F)$ is defined in the given case by equality (23).

We shall now indicate examples of composition algebras without zero divisors. As we have seen, over an algebraically closed field F there do not exist such algebras other than F. We shall next show that any field F can be extended to a field F_1 over which there already exist composition algebras without zero divisors.

LEMMA 12. Let F be an arbitrary field and $F_1 = F(\alpha)$ be a simple transcendental extension of the field F. Then the composition algebra $\boldsymbol{K}(\alpha)$ over the field F_1 does not contain zero divisors.

PROOF. It suffices to prove that the polynomial $x^2 - x - \alpha$ is irreducible over the field $F(\alpha)$. Suppose that this is not so, that is, $a^2 - a - \alpha = 0$ for some $a \in F(\alpha)$. Then $a = f \cdot g^{-1}$ where f, g are polynomials in α with coefficients from the field F. We also have $f^2 - fg = \alpha g^2$. If $\deg f = m$, $\deg g = n$, then we obtain that $\max\{2m, m + n\} = 2n + 1$. However, as is easy to see, this equality is impossible no matter what m, n. We have obtained a contradiction which proves the lemma.

LEMMA 13. Let A be a finite-dimensional algebra over a field F with identity element 1 and with involution $a \to \bar{a}$ such that $a + \bar{a}, a\bar{a} \in F$ for any $a \in A$; and let $F_1 = F(\alpha)$ be a simple transcendental extension of the field F. Then if A does not have zero divisors, the algebra $A_{F_1} = F_1 \otimes_F A$ and the algebra $A_1 = (A_{F_1}, \alpha)$, obtained from the algebra A_{F_1} by means of the Cayley–Dickson process, also do not contain nonzero divisors of zero.

PROOF. We shall prove first that there are no zero divisors in the algebra A_{F_1}. Let $ab = 0$ for some $a, b \in F_1$ where $a \neq 0$, $b \neq 0$. Then $a = \sum_i t_i \otimes a_i$, $b = \sum_j s_j \otimes b_j$, where $t_i, s_j \in F(\alpha)$ and $a_i, b_j \in A$. Let f be the common denominator of the quotients t_i, and g be the common denominator of the quotients s_j. Then $fa = \sum_{i=0}^n \alpha^i \otimes a_i'$, $gb = \sum_{j=0}^m \alpha^j \otimes b_j'$, where $a_i', b_j' \in A$ and $a_n' \neq 0$, $b_m' \neq 0$. We have $(fa)(gb) = 0$, whence $a_n'b_m' = 0$, and either $a_n' = 0$ or $b_m' = 0$. We have obtained a contradiction proving that either $a = 0$ or $b = 0$, that is, there are no zero divisors in the algebra A_{F_1}. Now let $x = a + vb$, $y = c + vd$ be nonzero elements from (A_{F_1}, α) such that $xy = 0$. Then

$$ac + \alpha d\bar{b} = 0, \qquad \bar{a}d + cb = 0. \tag{26}$$

Since there are no zero divisors in the algebra A_{F_1}, all the elements a, b, c, d must be nonzero. As in the previous case, there exist polynomials f, g, h, t

from the ring $F[\alpha]$ such that

$$fa = \sum_{i=0}^{n} \alpha^i \otimes a_i, \qquad gb = \sum_{i=0}^{m} \alpha^i \otimes b_i,$$

$$hc = \sum_{i=0}^{k} \alpha^i \otimes c_i, \qquad td = \sum_{i=0}^{l} \alpha^i \otimes d_i,$$

where all $a_i, b_i, c_i, d_i \in A$ and a_n, b_m, c_k, d_l are different from zero. The first of equalities (26) gives us

$$gt\left(\sum_{i=0}^{n} \alpha^i \otimes a_i\right)\left(\sum_{i=0}^{k} \alpha^i \otimes c_i\right) + \alpha fh\left(\sum_{i=0}^{l} \alpha^i \otimes d_i\right)\left(\sum_{i=0}^{m} \alpha^i \otimes \bar{b}_i\right) = 0,$$

whence

$$\deg g + \deg t + n + k = \deg f + \deg h + l + m + 1. \tag{27}$$

Analogously from the second equality (26) we obtain

$$\deg g + \deg h + n + l = \deg f + \deg t + k + m. \tag{28}$$

Now subtracting (28) from (27) we obtain

$$2(\deg t - \deg h) + 2(k - l) = 1,$$

which is impossible since 1 is not an even number.

This contradiction proves that the algebra (A_{F_1}, α) does not have zero divisors, which proves the lemma.

From Lemmas 12 and 13 follows

THEOREM 8. Let F be an arbitrary field. Then for any $n = 2, 4, 8$ there exists an infinite extension F_1 of the field F and a composition algebra A over the field F_1 of dimension n which does not contain nonzero divisors of zero.

Exercises

1. Prove that the basis elements v_1, v_2, v_3 of a Cayley–Dickson algebra satisfy the relations

$$(v_i v_j)v_k = \pm(v_1 v_2)v_3 + \varepsilon v_2 v_3, \qquad v_i(v_j v_k) = \pm(v_1 v_2)v_3 + \varepsilon v_2 v_3,$$

where (i, j, k) is an arbitrary permutation of the symbols 1, 2, 3; $\varepsilon = 0, \pm 1$. If F is a field of characteristic $\neq 2$, then the elements $e_1 = v_1 - \frac{1}{2}$, $e_2 = v_2$, $e_3 = v_3$ satisfy the relation

$$(e_i e_j)e_k = -e_i(e_j e_k) = (-1)^{\operatorname{sgn} \sigma}(e_1 e_2)e_3$$

for any permutation $\sigma = (i, j, k)$ of the symbols 1, 2, 3.

In Exercises 2–5 A is an alternative algebra with identity element 1 which contains a system of Cayley–Dickson matrix units such that $e_{11} + e_{22} = 1$.

2. Prove that A is decomposable into a direct sum of submodules:

$$A = A_{11} \oplus A_{12} \oplus A_{21} \oplus A_{22},$$

where $A_{ij} = \{a \in A \mid e_{ii}a = ae_{jj} = a\}$. In addition, the components A_{ij} are connected by the relations

$$\begin{aligned} A_{ii}^2 &\subseteq A_{ii}, & A_{ii}A_{jj} &= 0, \\ A_{ii}A_{ij} + A_{ij}A_{jj} &\subseteq A_{ij}, & A_{jj}A_{ij} &= A_{ij}A_{ii} = 0, \\ A_{ij}A_{ij} &\subseteq A_{ji}, & A_{ij}A_{ji} &\subseteq A_{ii}, \end{aligned}$$

where $i \neq j$.

3. Prove that for $i \neq j$

$$A_{ij} = A_{ij}^{(1)} + A_{ij}^{(2)} + A_{ij}^{(3)},$$

where $A_{ij}^{(k)} = A_{ii}e_{ij}^{(k)} = e_{ij}^{(k)}A_{jj}$, $k = 1, 2, 3$. In addition, the components $A_{ij}^{(k)}$ are connected by the relations

$$\begin{aligned} A_{ij}^{(k)}A_{ji}^{(k)} &\subseteq A_{ii}, & A_{ij}^{(k)}A_{ji}^{(l)} &= 0, & k &\neq l, \\ A_{ij}^{(k)}A_{ij}^{(k)} &= 0, & A_{ij}^{(k)}A_{ij}^{(k+1)} &\subseteq A_{ji}^{(k+2)}, & k &\bmod 3. \end{aligned}$$

4. Prove that the ring $\Omega = A_{11}$ is associative and commutative. *Hint*. Prove that the associator (x_{11}, y_{11}, z_{11}) is annihilated by the element $e_{12}^{(1)}$, and consequently also by the element $e_{12}^{(1)}e_{21}^{(1)} = e_{11}$; and prove that the commutator $[x_{11}, y_{11}]$ is annihilated by the element $e_{21}^{(3)}e_{21}^{(2)} = e_{12}^{(1)}$, and consequently also by the element $e_{12}^{(1)}e_{21}^{(1)} = e_{11}$.

We define "coordinate functions" $\pi_{ij}^{(k)}: A \to \Omega$ on A by setting $\pi_{ij}^{(k)}$ equal to zero on all components except $A_{ij}^{(k)}$, and on $A_{ij}^{(k)}$ they are defined in the following manner:

$$\begin{aligned} \pi_{11}(a_{11}) &= a_{11}, & \pi_{22}(a_{22}) &= e_{12}(a_{22}e_{21}), \\ \pi_{12}^{(k)}(a_{12}^{(k)}) &= a_{12}^{(k)}e_{21}^{(k)}, & \pi_{21}^{(k)}(a_{21}^{(k)}) &= e_{12}^{(k)}a_{21}^{(k)}. \end{aligned}$$

Furthermore, we define a linear mapping $\pi: A \to \boldsymbol{C}(\Omega)$ by setting

$$\pi(a) = \sum_{i=1}^{2} \pi_{ii}(a)e_{ii} + \sum_{k=1}^{3} (\pi_{12}^{(k)}(a)e_{12}^{(k)} + \pi_{21}^{(k)}(a)e_{21}^{(k)}).$$

5. Prove that $\pi(ab) = \pi(a)\pi(b)$. *Hint*. It suffices to consider the case when the elements a and b lie in the components $A_{ij}^{(k)}$. As a preliminary prove the equality $e_{12}(a_{22}e_{21}) = e_{12}^{(k)}(a_{22}e_{21}^{(k)})$, $k = 2, 3$.

6. (Zorn coordinatization theorem) Let A be an alternative algebra with identity element 1 which contains a system of Cayley–Dickson matrix units such that $e_{11} + e_{22} = 1$. Then A is a Cayley–Dickson matrix algebra over the ring $\Omega = A_{11}$.

7. Let $C(\Phi)$ be a Cayley–Dickson matrix algebra over an associative–commutative ring Φ. Then every one-sided ideal B of the algebra $C(\Phi)$ is a two-sided ideal and has the form $C(\Lambda)$, where Λ is an ideal of the ring Φ with $\Lambda = B \cap \Phi \cdot 1$.

LITERATURE

Albert [1, 8], Bruck and Kleinfeld [24], Brown [25], Jacobson [43, 45] [47, pp. 162–171], Dickson [53], Kaplansky [88], Linnik [106], McCrimmon [118], Freudenthal [239], Herstein [250], Zorn [258], Schafer [261, pp. 44–50].

Quadratic forms satisfying the weaker condition of composition $n(x^2) = (n(x))^2$ were studied by Gaynov [31–33].

In the works of Schafer [260, 263] and McCrimmon [115, 119, 120] the concept of an nth degree form admitting composition was introduced, and the theorem on quadratic forms admitting composition was generalized to nth degree forms.

CHAPTER 3

Special and Exceptional Jordan Algebras

In this chapter Φ is an associative and commutative ring with identity element 1 in which the equation $2x = 1$ is solvable. The solution of this equation (as is easy to see, it is unique) is denoted by $\frac{1}{2}$, and F denotes an arbitrary field of characteristic $\neq 2$.

1. DEFINITION AND EXAMPLES OF JORDAN ALGEBRAS

An algebra is called *Jordan* if it satisfies the identities

$$xy = yx, \qquad (x^2y)x = x^2(yx).$$

Let A be an associative Φ-algebra. We define on the additive Φ-module of the algebra A a new operation $\odot$ of multiplication, which is connected with the old multiplication by the formula

$$a \odot b = \tfrac{1}{2}(ab + ba).$$

After replacing the old multiplication with the new, a new algebra is obtained which is denoted by $A^{(+)}$. It is easy to verify that $A^{(+)}$ is Jordan. If J is a submodule of the algebra A which is closed with respect to the operation

$a \odot b = \frac{1}{2}(ab + ba)$, then J together with this operation is a subalgebra of $A^{(+)}$, and consequently a Jordan algebra. Such a Jordan algebra J is called *special*. The subalgebra A_0 of the algebra A generated by the set J is called an *associative enveloping algebra* for J. Nonspecial Jordan algebras are also called *exceptional*. We note that the algebra $A^{(+)}$ can also be defined for a nonassociative algebra A, however it is not always Jordan.

Let V be a vector space over a field F with a symmetric bilinear form $f = f(x, y)$ defined on V. We consider the direct sum $B = F \cdot 1 + V$ of the vector space V and the one-dimensional vector space $F \cdot 1$ with basis 1, and we define a multiplication on B by the rule

$$(\alpha \cdot 1 + x)(\beta \cdot 1 + y) = (\alpha\beta + f(x, y)) \cdot 1 + (\beta x + \alpha y),$$

where $\alpha, \beta \in F$ and $x, y \in V$. We note at once that the element 1 is an identity element for the algebra B, and that in view of the symmetry of the form f the multiplication on the algebra B is commutative. Let $a = \alpha \cdot 1 + x \in B$. Then

$$\begin{aligned} a^2 = (\alpha^2 + f(x, x)) \cdot 1 + 2\alpha x &= (\alpha^2 + f(x, x)) \cdot 1 + 2\alpha a - 2\alpha^2 \cdot 1 \\ &= (f(x, x) - \alpha^2) \cdot 1 + 2\alpha a. \end{aligned}$$

The identity $(a^2 b)a = a^2(ba)$ for the algebra B is now a consequence of the relations $(ab)a = a(ba)$ and $(1b)a = 1(ba)$, which are valid in any commutative algebra. Consequently the algebra B is Jordan. It is called the *Jordan algebra of the symmetric bilinear form f*. This algebra is special (see Exercise 1).

Now let U be some (not necessarily associative) algebra and $*$ be an involution on U. The set $H(U, *) = \{u \in U \mid u = u^*\}$ of symmetric with respect to $*$ elements is closed with respect to the operation $a \odot b = \frac{1}{2}(ab + ba)$, and is a subalgebra of the algebra $U^{(+)}$. If U is associative, then $U^{(+)}$ is Jordan, and $H(U, *)$ is a special Jordan algebra.

Let D be a composition algebra with involution $d \mapsto \bar{d}$ and D_n be the algebra of nth order matrices over D. As is easy to see, the mapping $S: X \to \bar{X}^t$, where $\bar{X}^t$ is the matrix obtained from the matrix X by applying the involution to each entry in X and then transposing, is an involution of the algebra D_n. If D is associative, then $H(D_n, S)$ is a special Jordan algebra. But if D is a Cayley–Dickson algebra, then the algebra $H(D_n, S)$ will only be Jordan for $n \leq 3$, and for $n = 3$ it will be exceptional. The remaining portion of this section is devoted to the proof of these two assertions.

For brevity we shall subsequently denote the algebra $H(D_n, S)$ by $H(D_n)$.

Let $\mathbf{C}$ be a Cayley–Dickson algebra over a field F, and consider the algebra $H(\mathbf{C}_3)$. We shall denote by $\mathscr{E}_{ij}$ the matrix for which there is a 1 in the intersection of the ith row and jth column and for which all remaining elements are zero. The elements $\mathscr{E}_{ij}$ are multiplied according to the usual formula: $\mathscr{E}_{ij}\mathscr{E}_{kl} = \delta_{jk}\mathscr{E}_{il}$ where δ_{jk} is the Kronecker symbol. The element

$\sum_{i=1}^{3} \mathscr{E}_{ii}$ is the identity element of the algebra $\boldsymbol{C}_3$, and we shall denote it by 1, the same as the identity element of the algebra $\boldsymbol{C}$. The mapping

$$c \mapsto \operatorname{diag}\{c, c, c\}$$

of the algebra $\boldsymbol{C}$ into $\boldsymbol{C}_3$ is a monomorphism. We shall also denote by c the diagonal matrix to which the element c is mapped under this monomorphism. Then the matrix $X = (x_{ij})$ can be written in the form $X = \sum x_{ij}\mathscr{E}_{ij}$. It is easy to verify that $(a\mathscr{E}_{ij})(b\mathscr{E}_{kl}) = \delta_{jk}(ab)\mathscr{E}_{il}$.

LEMMA 1. Let A be an arbitrary algebra and $a, b, c \in A$. If $(a, b, c)^+$ is the associator of the elements a, b, and c in the algebra $A^{(+)}$, then

$$\begin{aligned} 4(a, b, c)^+ = {} & (a, b, c) - (c, b, a) + (b, a, c) \\ & - (c, a, b) + (a, c, b) - (b, c, a) + [b, [a, c]]. \end{aligned} \tag{1}$$

The proof consists of the expansion of all associators and commutators and then comparison of both sides of the equality.

LEMMA 2. In an arbitrary alternative algebra A, the following identity is valid:

$$[(a, b, c), d] = (ab, c, d) + (bc, a, d) + (ca, b, d). \tag{2}$$

PROOF. We linearize the middle Moufang identity (see Section 2.3) to obtain

$$(ca)(bd) + (da)(bc) = [c(ab)]d + [d(ab)]c. \tag{3}$$

We now have

$$\begin{aligned} d(a, b, c) &= d[(ab)c] - d[a(bc)] \\ &= -(d, ab, c) + [d(ab)]c - d[a(bc)] \\ &= -(d, ab, c) + [d(ab)]c + (d, a, bc) - (da)(bc) \\ &\text{by (3)} \\ &= -(ab, c, d) - (bc, a, d) - [c(ab)]d + (ca)(bd) \\ &= -(ab, c, d) - (bc, a, d) + (c, a, b)d - [(ca)b]d + (ca)(bd) \\ &= -(ab, c, d) - (bc, a, d) + (c, a, b)d - (ca, b, d), \end{aligned}$$

which is what was to be proved.

We recall (see the proof of Lemma 2.8) that in a Cayley–Dickson algebra $\boldsymbol{C}$ the following relations are valid:

$$(a, b, c) = -(\bar{a}, b, c) = -(a, \bar{b}, c) = -(a, b, \bar{c}), \tag{4}$$

$$\overline{(a, b, c)} = -(a, b, c). \tag{5}$$

In addition to this, since F is a field of characteristic $\neq 2$, from the equality $\bar{a} = a$ it follows $a = \frac{1}{2}t(a) \in F$. In particular, if $A = (a_{ij}) \in H(\boldsymbol{C}_3)$, then $a_{ii} \in F$. Finally, it is easy to verify that for any $a, b, c \in \boldsymbol{C}$ the following relation is satisfied in the algebra $\boldsymbol{C}_3$:

$$(a\mathscr{E}_{ij}, b\mathscr{E}_{kl}, c\mathscr{E}_{pq}) = (a, b, c)\mathscr{E}_{ij}\mathscr{E}_{kl}\mathscr{E}_{pq}. \tag{6}$$

LEMMA 3. Let $A = (a_{ij}) \in H(\boldsymbol{C}_3)$. Then $[A^2, A] = 2a$, where $a = (a_{12}, a_{23}, a_{31})$.

PROOF. By (6) and the rule for multiplication of matrix units, we have

$$[A^2, A] = (A, A, A) = \sum (a_{ij}, a_{jk}, a_{kl})\mathscr{E}_{il}.$$

If $i = j$, $j = k$, or $k = l$, then $(a_{ij}, a_{jk}, a_{kl}) = 0$, since in this case one of the elements of the associator lies in F. If $i = k$, then $a_{jk} = \bar{a}_{ij}$ and $(a_{ij}, \bar{a}_{ij}, a_{kl}) = -(a_{ij}, a_{ij}, a_{kl}) = 0$. If $j = l$, then analogously $(a_{ij}, a_{jk}, a_{kl}) = 0$. Since there are all together three indices, the associator (a_{ij}, a_{jk}, a_{kl}) can only be nonzero in the case when i, j, and k are all different and $i = l$. Consequently

$$[A^2, A] = \sum_{i, j, k \neq} (a_{ij}, a_{jk}, a_{ki})\mathscr{E}_{ii}.$$

Furthermore, we have $(a_{ij}, a_{jk}, a_{ki}) = (a_{jk}, a_{ki}, a_{ij})$ and $(a_{ik}, a_{kj}, a_{ji}) = (\bar{a}_{ki}, \bar{a}_{jk}, \bar{a}_{ij}) = -(a_{ki}, a_{jk}, a_{ij}) = (a_{ij}, a_{jk}, a_{ki})$. Consequently, $(a_{ij}, a_{jk}, a_{ki}) = (a_{12}, a_{23}, a_{31}) = a$, and $[A^2, A] = 2a(\sum \mathscr{E}_{ii}) = 2a$. This proves the lemma.

THEOREM 1. $H(\boldsymbol{C}_3)$ is a Jordan algebra.

PROOF. (*McCrimmon*). Since the commutativity of the algebra $H(\boldsymbol{C}_3)$ is obvious, we must only verify that $(A^2 \odot B) \odot A = A^2 \odot (B \odot A)$ for any two matrices $A, B \in H(\boldsymbol{C}_3)$. This is equivalent to $(A, B, C)^+ = 0$ for $C = A^2$. It suffices to show that in the element $(A, B, C)^+$ the coefficients for $\mathscr{E}_{11}$ and $\mathscr{E}_{12}$ are zero, since, with a change of some of the indices, the same proof will also apply to the other coefficients. In view of (1), (2), (6), and Lemma 3, the coefficient for $\mathscr{E}_{11}$ of the element $(A, B, C)^+$ is a sum of associators, and so by (5) it is skew-symmetric. However, $(A, B, C)^+ \in H(\boldsymbol{C}_3)$ and its coefficient for $\mathscr{E}_{11}$ must be symmetric. Consequently this coefficient equals zero. We next consider the coefficient for $\mathscr{E}_{12}$ of the element $(A, B, C)^+$. Since the associator $(A, B, C)^+$ is linear in B, it suffices to show that the coefficient for $\mathscr{E}_{12}$ equals zero in the following four cases: $B = b_{ii}\mathscr{E}_{ii}$, $B = b_{13}\mathscr{E}_{13} + b_{31}\mathscr{E}_{31}$, $B = b_{23}\mathscr{E}_{23} + b_{32}\mathscr{E}_{32}$, $B = b_{12}\mathscr{E}_{12} + b_{21}\mathscr{E}_{21}$, where $b_{ij} = \bar{b}_{ji}$. In the first case $b_{ii} \in F \cdot 1$, and therefore in view of (1) the coefficient for $\mathscr{E}_{12}$ equals zero. If $B = b_{13}\mathscr{E}_{13} + b_{31}\mathscr{E}_{31}$, then in view of (6) the coefficient for $\mathscr{E}_{12}$ of the element (A, B, C) equals $\sum_{k,l}(a_{1k}, b_{kl}, c_{l2}) = (a_{11}, b_{13}, c_{32}) + (a_{13}, b_{31}, c_{12}) = (a_{13}, b_{31}, c_{12})$. Analogously the coefficients for $\mathscr{E}_{12}$ of the elements $-(C, B, A)$,

(B, A, C), $-(C, A, B)$, (A, C, B), $-(B, C, A)$ equal $-(c_{13}, b_{31}, a_{12})$, (b_{13}, a_{31}, c_{12}), 0, 0, $-(b_{13}, c_{31}, a_{12})$, respectively. The coefficient for $\mathscr{E}_{12}$ of the element $[B, [A, C]] = -2[b_{13}\mathscr{E}_{13} + b_{31}\mathscr{E}_{31}, a]$ equals zero. Adding these we obtain that the coefficient for $\mathscr{E}_{12}$ of the element $4(A, B, C)^{+}$ equals

$$\begin{aligned}&(a_{13}, b_{31}, c_{12}) - (c_{13}, b_{31}, a_{12}) + (b_{13}, a_{31}, c_{12}) - (b_{13}, c_{31}, a_{12})\\ &\quad = (a_{13}, b_{31}, c_{12}) + (\bar{b}_{31}, \bar{a}_{13}, c_{12}) - (c_{13}, b_{31}, a_{12}) - (\bar{b}_{31}, \bar{c}_{13}, a_{12}).\end{aligned}$$

This expression equals zero by (4) and the skew-symmetry of the associator. Analogous arguments also apply to the case $B = b_{23}\mathscr{E}_{32} + b_{32}\mathscr{E}_{32}$. Now let $B = b_{12}\mathscr{E}_{12} + b_{21}\mathscr{E}_{21}$. Using (1) we find the coefficient for $\mathscr{E}_{12}$ of the element $4(A, B, C)^{+}$. It equals

$$\begin{aligned}&(a_{12}, b_{21}, c_{12}) - (c_{12}, b_{21}, a_{12}) + (b_{12}, a_{21}, c_{12}) + (b_{12}, a_{23}, c_{32})\\ &\quad - (c_{12}, a_{21}, b_{12}) - (c_{13}, a_{31}, b_{12}) + (a_{12}, c_{21}, b_{12}) + (a_{13}, c_{31}, b_{12})\\ &\quad - (b_{12}, c_{21}, a_{12}) - (b_{12}, c_{23}, a_{32}) - 2[b_{12}, a].\end{aligned}$$

By (4) and the skew-symmetry of the associator it can be written as

$$-2(c_{32}, a_{23}, b_{12}) - 2(c_{21}, a_{12}, b_{12}) - 2(c_{13}, a_{31}, b_{12}) - 2[b_{12}, (a_{12}, a_{23}, a_{31})]. \tag{7}$$

Now $c_{ij} = \sum a_{ik}a_{kj}$, and therefore $(c_{ij}, a_{ji}, b_{12}) = (a_{ii}a_{ij}, a_{ji}, b_{12}) + (a_{ij}a_{jj}, a_{ji}, b_{12}) + (a_{ik}a_{kj}, a_{ji}, b_{12})$ where i, j, k are distinct. Since $a_{ii} \in F \cdot 1$, by (4) $(a_{ii}a_{ij}, a_{ji}, b_{12}) = a_{ii}(a_{ij}, a_{ji}, b_{12}) = a_{ii}(a_{ij}, \bar{a}_{ij}, b_{12}) = 0$. Analogously $(a_{ij}a_{jj}, a_{ji}, b_{12}) = 0$. Consequently (7) takes the form $-2(a_{31}a_{12}, a_{23}, b_{12}) - 2(a_{23}a_{31}, a_{12}, b_{12}) - 2(a_{12}a_{23}, a_{31}, b_{12}) - 2[b_{12}, (a_{12}, a_{23}, a_{31})]$. However, this expression equals zero by (2).

This proves the theorem.

It is now convenient for us to introduce the following notation for the elements of the algebra $H(\mathbf{C}_3)$:

$$\begin{aligned}&x_{ii} = x\mathscr{E}_{ii}, && x = \alpha \cdot 1, && \alpha \in F,\\ &x_{ij} = x\mathscr{E}_{ij} + \bar{x}\mathscr{E}_{ji}, && x \in \mathbf{C}, && i \neq j,\\ &e_{ii} = \mathscr{E}_{ii}, && e_{ij} = \mathscr{E}_{ij} + \mathscr{E}_{ji}, && i \neq j.\end{aligned}$$

THEOREM 2 (*Albert*). $H(\mathbf{C}_3)$ is an exceptional algebra.

PROOF. The proof is by contradiction. We assume there exists an associative algebra A such that $H(\mathbf{C}_3)$ is a subalgebra of the algebra $A^{(+)}$. We shall denote the multiplication in the algebra $\mathbf{C}$ by a dot $\cdot$, and multiplication in the algebra $H(\mathbf{C}_3)$ by a circle with a dot inside $\odot$. When multiplying elements in the algebra A, we shall not place anything between them.

Let $X, Y \in H(\mathbf{C}_3)$. Then $X \odot Y = \frac{1}{2}(XY + YX)$. This relation connects multiplication in A with multiplication in $H(\mathbf{C}_3)$, and as a final result with multiplication in $\boldsymbol{C}$. Using it we shall find a subalgebra in A isomorphic to the Cayley–Dickson algebra $\boldsymbol{C}$. In view of the nonassociativity of the algebra $\boldsymbol{C}$, this will prove the theorem.

We indicate some relations we need in the algebra A:

$$e_{ii}^2 = e_{ii}, \qquad e_{ij}^2 = e_{ii} + e_{jj}, \tag{8}$$

$$e_{ii}x_{ij} + x_{ij}e_{ii} = e_{jj}x_{ij} + x_{ij}e_{jj} = x_{ij}, \tag{9}$$

$$e_{kk}x_{ij} + x_{ij}e_{kk} = 0, \qquad k \neq i, j, \tag{10}$$

$$x_{12}y_{23} + y_{23}x_{12} = (x \cdot y)_{13}. \tag{11}$$

Relations (8)–(10) are obvious. We shall prove equality (11). We have

$$\begin{aligned} x_{12}y_{23} + y_{23}x_{12} &= 2x_{12} \odot y_{23} \\ &= 2\begin{pmatrix} 0 & x & 0 \\ \bar{x} & 0 & 0 \\ 0 & 0 & 0 \end{pmatrix} \odot \begin{pmatrix} 0 & 0 & 0 \\ 0 & 0 & y \\ 0 & \bar{y} & 0 \end{pmatrix} = \begin{pmatrix} 0 & 0 & x \cdot y \\ 0 & 0 & 0 \\ \bar{y} \cdot \bar{x} & 0 & 0 \end{pmatrix} = (x \cdot y)_{13}. \end{aligned}$$

Analogously we obtain the relations

$$x_{12}y_{13} + y_{13}x_{12} = (\bar{x} \cdot y)_{23}, \tag{12}$$

$$x_{13}y_{23} + y_{23}x_{13} = (x \cdot \bar{y})_{12}. \tag{13}$$

By (10) for $k \neq i, j$ we have

$$0 = e_{kk}(e_{kk}x_{ij} + x_{ij}e_{kk}) = e_{kk}x_{ij} + e_{kk}x_{ij}e_{kk}.$$

However, by (8) and (10)

$$2e_{kk}x_{ij}e_{kk} = (e_{kk}x_{ij} + x_{ij}e_{kk})e_{kk} + e_{kk}(e_{kk}x_{ij} + x_{ij}e_{kk}) - (e_{kk}^2 x_{ij} + x_{ij}e_{kk}^2) = 0,$$

and therefore

$$e_{kk}x_{ij} = x_{ij}e_{kk} = 0, \qquad k \neq i, j. \tag{14}$$

We now map $\boldsymbol{C}$ into A in the following manner:

$$\sigma : x \mapsto e_{11}x_{12}e_{12}.$$

It is clear that this mapping is a homomorphism of the vector spaces. We show that $(x \cdot y)^\sigma = x^\sigma y^\sigma$. By (13) and (14) we have

$$(x \cdot y)^\sigma = e_{11}(x \cdot y)_{12}e_{12} = e_{11}(x_{13}\bar{y}_{23} + \bar{y}_{23}x_{13})e_{12} = e_{11}x_{12}\bar{y}_{23}e_{12}.$$

We then change this expression by means of (11) and (14):

$$e_{11}x_{13}\bar{y}_{23}e_{12} = e_{11}(x_{12}e_{23} + e_{23}x_{13})\bar{y}_{23}e_{12} = e_{11}x_{12}e_{23}\bar{y}_{23}e_{12},$$

and then by means of (12):

$$e_{11}x_{12}e_{23}\bar{y}_{23}e_{12} = e_{11}x_{12}e_{23}(y_{12}e_{13} + e_{13}y_{12})e_{12} = e_{11}x_{12}e_{23}e_{13}y_{12}e_{12},$$

since by (9) and (14) $y_{12}e_{13}e_{12} = y_{12}(e_{33}e_{13} + e_{13}e_{33})e_{12} = 0$. Thus

$$(x \cdot y)^\sigma = e_{11}x_{12}e_{23}e_{13}y_{12}e_{12}. \tag{15}$$

However by (9), (13), and (14), $e_{23}e_{13}y_{12} = e_{23}(e_{11}e_{13} + e_{13}e_{11})y_{12} = e_{23}e_{13}e_{11}y_{12} = (e_{12} - e_{13}e_{23})e_{11}y_{12} = e_{12}e_{11}\,y_{12}$, and therefore it follows from (15) that

$$(x \cdot y)^\sigma = e_{11}x_{12}e_{12}e_{11}y_{12}e_{12} = x^\sigma y^\sigma.$$

This proves the mapping introduced is a homomorphism. By Lemma 2.3, however, the Cayley–Dickson algebra is simple, and therefore our homomorphism is either an isomorphism or maps the algebra $\boldsymbol{C}$ to zero. But the latter is false, since by (8) and (14)

$$1^\sigma = e_{11}e_{12}e_{12} = e_{11}(e_{11} + e_{22}) = e_{11} \neq 0.$$

Since the algebra $\boldsymbol{C}$ is not associative and cannot be mapped isomorphically into an associative algebra, we have arrived at a contradiction. This proves the theorem.

Exercises

1. Let $B(f) = F \cdot 1 + M$ be the Jordan algebra of a symmetric bilinear form $f: M \times M \to F$. Prove that the Clifford algebra $C(f)$ of the form f (see S. Lang, "Algebra," p. 441) is an associative enveloping algebra for $B(f)$.

2. Prove that if the form f is nondegenerate and $\dim_F M > 1$, then the Jordan algebra $B(f) = F \cdot 1 + M$ is simple.

3. Prove that for a Cayley–Dickson algebra $\boldsymbol{C}$ the algebra $\boldsymbol{C}^{(+)}$ is isomorphic to a Jordan algebra of a symmetric nondegenerate bilinear form.

4. Prove that for any elements a, b, c of an associative Φ-algebra the following equality holds:

$$\tfrac{1}{2}(abc + cba) = (a \odot b) \odot c + (b \odot c) \odot a - (c \odot a) \odot b. \tag{16}$$

5. (*McCrimmon*) Let A be an associative Φ-algebra and J be an ideal of the algebra $A^{(+)}$. Then $A^\# b^2 A^\# \subseteq J$ for any $b \in J$, and if $b^2 = 0$ for all $b \in J$, then $(A^\# b A^\#)^3 = (0)$ for any $b \in J$.

6. (*Herstein*) If A is a simple associative Φ-algebra, then the algebra $A^{(+)}$ is also simple.

7. (*McCrimmon*) Let A be an associative Φ-algebra with involution $*$ and J be an ideal in $H(A, *)$. If $bcb \neq 0$ for some $b, c \in J$, then the containment $B \cap H(A, *) \subseteq J$ holds for the ideal $B = A^\# bcbA^\#$ of the algebra A; and if $bcb = 0$ for all $b, c \in J$, then A contains a nonzero nilpotent ideal.

SKETCH OF PROOF. Prove that for any $x, y \in A$

$$xbcbx^* \in J, \qquad xbcby^* + ybcbx^* \in J.$$

The first part of the assertion follows from this. In the case when $bcb = 0$ for all $b, c \in J$, assume first that $bhb \neq 0$ for some $b \in J, h \in H(A, *)$, and prove that for any $x \in A$

$$xbhb + bhbx^* \in \mathbf{J}.$$

Hence $(xbhb + bhbx^*)^3 = 0$ and $0 = xbhb(xbhb + bhbx^*)^3 = (xbhb)^4$, that is, the left ideal $I = A^{\#}bhb$ is a nil-algebra of bounded index. By a well-known result from the theory of associative PI-rings (e.g., see [46, p. 335]), this implies the existence of a nilpotent ideal in I. It is not difficult to then find such an ideal also in A. If $bhb = 0$ for all $h \in H(A, *)$, then prove that the left ideal $A^{\#}b$ satisfies the identity $x^3 = 0$, which likewise implies the existence of a nilpotent ideal in A.

8. (*Herstein*) If A is an associative Φ-algebra with involution $*$ which does not contain proper $*$-*ideals* (i.e., ideals I such that $I^* \subseteq I$), then the Jordan algebra $H(A, *)$ is simple.

9. The Jordan algebra $H(D_n)$, where D is an associative composition algebra, is simple.

10. Prove that $H(\mathbf{C}_3)$ is a simple algebra.

11. Prove that the algebra $H(\mathbf{C}_2)$ is isomorphic to a Jordan algebra of a symmetric nondegenerate bilinear form, and is therefore simple and special.

12. Let $u = v_1$, $v = v_2$, $w = v_3$ be generators of the Cayley–Dickson algebra $\mathbf{C}$ (see Section 2.4). Prove that the elements $X = e_{12}$, $Y = e_{23}$, $Z = u_{12} + v_{13} + w_{23}$ generate the algebra $H(\mathbf{C}_3)$.

13. Prove that the algebra $H(\mathbf{C}_n)$ is not Jordan for $n \geq 4$.

2. FREE SPECIAL JORDAN ALGEBRAS

Consider the free associative Φ-algebra Ass$[X]$ from the set of free generators $X = \{x_\alpha\}$. We shall call the subalgebra generated by the set X in the algebra Ass$[X]^{(+)}$ the *free special Jordan algebra from the set of free generators* X, and we shall denote it by SJ$[X]$. As we shall see at the end of the section, special Jordan algebras do not form a variety since for card$(X) \geq 3$ there exist homomorphic images of the algebra SJ$[X]$ which are exceptional algebras. However, all special Jordan algebras from a set of generators with cardinality λ are homomorphic images of the algebra SJ$[X]$ where card$(X) = \lambda$. It is also in this sense that we speak about the freedom of the algebras SJ$[X]$.

PROPOSITION 1. Let J be a special Jordan algebra. Then every mapping of X into J can be uniquely extended to a homomorphism of $\mathrm{SJ}[X]$ into J.

PROOF. Let $\sigma: X \to J$ be some mapping and A be an associative enveloping algebra for J. Then the mapping σ can be extended to a homomorphism $\bar{\sigma}: \mathrm{Ass}[X] \to A$. It is now clear that the restriction of $\bar{\sigma}$ to $\mathrm{SJ}[X]$ is a homomorphism of $\mathrm{SJ}[X]$ into J which extends σ. This proves the proposition.

An element of the free associative algebra $\mathrm{Ass}[X]$ is called a *Jordan polynomial* (*j-polynomial*) if it belongs to $\mathrm{SJ}[X]$, that is, can be expressed from elements of the set X by means of the operations $+$ and $\odot$. For example, the polynomials $x_1x_2x_1$ and $x_1x_2x_3 + x_3x_2x_1$ are j-polynomials by (16). There is still not a single appropriate criterion known that enables one to recognize from the expression of a polynomial in $\mathrm{Ass}[X]$ whether or not it is a j-polynomial. Cohn's theorem, which will be formulated later, gives such a criterion for $\mathrm{card}(X) \leq 3$.

There is yet one more Jordan algebra connected with the free associative algebra $\mathrm{Ass}[X]$. For this we define an involution $*$ on $\mathrm{Ass}[X]$, which is defined on monomials by the rule

$$(x_{i_1}x_{i_2}\cdots x_{i_k})^* = x_{i_k}\cdots x_{i_2}x_{i_1},$$

and which is extended to polynomials by linearity: if $f = \sum \alpha_s u_s$ where $\alpha_s \in \Phi$ and u_s are monomials, then $f^* = \sum \alpha_s u_s^*$. We denote by $H[X]$ the Jordan algebra $H(\mathrm{Ass}[X], *)$ of symmetric elements with respect to $*$ of the algebra $\mathrm{Ass}[X]$.

THEOREM 3 (*Cohn*). The containment $H[X] \supseteq \mathrm{SJ}[X]$ is valid for any set X. Moreover, equality holds for $\mathrm{card}(X) \leq 3$, but this is a strict containment for $\mathrm{card}(X) > 3$.

PROOF. The generators $x_\alpha \in X$ are symmetric with respect to $*$ and are in $H[X]$. In addition, if $a, b \in H[X]$, then also $a \odot b \in H[X]$ since

$$(a \odot b)^* = \tfrac{1}{2}[(ab)^* + (ba)^*] = \tfrac{1}{2}(b^*a^* + a^*b^*) = \tfrac{1}{2}(ab + ba) = a \odot b.$$

Therefore $H[X] \supseteq \mathrm{SJ}[X]$.

Let $\mathrm{card}(X) = 3$, that is, $X = \{x_1, x_2, x_3\}$. For the proof of the fact that $H[X] = \mathrm{SJ}[X]$, it suffices to show that elements from $H[X]$ of the form $u + u^*$, where u is a monomial, are j-polynomials. In fact, if $f = \sum u_i$ is symmetric with respect to $*$, then $f = \frac{1}{2}(f + f^*) = \frac{1}{2}\sum(u_i + u_i^*)$.

Let $u = x_{i_1}^{l_1}x_{i_2}^{l_2}\cdots x_{i_h}^{l_h}$, where $i_r \neq i_{r+1}$ for $r = 1, 2, \ldots, h-1$. We call the number h the *height* of the monomial u.

To prove that $u + u^* \in \mathrm{SJ}[X]$, we carry out a double induction. The first induction is on the length of the monomial u, with the initial case being obvious. Assume the assertion has been proved for monomials of length $<k$. We consider the set of monomials of length k. Among them are all together three monomials of height 1. These are x_1^k, x_2^k, x_3^k. For them our assertion is obvious, and so we have the basis for a second induction. Assume that the assertion is valid for all monomials of length k and height $<h$. We first consider the case when a monomial u begins and ends with the same generating element, namely $u = a^p v a^q$ where $a \in \{x_1, x_2, x_3\}$ and v is a monomial. In this case

$$\begin{aligned} u + u^* &= a^p v a^q + a^q v a^p \\ &= a^p(va^q + a^q v^*) + (a^q v^* + va^q)a^p - (a^{p+q}v^* + va^{p+q}). \end{aligned}$$

We observe that the sum of the first two summands is a j-polynomial by our first induction assumption, and since the monomial $a^{p+q}v^*$ has height $h-1$, the last summand is a j-polynomial by our second assumption.

A second case which it is necessary to consider is the following: $u = a^p b^r a^q v$ where $a, b \in \{x_1, x_2, x_3\}$, $a \neq b$, and v is a monomial. We have

$$\begin{aligned} u + u^* &= a^p b^r a^q v + v^* a^q b^r a^p \\ &= a^p b^r(a^q v + v^* a^q) + (v^* a^q + a^q v)b^r a^p - (a^p b^r v^* a^q + a^q v b^r a^p), \end{aligned}$$

where in view of (16) the sum of the first two summands is a Jordan polynomial, and in the last parentheses we have our first case.

A third case is $u = a^p v a^q b^r$, where $a, b \in \{x_1, x_2, x_3\}$, $a \neq b$, and v is a monomial. In this case

$$\begin{aligned} u + u^* &= a^p v a^q b^r + b^r a^q v^* a^p \\ &= a^p(va^q + a^q v^*)b^r + b^r(a^q v^* + va^q)a^p - (a^{p+q}v^* b^r + b^r v a^{p+q}). \end{aligned}$$

In view of (16) and our first induction assumption, the sum of the first two terms is a j-polynomial, while the subtracted term satisfies the conditions of our second induction assumption.

It only remains to consider as a fourth case two possibilities for the monomial u: $u = a^p b^r c^s$ and $u = a^p c^q b^t v a^l c^r b^s$, where v is a monomial which is possibly missing, and a, b, $c \in \{x_1, x_2, x_3\}$ are pairwise distinct. If $u = a^p b^r c^s$, then by (16) $u + u^* = a^p b^r c^s + c^s b^r a^p \in \mathrm{SJ}[X]$. Now let $u = a^p c^q b^t v a^l c^r b^s$. Then

$$\begin{aligned} u + u^* &= a^p c^q b^t v a^l c^r b^s + b^s c^r a^l v^* b^t c^q a^p \\ &= a^p(c^q b^t v a^l c^r + c^r a^l v^* b^t c^q)b^s + b^s(c^r a^l v^* b^t c^q + c^q b^t v a^l c^r)a^p \\ &\quad - (a^p c^r a^l v^* b^t c^q b^s + b^s c^q b^t v a^l c^r a^p). \end{aligned}$$

Because of (16) and the first induction assumption, the sum of the first two terms is a j-polynomial, and by our already considered second case, the third is likewise a j-polynomial.

Thus the equality $H[X] = \mathrm{SJ}[X]$ is proved in the case when $\mathrm{card}(X) = 3$. Of course from this also follows the validity of the equality $H[X] = \mathrm{SJ}[X]$ for $\mathrm{card}(X) = 2$. The proof we have given belongs to Shirshov. It is constructive and gives an algorithm which allows one to find from the expression for a symmetric polynomial in three variables its representation in the form of a j-polynomial.

We shall now prove that $H[X] \neq \mathrm{SJ}[X]$ for $\mathrm{card}(X) > 3$. For this it suffices to show that the element $f(x_1, x_2, x_3, x_4) = x_1x_2x_3x_4 + x_4x_3x_2x_1$ is not a Jordan polynomial from x_1, x_2, x_3, x_4.

We consider the free Φ-module E from generators e_1, e_2, e_3, e_4 and the exterior algebra $\bigwedge(E)$ of this module. (For its definition and properties see the book by S. Lang, "Algebra," p. 474. Mir, 1968.) The Φ-module of the algebra $\bigwedge(E)$ is also free with basis $\{1, e_{i_1} \wedge \cdots \wedge e_{i_s}; i_1 < \cdots < i_s; s = 1, 2, 3, 4\}$, and in addition $e_i \odot e_j = 0$ in $\bigwedge(E)$. Let $\sigma: \mathrm{Ass}[X] \to \bigwedge(E)$ be the homomorphism such that $\sigma(x_i) = e_i$ for $i = 1, 2, 3, 4$, and $\sigma(x_i) = 0$ for $i > 4$. All Jordan polynomials are mapped to zero under this homomorphism. However, $\sigma[f(x_1, x_2, x_3, x_4)] = 2e_1 \wedge e_2 \wedge e_3 \wedge e_4 \neq 0$. This means $f(x_1, x_2, x_3, x_4)$ is not a Jordan polynomial, and therefore the theorem is completely proved.

REMARK. If the equation $2x = 1$ is not solvable in Φ, the operation $\odot$ cannot be defined. In this case the elements which can be obtained from elements of the set X by means of the operations of addition, squaring, and the quadratic multiplication $\{xyx\} = xyx$ are called the Jordan polynomials of the free associative algebra $\mathrm{Ass}[X]$. If $\frac{1}{2} \in \Phi$, then, as is easy to see, this definition gives us the usual Jordan polynomials. Tracing the proof of Cohn's theorem, it is possible to observe that, for the set $\mathrm{SJ}[X]$ of Jordan polynomials in the new sense, the equality $H[X] = \mathrm{SJ}[X]$ is true for $\mathrm{card}(X) \leq 3$ for any ring of operators Φ.

THEOREM 4. If $\mathrm{card}(X) > 1$, then the free Jordan algebra $\mathrm{SJ}[X]$ is not isomorphic to an algebra $A^{(+)}$ for any associative algebra A.

PROOF. Let us assume that $\mathrm{SJ}[X]$ is isomorphic to an algebra $A^{(+)}$, where A is an associative algebra. We can then assume that an associative operation (denoted by a dot) which distributes with addition is defined on the set $\mathrm{SJ}[X]$, such that for any $a, b \in \mathrm{SJ}[X]$

$$a \odot b = \tfrac{1}{2}(a \cdot b + b \cdot a), \tag{17}$$

and that $A = \langle \mathrm{SJ}[X], +, \cdot \rangle$. By (17) the set X is also a set of generators for A. Let $\sigma: \mathrm{Ass}[X] \to A$ be the homomorphism such that $x_\alpha^\sigma = x_\alpha$ for all $x_\alpha \in X$. We consider the kernel I of this homomorphism. By (17) the restriction of σ to $\mathrm{SJ}[X]$ is the identity mapping. Therefore

$$I \cap \mathrm{SJ}[X] = (0). \tag{18}$$

In addition, for any $f(x) \in \mathrm{Ass}[X]$ there can be found a $j(x) \in \mathrm{SJ}[X]$ such that $f(x) - j(x) \in I$. In particular, this is true for the polynomial $f(x) = x_1x_2$. But then by (16)

$$[x_1x_2 - j(x)][x_2x_1 - j(x)]$$
$$= x_1x_2^2x_1 - x_1x_2j(x) - j(x)x_2x_1 + j(x)^2 \in I \cap \mathrm{SJ}[X].$$

By (18) this means $[x_1x_2 - j(x)][x_2x_1 - j(x)] = 0$, and so we obtain that either $j(x) = x_1x_2$ or $j(x) = x_2x_1$. But neither one is possible, which proves the theorem.

We now turn to the study of homomorphic images of free special Jordan algebras.

LEMMA 4. Let I be an ideal of a special Jordan algebra J with associative enveloping algebra A, and let $\hat{I}$ be the ideal of the algebra A generated by the set I where

$$\hat{I} \cap J = I.$$

Then the quotient algebra J/I is special.

PROOF. We first note an obvious isomorphism: for any ideal B of the algebra A

$$(A/B)^+ \cong A^{(+)}/B^{(+)}.$$

Then by the second homomorphism theorem

$$J/I = J/J \cap \hat{I}^{(+)} \cong J + \hat{I}^{(+)}/\hat{I}^{(+)} \subseteq A^{(+)}/\hat{I}^{(+)},$$

which by our previous remark gives us that the algebra J/I is special.

LEMMA 5 (*Cohn*). Let I be an ideal of the free special Jordan algebra $\mathrm{SJ}[X]$ and $\hat{I}$ be the ideal generated by the set I in $\mathrm{Ass}[X]$. The quotient algebra $\mathrm{SJ}[X]/I$ is special if and only if $\hat{I} \cap \mathrm{SJ}[X] = I$.

PROOF. If the condition $\hat{I} \cap \mathrm{SJ}[X] = I$ is satisfied, then the quotient algebra $\mathrm{SJ}[X]/I$ is special by Lemma 4.

Now let us assume that the quotient algebra $\mathrm{SJ}[X]/I$ is special and that A is an associative enveloping algebra for it. We denote by τ the canonical homomorphism of the algebra $\mathrm{SJ}[X]$ onto $\mathrm{SJ}[X]/I$. The algebra A is generated by the elements $a_\alpha = \tau(x_\alpha)$, where $x_\alpha \in X$. Let σ be the homomorphism of the algebra $\mathrm{Ass}[X]$ onto A such that $\sigma(x_\alpha) = a_\alpha$. Then the restriction of σ to $\mathrm{SJ}[X]$ is a homomorphism of $\mathrm{SJ}[X]$ to $\mathrm{SJ}[X]/I$ which coincides with τ on the generators, and consequently equals τ. But then $\mathrm{Ker}\,\sigma \cap \mathrm{SJ}[X] = \mathrm{Ker}\,\tau$. We now note that $\hat{I} \subseteq \mathrm{Ker}\,\sigma$ and $\mathrm{Ker}\,\tau = I$. Consequently $\hat{I} \cap \mathrm{SJ}[X] \subseteq I$, and since the reverse containment is obvious, we have $\hat{I} \cap \mathrm{SJ}[X] = I$. This proves the lemma.

THEOREM 5 (*Cohn*). Let $\mathrm{SJ}[x, y, z]$ be the free special Jordan algebra from generators x, y, z, and let I be the ideal in $\mathrm{SJ}[x, y, z]$ generated by the element $k = x^2 - y^2$. Then the quotient algebra $\mathrm{SJ}[x, y, z]/I$ is exceptional.

PROOF. The element $v = kxyz + zyxk$ is in $\hat{I} \cap \mathrm{SJ}[x, y, z]$, where $\hat{I}$ is the ideal of the algebra $\mathrm{Ass}[x, y, z]$ generated by the set I. We shall show that $v \notin I$, and then by Lemma 5 everything will be proved. Assume that $v \in I$. Then there exists a Jordan polynomial $j(x, y, z, t)$, each monomial of which contains t, such that $v = j(x, y, z, k)$. It is clear we can consider that all the monomials in $j(x, y, z, t)$ have degree 4 and are linear in z. Comparing the degrees of the elements v and $j(x, y, z, x^2 - y^2)$ in the separate variables, we conclude that $j(x, y, z, t)$ is linear in t, and consequently in the other variables as well. The polynomial $j(x, y, z, t)$ is symmetric and is in $H[x, y, z, t]$. Therefore it is a linear combination of the 24 quadruples of the form $\{xyzt\} = xyzt + tzyx$ with all permutations of the elements x, y, z, t. However, the form of the element v indicates that in this linear combination there are nonzero coefficients only for those quadruples which end (or begin) with a z:

$$\begin{aligned} j(x, y, z, t) = \alpha_1\{txyz\} &+ \alpha_2\{xtyz\} + \alpha_3\{tyxz\} \\ &+\alpha_4\{ytxz\} + \alpha_5\{xytz\} + \alpha_6\{yxtz\}. \end{aligned}$$

Substituting $x^2 - y^2$ for t in this equality, we obtain the relation

$$\begin{aligned} \{x^3yz\} - \{y^2xyz\} = \alpha_1\{x^3yz\} &- \alpha_1\{y^2xyz\} + \alpha_2\{x^3yz\} \\ &- \alpha_2\{xy^3z\} + \alpha_3\{x^2yxz\} - \alpha_3\{y^3xz\} \\ &+ \alpha_4\{yx^3z\} - \alpha_4\{y^3xz\} + \alpha_5\{xyx^2z\} \\ &- \alpha_5\{xy^3z\} + \alpha_6\{yx^3z\} - \alpha_6\{yxy^2z\}. \end{aligned}$$

Comparing the coefficients for $\{y^2xyz\}$, we conclude that $\alpha_1 = 1$. We next compare the coefficients for $\{x^3yz\}$ and obtain $\alpha_2 = 0$. Comparison of the coefficients for $\{x^2yxz\}$ gives us $\alpha_3 = 0$. Hence, it follows that as the coefficient for $\{y^3xz\}$, $\alpha_4 = 0$. Furthermore, comparing the coefficients for $\{xyx^2z\}$ and $\{yxy^2z\}$, we obtain that $\alpha_5 = \alpha_6 = 0$. But this means that

$$j(x, y, z, t) = \{xyzt\}.$$

However, as we saw in the proof of Theorem 3, $\{xyzt\}$ is not a Jordan polynomial. This proves the theorem.

THEOREM 6. All homomorphic images of the free special Jordan algebra $\mathrm{SJ}[x, y]$ from two generators are special.

PROOF. Let I be an arbitrary ideal of the algebra $\mathrm{SJ}[x, y]$ and $\hat{I}$ be the ideal generated by the set I in $\mathrm{Ass}[x, y]$. By Lemma 5 it suffices to show that $\hat{I} \cap \mathrm{SJ}[x, y] = I$. Let $u \in \hat{I}$. Then $u = \sum v_i k_i w_i$, where $k_i \in I$ and v_i, w_i are monomials. Let us assume that $u \in \mathrm{SJ}[x, y]$. For the proof that $u \in I$, it suffices to show the inclusion $vkw + w^*kv^* \in I$ is valid for all $k \in I$ and any monomials v, w. But this inclusion, in fact, holds, since by Theorem 3 $vzw + w^*zv^*$ is a Jordan polynomial from x, y, z. This proves the theorem.

As it well known,[†] in order to extend a class of algebras $\mathfrak{K}$ to a variety, it is necessary to first close $\mathfrak{K}$ with respect to forming direct products, next subalgebras, and finally homomorphic images. The smallest variety Var $\mathfrak{K}$ containing the class of algebras $\mathfrak{K}$ is thus obtained. It is easy to observe that the class $\mathfrak{S}$ of all special Jordan Φ-algebras is closed both with respect to direct products and with respect to subalgebras. Therefore, the variety Var $\mathfrak{S}$ consists of all possible homomorphic images of special Jordan algebras. Furthermore, as we shall show, it is distinct from the variety of all Jordan algebras. It is obvious that the free special Jordan algebras are the free algebras for this variety.

We shall now prove a theorem on imbedding for special Jordan algebras. For its proof we establish two lemmas.

LEMMA 6. In the free associative algebra $\mathrm{Ass}[x, y]$, an element belongs to the subalgebra $\langle Z \rangle$ generated by the set $Z = \{z_i = xy^i x : i = 1, 2, \ldots\}$ if and only if it is a linear combination of monomials of the form

$$xy^{j_1}x^2y^{j_2}x^2 \cdots x^2y^{j_n}x, \qquad j_k \geq 1. \tag{19}$$

This subalgebra is a free associative algebra with set of free generators Z.

The proof of the first assertion of the lemma is obvious. For the proof of the second assertion we consider the homomorphism

$$\phi : \mathrm{Ass}[x_1, \ldots, x_n, \ldots] \to \langle Z \rangle$$

sending x_i to z_i. It is easy to see that ϕ is an isomorphism, since

$$f(z_1, z_2, \ldots, z_n) = 0$$

implies $f = 0$. We shall now denote the algebra $\langle Z \rangle$ by $\mathrm{Ass}[Z]$.

LEMMA 7. Let I be an ideal of the algebra $\mathrm{Ass}[Z]$ and $\tilde{I}$ be the ideal generated by the set I in $\mathrm{Ass}[x, y]$. Then

$$\tilde{I} \cap \mathrm{Ass}[Z] = I.$$

[†] For example, see A. I. Mal'cev, "Algebraic Systems," p. 339. Nauka, 1970.

PROOF. Every element u from $\tilde{I}$ is representable in the form

$$u = \sum_i v_i k_i w_i, \tag{20}$$

where $k_i \in I$ and v_i, w_i are monomials. If $u \in \mathrm{Ass}[Z]$, then u is a linear combination of monomials of the form (19). The elements k_i can also be represented analogously. We note that if k and vkw are monomials of the indicated form, then the monomials v, w likewise have this same form. Therefore, if $u \in \mathrm{Ass}[Z]$, then all the terms $v_i k_i w_i$ on the right side of equality (20) for which either v_i or w_i does not belong to $\mathrm{Ass}[Z]$ must cancel. But in this case $u \in I$. This proves the lemma.

THEOREM 7 (*Shirshov*). Every special Jordan algebra with not more than a countable number of generators is imbeddable in a special Jordan algebra with two generators.

PROOF. Let J be a special Jordan algebra with not more than a countable number of generators. Then J is isomorphic to some quotient algebra $\mathrm{SJ}[Z]/I$ of the algebra $\mathrm{SJ}[Z]$. We note that by Lemma 5 the relation

$$\hat{I} \cap \mathrm{SJ}[Z] = I$$

holds for the ideal $\hat{I}$ of the algebra $\mathrm{Ass}[Z]$ generated by the set I. We now consider the ideal $\tilde{\hat{I}}$ of the algebra $\mathrm{Ass}[x, y]$ generated by the set $\hat{I}$. In view of Lemma 7

$$\tilde{\hat{I}} \cap \mathrm{Ass}[Z] = \hat{I},$$

whence $\mathrm{SJ}[Z] \cap \tilde{\hat{I}} = \mathrm{SJ}[Z] \cap (\mathrm{Ass}[Z] \cap \tilde{\hat{I}}) = \mathrm{SJ}[Z] \cap \hat{I} = I$. The image of the algebra $\mathrm{SJ}[Z]$ in the quotient algebra $\bar{A} = \mathrm{Ass}[x, y]/\tilde{\hat{I}}$ is the subalgebra $\mathrm{SJ}[Z]/\mathrm{SJ}[Z] \cap \tilde{\hat{I}} = \mathrm{SJ}[Z]/I$ isomorphic to J. Since $xy^i x = 2(y^i \odot x) \odot x - y^i \odot x^2$, this subalgebra is in the subalgebra of the algebra $\bar{A}^{(+)}$ generated by the elements $\bar{x} = x + \tilde{\hat{I}}$ and $\bar{y} = y + \tilde{\hat{I}}$.

This proves the theorem.

Exercises

1. (*Cohn*) Prove that $\mathrm{H}[X]$ is the subalgebra of the algebra $\mathrm{Ass}[X]^{(+)}$ generated by the set X and quadruples of the form $\{x_\alpha x_\beta x_\gamma x_\delta\}$.

2. (*Cohn*) Prove that the quotient algebra $\mathrm{SJ}[x, y, z]/I$ is exceptional, where I is the ideal generated by the element $x \odot y$.

3. Prove that the quotient algebra $\mathrm{Ass}[x, y, z]^{(+)}/I$ is exceptional, where I is the ideal generated by the element $x^2 - y^2$.

4. (*Shirshov*) Every special Jordan algebra is isomorphically imbeddable in a special Jordan algebra, each countable subset of which is contained in a subalgebra with two generators.

3. SHIRSHOV'S THEOREM

In the previous section we established that the analog of the Poincaré–Birkhoff–Witt theorem is not valid for Jordan algebras in that not every Jordan algebra has an associative enveloping algebra. We saw that even Jordan algebras generated by three elements can be exceptional. For Jordan algebras with two generators, however, the situation is radically different, since in 1956 Shirshov proved every Jordan algebra from two generators is special.

This result had a noticeable influence on the subsequent development of the theory. For example, it follows from this result that all identities in two variables which are valid for all special Jordan algebras are also true for all Jordan algebras. Since the question of the validity of an identity in all special Jordan algebras is not difficult—it is only necessary to verify that it is valid in a free special Jordan algebra—this theorem gives an algorithm for the recognition of identities in two variables in the variety of Jordan algebras.

In 1960 Macdonald proved by a similar method that all identities in three variables which are linear in one of them and valid for all special Jordan algebras are also valid for all Jordan algebras. Glennie showed that the results of Shirshov and Macdonald cannot be strengthened. He found identities of degrees 8 and 9 which are satisfied in special Jordan algebras but are not satisfied in the simple exceptional Jordan algebra $H(C_3)$. These identities depend on three variables and have degree ≥ 2 in each of them.

The question of the description of the identities which are valid in all special but not all Jordan algebras still remains open.

The aim of the present section is to prove Shirshov's theorem and a series of corollaries to it. Even though Macdonald's result is not a formal corollary to Shirshov's theorem, it follows from its proof.

We shall begin with some preliminary results.

Let A be an arbitrary Φ-algebra. For every element $a \in A$ we define two mappings from the algebra A to itself: $R_a : x \mapsto xa$ and $L_a : x \mapsto ax$. These mappings are endomorphisms of the Φ-module A. The first endomorphism is called the *operator of right multiplication by the element a*, and the second is the *operator of left multiplication by the element a.* The subalgebra of the algebra of endomorphisms of the Φ-module A generated by all possible operators R_a and L_a, where $a \in A$, is called the *multiplication algebra of the algebra A* and is denoted by $M(A)$. The subalgebra of the algebra $M(A)$

generated by all the operators of right (left) multiplications is called the *right (left) multiplication algebra of the algebra* A and is denoted by $R(A)$ ($L(A)$, respectively). If the algebra A is commutative or anticommutative, then all these algebras coincide: $M(A) = R(A) = L(A)$.

Let A be a subalgebra of an algebra B. We shall denote by $R^B(A)$ the subalgebra generated in $R(B)$ by the operators R_a where $a \in A$. The algebras $M^B(A)$, $L^B(A)$ are defined analogously.

We consider the multiplication algebras of Jordan algebras. The identities satisfied by Jordan algebras imply a series of relations for the multiplication operators. By commutativity, the basic Jordan identity is equivalent to the identity $(yx)x^2 = (yx^2)x$, which in turn is equivalent to the operator relation

$$[R_x, R_{x^2}] = 0. \tag{21}$$

We carry out a complete linearization of the identity $(x^2y)x = x^2(yx)$. Changing notation for the unknowns and using commutativity, we obtain the identity

$$[(xy)z]t + [(xt)z]y + x[(yt)z] = (xy)(zt) + (xz)(yt) + (xt)(yz). \tag{22}$$

We now note that in the right side of this equality all variables appear symmetrically, and, therefore, the left side must not depend on the positions of the variables x and z. Taking this property into account, we obtain yet the identities

$$[(xy)z]t + [(xt)z]y + x[(yt)z] = [x(yz)]t + [x(yt)]z + [x(zt)]y, \tag{23}$$

$$[x(yz)]t + [x(yt)]z + [x(zt)]y = (xy)(zt) + (xz)(yt) + (xt)(yz). \tag{24}$$

Identities (22)–(24) are equivalent to the operator relations

$$R_yR_zR_t + R_tR_zR_y + R_{(yt)z} = R_yR_{zt} + R_zR_{yt} + R_tR_{yz}, \tag{25}$$

$$R_yR_zR_t + R_tR_zR_y + R_{(yt)z} = R_{yz}R_t + R_{yt}R_z + R_{zt}R_y, \tag{26}$$

$$[R_{yz}, R_t] + [R_{yt}, R_z] + [R_{zt}, R_y] = 0. \tag{27}$$

PROPOSITION 2. If the subalgebra A of a Jordan algebra B is generated by the set A_0, then the algebra $R^B(A)$ is generated by the set of operators $\{R_a, R_{ab} \mid a, b \in A_0\}$.

PROOF. It is obvious that the algebra $R^B(A)$ is generated by operators of the form $R_{\bar{v}}$, where $v = v(x_1, \ldots, x_n)$ is some nonassociative monomial and $\bar{v} = v(a_1, \ldots, a_n)$ with $a_i \in A_0$. It suffices to show that for any monomial v the operator $R_{\bar{v}}$ is in the subalgebra A_0^* generated by the set $\{R_a, R_{ab} \mid a, b \in A_0\}$. We shall carry out an induction on the degree of the monomial v. If $d(v) \leq 2$, then it is obvious $R_{\bar{v}} \in A_0^*$. If $d(v) \geq 3$, then $v = (v_1v_2)v_3$ where the v_i

are monomials of lesser length. By (25)

$$R_{\bar{v}} = -R_{\bar{v}_1}R_{\bar{v}_3}R_{\bar{v}_2} - R_{\bar{v}_2}R_{\bar{v}_3}R_{\bar{v}_1} + R_{\bar{v}_1}R_{\bar{v}_2\bar{v}_3} + R_{\bar{v}_2}R_{\bar{v}_1\bar{v}_3} + R_{\bar{v}_3}R_{\bar{v}_1v_2}.$$

By the induction assumption $R_{\bar{v}_i}, R_{\bar{v}_i\bar{v}_j} \in A_0^*$, and therefore $R_{\bar{v}} \in A_0^*$ also. This proves the proposition.

A subalgebra A of a Jordan algebra B is called *strongly associative* if, for any elements $a, a' \in A$ and $b \in B$, the equality $(a, b, a') = 0$ holds. By the relation $(a, b, a') = b[R_a, R_{a'}]$, the condition of strong associativity for a subalgebra A in an algebra B is equivalent to the commutativity of the algebra $R^B(A)$.

THEOREM 8. Every singly generated subalgebra of a Jordan algebra is strongly associative.

PROOF. Let the subalgebra A of the Jordan algebra B be generated by the element a. Then the algebra $R^B(A)$ is generated by the elements R_a and R_{a^2}, which commute by (21). Consequently the algebra $R^B(A)$ is commutative. This proves the theorem.

COROLLARY. Every Jordan algebra is power associative.

It also follows from Theorem 8 that in every Jordan algebra the identity

$$(x^n y)x^m = x^n(yx^m) \tag{28}$$

is valid.

Along with the usual multiplication, in every Jordan algebra we define the *Jordan triple product*:

$$\{xyz\} = (xy)z + (zy)x - (xz)y.$$

If the Jordan algebra A is special and imbedded in the algebra $P^{(+)}$ for some associative algebra P, then it is not difficult to verify for any $a, b, c \in A$

$$\{abc\} = \tfrac{1}{2}(abc + cba),$$

where the associative product of the elements x and y is denoted by xy. In particular, the Jordan triple product $\{aba\}$ in $P^{(+)}$ equals the associative product aba.

For any elements, a, b of a Jordan algebra A we now define the mapping $U_{a,b}: x \mapsto \{axb\}$, and we set $U_a = U_{a,a}$. It is clear that $U_{a,b}$ and U_a are elements of the algebra $R(A)$ of right multiplications:

$$U_{a,b} = R_aR_b + R_bR_a - R_{ab}, \tag{29}$$

$$U_a = 2R_a^2 - R_{a^2}. \tag{30}$$

Sometimes we shall also write $U(a, b)$ and $U(a)$.

LEMMA 8. Let A be a strongly associative subalgebra of a Jordan algebra B. Then for any $a, a' \in A$ and $b \in B$ the following relations are valid:

$$U_a U_{a',b} = 2R_a U_{aa',b} - U_{a^2a',b}, \tag{31}$$

$$U_{a',b} U_a = 2U_{aa',b} R_a - U_{a^2a',b}. \tag{31'}$$

PROOF. We first note that for any $x, y \in B$

$$U_{x^2,y} = 2U_{x,y}R_x - R_y U_x = 2R_x U_{x,y} - U_x R_y. \tag{32}$$

For the proof of this relation it is necessary to write the operators U_x, $U_{x,y}$, $U_{x^2,y}$ according to formulas (29) and (30) and use identities (25)–(27).

Linearizing equality (32) in x, we obtain

$$U_{xz,y} = U_{x,y}R_z + U_{z,y}R_x - R_y U_{x,z} = R_z U_{x,y} + R_x U_{z,y} - U_{x,z}R_y. \tag{33}$$

We now replace x by x^2 in the second of equalities (33), and we obtain the relation

$$U_{x^2z,y} = R_z U_{x^2,y} + R_{x^2} U_{z,y} - U_{x^2,z} R_y.$$

By (32) and the definition of the operator U_x, this relation can be changed in the following manner:

$$\begin{aligned} U_{x^2z,y} &= 2R_z R_x U_{x,y} - R_z U_x R_y + (2R_x^2 - U_x)U_{z,y} - 2U_{x,z}R_x R_y + R_z U_x R_y \\ &= 2[R_z, R_x]U_{x,y} + 2R_x R_z U_{x,y} + 2R_x^2 U_{x,z} - U_x U_{x,z} \\ &\quad + 2[R_x, U_{x,z}]R_y - 2R_x U_{x,z} R_y. \end{aligned}$$

Consequently, by (33)

$$U_{x^2z,y} = 2[R_z, R_x]U_{x,y} + 2R_x U_{xz,y} + 2[R_x, U_{x,z}]R_y - U_x U_{z,y}.$$

Setting here $x = a$, $z = a'$, $y = b$ we obtain relation (31). Relation (31′) is proved analogously, which proves the lemma.

COROLLARY. The following identities are valid in every Jordan algebra:

$$\{x^k\{x^n y x^n\}z\} = 2\{x^{n+k}(yx^n)z\} - \{x^{2n+k}yz\}, \tag{34}$$

$$\{xty\}z + \{zty\}x = \{(xz)ty\} + \{x(ty)z\}, \tag{35}$$

$$\{xtz\}y + \{(xz)ty\} = \{x(tz)y\} + \{z(tx)y\}, \tag{36}$$

$$\{x^n y x^n\}x^k = \{x^n y x^{n+k}\}. \tag{37}$$

PROOF. The validity of identity (34) follows from (31) by Theorem 8. Furthermore, the operator relations (33) give us identities (35) and (36). For the proof of identity (37) we set $z = 1$ in (34), formally adjoining beforehand, if necessary, an identity element 1 to the Jordan algebra. By (28) we obtain

$$\{x^n y x^n\} x^k = 2x^{n+k}(yx^n) - x^{2n+k} y = \{x^n y x^{n+k}\}.$$

This proves the corollary.

We consider the free associative Φ-algebra $A = \mathrm{Ass}[x, y, z]$ from generators x, y, z. The algebra A is a free module over Φ for which the set S of all associative words from x, y, z constitutes a basis. For an associative word $\alpha \in S$ we shall denote by $d(\alpha)$ the length of the word α, and by $h(\alpha)$ its height which was defined for the proof of Theorem 3. We also recall that an involution $*$ is defined on the algebra A such that if $\alpha = x_1 x_2 \cdots x_n$, where $x_i \in \{x, y, z\}$, then $\alpha^* = x_n x_{n-1} \cdots x_1$.

Let S_1 be the set of words from S for which the degree in z is not greater than 1. We define by induction on the height of a word $\alpha \in S_1$ a mapping $\alpha \to \alpha'$ of the set S_1 into the free Jordan algebra $J = \mathrm{J}[x, y, z]$, setting

(1) $\alpha' = \alpha$, if $h(\alpha) = 1$;
(2) $(\alpha\beta)' = \alpha' \cdot \beta'$, if $h(\alpha\beta) = 2$, $h(\alpha) = h(\beta) = 1$;
(3) $(x^k \beta x^n)' = x^k \cdot (\beta x^n)' + (x^k \beta)' \cdot x^n - x^{k+n} \cdot \beta'$,
where $h(x^k \beta x^n) = h(\beta) + 2$;
(4) $(y^k \beta x^n)' = 2\{y^k \beta' x^n\} - (x^n \beta y^k)'$,
where $\beta \neq z$, $h(y^k \beta x^n) = h(\beta) + 2$

(we note that in this case $h(x^n \beta y^k) < h(y^k \beta x^n)$, so that it is possible to assume $(x^n \beta y^k)'$ is defined);

(5) $(y^m z x^k)' = \{y^m z x^k\}$;
(6) $(x^k \beta z)' = (z\beta^* x^k)' = 2x^k \cdot (\beta z)' - (\beta z x^k)'$,
where $h(x^k \beta z) = h(\beta) + 2$.

(We note that $(\beta z x^k)'$ is already defined by (3)–(5).)

Changing the roles of x and y, we obtain four more rules (3′)–(6′) for this inductive definition.

We denote by A_1 the Φ-submodule in A generated by the set S_1. It is clear that the mapping $\alpha \mapsto \alpha'$ can be extended to a homomorphism of A_1 into J as Φ-modules.

Furthermore, for any elements $a, b \in A$ we define the operations

$$[a] = \tfrac{1}{2}(a + a^*), \tag{38}$$

$$a \circ b = a \odot [b] = \tfrac{1}{4}(ab + ab^* + ba + b^* a), \tag{39}$$

where, as usual, by $\odot$ is denoted the multiplication in the algebra $A^{(+)}$. We note that generally speaking the operation $\circ$ is not commutative. By Theorem 3, the mapping $a \to [a]$ is an endomorphism of the algebra A

onto the free special Jordan algebra $\mathrm{SJ}[x, y, z]$. In addition, if $a \in \mathrm{SJ}[x, y, z]$, then $[a] = a$.

LEMMA 9. The equality $[a \circ b] = [a] \odot [b]$ is valid for any $a, b \in A$.

PROOF. By (38) and (39) $[a \circ b] = \frac{1}{2}\{a \odot [b] + (a \odot [b])^*\} = \frac{1}{2}(a \odot [b] + a^* \odot [b]) = [a] \odot [b]$. This proves the lemma.

SHIRSHOV'S THEOREM. Every Jordan algebra from two generators is special.

PROOF. In view of Theorem 6, it suffices to prove that the algebras $\mathrm{J}[x, y]$ and $\mathrm{SJ}[x, y]$ are isomorphic. We denote by ϕ the restriction of the mapping $a \to a'$ to the set $\mathrm{SJ}[x, y]$. It is clear that ϕ maps $\mathrm{SJ}[x, y]$ into $\mathrm{J}[x, y]$ and is a homomorphism of the Φ-modules. We shall show that ϕ is an isomorphism of $\mathrm{SJ}[x, y]$ onto $\mathrm{J}[x, y]$.

By the definition of the mapping $'$, by induction on the height of the word $\alpha \in S_1$ it is easy to show that $(\alpha^*)' = \alpha'$. Therefore, for any $a \in A_1$

$$(a^*)' = a', \tag{40}$$

and consequently

$$[a]' = a'. \tag{41}$$

By (41) and Lemma 9, for any $a, b \in \mathrm{SJ}[x, y]$ we now have $\phi(a \odot b) = (a \odot b)' = ([a] \odot [b])' = [a \circ b]' = (a \circ b)'$. On the other hand, $\phi(a) \cdot \phi(b) = a' \cdot b'$. This means in order for the mapping ϕ to be a homomorphism it is necessary and sufficient that the relation

$$(a \circ b)' = a' \cdot b'$$

be satisfied for any $a, b \in \mathrm{SJ}[x, y]$. We shall prove this relation later in a basic lemma, but now we complete the proof of the theorem. We have a homomorphism $\phi: \mathrm{SJ}[x, y] \to \mathrm{J}[x, y]$ such that $\phi(x) = x$, $\phi(y) = y$. We note that if the elements $m, n \in \mathrm{J}[x, y]$ have inverse images, that is, $m = p'$, $n = q'$ for some $p, q \in \mathrm{SJ}[x, y]$, then the element $m \cdot n$ also has an inverse image, the element $p \odot q$, since

$$\phi(p \odot q) = \phi(p) \cdot \phi(q) = p' \cdot q' = m \cdot n.$$

Hence it follows that ϕ is a surjective homomorphism. However, there is the canonical homomorphism $\pi: \mathrm{J}[x, y] \to \mathrm{SJ}[x, y]$ for which $\pi(x) = x$, $\pi(y) = y$. The mappings $\pi\phi$ and $\phi\pi$ are the identity mappings on $\mathrm{SJ}[x, y]$ and $\mathrm{J}[x, y]$. Consequently π and ϕ are isomorphisms, which is what was to be proved.

LEMMA 10. Let $\pi : J[x, y, z] \to SJ[x, y, z]$ be the canonical homomorphism. Then the equality

$$\pi(a') = [a]$$

is valid for any $a \in A_1$.

PROOF. It is obvious that it suffices to prove the assertion of the lemma in the case $a = \alpha$ is a word from x, y, z. If $h(\alpha) = 1$, then everything is clear. We now assume that the lemma is true for all words of lesser height, and we consider the possible cases:

1. $\alpha = \alpha_1\alpha_2$, $h(\alpha) = 2$, $h(\alpha_1) = h(\alpha_2) = 1$. Then $\pi(\alpha') = \pi(\alpha_1 \cdot \alpha_2) = \alpha_1 \odot \alpha_2 = [\alpha_1\alpha_2] = [\alpha]$.

2. $\alpha = x^k\beta x^n$. In this case by the definition of the mapping $\alpha \to \alpha'$ and the induction assumption

$$\begin{aligned}\pi(\alpha') &= \pi(x^k \cdot (\beta x^n)' + (x^k\beta)' \cdot x^n - x^{k+n} \cdot \beta') \\ &= x^k \odot [\beta x^n] + [x^k\beta] \odot x^n - x^{k+n} \odot [\beta] \\ &= \tfrac{1}{2}(x^k\beta x^n + x^n\beta^* x^k) = [x^k\beta x^n] = [\alpha].\end{aligned}$$

3. $\alpha = y^k\beta x^n$. Then we have

$$\begin{aligned}\pi(\alpha') &= 2\{y^k[\beta]x^n\} - [x^n\beta y^k] \\ &= y^k[\beta]x^n + x^n[\beta]y^k - [x^n\beta y^k] \\ &= \tfrac{1}{2}(y^k\beta x^n + x^n\beta^* y^k) = [y^k\beta x^n] = [\alpha].\end{aligned}$$

4. $\alpha = x^k\beta z$. By a case already considered, we have

$$\pi(\alpha') = 2x^k \odot [\beta z] - [\beta z x^k] = \tfrac{1}{2}(x^k\beta z + z\beta^* x^k) = [x^k\beta z] = [\alpha].$$

The remaining cases are considered analogously. This proves the lemma.

COROLLARY. Let $a, b, ab \in A_1$ and also $a' \cdot b' = c'$ for some $c \in A_1$. Then $a' \cdot b' = (a \circ b)'$.

PROOF. If we apply the homomorphism π to the equality $a' \cdot b' = c'$, we obtain $[a] \odot [b] = [c]$. Then, by Lemma 9 and equality (41), we obtain

$$a' \cdot b' = c' = [c]' = ([a] \odot [b])' = [a \circ b]' = (a \circ b)'.$$

This proves the corollary.

BASIC LEMMA (*Shirshov*). The relation

$$(a \circ b)' = a' \cdot b'$$

is valid for any $a, b \in A_1$ such that $ab \in A_1$.

PROOF. Let $f, g \in J[x, y, z]$. We shall write $f \equiv g$ if $f - g = h'$ for some $h \in A_1$. By the corollary to Lemma 10 it suffices for us to show that the relation $a' \cdot b' \equiv 0$ is valid for any $a, b \in A_1$ such that $ab \in A_1$. It is also clear that it is sufficient to prove this relation in the case when a and b are associative words from S_1. We shall say that a word α is greater than a word β if α contains z but β does not, or if both α and β do not contain z and $h(\alpha) > h(\beta)$.

Thus, let $a = \alpha$, $b = \beta$ be associative words that satisfy the conditions of the basic lemma. We formulate three induction assumptions:

(1) the lemma is valid for words α_1 and β_1 if $h(\alpha_1) + h(\beta_1) < h(\alpha) + h(\beta)$;

(2) the lemma is valid for words α_1 and β_1 if $h(\alpha_1) + h(\beta_1) = h(\alpha) + h(\beta)$, $d(\alpha_1) + d(\beta_1) < d(\alpha) + d(\beta)$.

The basis for the first induction assumption is the obvious validity of the lemma for a sum of heights equal to 2. The first induction assumption serves as basis for the second induction assumption, since a decrease of the sum of degrees leads as a final result to a decrease of the sum of heights.

(3) the lemma is valid for words α_1 and β_1 if $h(\alpha_1) + h(\beta_1) = h(\alpha) + h(\beta)$, $d(\alpha_1) + d(\beta_1) = d(\alpha) + d(\beta)$, and the height of the smaller of the words α_1, β_1 is less than the height of the smaller of the words α, β.

This last induction assumption will be justified later, where it will be shown that the lemma is valid if the height of the smaller of the words α, β is less than 3, but now we complete the proof of the basic lemma.

Let β be less than α and $h(\beta) > 2$. Then, as is easy to see from the definition of the mapping $'$,

$$\beta' = \sum_i \sigma_i (c_i' \cdot d_i') \cdot e_i' + \sum_j \sigma_j x^{k_j} \cdot y^{s_j},$$

where $\sigma_k \in \Phi$; c_i, d_i, $e_i \in S$; $h(c_i) + h(d_i) + h(e_i) \leq h(\beta)$. By the linearity of all the operations, it suffices to prove the validity of the lemma by considering, instead of β', the elements $\beta_1 = (c' \cdot d') \cdot e'$ or $\beta_2 = x^k \cdot y^s$.

The basis for the third induction assumption gives us

$$\alpha' \cdot \beta_2 = \alpha' \cdot (x^k y^s)' \equiv 0.$$

Furthermore, in view of identity (22) and induction assumption (3),

$$\begin{aligned}
\alpha' \cdot \beta_1 &= \alpha' \cdot ((c' \cdot d') \cdot e') \\
&= -((\alpha' \cdot c') \cdot e') \cdot d' - ((\alpha' \cdot d') \cdot e') \cdot c' + (\alpha' \cdot c') \cdot (d' \cdot e') \\
&\quad + (\alpha' \cdot d') \cdot (c' \cdot e') + (\alpha' \cdot e') \cdot (c' \cdot d') \\
&= -\{((\alpha \circ c) \circ e) \circ d - ((\alpha \circ d) \circ e) \circ c + (\alpha \circ c) \circ (d \circ e) \\
&\quad + (\alpha \circ d) \circ (c \circ e) + (\alpha \circ e) \circ (c \circ d)\}' \equiv 0.
\end{aligned}$$

Thus, modulo induction assumption (3) the lemma is proved.

Before proceeding to the basis for induction assumption (3), we shall establish a useful relation. Let $a, c \in \mathrm{SJ}[x, y, z]$ and $b \in \mathrm{Ass}[x, y, z]$. Then the equality

$$\{abc\} = (b \circ a) \circ c + (b \circ c) \circ a - b \circ (c \circ a)$$

holds. In fact, by the definition of the operation $\circ$ we have the equality $x \circ y = x \odot [y]$. Hence, since $[a] = a$, $[c] = c$, and $[a \circ c] = a \odot c$, we obtain

$$\begin{aligned}\{abc\} &= (b \odot a) \odot c + (b \odot c) \odot a - b \odot (c \odot a) \\ &= (b \circ a) \circ c + (b \circ c) \circ a - b \circ (c \circ a).\end{aligned}$$

If the induction assumptions apply to the pairs of elements $(b \circ a, c)$, $(b \circ c, a)$, and $(b, c \circ a)$, then from the obtained relation follows the equality

$$\{abc\}' = \{a'b'c'\}. \tag{42}$$

In fact, in this case we have

$$\begin{aligned}\{abc\}' &= ((b \circ a) \circ c + (b \circ c) \circ a - b \circ (a \circ c))' \\ &= (b' \cdot a') \cdot c' + (b' \cdot c') \cdot a' - b' \cdot (a' \cdot c') = \{a'b'c'\}.\end{aligned}$$

The proof of the basis for induction assumption (3) is divided into a series of cases. Everywhere below $h(\alpha) = h(\gamma) + 2$.

CASE 1

$$\alpha = x^s \gamma x^m, \qquad \beta = x^t.$$

We shall assume that $s \geq m$. This does not restrict generality, since otherwise, in view of (40), α^* can be considered in place of α. We shall carry out an induction on $s + m$. If $s + m = 1$, then one of the numbers s, m equals 0, and we satisfy the conditions for induction assumption (1). Thus we have a basis for the induction. Now we shall perform the induction step. Using the induction assumptions (1) and (2) and formula (42), as well as identities (37) and (28), we obtain

$$\begin{aligned}\alpha' \cdot \beta' &= (x^s \gamma x^m)' \cdot x^t = \{x^m (x^{s-m}\gamma) x^m\}' \cdot x^t \\ &= \{x^m (x^{s-m}\gamma)' x^m\} \cdot x^t = \{x^m (x^{s-m}\gamma)' x^{m+t}\} \\ &= ((x^{s-m}\gamma)' \cdot x^m) \cdot x^{m+t} + ((x^{s-m}\gamma)' \cdot x^{m+t}) \cdot x^m - (x^{s-m}\gamma)' \cdot x^{2m+t} \\ &\equiv 2((x^{s-m}\gamma)' \cdot x^m) \cdot x^{m+t} = 2((x^{s-m}\gamma) \circ x^m)' \cdot x^{m+t} \\ &= (x^{s-m}\gamma x^m + x^s \gamma)' \cdot x^{m+t} \equiv (x^{s-m}\gamma x^m)' \cdot x^{m+t},\end{aligned}$$

whence by the induction assumption $\alpha' \cdot \beta' \equiv 0$.

CASE 2

$$\alpha = x^m \gamma y^n, \qquad \beta = x^t.$$

(a) $\gamma = z$. In this case by the definitions of the operation $\circ$ and the mapping $'$ and by (35) and (28), we obtain

$$\begin{aligned}\alpha' \cdot \beta' - (\alpha \circ \beta)' &= (x^m z y^n)' \cdot x^t - \tfrac{1}{2}(x^{m+t} z y^n)' - \tfrac{1}{2}(x^m z y^n x^t)' \\ &= \{x^m z y^n\} \cdot x^t - \tfrac{1}{2}\{x^{m+t} z y^n\} - \tfrac{1}{2}x^m \cdot (x y^n x^t)' \\ &\quad - \tfrac{1}{2}\{x^m z y^n\} \cdot x^t + \tfrac{1}{2}x^{m+t} \cdot (z \cdot y^n) \\ &= \tfrac{1}{2}\{x^m z y^n\} \cdot x^t - \tfrac{1}{2}\{x^{m+t} z y^n\} + \tfrac{1}{2}\{x^t z y^n\} \cdot x^m \\ &\quad - x^m \cdot (x^t \cdot (z \cdot y^n)) + \tfrac{1}{2}x^{m+t} \cdot (z \cdot y^n) \\ &= \tfrac{1}{2}\{x^m (z \cdot y^n) x^t\} - x^m \cdot (x^t \cdot (z \cdot y^n)) + \tfrac{1}{2}x^{m+t} \cdot (z \cdot y^n) = 0.\end{aligned}$$

We can reason analogously in the case when the word γ is missing. In fact this only simplifies the proof.

(b) $\gamma \neq z$, γ is not the empty word. In this case $h(y^n \gamma x^m) < h(x^m \gamma y^n)$, and therefore by induction assumption (1),

$$\begin{aligned}\alpha' \cdot \beta' &= (x^m \gamma y^n)' \cdot x^t \\ &= 2\{x^m \gamma' y^n\} \cdot x^t - (y^n \gamma x^m)' \cdot x^t \equiv 2\{x^m \gamma' y^n\} \cdot x^t. \qquad (43)\end{aligned}$$

On the other hand, if $h(x^m \delta x^n) = h(\delta) + 2 = h(\alpha) + h(\beta)$, then by the definition of the mapping $'$ and induction assumption (1), we have

$$(x^m \delta)' \cdot x^n = (x^m \delta x^n)' - x^m \cdot (\delta x^n)' + x^{m+n} \cdot \delta' \equiv -x^m \cdot (\delta x^n)'. \qquad (44)$$

Consequently, $\alpha' \cdot \beta' = (x^m \gamma y^n)' \cdot x^t \equiv -x^m (\gamma y^n x^t)'$, whence by the definition of the operation $\circ$ and induction assumption (1), we obtain

$$\begin{aligned}\alpha' \cdot \beta' &\equiv -2x^m \cdot ((\gamma y^n) \circ x^t)' + x^m \cdot (x^t \gamma y^n)' \\ &= -2x^m \cdot (x^t \cdot (\gamma y^n)') + 2x^m \cdot \{x^t \gamma' y^n\} - x^m \cdot (y^n \gamma x^t)' \\ &\equiv -2x^m \cdot (x^t \cdot (\gamma y^n)') + 2x^m \cdot \{x^t \gamma' y^n\}.\end{aligned}$$

Combining the obtained relation with (43), by (35) and (28) we obtain

$$\begin{aligned}\alpha' \cdot \beta' &\equiv \{x^m \gamma' y^n\} \cdot x^t + \{x^t \gamma' y^n\} \cdot x^m - x^m \cdot (x^t \cdot (\gamma y^n)') \\ &= \{x^{m+t} \gamma' y^n\} + \{x^m (\gamma' \cdot y^n) x^t\} - x^m \cdot (x^t \cdot (\gamma y^n)') \\ &\equiv \{x^m (\gamma \circ y^n)' x^t\} - x^m \cdot (x^t \cdot (\gamma y^n)') \\ &= \tfrac{1}{2}\{x^m (y^n \gamma)' x^t\} - \tfrac{1}{2}x^{m+t} \cdot (\gamma y^n)' \equiv x^m \cdot (x^t \cdot (y^n \gamma)') \\ &= x^m \cdot ((y^n \gamma) \circ x^t)' = \tfrac{1}{2}x^m \cdot (x^t y^n \gamma + y^n \gamma x^t)' \equiv \tfrac{1}{2}x^m \cdot (x^t y^n \gamma)'.\end{aligned}$$

Since $h(\alpha) = h(\gamma) + 2$, then either $\gamma = y^k \gamma_1$ or $\gamma = \gamma_2 x^k$. Thus, either by the first induction assumption or by Case 1, we have $\alpha' \cdot \beta' \equiv \tfrac{1}{2}x^m \cdot (x^t y^n \gamma)' \equiv 0$.

CASE 3

$$\alpha = x^m \gamma z, \qquad \beta = x^n.$$

If γ is the empty word, then according to (28) we have

$$\begin{aligned}\alpha' \cdot \beta' - (\alpha \circ \beta)' &= (x^m \cdot z) \cdot x^n - \tfrac{1}{2}(x^m z x^n + x^{m+n} z)' \\ &= (x^m \cdot z) \cdot x^n - \tfrac{1}{2}(x^m \cdot z) \cdot x^n - \tfrac{1}{2}x^m \cdot (z \cdot x^n) \\ &\quad + \tfrac{1}{2}x^{m+n} \cdot z - \tfrac{1}{2}x^{m+n} \cdot z = 0.\end{aligned}$$

If γ is not the empty word, then by (44) and the first two cases we obtain

$$\alpha' \cdot \beta' = (x^m \gamma z)' \cdot x^n \equiv -x^m \cdot (\gamma z x^n)' \equiv 0.$$

CASE 4

$$\alpha = x^m \gamma x^n, \qquad \beta = y^t.$$

We first note that if the word $\delta \in A_1$ and $h(x^k\,\delta y^n) = h(\delta) + 2$, then by the definition of the mapping $'$

$$2\{x^k\,\delta' y^n\} = (x^k\,\delta y^n)' + (y^n\,\delta x^k)' \equiv 0. \tag{45}$$

Furthermore, by (40) we can assume without loss of generality that $m \geq n$. Now by induction assumption (2) and relations (42), (36), and (45), we obtain

$$\begin{aligned}\alpha' \cdot \beta' &= (x^m \gamma x^n)' \cdot y^t = \{x^n(x^{m-n}\gamma)x^n\}' \cdot y^t = \{x^n(x^{m-n}\gamma)' x^n\} \cdot y^t \\ &= -\{x^{2n}(x^{m-n}\gamma)' y^t\} + 2\{x^n(x^n \cdot (x^{m-n}\gamma)')y^t\} \\ &= -\{y^t(x^{m-n}\gamma)' x^{2n}\} + 2\{x^n((x^{m-n}\gamma) \circ x^n)' y^t\} \\ &= -\{y^t(x^{m-n}\gamma)' x^{2n}\} + \{y^t(x^m\gamma)' x^n\} + \{y^t(x^{m-n}\gamma x^n)' x^n\} \\ &\equiv \{y^t(x^{m-n}\gamma x^n)' x^n\}.\end{aligned}$$

For convenience we denote $m - n$ by k. For completion of the proof it now suffices to prove the relation

$$\{y^t(x^k \gamma x^n)' x^m\} \equiv 0. \tag{46}$$

We carry out an induction on $k + n$. If $k + n = 1$, then either $k = 0$ or $n = 0$, and (46) follows from (45). We now assume that (46) is true for $k' + n' < k + n$. It is possible to assume that $k \geq n$. By induction assumption (2) and relations (42), (34), and (45) we obtain

$$\begin{aligned}\{y^t(x^k\gamma x^n)' x^m\} &= \{y^t\{x^n(x^{k-n}\gamma)x^n\}' x^m\} = \{y^t\{x^n(x^{k-n}\gamma)' x^n\} x^m\} \\ &= 2\{x^{n+m}(x^n \cdot (x^{k-n}\gamma)')y^t\} - \{x^{2n+m}(x^{k-n}\gamma)' y^t\} \\ &= 2\{y^t((x^{k-n}\gamma) \circ x^n)' x^{n+m}\} - \{y^t(x^{k-n}\gamma)' x^{2n+m}\} \\ &\equiv \{y^t(x^k\gamma)' x^{n+m}\} + \{y^t(x^{k-n}\gamma x^n)' x^{n+m}\} \\ &\equiv \{y^t(x^{k-n}\gamma x^n)' x^{n+m}\},\end{aligned}$$

whence by the induction assumption (46) is valid. Thus the lemma is also true in Case 4.

CASE 5

$$\alpha = x^k\gamma z, \qquad \beta = y^n.$$

In this case by definition of the mapping $'$ we have

$$2\alpha' \cdot \beta' = 2(x^k\gamma z)' \cdot y^n = (y^n x^k \gamma z)' + (x^k\gamma z y^n)' \equiv 0,$$

and everything is proved.

CASE 6

$$\alpha = x^k\gamma x^n, \qquad \beta = y^t x^m.$$

Applying cases already considered, in view of (46) we obtain

$$\begin{aligned}\alpha' \cdot \beta' &= (x^k\gamma x^n)' \cdot (y^t x^m)' = (x^k\gamma x^n)' \cdot (y^t \gamma x^m)\\ &= -\{y^t(x^k\gamma x^n)' x^m\} + ((x^k\gamma x^n)' \cdot y^t) \cdot x^m + ((x^k\gamma x^n)' \cdot x^m) y^t \equiv 0,\end{aligned}$$

which is what was to be proved.

CASE 7

$$\alpha = x^k\gamma y^n, \qquad \beta = y^t x^m.$$

Again applying cases considered earlier, by (45) we obtain

$$\begin{aligned}\alpha' \cdot \beta' &= (x^k\gamma y^n)' \cdot (y^t \cdot x^m)\\ &= -\{y^t(x^k\gamma y^n)' x^m\} + ((x^k\gamma y^n)' \cdot x^m) \cdot y^t + ((x^k\gamma y^n)' \cdot y^t) \cdot x^m \equiv 0,\end{aligned}$$

which is what was to be proved.

The case when $\alpha = x^k\gamma z$, $\beta = y^t x^m$ is considered analogously.

The cases considered together with those obtained from them by means of interchanging the roles of the generators x and y or replacing the word α by the word α^* justify induction assumption (3). With this the proof of the basic lemma is completed.

COROLLARY 1. Let J_1 be the submodule of the algebra $\mathrm{J}[x, y, z]$ consisting of elements of degree ≤ 1 in z and

$$\pi : \mathrm{J}[x, y, z] \to \mathrm{SJ}[x, y, z]$$

be the canonical homomorphism. Then $\operatorname{Ker} \pi \cap J_1 = (0)$.

PROOF. Let $SJ_1 = A_1 \cap \mathrm{SJ}[x, y, z]$. Then as is easy to see $\pi(J_1) = SJ_1$. We denote by ϕ the restriction of the mapping $a \mapsto a'$ to the set SJ_1. By Lemma 10 for any $a \in SJ_1$ we have $\pi(\phi(a)) = \pi(a') = [a] = a$. Hence it follows that for the proof of the corollary it only remains to show that ϕ maps SJ_1 onto J_1. This can be established just as in the proof of Shirshov's theorem.

COROLLARY 2 (*Shirshov–Macdonald*). If a nonassociative polynomial $f(x_1, x_2, x_3)$ of degree ≤ 1 in x_3 is an identity in every special Jordan algebra, then it is an identity for all Jordan algebras.

PROOF. It suffices to show that $f(x, y, z) = 0$ in the free Jordan algebra $\mathrm{J}[x, y, z]$. By assumption $f(x, y, z) \in \operatorname{Ker} \pi \cap J_1$. Therefore, by Corollary 1 $f(x, y, z) = 0$. This proves the corollary.

The assertion of this corollary is due to Shirshov when the degree of the polynomial f in x_3 equals zero, and when this degree equals one is due to Macdonald. The assertion of the corollary is not true for degree 2 in x_3. This was shown by Glennie (see Exercises).

Macdonald's result can be restated in the following form.

COROLLARY 3. Let $J_2 = \mathrm{J}[x, y]$, $J_3 = \mathrm{J}[x, y, z]$, $SJ_2 = \mathrm{SJ}[x, y]$, and $SJ_3 = \mathrm{SJ}[x, y, z]$. Then the mapping $R_u \mapsto R_{u^\pi}$ produces an isomorphism

$$\Pi : R^{J_3}(J_2) \to R^{SJ_3}(SJ_2).$$

PROOF. We define a mapping Π in the following manner:

$$\Pi\left(\sum_i R_{u_{i1}} R_{u_{i2}} \cdots R_{u_{ik_i}}\right) = \sum_i R_{u_{i1}^\pi} R_{u_{i2}^\pi} \cdots R_{u_{ik_i}^\pi}.$$

It is obvious that Π is a homomorphism, and it is only necessary to verify Π is one-to-one. Let

$$\sum_i R_{u_{i1}} R_{u_{i2}} \cdots R_{u_{ik_i}} \in \operatorname{Ker} \Pi.$$

Then

$$0 = z^\pi \sum_i R_{u_{i1}^\pi} R_{u_{i2}^\pi} \cdots R_{u_{ik_i}^\pi} = \left(z \sum_i R_{u_{i1}} R_{u_{i2}} \cdots R_{u_{ik_i}}\right)^\pi,$$

and by Corollary 1

$$z \sum_i R_{u_{i1}} R_{u_{i2}} \cdots R_{u_{ik_i}} = 0.$$

Since the algebra $\mathrm{J}[x, y, z]$ is free, we finally have

$$\sum_i R_{u_{i1}} R_{u_{i2}} \cdots R_{u_{ik_i}} = 0,$$

which proves the mapping Π is an isomorphism. This proves the corollary.

McCrimmon obtained an interesting result in this direction. He proved that if a nonassociative polynomial $f(x, x^{-1}, y, y^{-1}, z)$ of degree ≤ 1 in z vanishes in every special Jordan algebra when substituting for x and y

their inverse elements, for x^{-1} and y^{-1} their inverses†, and for z any element of the special Jordan algebra, then the polynomial $f(x, x^{-1}, y, y^{-1}, z)$ has this same property in the class of all Jordan algebras.

We note two useful identities, which by Corollary 2 are valid in all Jordan algebras:

$$\{xyx\}^2 = \{x\{yx^2y\}x\}, \tag{47}$$

$$zU(xU_y) = zU_yU_xU_y. \tag{48}$$

The last identity goes by the name of *Macdonald's identity*.

We denote by $\mathfrak{Z}_n$ the kernel of the canonical homomorphism of the free Jordan algebra $J[x_1, x_2, \ldots, x_n]$ onto the free special Jordan algebra $SJ[x_1, x_2, \ldots, x_n]$. Shirshov's theorem states that $\mathfrak{Z}_2 = (0)$. However $\mathfrak{Z}_3$ already turns out to be different from zero. This was first established by Albert and Paige who showed that the algebra $H(\boldsymbol{C}_3)$ is generated by three elements and is not the homomorphic image of any special Jordan algebra. Each nonzero element from $\mathfrak{Z}_n$ is called an *s-identity*. Such a name stems from the fact that if $f \in \mathfrak{Z}_n$, then f is an identity satisfied in all special but not all Jordan algebras. The existence of s-identities in three variables x, y, z follows from the theorem of Albert and Paige. However their proof does not determine a single concrete s-identity. Analyzing the proof of Albert and Paige, Glennie found a series of s-identities in three variables of degrees 8 and 9. He also showed that there do not exist s-identities of degree ≤ 5 in any number of variables. Later, with the aid of IBM, Glennie also established that s-identities of degree ≤ 7 do not exist in the case of a field of characteristic 0.

Exercises

1. Prove that the following two elements of the algebra $J[x, y, z]$ are s-identities:

$$\begin{aligned}&2\{\{y\{xzx\}y\}z(xy)\} - \{y\{x\{z(xy)z\}x\}y\}\\&\quad - 2\{(xy)z\{x\{yzy\}x\}\} + \{x\{y\{z(xy)z\}y\}x\},\end{aligned}$$

$$\begin{aligned}&2\{xzx\} \cdot \{y\{zy^2z\}x\} - 2\{yzy\} \cdot \{x\{zx^2z\}y\}\\&\quad - \{x\{z\{x\{yzy\}y\}z\}x\} + \{y\{z\{y\{xzx\}x\}z\}y\}.\end{aligned}$$

Hint. Refute the first identity with the following elements of the algebra $H(\boldsymbol{C}_3)$: $X = e_{11} + e_{33}$, $Y = e_{12} + e_{23}$, $Z = u_{12} + v_{13} + w_{23}$, where $(u, v, w) \neq 0$; the second identity with the elements $X = e_{12}$, $Y = e_{23}$, $Z = u_{12} + v_{13} + w_{23}$, where $(u, v, w) \neq 0$.

† For Jordan algebras the definition of invertibility of elements is somewhat different from the usual. It will be given in Chapter 14.

2. (*Albert–Paige*) The algebra $H(C_3)$ is not the homomorphic image of a special Jordan algebra and does not belong to the variety Var $\mathfrak{S}$.

3. The free Jordan algebra $J[X]$, where $\operatorname{card}(X) > 2$, is exceptional.

LITERATURE

Albert and Paige [9], Glennie [35, 36], Jacobson [47], Jacobson and McCrimmon [49], Cohn [101], Macdonald [113], McCrimmon [121, 123, 125, 126], Shirshov [274, 275, 279].

CHAPTER 4

Solvability and Nilpotency of Jordan Algebras

An associative algebra is called *nilpotent* if for some natural number n the product of any n elements from the algebra equals zero. When speaking about the product of n elements from a nonassociative algebra, we must also indicate the order in which the products are taken, or what is equivalent, the arrangement of parentheses. It can turn out that products of any n elements from the algebra equal zero for one arrangement of parentheses, and are different from zero for another. Thus, for nonassociative algebras, and in particular for Jordan algebras, there appear different variations of the concept of nilpotency. The interconnection between these variations is studied in this chapter.

1. DEFINITIONS AND EXAMPLES

A nonassociative algebra A is called *nilpotent* if there exists a natural number n such that the product of any n elements from the algebra A, with any arrangement of parentheses, equals zero. The least such number is called the *index of nilpotency* of the algebra A. The algebra A is called *right nilpotent* if there exists a natural number n such that $a_1 R_{a_2} R_{a_3} \cdots R_{a_n} = 0$, and *left nilpotent* if $a_1 L_{a_2} L_{a_3} \cdots L_{a_n} = 0$ for any elements $a_1, \ldots, a_n$. The minimal such number n is called, respectively, the *index of right nilpotency or left nilpotency of the algebra A*.

In an algebra A we define inductively a series of subsets by setting $A^1 = A^{\langle 1 \rangle} = A$ and $A^n = A^{n-1}A + A^{n-2}A^2 + \cdots + AA^{n-1}$, $A^{\langle n \rangle} = A^{\langle n-1 \rangle}A$. The subset A^n is called the nth *power* of the algebra A. The chain of subsets

$$A^1 \supseteq A^2 \supseteq \cdots \supseteq A^n \supseteq \cdots$$

is a chain of ideals of the algebra A. It is also easy to see that the chain

$$A^{\langle 1 \rangle} \supseteq A^{\langle 2 \rangle} \supseteq \cdots \supseteq A^{\langle n \rangle} \supseteq \cdots$$

is a chain of right ideals of this algebra, and if A is commutative or anticommutative, then this chain consists of ideals. The algebra A is nilpotent if $A^n = (0)$ for some n, and right nilpotent if $A^{\langle n \rangle} = (0)$.

PROPOSITION 1. Let A be a commutative (anticommutative) algebra. Then $A^{2^n} \subseteq A^{\langle n \rangle}$ for any $n \geq 1$. In other words, if the algebra A is right nilpotent of index n, then it is nilpotent of index not greater than 2^n.

PROOF. By the commutativity (anticommutativity) of the algebra A,

$$A^{2^n} = \sum_{\substack{i+j=2^n \\ i \geq j}} A^i A^j \subseteq A^{2^{n-1}}A.$$

A trivial induction now gives $A^{2^n} \subseteq A^{\langle n \rangle}$, which also proves the proposition.

We define one more chain of subsets in an algebra A, setting $A^{(1)} = A^2$ and $A^{(n)} = (A^{(n-1)})^2$. We shall call the algebra A *solvable* if $A^{(n)} = (0)$ for some n. The minimal such n is called the *index of solvability* of the algebra A. The terms of the chain

$$A \supseteq A^{(1)} \supseteq A^{(2)} \supseteq \cdots \supseteq A^{(n)} \supseteq \cdots,$$

beginning with $A^{(2)}$, are not, generally speaking, ideals in A.

We say that an algebra has some property locally if every finitely generated subalgebra has this property. Thus are introduced the concepts of *local nilpotency*, *local solvability*, etc.

The basic result of the present chapter is a theorem of Zhevlakov asserting that every locally solvable Jordan algebra is locally nilpotent. This theorem will be proved in Section 3. In concluding this section we present two examples showing that solvable Jordan algebras can be nonnilpotent.

EXAMPLE 1 (*Zhevlakov*). Let $X = \{x_1, x_2, \ldots, x_n, \ldots\}$ be a countable set of symbols. We consider the set of words in the alphabet X, and call a word $x_{i_1} x_{i_2} \cdots x_{i_n}$ *regular* if $i_1 < i_2 < \cdots i_n$. We linearly order the set of regular words lexicographically, setting $x_i > x_j$ if $i < j$, and if $u = vv_1$, then we consider $u > v$.

We now consider a field F of characteristic $\neq 2, 3$ and an algebra A over F with basis from the regular words of the alphabet X, and we define a multiplication on the basis by the following rules: if u, v are regular words, then

1. $u * v = v * u$;
2. $u * v = 0$, if $\min(d(u), d(v)) > 1$;
3. $u * x_i = 0$, if x_i is in the listing of u;
4. $u * x_i = 0$, if $d(u) > 1$ and $x_i > u$;
5. $x_i * x_j = x_i x_j$, if $i < j$;
6. $u * x_i = (-1)^\sigma \langle u x_i \rangle$, if 1–5 are not applicable.

Here $\langle u x_i \rangle$ is the regular word formed by all letters of the word u and the letter x_i, and if $u = x_{j_1} x_{j_2} \cdots x_{j_n}$, then σ is the number of inversions in the permutation $(j_1, j_2, \ldots, j_n, i)$.

The Jacobi identity is valid in the algebra A:

$$(a * b) * c + (b * c) * a + (c * a) * b = 0 \tag{1}$$

for any $a, b, c \in A$. We shall prove this. It is clear that if from the three words a, b, c two have length >1 or have a common symbol from X, then (1) is valid. Analogously, if $d(a) > 1$ and $d(b) = d(c) = 1$, but either $b > a$ or $c > a$, then (1) is also true. This leaves two cases:

(a) $a = x_{i_1} x_{i_2} \cdots x_{i_n}$, $b = x_j$, $c = x_k$, and $a > b > c$;
(b) $a = x_i$, $b = x_j$, $c = x_k$, and $a > b > c$

which are verified by direct computations.

We shall now prove that A is a special Jordan algebra. In view of (1), for any $a, b \in A$ the following relation holds in the multiplication algebra $R(A)$:

$$R_{a*b} = -R_a R_b - R_b R_a.$$

We now define on the vector space of the algebra $R(A)$ another multiplication: $w_1 \cdot w_2 = -2 w_1 w_2$. It is easy to see that the new algebra $\overline{R(A)}$ obtained in this manner is also associative. In addition, for $a, b \in A$ the following relation is valid in $\overline{R(A)}$:

$$R_{a*b} = \tfrac{1}{2}(R_a \cdot R_b + R_b \cdot R_a),$$

which shows that the mapping $\phi: A \to \overline{R(A)}$ sending $a \in A$ to $R_a \in \overline{R(A)}$ is a homomorphism of the algebra A into $\overline{R(A)}^{(+)}$. It is easy to see that the kernel of ϕ equals zero, that is, ϕ is an isomorphism. This proves the algebra A is Jordan and special.

As is obvious from the multiplication table, the algebra A is solvable of index 2, that is, $A^{(2)} = (0)$, and in addition the identity $x^3 = 0$ is satisfied in

A. However, A is not nilpotent, since the product

$$(\cdots((x_1 * x_2) * x_3)\cdots) * x_n$$

does not equal zero for any n.

EXAMPLE 2 (*Shestakov*). Let M be a vector space over a field F with basis $\{x_1, x_2, \ldots, x_n, \ldots\}$, $\bigwedge(M)$ be its exterior algebra, and $\bigwedge^0(M)$ be the subalgebra of the algebra $\bigwedge(M)$ generated by the set M. We consider the space $A = \bigwedge^0(M) \oplus M$, and define a multiplication on A by the rule

$$(u + x)(v + y) = v \wedge x + u \wedge y,$$

where $u, v \in \bigwedge^0(M)$; $x, y \in M$. It is easy to see that the algebra A is Jordan, solvable of index 2, satisfies the identity $x^3 = 0$, but is not nilpotent.

Exercises

1. Prove that every right nilpotent (left nilpotent) algebra is solvable.
2. Let A be a Jordan algebra. Prove that

$$A^{2(n-1)} \subseteq A^{\langle n \rangle}.$$

3. Let A be an algebra, and set $A^{(1)n} = A^n$, $A^{(i+1)n} = (A^{(i)n})^n$. We call the algebra *n-solvable* if $A^{(i)n} = (0)$ for some i. Prove that if $2^k > n$, then

$$A^{(ki)} \subseteq A^{(i)n} \subseteq A^{(i)}.$$

Deduce from this that the concepts of n-solvability are equivalent to each other.

4. (*Zhevlakov*) Let A be an algebra over an arbitrary field F with basis $X = \{x_1, x_2, \ldots, x_n, \ldots,\}$ and multiplication on the basis

$$x_i x_j = \begin{cases} x_{i-j}, & \text{if } i > j, \\ 0, & \text{if } i \leq j. \end{cases}$$

Prove that the algebra A is simple and locally right nilpotent.

2. THE NORMAL FORM FOR ELEMENTS OF THE MULTIPLICATION ALGEBRA

Let $J = \mathrm{J}[X]$ be the free Jordan Φ-algebra from a countable set $X = \{x_1, x_2, \ldots, x_n, \ldots,\}$ of free generators. As we saw in Section 1.3, the variety of Jordan Φ-algebras (under the presence of an element $\frac{1}{2}$ in Φ) is homogeneous, and the concepts of homogeneity and degree are well defined for elements of the algebra J. We shall denote by $d(u)$ the degree of a monomial $u \in J$.

We shall call every expression of the form

$$w = R_{a_1} R_{a_2} \cdots R_{a_n},$$

where all the a_i are monomials from the algebra J, a *word* in the right multiplication algebra $R(J)$. We shall call the number n the *length of the word* w, and the number $\sum_{i=1}^{n} d(a_i)$ its *degree*. We shall call the sequence $(l_1, l_2, \ldots, l_n, \ldots)$, where l_i is the total degree of the monomials $a_1, a_2, \ldots, a_n$ in x_i, the *composition* of the word w.

We now assume that the set of all monomials of the algebra J is linearly ordered in an arbitrary manner. In this case we shall call words of the form

$$R_{b_1} R_{c_1} R_{b_2} R_{c_2} \cdots R_{b_n} \hat{R}_{c_n}$$

(R_{c_n} can be present or not), where $b_1 < b_2 < \cdots < b_n$ and $c_1 < c_2 < \cdots < c_n$, *normal.*

LEMMA 1 (*Zhevlakov's lemma on normal form*). Every word $w \in R(J)$ can be represented in the form of a linear combination of normal words with the same composition as w and with coefficients of the form $k/2^m$, where $k, m \in \boldsymbol{Z}$.

PROOF. Words of length 1 and 2 are normal. We shall carry out an induction on the length of the word w. Let the length of w equal n. We introduce on words from $R(J)$ of length n the lexicographic ordering, setting

$$R_{u_1} R_{u_2} \cdots R_{u_n} > R_{v_1} R_{v_2} \cdots R_{v_n}$$

if for some $l \in \{1, 2, \ldots, n\}$ we have $u_i = v_i$ for $i < l$ and $u_l > v_l$.

Assume that the word $w = R_{a_1} R_{a_2} \cdots R_{a_n}$ is not normal. Then it contains a subword of the form $R_a R_b R_a$, or a subword of the form $R_a R_b R_c$ where $a > c$. In the first case, by relation (3.26) the word w can be expressed by words of smaller length with the same composition,

$$w = w' R_a R_b R_a w'' = \tfrac{1}{2} w' R_{a^2} R_b w'' + w' R_{ab} R_a w'' - \tfrac{1}{2} w' R_{a^2 b} w'',$$

and therefore by the induction assumption is representable in the necessary form.

In the second case, the same relation permits us to represent the word w in the following manner:

$$\begin{aligned} w &= w' R_a R_b R_c w'' \\ &= -w' R_c R_b R_a w'' + w'(R_{ab} R_c + R_{ac} R_b + R_{bc} R_a - R_{(ac)b}) w''. \end{aligned}$$

The word w is represented in the form of a linear combination of words with the same composition and to which our induction assumption applies, and the word $w' R_c R_b R_a w''$ which is lexicographically less than the word w. In view of the fact that the number of words in $R(J)$ with a fixed composition is finite, the described process for elimination of disorders is finite. This means that by a finite number of steps the word w is representable in the form of a linear combination of normal words, which proves the lemma.

Let $J_n = \mathrm{J}[x_1, x_2, \ldots, x_n]$ be the free Jordan algebra from generators $x_1, x_2, \ldots, x_n$. It is in a natural manner imbeddable in J, and henceforth we shall consider it a subalgebra of the algebra J.

LEMMA 2 (*Zhevlakov*). For every natural number $k \geq 1$ there exists a number $h(n, k)$ such that any word $w \in R^J(J_n)$ of degree $h(n, k)$ is representable in the form of a linear combination words having the same composition and each of which contains an operator of right multiplication by a monomial of degree $\geq k$.

PROOF. We set by definition

$$h(n, k) = 1 + 2 \sum_{j=1}^{k-1} jP_{n,j},$$

where $P_{n,j}$ is the number of different monomials from J_n of degree j. We shall prove that the number $h(n, k)$ satisfies the conditions of the lemma. Let $w \in R^J(J_n)$ have degree $h(n, k)$. We order the monomials from J so that the monomials of greater degree are larger, and we represent w in the form of a linear combination of normal words:

$$w = \sum \alpha_l w_l.$$

We shall show that in this decomposition every normal word w_l contains an operator of multiplication by a monomial of degree $\geq k$. Suppose the opposite. Let

$$w_l = R_{b_1} R_{c_1} R_{b_2} R_{c_2} \cdots R_{b_m} \hat{R}_{c_m}$$

and the degrees of all the monomials b_i and c_i be strictly less than k. Since $b_1 < b_2 < \cdots < b_m$, all the monomials are different. Therefore

$$m \leq \sum_{j=1}^{k-1} P_{n,j},$$

and the total degree of the monomials b_i does not exceed the number

$$\sum_{j=1}^{k-1} jP_{n,j}.$$

Analogous arguments are also valid for the monomials c_i. Consequently the degree of the word w_l does not exceed

$$2 \sum_{j=1}^{k-1} jP_{n,j}.$$

But this number is strictly less than $h(n, k)$. This contradiction proves the lemma.

THEOREM 1 (*Zhevlakov*). Let B be a Jordan algebra and A be a locally nilpotent subalgebra of B. Then the algebra $R^B(A)$ is locally nilpotent.

PROOF. Let $\{w_1, w_2, \ldots, w_k\}$ be an arbitrary finite subset in $R^B(A)$. In the listings of $w_1, w_2, \ldots, w_k$ there is only a finite number of operators $R_{a_1}, R_{a_2}, \ldots, R_{a_n}$ with $a_i \in A$. It suffices to show that these operators generate a nilpotent subalgebra. We consider the subalgebra $A_0 \subseteq A$ generated by the elements $a_1, a_2, \ldots, a_n$. It is nilpotent of some index N. We shall show that any product of $h = h(n, N)$ operators from the set $\{R_{a_1}, R_{a_2}, \ldots, R_{a_n}\}$ equals zero. Let

$$R_{a_{i_1}} R_{a_{i_2}} \cdots R_{a_{i_h}}$$

be such a product. We consider the word

$$w = R_{x_{i_1}} R_{x_{i_2}} \cdots R_{x_{i_h}}$$

from $R^J(J_n)$. By Lemma 2 the word w is representable in the form of a linear combination of words each of which contains an operator of right multiplication by a monomial of degree $\geq N$:

$$w = \sum \alpha_j w'_j R_{u_j(x_1, \ldots, x_n)} w''_j, \qquad d(u_j) \geq N.$$

In particular, the action of the right and left sides of this equality on the element x_{n+1} is the same:

$$x_{n+1} w = x_{n+1} \sum \alpha_j w'_j R_{u_j(x_1, \ldots, x_n)} w''_j.$$

This equality is satisfied in the free Jordan algebra J. It remains valid when in place of the generators $x_1, x_2, \ldots, x_n$ are substituted the elements $a_1, a_2, \ldots, a_n$, and in place of x_{n+1} an arbitrary element $b \in B$. In view of the nilpotency of the algebra A_0, we obtain

$$bR_{a_{i_1}} R_{a_{i_2}} \cdots R_{a_{i_h}} = 0, \qquad \text{that is,} \quad R_{a_{i_1}} R_{a_{i_2}} \cdots R_{a_{i_h}} = 0 \quad \text{in} \quad R(B).$$

This proves the theorem.

Exercises

In all of the exercises A and B are Jordan algebras, and A is also a subalgebra of the algebra B.

1. Prove that when an algebra A is finitely generated, it follows the algebra $R^B(A)$ is finitely generated.

2. Prove that the algebra A is locally nilpotent if and only if its right multiplication algebra $R(A)$ is locally nilpotent.

3. Prove that if A is finite-dimensional, then $R^B(A)$ is also finite-dimensional.

4. Prove that all the algebras from the set $\{R^C(A)\}$, where A is fixed and C runs through the class of all Jordan extension algebras of A, are the homomorphic images of some algebra $R^*(A)$ from this set. If A is the algebra with zero multiplication, what kind of algebra is $R^*(A)$? *Hint*. For the proof of the existence of the algebra $R^*(A)$, use the construction for the free products of algebras (e.g., see A. I. Mal'cev, "Algebraic Systems," p. 306, Nauka, 1970).

3. ZHEVLAKOV'S THEOREM

In a Jordan algebra A, instead of the chain of subsets $A^{(i)}$ it is convenient for many purposes to consider the chain

$$A \supseteq A^{[1]} \supseteq A^{[2]} \supseteq \cdots \supseteq A^{[n]} \supseteq \cdots,$$

where $A^{[1]} = A^3$ and $A^{[i+1]} = (A^{[i]})^3$. In a Jordan algebra A this chain is a chain of ideals, since the following assertion is valid.

LEMMA 3. Let B and C be ideals of a Jordan algebra A. Then $(BC)C + BC^2$ is also an ideal in A. In particular, if I is an ideal in A, then I^3 is also an ideal.

PROOF. Let $b \in B$, $c, d \in C$, $a \in A$. Then by identity (3.22)

$$\begin{aligned}[(bc)d]a = &-[(ba)d]c - b[(ac)d] + (bc)(ad) \\ &+ (bd)(ac) + (ba)(cd) \in (BC)C + BC^2,\end{aligned}$$

and furthermore by (3.24)

$$\begin{aligned}[b(cd)]a = &-[b(ac]d - [b(ad)]c + (bc)(ad) \\ &+ (bd)(ac) + (ba)(cd) \in (BC)C + BC^2.\end{aligned}$$

We see that $(BC)C + BC^2$ is an ideal in A, and this proves the lemma.

LEMMA 4. In an arbitrary algebra A the containment

$$A^{(2i)} \subseteq A^{[i]} \subseteq A^{(i)}$$

is valid, so that "cubic solvability" is equivalent to the usual solvability. In a Jordan algebra A every nonzero solvable ideal I contains a nonzero nilpotent ideal of the algebra A.

PROOF. The first assertion of the lemma is proved by an obvious induction. We now assume that there is a solvable ideal I in the Jordan algebra A. If n is its index of solvability, then by the first assertion of the lemma $I^{[n]} = (0)$. In the chain

$$I \supseteq I^{[1]} \supseteq \cdots \supseteq I^{[n]} = (0),$$

however, all of the terms are ideals, and for some term Q of this chain $Q^3 = (0)$, but $Q \neq (0)$. This proves the lemma.

LEMMA 5 (*Zhevlakov*). For any two natural numbers m and n there exists a number $f(n, m)$ such that for every Jordan algebra A with n generators

$$A^{f(n,m)} \subseteq A^{[m]}.$$

PROOF. It suffices to prove the lemma in the case when $A = J_n$ is the free Jordan algebra with n generators. This same number $f(n, m)$ will also serve in the general case. Since $A^3 = A^{[1]}$ we have $f(n, 1) = 3$, which gives us a basis for an induction on m.

Let us assume that the number $f(n, m-1)$ has been found. Let

$$M = \sum_{j=1}^{f(n, m-1)} P_j$$

be the number of different monomials of degree $\leq f(n, m-1)$ in J_n, and let h be the function from Lemma 2. We shall prove that the number 2^N can be taken as $f(n, m)$, where

$$N = f(n, m-1) + (M+2)h(n, f(n, m-1)).$$

By Proposition 1 it suffices for us to prove that

$$A^{\langle N \rangle} \subseteq A^{[m]}.$$

We consider an arbitrary element from $A^{\langle N \rangle}$. It is a sum of monomials of the form aw, where $a \in A$ and w is a word from $R(A)$ of length $N-1$. We separate the word w into subwords

$$w = w_0 w_1 w_2 \cdots w_{M+2},$$

so that the length of the word w_0 equals $f(n, m-1) - 1$ and the lengths of the remaining words are equal to the number $h(n, f(n, m-1))$. Applying Lemma 2 to the words $w_1, w_2, \ldots, w_{M+2}$, we can assume that the word $w_1 w_2 \cdots w_{M+2}$ contains $M+2$ operators $R_{u_1}, R_{u_2}, \ldots, R_{u_{M+2}}$ where $d(u_i) \geq f(n, m-1)$, and consequently $u_i \in A^{[m-1]}$. We also note that $u_0 = aw_0 \in A^{[m-1]}$, since $d(aw_0) = f(n, m-1)$. We must now prove that every monomial of the form

$$b = u_0 w_0' R_{u_1} w_1' R_{u_2} \cdots w_{M+1}' R_{u_{M+2}} w_{M+2}'$$

lies in $A^{[m]}$, where

$$w_s' = R_{a_{s1}} R_{a_{s2}} \cdots R_{a_{sk_s}}, \qquad a_{ij} \in A.$$

Also, without restricting generality, we can assume $a_{ij} \notin A^{[m-1]}$, and in particular $d(a_{ij}) < f(n, m-1)$.

We show that all w'_s, where $s = 1, 2, \ldots, M+1$, can be considered as either empty or consisting of one operator. In fact, if $w'_1 = w''_1 R_{a_1} R_{a_2}$, then according to (3.26) the subword $R_{a_1} R_{a_2} R_{u_2}$ can be replaced by the sum

$$-R_{u_2} R_{a_2} R_{a_1} - R_{(a_1 u_2) a_2} + R_{a_1 a_2} R_{u_2} + R_{a_1 u_2} R_{a_2} + R_{a_2 u_2} R_{a_1}.$$

As a result the word $w'_0 R_{u_1} w'_1 R_{u_2} \cdots R_{u_{M+2}} w'_{M+2}$ is represented in the form of a linear combination of words with the same form but with a smaller length subword between the first and second operators of the form R_u, where $u \in A^{[m-1]}$. Consecutive applications of such transformations enable us to shorten the length of this subword to the needed size. We then perform these same operations with the words $w'_2, \ldots, w'_{M+1}$.

We consider the case when one of the words $w'_1, w'_2, \ldots, w'_{M+1}$ is empty, for example, w'_s. Then since $A^{[m-1]}$ and $A^{[m]}$ are ideals in A, we have

$$b = u_0 w'_0 R_{u_1} w'_1 \cdots w'_{s-1} R_{u_s} R_{u_{s+1}} \cdots \in (A^{[m-1]})^3 = A^{[m]}.$$

It remains to consider the case when the words $w'_1, w'_2, \ldots, w'_{M+1}$ consist of one operator: $w'_i = R_{a_i}$, $i = 1, 2, \ldots, M+1$, and $d(a_i) < f(n, m-1)$. We have

$$b = u_0 w'_0 R_{u_1} R_{a_1} \cdots R_{u_{M+1}} R_{a_{M+1}} R_{u_{M+2}} w'_{M+2}.$$

We transform b by identity (3.25). Then by what has been proved above we obtain

$$\begin{aligned} b &= u_0 w'_0 \cdots R_{u_i} R_{a_i} R_{u_{i+1}} R_{a_{i+1}} R_{u_{i+2}} \cdots w'_{M+2} \\ &= -u_0 w'_0 \cdots R_{u_i} R_{a_{i+1}} R_{u_{i+1}} R_{a_i} R_{u_{i+2}} \cdots w'_{M+2} + b', \end{aligned}$$

where $b' \in A^{[m]}$. This means interchanging the positions of the operators R_{a_i} and $R_{a_{i+1}}$ changes the sign of the element b modulo terms from $A^{[m]}$. We now note that among the monomials $a_1, a_2, \ldots, a_{M+1}$ there are found two the same. In fact, $d(a_i) < f(n, m-1)$, and M is the number of different monomials from J_n of degree $< f(n, m-1)$. The presence of two identical monomials now implies that $2b \in A^{[m]}$, whence it follows that $b \in A^{[m]}$. With this the proof of the lemma is completed.

THEOREM 2 (*Zhevlakov*). Every solvable finitely generated Jordan algebra is nilpotent.

PROOF. Let the Jordan algebra A be solvable of index m and have n generators. By Lemma 4 we have $A^{[m]} = (0)$, and by Lemma 5 it follows from this that $A^{f(n,m)} = (0)$. This proves the theorem.

COROLLARY. Every locally solvable Jordan algebra is locally nilpotent.

Another way of proving Theorem 2, which was proposed by Shestakov, is deduced in the exercises to this section.

Exercises

1. Let A be an algebra and I be an ideal in A. Prove that if the ideal I is nilpotent, then the algebra $R^A(I)$ is also nilpotent.

2. Let A be a Jordan algebra, I be an ideal of the algebra A such that $A^2 \subseteq I$, and $A = I + \Phi \cdot v$ where $v \in A$. Prove that

$$[R(A)]^3 \subseteq R^A(I) + R^A(I) \cdot R(A).$$

3. Prove that under the conditions of Exercise 2 the nilpotency of the ideal I implies the nilpotency of the algebra A.

4. Let A be a finitely generated Jordan algebra. Prove that the nilpotency of A^2 implies the nilpotency of the whole algebra A.

5. Prove that if the Jordan algebra A is finitely generated, then the algebra A^2 is also finitely generated. *Hint*. First prove this assertion for a free Jordan algebra J with the same number of generators by establishing for J the existence of a number N such that $J^N \subseteq (J^2)^3$, and then use the homogeneity of the variety of Jordan algebras.

6. Using Exercises 4 and 5, prove Zhevlakov's theorem.

4. JORDAN NIL-ALGEBRAS

This section is devoted to the proof of the following theorem.

THEOREM 3 (*Albert–Zhevlakov*). Let A be a Jordan nil-algebra over a ring Φ containing an element $\frac{1}{2}$. Then if A satisfies the maximal condition for Φ-subalgebras, A is nilpotent.

PROOF. All subalgebras of A generated by one element are nilpotent, and therefore by the maximal condition there can be found a maximal nilpotent subalgebra B in A. In every algebra satisfying the maximal condition for subalgebras, all subalgebras are finitely generated. Consequently B is finitely generated. What is more, since B is nilpotent it is a finitely generated Φ-module, $B = \Phi \cdot e_1 + \cdots + \Phi \cdot e_n$. Hence it follows that in the right multiplication algebra $R(A)$ of the algebra A the subalgebra $R^A(B)$ is generated by the operators $R_{e_1}, \ldots, R_{e_n}$, and by Theorem 1 is nilpotent of some index m.

Let us assume that $B \neq A$. Then we shall find an element $a \in A \backslash B$ such that $aB \subseteq B$. We take an arbitrary element $a_0 \in A \backslash B$. If it is not the desired one, then there exists an element $b_1 \in B$ such that $a_1 = a_0 b_1 \notin B$. If a_1 is also not the desired one, then there exists $b_2 \in B$ such that $a_2 = a_1 b_2 \notin B$, etc. Since for any $b_1, \ldots, b_m \in B$ the element $a_0 R_{b_1} \cdots R_{b_m}$ equals zero, and consequently belongs to B, among the elements $a_1, \ldots, a_{m-1}$ must necessarily be found the desired element a.

Two cases are possible:

1. $a^2 B \subseteq B$. By Proposition 3.2 the operator R_{a^t} for any $t \geq 1$ lies in the subalgebra generated by the operators R_a and R_{a^2}, and therefore in this case $a^t B \subseteq B$ for any $t \geq 1$. In view of the nilpotency of the element a, there can be found a $p \geq 1$ such that $a^p \notin B$ but $(a^p)^2 \in B$. We denote the element a^p by s. The element s has the properties

$$s \notin B, \qquad s^2 \in B, \qquad sB \subseteq B. \tag{2}$$

2. There exists an element $c \in B$ such that $a^2 c \notin B$. In this case we shall show that the element $a^2 c$ satisfies properties (2), and it can be taken as s. By identity (3.23) for any $b \in B$

$$(a^2 c)b = -2[(ab)c]a + [a(ac)]b + [a(ab)]c + [a(bc)]a \in B. \tag{3}$$

In addition, (3.23) gives

$$\begin{aligned}(a^2 c)^2 = &-2\{[a(a^2 c)]c\}a + [a(ac)](a^2 c)\\ &+\{a[(a^2 c)a]\}c + \{a[(a^2 c)c]\}a.\end{aligned}$$

The second and fourth terms of the right side of this equality lie in B according to (3). We consider the first term. By (3)

$$\{[a(a^2 c)]c\}a = \{[(a^2 c)a]c\}a = \{[a^2(ca)]c\}a \in B.$$

Analogously we have

$$\{a[(a^2 c)a]\}c = \{a[a^2(ca)]\}c = \{a^2[(ca)a]\}c \in B.$$

Thus $(a^2 c)^2 \in B$ and the element $a^2 c$ satisfies conditions (2). We denote it by s.

We consider the set $B_1 = B + \Phi \cdot s$. By the properties of the elements s, B_1 is a subalgebra of the algebra A which strictly contains B. Since $B_1^2 \subseteq B$, it is a solvable finitely generated algebra. By Zhevlakov's theorem B_1 is nilpotent, which contradicts the maximality of B. This proves the theorem.

COROLLARY (*Albert*). Every finite-dimensional Jordan nil-algebra over a field F of characteristic $\neq 2$ is nilpotent.

This result of Albert can be strengthened. As shown by Shestakov, every Jordan Φ-algebra which is generated as a Φ-module by a finite set of nilpotent elements is nilpotent.

5. THE LOCALLY NILPOTENT RADICAL

A locally nilpotent ideal $\mathscr{L}(A)$ such that the quotient algebra $A/\mathscr{L}(A)$ does not contain nonzero locally nilpotent ideals is called the *locally nilpotent radical* of the algebra A. Since the homomorphic image of a locally nilpotent algebra is a locally nilpotent algebra, the ideal $\mathscr{L}(A)$ contains any other locally nilpotent ideal of the algebra A.

In this section we shall show that in every Jordan algebra the locally nilpotent radical exists, and shall prove an important theorem of Skosyrskiy on the relation between the locally nilpotent radical of a special Jordan algebra and the locally nilpotent radical of an associative enveloping algebra for it.

LEMMA 6. Let J be a finitely generated Jordan algebra. Then the ideals $J^{[k]}$ of A are also finitely generated algebras.

PROOF. It suffices to prove that J^3 is a finitely generated algebra. In addition, it is obvious one can assume that J is a free Jordan algebra. To be specific, let J have n generators. Then by Lemma 5 $J^{f(n,2)} \subseteq J^{[2]}$, that is, every monomial u of degree $\geq f(n, 2)$ is representable in the form

$$u = \sum_i (v_i w_i) t_i,$$

where v_i, w_i, t_i are monomials from J^3. By the homogeneity of the variety of Jordan algebras, it is possible to assume that $d(v_i) + d(w_i) + d(t_i) = d(u)$ for any i. This means that as generators of the algebra J^3 can be taken all monomials v of the algebra J for which $3 \leq d(v) \leq f(n, 2)$. This proves the lemma.

LEMMA 7. If both the ideal I of a Jordan algebra J and the quotient algebra J/I are locally nilpotent, then the algebra J is also locally nilpotent.

PROOF. Let J_0 be a finitely generated subalgebra in J. By the local nilpotency of the quotient algebra J/I, we have $J_0^{[k]} \subseteq I$ for some k. Since I is locally nilpotent and the subalgebra $J_0^{[k]}$ is finitely generated, there exists a number m such that $(0) = (J_0^{[k]})^{[m]} = J_0^{[k+m]}$. Thus the subalgebra J_0 is solvable, and by Zhevlakov's theorem it is nilpotent. This proves the lemma.

THEOREM 4 (*Zhevlakov*). Every Jordan algebra has a locally nilpotent radical.

PROOF. In a Jordan algebra J we consider the sum $\mathscr{L}(J)$ of all locally nilpotent ideals. If L_1 and L_2 are two such ideals, then by the second homomorphism theorem

$$L_1 + L_2/L_1 \cong L_2/L_2 \cap L_1,$$

whence by Lemma 7 it follows that $L_1 + L_2$ is also a locally nilpotent ideal. It is now easy to see that the sum of a finite number of locally nilpotent ideals is a locally nilpotent ideal. Let $l_1, l_2, \ldots, l_n$ be a finite number of elements from $\mathscr{L}(J)$. All these elements lie in the sum of a finite number of locally nilpotent ideals, and therefore generate a nilpotent subalgebra. Consequently the ideal $\mathscr{L}(J)$ is locally nilpotent. In view of Lemma 7, the quotient algebra $J/\mathscr{L}(J)$ does not contain nonzero locally nilpotent ideals. This means $\mathscr{L}(J)$ is the locally nilpotent radical of the algebra J, which proves the theorem.

We now turn to a study of the relation between the locally nilpotent radical of a special Jordan algebra and the locally nilpotent radical of an associative enveloping algebra for it. We note that in every associative algebra the relation

$$x(y \odot z) = (x \odot y)z + (x \odot z)y - \{yxz\} \tag{4}$$

is valid.

LEMMA 8. In the free associative algebra $\mathrm{Ass}[x_0, x_1, \ldots, x_n]$, for any homogeneous Jordan polynomial $j(x_1, \ldots, x_n)$ there exist homogeneous Jordan polynomials $j_i(x_0, x_1, \ldots, x_n)$, $i = 0, 1, \ldots, n$, each linear in x_0 and such that

$$x_0 j(x_1, \ldots, x_n) = j_0(x_0, x_1, \ldots, x_n) + \sum_{i=1}^{n} j_i(x_0, x_1, \ldots, x_n)x_i, \tag{5}$$

where also $d(j_0) - 1 = d(j_i) = d(j)$.

PROOF. It is obviously possible to assume that j is a Jordan monomial. We shall carry out an induction on the degree $d(j)$ of this monomial. If $d(j) = 1$, then the assertion is evident. Let $d(j) = m \geq 2$. Then $j = j_1 \odot j_2$, where $d(j_k) < m$ for $k = 1, 2$. By relation (4)

$$x_0 j = x_0(j_1 \odot j_2) = (x_0 \odot j_1)j_2 + (x_0 \odot j_2)j_1 - \{j_1 x_0 j_2\}. \tag{6}$$

Now by the induction assumption there exists a representation of the form (5) for the polynomial $x_0 j_2$. Substituting in it the polynomial $x_0 \odot j_1$ in place of x_0, we obtain an analogous representation for $(x_0 \odot j_1)j_2$. In the same way, the existence of the desired representation for $x_0 j_1$ implies the existence of such a representation for $(x_0 \odot j_2)j_1$. By (6) it is now clear that $x_0 j$ is also representable in the required form, which proves the lemma.

COROLLARY. Let $j'(x)$, $j''(x)$ be homogeneous Jordan polynomials in $\mathrm{Ass}[x_1, \ldots, x_n]$. Then there exist homogeneous Jordan polynomials $j_i(x)$, $i = 0, 1, \ldots, n$, such that

$$j'(x)j''(x) = j_0(x) + \sum_{i=1}^{n} j_i(x)x_i, \tag{7}$$

and moreover $d(j_0) = d(j_i) + 1 = d(j') + d(j'')$.

PROOF. By Lemma 8

$$x_0 j''(x_1, \ldots, x_n) = j_0''(x_0, x_1, \ldots, x_n) + \sum_{i=1}^{n} j_i''(x_0, x_1, \ldots, x_n)x_i.$$

Substituting in this expression $j'(x)$ in place of x_0, we obtain relation (7). This proves the corollary.

LEMMA 9. For any homogeneous Jordan polynomial $j(x_0, x_1, \ldots, x_n)$ linear in x_0 and any monomial $v(x_1, \ldots, x_n)$ there can be found homogeneous Jordan polynomials $j_i(x_0, x_1, \ldots, x_n)$ linear in x_0 and multilinear monomials $v_i(x_1, \ldots, x_n)$ such that

$$j(x_0, x_1, \ldots, x_n)v(x_1, \ldots, x_n) = j_0(x_0, x_1, \ldots, x_n) + \sum_i j_i(x_0, x_1, \ldots, x_n)v_i(x_1, \ldots, x_n),$$

and where $d(j) + d(v) = d(j_0) = d(j_i) + d(v_i)$.

PROOF. We shall carry out an induction on $d(v)$. If $d(v) = 1$, then everything is clear. Let $d(v) = m > 1$, and suppose everything is proved for monomials of lesser length. We introduce the following notation: we shall write $p \equiv q$ if the difference $p - q$ satisfies the conditions of our induction assumption. By relations (5) and (7) it is clear that

$$j(x_0, x_1, \ldots, x_n)x_{i_1} \cdots (x_{i_s} \odot x_{i_{s+1}}) \cdots x_{i_m} \equiv 0.$$

Therefore

$$j(x_0, x_1, \ldots, x_n)x_{i_1} \cdots x_{i_s}x_{i_{s+1}} \cdots x_{i_m} \equiv -j(x_0, x_1, \ldots, x_n)x_{i_1} \cdots x_{i_{s+1}}x_{i_s} \cdots x_{i_m}.$$

It is now obvious that if there are two identical variables in the monomial $v(x_1, \ldots, x_n) = x_{i_1} \cdots x_{i_m}$, then

$$j(x_0, x_1, \ldots, x_n)v(x_1, \ldots, x_n) \equiv 0,$$

which is what was to be proved.

COROLLARY. For every monomial $v(x_1, \ldots, x_n)$ of the algebra $\mathrm{Ass}[x_1, \ldots, x_n]$ there exist homogeneous Jordan polynomials $j_i(x_1, \ldots, x_n)$ and multilinear monomials $v_i(x_1, \ldots, x_n)$ such that

$$v(x_1, \ldots, x_n) = j_0(x_1, \ldots, x_n) + \sum_i j_i(x_1, \ldots, x_n)v_i(x_1, \ldots, x_n),$$

where $d(v) = d(j_0) = d(j_i) + d(v_i)$.

PROOF. It suffices to employ Lemma 9 for $j(x_0, x_1, \ldots, x_n) = x_0$, and then to set x_0 equal to the identity element 1 of the algebra $\mathrm{Ass}[x_1, \ldots, x_n]^{\#}$.

PROPOSITION 2. If a special Jordan algebra J from n generators is nilpotent of index m, then an associative enveloping algebra A for J is also nilpotent with index of nilpotency not greater than $m + n$.

PROOF. If $a_1, \ldots, a_n$ are generators of the algebra J, then they also generate the algebra A. By the corollary to Lemma 9, if the associative monomial $v(x_1, \ldots, x_n)$ has degree not less than $m + n$, then $v(a_1, \ldots, a_n) = 0$. Consequently $A^{m+n} = (0)$, which proves the proposition.

COROLLARY. If J is a special locally nilpotent Jordan algebra, then every associative enveloping algebra for J is locally nilpotent.

THEOREM 5 (*Skosyrskiy*). Let J be a special Jordan algebra and A be an associative enveloping algebra for J. Then their locally nilpotent radicals satisfy the relation

$$\mathscr{L}(J) = J \cap \mathscr{L}(A).$$

PROOF. The containment $J \cap \mathscr{L}(A) \subseteq \mathscr{L}(J)$ is obvious. For the proof of the reverse containment it suffices to show that the set $I = \mathscr{L}(J)A^{\#}$ is a locally nilpotent ideal in A. It is clear that I is a right ideal. For the proof that I is a left ideal it suffices to note that for any elements $x \in J$, $l \in \mathscr{L}(J)$,

$$xl = -lx + 2l \odot x. \tag{8}$$

Since $2l \odot x \in \mathscr{L}(J)$ and the algebra A is generated by the set J, we have $A\mathscr{L}(J) \subseteq \mathscr{L}(J)A^{\#}$. Hence it follows that I is a left ideal in A, and consequently also an ideal.

We shall show that this ideal is locally nilpotent. We fix some set $\{a_\alpha\}$ of generators of the algebra J, and consequently also of the algebra A. It suffices to prove that any finite set M of elements of the form $lv(a_1, \ldots, a_k)$, where l runs through a finite set $L_0 \subseteq \mathscr{L}(J)$ and v is some set of words (maybe empty), generates a nilpotent subalgebra in A. We note that by Lemma 9, all the words v can be considered multilinear.

For each natural number p we consider the set L_p of elements of the form $j(l, a_1, \ldots, a_k)$, where $l \in L_0$ and $j(x_0, x_1, \ldots, x_k)$ is a linear in x_0 Jordan monomial of degree $p+1$ with coefficients equal to the identity element. It is clear that $L_p \subseteq \mathscr{L}(J)$. The sets L_p are finite, and therefore by Proposition 2 generate nilpotent subalgebras in A. Let t be the index of nilpotency of the subalgebra generated by the set L_{2k}. We shall show that a product of any t elements from M also equals zero. We consider some such product:

$$b = l_1 v_1(a_1, \ldots, a_k) l_2 v_2(a_1, \ldots, a_k) \cdots l_t v_t(a_1, \ldots, a_k).$$

By means of relation (8) the expression $v_1(a_1, \ldots, a_k) l_2$ can be transformed into a sum of expressions of the form $l'_2 v'_1(a_1, \ldots, a_k)$, where $l'_2 \in L_k$ and v'_1 is a multilinear monomial. By Lemma 9, each product of the form

$$l'_2 v'_1(a_1, \ldots, a_n) v_2(a_1, \ldots, a_k)$$

can be represented in the form of a sum of elements of the form $l''_2 v''_1(a_1, \ldots, a_k)$, where $l''_2 \in L_{2k}$ and v''_1 is a multilinear monomial. Then we repeat this process, but now with the product $v''_1(a_1, \ldots, a_k) l_3$, etc. We finally obtain a representation of the product b in the form of a sum of products of the form

$$l_1 l''_2 l''_3 \cdots l''_t v_t(a_1, \ldots, a_k),$$

where $l''_i \in L_{2k}$. Each such product equals zero, since t is the index of nilpotency of the subalgebra generated by the set L_{2k}. Consequently $b = 0$ also, which proves the theorem.

COROLLARY. If the Jordan algebra J is special, then the quotient algebra $J/\mathscr{L}(J)$ is also special.

PROOF. Let I be the ideal of an associative enveloping algebra A generated by the set $\mathscr{L}(J)$. By Skosyrskiy's theorem we have the equality $I \cap J = \mathscr{L}(J)$, and so that the quotient algebra $J/\mathscr{L}(J)$ is special follows from Lemma 3.4. This proves the corollary.

LITERATURE

Jacobson [47], Zhevlakov [69], Zhevlakov and Shestakov [80], Lyu Shao-syue [112], Skosyrskiy [194], Slin'ko [208], Tsai [257], Shestakov [266, 267].

CHAPTER 5

Algebras Satisfying Polynomial Identities

The theory of associative algebras with polynomial identities (PI-algebras) is adequately developed at present, and contains a variety of topics. This chapter is devoted to one topic that arises in connection with the well-known Kurosh problem: "Is every algebraic algebra locally finite?"

This problem was solved affirmatively for associative PI-algebras by Levitzki and Kaplansky, who used the well-developed structure theory of associative rings for its solution. These methods prove to be ineffective in the nonassociative case, in view of the lack of suitable structure results.

Shirshov approached the problem from a combinatorial point of view. The methods he developed allowed him to obtain not only fundamentally new results, such as the theorem on height, but also to solve the Kurosh problem affirmatively for alternative and special Jordan PI-algebras.

This chapter is devoted to a presentation of these results.

1. SHIRSHOV'S LEMMA

We consider associative words formed from elements of some finite ordered set of symbols:

$$R = \{x_i\}, \qquad i = 1, 2, \ldots, k; \qquad x_i > x_j \quad \text{if} \quad i > j.$$

We shall call a word α x_k-*indecomposable* if it has the form $\alpha = x_k x_k \cdots x_k x_{i_1} x_{i_2} \cdots x_{i_s}$, where $s \geq 1$ and $i_t \neq k$ for $t = 1, 2, \ldots, s$. We shall call a representation of a word β in the form of a product of some number of x_k-indecomposable words an x_k-*factorization* of the word β. It is easy to see that there exists an x_k-factorization for a word β, which is also unique, if and only if the word β begins with the symbol x_k and ends in a symbol different from x_k.

We introduce a partial ordering on the set of all associative words from elements of the set R: for words α and β having equal length we set $\alpha > \beta$ if this relation is satisfied in the lexicographic sense. We introduce a linear order on the set T of all x_k-indecomposable words: for words $\alpha, \beta \in T$ (not necessarily of equal length) we set $\alpha > \beta$ if α is lexicographically greater than β or if α is an initial segment of the word β.

We shall call an associative word γ *n-partitionable* if it can be represented in the form of a product of n of its subwords such that for any nonidentity permutation of these subwords an associative word is obtained which is strictly less than γ.

For example, the word $x_3x_1x_2x_2x_1x_1x_2x_1x_1x_1$ is 3-partitionable and allows several 3-partitionings:

$$(x_3x_1)(x_2x_2x_1x_1)(x_2x_1x_1x_1), \qquad (x_3x_1x_2)(x_2x_1x_1)(x_2x_1x_1x_1),$$
$$(x_3)(x_1x_2x_2x_1x_1x_2)(x_1x_1x_1),$$

etc., but the word $x_1x_2x_1x_3x_2x_1x_2x_3x_2$ is not 2-partitionable.

Words allowing x_k-factorization can be considered as words formed from elements of the set T. In this case we shall call them T-*words* (as opposed to R-*words*). Analogously we shall use the terms T-*length* and R-*length* to emphasize which set of symbols is considered as the set of generators. We introduce a partial ordering $\prec$ on the set of all associative T-words: for T-words α and β having equal T-length, $\alpha \prec \beta$ if as a T-word α is less than β in the lexicographic sense. It is now also meaningful to talk about n-partitionable T-words. Therefore, where this could subsequently be ambiguous, we shall use the terms n_R-*partitionable* and n_T-*partitionable*.

LEMMA 1. Let α be an associative T-word. Then if it is n_T-partitionable, it is also n_R-partitionable.

PROOF. Let $\alpha = \alpha_1\alpha_2 \cdots \alpha_n$ be an n_T-partitioning of the word α. Then $\alpha, \alpha_1, \ldots, \alpha_n$ are words allowing x_k-factorization. It follows from the definition of n-partitionable that $\alpha \succ_T \alpha_{i_1}\alpha_{i_2} \cdots \alpha_{i_n}$ for any nonidentity permutation $(i_1 i_2 \cdots i_n)$. It is easy to see that for the R-words α and $\alpha_{i_1}\alpha_{i_2} \cdots \alpha_{i_n}$ we also have $\alpha > \alpha_{i_1}\alpha_{i_2} \cdots \alpha_{i_n}$. Therefore the given n_T-partitioning will also be an n_R-partitioning, which proves the lemma.

LEMMA 2. The $(n-1)_T$-partitionability of a word α implies n_R-partitionability of the word αx_k.

PROOF. From Lemma 1 follows the existence of the following $(n-1)_R$-partitioning of the word α:

$$\alpha = (x_k x_{i_1} \cdots x'_{i_1})(x_k x_{i_2} \cdots x'_{i_2}) \cdots (x_k x_{i_{n-1}} \cdots x'_{i_{n-1}}),$$

where $x_i, x'_i \in R$, $x'_{i_t} \neq x_k$ for $t = 1, 2, \ldots, n-1$.

We shall show that the following n_R-partitioning holds for the word αx_k:

$$\alpha x_k = (x_k)(x_{i_1} \cdots x'_{i_1} x_k)(x_{i_2} \cdots x'_{i_2} x_k) \cdots (x_{i_{n-1}} \cdots x'_{i_{n-1}} x_k).$$

In fact, any permutation of the subwords of the word αx_k which retains (x_k) in the first position changes the word αx_k into a word $\alpha' x_k$, where the α' is obtained by means of some permutation of the subwords in the given $(n-1)_R$-partitioning of the word α. Therefore $\alpha > \alpha'$, and $\alpha x_k > \alpha' x_k$. If permutations are considered which displace the symbol x_k from the first position, then it is obvious that the words obtained in this manner will begin with a strictly less number of symbols x_k in comparison with the word αx_k. Therefore they will be strictly less than the word αx_k. This proves the lemma.

LEMMA 3 (*Shirshov*). For any three natural numbers k, s, n, there can be found a natural number $N(k, s, n)$ such that in any associative word from k ordered symbols with length $N(k, s, n)$ there either occur s consecutive equal subwords or there can be found an n-partitionable subword.

PROOF. We shall consider associative words from elements of the set $R = \{x_1, x_2, \ldots, x_k\}$. It is easy to see that the number $N(k, s, 1)$ exists for any k and s. This gives us a basis for induction on n. We assume that a number $N(k, s, n-1)$ exists no matter what k and s. We note that the number $N(1, s, n)$ also exists, which gives us a basis for a subsequent induction on k. Thus, letting the number $N(k-1, s, n)$ exist, we prove that the number $N(k, s, n)$ exists.

We consider an arbitrary associative word α of length

$$[s + N(k-1, s, n)][N(k^{N(k-1, s, n)+s}, s, n-1) + 1].$$

If at the beginning of the word α there is some number of symbols x_i different from the symbol x_k, and this number is not less than the number $N(k-1, s, n)$, then the induction assumption is satisfied with respect to a subword α' which is at the beginning of the word α and depends only on $k-1$ generators. Therefore, we can assume that the length of such a subword α' is less than $N(k-1, s, n)$. At the end of the word α there can be found a subword $\alpha'' = x_k x_k \cdots x_k$. We can assume that the length of the word α'' is less than s, since in the opposite case the conclusion of the lemma is realized. Discarding, if they exist, the words α' and α'', we obtain a subword α_1 for which the length is greater than the number:

$$[s + N(k-1, s, n)] \cdot N(k^{N(k-1,s,n)+s}, s, n-1).$$

For the word α_1 there exists an x_k-factorization $\alpha_1 = \alpha_{11} \cdots \alpha_{1m}$. We can obviously assume that the length of each x_k-indecomposable word α_{1i} is less than the number $s + N(k-1, s, n)$, since in the opposite case in such a word there would be found either s consecutive symbols x_k, or a subword of length $N(k-1, s, n)$ which did not contain the symbol x_k. It is easy to see that under the indicated bound on the length there do not exist more than $k^{N(k-1,s,n)+s}$ different x_k-indecomposable words. We shall consider the word α_1 as a T-word. Since its T-length is strictly greater than $N(k^{N(k-1,s,n)+s}, s, n-1)$, there can be found in the word α_1 either s consecutive equal subwords or an $(n-1)_T$-partitionable subword β.

If the second assumption is realized, then by the strict inequality of the T-length of the T-word α_1, we can consider that the symbol x_k follows the subword β. By Lemma 2 the subword βx_k is n_k-partitionable. Thus, in any case the conclusion of the lemma is realized. Therefore we set

$$N(k, s, n) = [N(k-1, s, n) + s][N(k^{N(k-1,s,n)+s}, s, n-1) + 1],$$

which proves the lemma.

LEMMA 4. Let α be an associative word of length m which is not representable in the form β^t, where β is a proper subword of the word α. Then for any natural number $n \leq m$, the word α^{2n} contains an n-partitionable subword.

PROOF. It is possible to obtain from the word α, by means of cyclic permutations of the generators, m words $\alpha = \alpha_0, \alpha_1, \ldots, \alpha_{m-1}$. Since the word α is not representable in the form of a power of a subword, it is easy to see all the obtained words $\alpha_0, \alpha_1, \ldots, \alpha_{m-1}$ are different. We assume that in the lexicographic sense $\alpha_{i_0} > \alpha_{i_1} > \cdots > \alpha_{i_{m-1}}$. It is obvious that each of the words α_i can be represented in the form $\alpha_i = u_i v_i$, where $v_i u_i = \alpha$. We now consider the word

$$\alpha^{2n} = v_{i_0} u_{i_0} v_{i_0} u_{i_0} v_{i_1} u_{i_1} v_{i_1} u_{i_1} \cdots v_{i_{n-1}} u_{i_{n-1}} v_{i_{n-1}} u_{i_{n-1}}.$$

We set $\alpha'_{i_k} = u_{i_k} v_{i_k} u_{i_k} v_{i_k}$ for $k = 0, 1, \ldots, n-2$, and $\alpha'_{i_{n-1}} = u_{i_{n-1}} v_{i_{n-1}} u_{i_{n-1}}$, $\gamma = v_{i_0}$. Then the word α^{2n} is represented in the form $\alpha^{2n} = \gamma \alpha'_{i_0} \alpha'_{i_1} \cdots \alpha'_{i_{n-1}}$. Since the beginning of each of the words α'_j coincides with the word α_j and $\alpha_{i_0} > \alpha_{i_1} > \cdots > \alpha_{i_{n-1}}$, it is easy to see that the word $\alpha'_{i_0} \alpha'_{i_1} \cdots \alpha'_{i_{n-1}}$ is n-partitionable. This proves the lemma.

2. ASSOCIATIVE PI-ALGEBRAS

We now begin the study of algebras satisfying a polynomial identity. Let Φ be an arbitrary associative–commutative ring with identity element 1. We consider the free associative Φ-algebra $\mathrm{Ass}[X]$ from the set of generators $X = \{x_1, x_2, \ldots\}$. Elements of the algebra $\mathrm{Ass}[X]$ are associative polynomials from the noncommuting variables $x_1, x_2, \ldots$ with coefficients from Φ. We call an element $f(x_1, x_2, \ldots, x_n) \in \mathrm{Ass}[X]$ *admissible* if even one of the coefficients for the terms of highest degree of the polynomial $f(x_1, x_2, \ldots, x_n)$ equals the identity element. Now let A be an associative Φ-algebra and M be some Φ-submodule of the algebra A. We shall say that M satisfies an *admissible polynomial identity* if there exists an admissible element $f(x_1, x_2, \ldots, x_n) \in \mathrm{Ass}[X]$ such that

$$f(m_1, m_2, \ldots, m_n) = 0$$

for all $m_1, m_2, \ldots, m_n \in M$.

If the algebra A satisfies an admissible polynomial identity f, then we shall call the algebra A a *PI-algebra* for short.

We note that if a submodule M of an algebra A satisfies an admissible polynomial identity f, then M also satisfies an admissible multilinear identity of the same degree. Actually, by Theorem 1.7 the complete linearization of the homogeneous component of the element f containing the singled-out term of highest degree (with coefficient 1) will be an identity in M. This identity will also be the desired admissible multilinear identity. It is clear that every such admissible multilinear identity can be written in the form

$$x_1 x_2 \cdots x_n - \sum_{(i_1, i_2, \ldots, i_n)} \alpha_{i_1 i_2 \cdots i_n} x_{i_1} x_{i_2} \cdots x_{i_n}, \tag{1}$$

where $\alpha_{i_1 i_2 \cdots i_n} \in \Phi$ and the summation on the right is extended over all permutations of the symbols $1, 2, \ldots, n$ different from the permutation $(1\ 2 \cdots n)$.

We consider some elementary examples of associative PI-algebras.

1. Any commutative algebra A is a PI-algebra since it satisfies the identity $f(x_1, x_2) = x_1 x_2 - x_2 x_1$.

2. The algebra Φ_2 of matrices of dimension 2×2 with elements from Φ satisfies the Hall identity

$$f(x_1, x_2, x_3) = (x_1 x_2 - x_2 x_1)^2 x_3 - x_3 (x_1 x_2 - x_2 x_1)^2.$$

3. A nil-algebra for which the indices of nilpotency of all the elements are bounded overall by some number n satisfies the identity $f(x) = x^n$. We shall call such an algebra a *nil-algebra of bounded index*.

We now become acquainted with more general examples of PI-algebras.

The polynomial

$$[x_1, \ldots, x_n] = \sum_{\sigma \in S_n} (-1)^{\operatorname{sgn} \sigma} x_{\sigma(1)} \cdots x_{\sigma(n)},$$

where σ runs through the symmetric group S_n and the number $(-1)^{\operatorname{sgn} \sigma}$ equals 1 or -1 depending on whether the permutation σ is even or odd, is called the *standard polynomial of degree n* in the algebra Ass$[X]$. We note that the standard polynomial of degree 2 will simply be the commutator $[x_1, x_2] = x_1 x_2 - x_2 x_1$. In this sense the class of algebras which satisfy a standard polynomial is a generalization of the class of commutative algebras.

The following lemma shows that every finite-dimensional algebra is a PI-algebra.

LEMMA 5. If the algebra A is an n-generated Φ-module, then A satisfies the identity $[x_1, \ldots, x_{n+1}]$.

PROOF. From the definition of a standard polynomial it is obvious that it is linear in all of its variables and reduces to zero if any two of the arguments are equal. Let $u_1, \ldots, u_n$ be generators of the Φ-module A and $a_1, \ldots, a_{n+1}$ be arbitrary elements of the algebra A. We can represent each of the elements a_i in the form of a linear combination of the elements $u_1, \ldots, u_n$ with coefficients from Φ. By the multilinearity of the polynomial $[x_1, \ldots, x_{n+1}]$, the expression $[a_1, \ldots, a_{n+1}]$ will be a linear combination of terms of the form $[u_{i_1}, \ldots, u_{i_{n+1}}]$, where the u_{i_k} are elements from the set $\{u_1, \ldots, u_n\}$. But then some two arguments in the expression $[u_{i_1}, \ldots, u_{i_{n+1}}]$ necessarily coincide, and therefore it must be equal to zero. Hence $[a_1, a_2, \ldots, a_{n+1}] = 0$, which proves the lemma.

If an algebra A over a field F has dimension n, then each element from A satisfies a polynomial over F of degree $n + 1$. This property is the basis for the definition of an algebraic algebra. An element a of a Φ-algebra A is called *algebraic* over Φ if there exists a natural number n such that $a^n = \sum_{i<n} \alpha_i a^i$ where $\alpha_i \in \Phi$. The smallest such number n is called the *algebraic degree* of the element a. The algebra A is called *algebraic* over Φ if each element of the algebra A is algebraic over Φ. If in addition the algebraic degrees of the elements of the algebra A are bounded overall, then A is called an *algebraic algebra of bounded degree* over Φ.

LEMMA 6. Let each element of a Φ-submodule M of an algebra A be algebraic over Φ, and also let the algebraic degrees of all elements from M be bounded overall. Then M satisfies an admissible polynomial identity.

PROOF. Let n be the bound on the algebraic degree of elements from M. Then for any $a, b \in M$ the element $[a^n, b]$ is a linear combination of the elements $[a^{n-1},b],\ldots,[a,b]$ with coefficients from Φ. Therefore the elements $[a^n,b],\ldots,[a,b]$ reduce the standard polynomial $[x_1,\ldots,x_n]$ to zero. Consequently the submodule M satisfies the identity

$$[[x^n,y],[x^{n-1},y],\ldots,[x,y]],$$

which as is easy to see is admissible. This proves the lemma.

COROLLARY. Every algebraic algebra of bounded degree over Φ is a PI-algebra.

Now let A be an associative algebra generated by a finite set $\{a_1,a_2,\ldots,a_k\}$. For each polynomial $f=f(x_1,\ldots,x_k)$ of the free associative algebra $\mathrm{Ass}[x_1,\ldots,x_k]$, we shall denote by $\bar{f}=f(a_1,\ldots,a_k)$ the image of the element f under the homomorphism of the algebra $\mathrm{Ass}[x_1,\ldots,x_k]$ onto A which sends x_i to a_i.

Let $Y=\{f_1,f_2,\ldots,f_l\}$ be some finite set of homogeneous polynomials from $\mathrm{Ass}[x_1,\ldots,x_k]$, $\bar{Y}=\{\bar{f}_1,\bar{f}_2,\ldots,\bar{f}_l\}$, and v be a monomial from $\mathrm{Ass}[x_1,\ldots,x_k]$. Let us assume that there exists a natural number q and associative monomials $u_i(x_1,\ldots,x_l)$, with the maximum of their heights equal to q, such that in the algebra A

$$\bar{v}=\sum_i u_i(\bar{f}_1,\bar{f}_2,\ldots,\bar{f}_l),$$

and where also each element $u_i(f_1,f_2,\ldots,f_l)$ has the same type as v. We call the least number q with this property the *height of the monomial* $\bar{v}$ with respect to the set $\bar{Y}$.

In the case when the heights with respect to $\bar{Y}$ of all the monomials of the algebra A are bounded overall by some natural number h, we shall say that A is an *algebra of bounded height* (*with respect to the set* $\bar{Y}$) and that h is the *height of the algebra* A.

As an example we consider in the free associative algebra $\mathrm{Ass}[x_1,x_2]$ the monomial $x_1x_1x_2x_1x_1x_2x_1x_1x_2$. This monomial has height 6 with respect to the set $\{x_1,x_2\}$, and height 1 with respect to the singleton set $\{x_1x_1x_2\}$.

Finitely generated commutative algebras serve as an obvious example of algebras of bounded height. In fact, if A is a commutative algebra generated by the elements $\{a_1,\ldots,a_k\}$, then for any monomial v of type $[i_1,i_2,\ldots,i_k]$,

$$\bar{v}=\alpha a_1^{i_1}a_2^{i_2}\cdots a_k^{i_k},\qquad \alpha\in\Phi,$$

that is, the height of every monomial of the algebra A with respect to the set $\{a_1,\ldots,a_k\}$ does not exceed k.

As it turns out, every finitely generated PI-algebra is an algebra of bounded height. This fact goes by the name of the theorem on height.

THEOREM ON HEIGHT (*Shirshov*). Let an admissible polynomial identity of degree n be satisfied in the associative Φ-algebra A from set of generators $\{a_1, \ldots, a_k\}$. Then the algebra A has bounded height with respect to the set $\bar{Y}$, where Y is the set of all associative words from $\mathrm{Ass}[x_1, \ldots, x_k]$ of length less than n.

PROOF. As was already noted, we can assume that A satisfies an identity of form (1). For the proof of the theorem it is sufficient to prove the existence of a number $M = M(n, k)$ such that each associative word s from elements $x_1, \ldots, x_k$ of height $\geq M$ with respect to Y contains an n-partitionable subword. Actually, in the algebra A we then have $\bar{s} = \sum_i \alpha_i \bar{s}_i$, where $\alpha_i \in \Phi$ and the s_i are words from $x_1, \ldots, x_k$ with the same composition as s, but strictly less than s. If some one of the words s_i has height $\geq M$, then we continue the process. In view of its strict monotonicity, we finally obtain that $\bar{s} = \sum_i \alpha_i \bar{s}_i$, where each of the words s_i has height less than M with respect to Y.

According to Lemma 3 there exists a number $N = N(k, 2n, n)$ such that each associative word s of length N from $x_1, \ldots, x_k$, which does not contain n-partitionable subwords, contains a subword of the form v^{2n}. Moreover, it is possible to assume that the word v is not represented in the form $(v_1)^k$ with $k > 1$. Then by Lemma 4, the length $d(v)$ of the word v is less than n. From the reasoning just mentioned it follows easily that each word from $x_1, \ldots, x_k$ of height $\geq N + 2$ with respect to Y, which does not contain n-partitionable subwords, contains a subword v_1 of the form $v_1 = v^n v'$, where $n > d(v) \geq d(v')$ and the word v' is not an initial segment of the word v. Since the set of subwords of the indicated form is finite, for a sufficiently large natural number M each word of height $\geq M$ with respect to Y, which does not contain n-partitionable subwords, contains n equal subwords of the form $v_1 = v^n v'$ (not necessarily consecutively). There exists for each such word, however, a subword which is n-partitionable in one of the following ways:

$$(v^n v' u_1 v)(v^{n-1} v' u_2 v^2) \cdots (v v' u_n),$$
$$(v' u_1 v^{n-1})(v v' u_2 v^{n-2}) \cdots (v^{n-1} v' u_n),$$

depending on which of the words v, v' is lexicographically greater. Consequently each word for which the height with respect to Y is greater than M contains an n-partitionable subword.

This proves the theorem.

COROLLARY 1 (*Kaplansky*). A finitely generated, associative, algebraic PI-algebra A over a field F is finite-dimensional.

PROOF. We denote by V the set of all products of less than n generators of the algebra A, where n is the degree of the identity. Let m be the maximum of the algebraic degrees of the elements of the set V, and h be the height of the algebra A with respect to V. Then the algebra A is generated as a linear space over F by the finite set of all products of less than $(m-1) \cdot n \cdot h$ generators.

Analogously, from the theorem on height is deduced

COROLLARY 2 (*Levitzki*). A finitely generated associative PI-algebra over a ring Φ, each element of which is nilpotent, is nilpotent.

In particular, every associative nil-algebra of bounded index is locally nilpotent.

We shall say that an algebra A has *locally bounded height* if each finitely generated subalgebra of A is an algebra of bounded height.

COROLLARY 3. Every associative PI-algebra has locally bounded height.

3. ALGEBRAIC ELEMENTS AND LOCAL FINITENESS

In the previous section we saw that the theorems of Kaplansky and Levitzki, which were obtained by their authors using completely different methods, prove to be corollaries of a single result. The generality of these theorems becomes even more apparent after consideration of the following concepts.

In what follows we fix some ideal Z of the ring Φ.

We shall call an element a of a power-associative Φ-algebra A *algebraic over* Z if there exist elements $z_i \in Z$ and a natural number m such that $a_m = \sum_{i=1}^{m-1} z_i a^i$.

We shall call a finitely generated Φ-algebra A *finite over* Z if there exist elements $a_1, a_2, \ldots, a_k$ in A such that for some natural number m every element $c \in A^m$ admits representation in the form $c = \sum_{i=1}^{k} z_i a_i$, where $z_i \in Z$. If each finitely generated Φ-subalgebra A in the Φ-algebra B is finite over Z, then we shall call the algebra B *locally finite over* Z.

In particular, when $Z = (0)$ the algebraic over Z elements are simply the nilpotent elements, and local finiteness over Z in this case turns into local nilpotency. When $Z = \Phi$ and Φ is a field, algebraic over Z elements and local finiteness over Z turn into the usual algebraic elements and local finite-dimensionality over Φ.

An immediate corollary of the theorem on height is

THEOREM 1 (*Shirshov*). If, in a finitely generated associative Φ-algebra A, with admissible polynomial identity of degree n, all of the products of less than n generators is algebraic over Z, then the algebra A is finite over Z.

This theorem contains the theorems of Kaplansky and Levitzki as particular cases. Its principal difference from them is that the algebraic condition is not imposed on all elements, but only on a finite number of them.

The following lemma is needed for the subsequent study of the concept of local finiteness.

LEMMA 7. Let the algebra A be finitely generated as a Φ-module. Then every finitely generated subalgebra B of the algebra A is also a finitely generated Φ-module.

PROOF. Let $R = \{b_1, \ldots, b_k\}$ be a set of generators of the algebra B. By assumption there exist elements $a_1, \ldots, a_n \in A$ such that $A = \Phi a_1 + \cdots + \Phi a_n$. In particular,

$$b_i = \sum_{l=1}^{n} \rho_i^l a_l, \qquad i = 1, 2, \ldots, k,$$

$$a_i a_j = \sum_{l=1}^{n} \rho_{ij}^l a_l, \qquad i, j = 1, 2, \ldots, n.$$

Let Φ_0 be the subring of the ring Φ generated by the set $\{1, \rho_i^l, \rho_{ij}^l\}$. By Hilbert's theorem (S. Lang, "Algebra," p. 169, Mir, 1968) Φ_0 is a Noetherian ring. We consider the Φ_0-module A_0:

$$A_0 = \Phi_0 a_1 + \Phi_0 a_2 + \cdots + \Phi_0 a_n.$$

It is easy to see that A_0 is a Φ_0-algebra, and moreover $R \subseteq A_0$. We consider the Φ_0-subalgebra $B_0 \subseteq A_0$ generated by the set R. Since the Φ_0-module A_0 is Noetherian (as a finitely generated module over the Noetherian ring Φ_0), then B_0 is a finitely generated module over Φ_0. For completion of the proof it only remains to note that $B = \Phi B_0$. This proves the lemma.

Lemma 7 is a particular case of the following assertion.

LEMMA 8. Let the algebra A be finite over Z. Then A is locally finite over Z.

PROOF. By assumption there exist elements $a_1, \ldots, a_n \in A$ and a natural number m such that $A^m \subseteq Za_1 + \cdots + Za_n$. Let B be an arbitrary finitely generated subalgebra of the algebra A. For the proof of the finiteness of B over Z, it is obviously sufficient to show that B is finitely generated as a Φ-module, and that the algebra $\bar{B} = B/ZB$ is nilpotent. It is clear that A is a finitely generated Φ-module. By Lemma 7, B also is a finitely generated module over Φ. Thus it remains to prove the nilpotency of the algebra $\bar{B} = B/ZB$. We denote the multiplication algebra $M^A(B)$ by B^*. It is obviously sufficient to show that the algebra $\bar{B}^* = B^*/ZB^*$ is nilpotent. (In fact, if

$(\bar{B}^*)^k = (0)$, then it is easy to see $(\bar{B})^{2^k} = (0)$.) Let W be an arbitrary element from B^*. We consider the elements $a_i W^m \in A^m$, $i = 1, 2, \ldots, n$. We have $a_i W^m = s_{i1}a_1 + \cdots + s_{in}a_n$, where $s_{ij} \in Z$; $i, j = 1, 2, \ldots, n$. We denote by S the matrix of dimension $n \times n$ composed from the elements s_{ij}. Further, if the element $c \in A^m$ is represented in the form $c = z_1 a_1 + \cdots + z_n a_n$ where $z_i \in Z$, then we associate with it the tuple $[c] = (z_1, \ldots, z_n)$. Finally, we denote by $\hat{a}$ the column

$$\begin{pmatrix} a_1 \\ \vdots \\ a_n \end{pmatrix}.$$

As is easy to see, with this notation $cW^m = [c] \cdot S \cdot \hat{a}$, where the usual product of matrices is on the right side. Also, $cW^{km} = [c] \cdot S^k \cdot \hat{a}$ for every natural number k. By the Hamilton–Cayley theorem (S. Lang, "Algebra," p. 446, Mir, 1968), the matrix S is a root of its characteristic polynomial, that is, $S^n = \sum_{k=0}^{n-1} \sigma_k S^k$ where $\sigma_k \in Z$. Now let x be an arbitrary element of the algebra. We consider

$$\begin{aligned} xW^{m(n+1)} &= (xW^m)W^{mn} = [xW^m] \cdot S^n \cdot \hat{a} = \sum_{k=0}^{n-1} \sigma_k [xW^m] \cdot S^k \cdot \hat{a} \\ &= \sum_{k=0}^{n-1} \sigma_k (xW^m) W^{mk} = x\left(\sum_{k=0}^{n-1} \sigma_k W^{m(k+1)}\right), \end{aligned}$$

whence since the element x was arbitrary we obtain

$$W^{m(n+1)} = \textstyle\sum_{k=0}^{n-1} \sigma_k W^{m(k+1)}, \qquad \sigma_k \in Z.$$

Thus $W^{m(n+1)} \in ZB^*$ for any element $W \in B^*$, that is, the algebra $\bar{B}^*$ satisfies the admissible polynomial identity $x^{m(n+1)} = 0$. The algebra $\bar{B}^*$ is associative, and since B is finitely generated over Φ, has a finite number of generators. By Corollary 2 to the theorem on height $\bar{B}^*$ is nilpotent. By the same token this proves the lemma.

In view of Lemma 8, from Theorem 1 easily follows

THEOREM 2. *If, in an associative algebra A, with admissible polynomial identity of degree n, all products of generators containing less than n factors are algebraic over Z, then the algebra A is locally finite over Z.*

Exercises

Everywhere below Φ is an arbitrary associative–commutative ring with identity element 1, and Z is some ideal of the ring Φ.

1. Let A be an associative Φ-algebra and I be an ideal in A. Prove that if the ideal I and the quotient algebra $\bar{A} = A/I$ are locally finite over Z, then A is also locally finite over Z.

2. Prove that the sum of two locally finite over Z two-sided ideals of an associative Φ-algebra A is again a locally finite over Z ideal.

3. Prove that in every associative algebra A there exists a unique locally finite over Z two-sided ideal $\mathscr{L}_Z(A)$ which contains all locally finite over Z two-sided ideals of the algebra A and is such that in the quotient algebra $\bar{A} = A/\mathscr{L}_Z(A)$ there are no nonzero locally finite over Z two-sided ideals.

We shall call the ideal $\mathscr{L}_Z(A)$ the *locally finite over Z radical* of the algebra A.

4. Prove that the locally finite over Z radical $\mathscr{L}_Z(A)$ of an associative algebra A contains all locally finite over Z one-sided ideals of the algebra A.

4. SPECIAL JORDAN PI-ALGEBRAS

We again consider the free associative Φ-algebra Ass$[X]$ from set of generators $X = \{x_1, x_2, \ldots\}$. We shall call an element h of the algebra Ass$[X]$ a *j-polynomial* if h can be expressed from elements of the set X by means of the operations of addition, multiplication by elements from Φ, squaring, and the "quadratic multiplication" $xU_y = yxy$.

We denote by $j[X]$ the set of all j-polynomials. As is easy to see, the set $j[X]$ is closed with respect to the Jordan multiplication $a \circ b = ab + ba$. In addition, if $a, b, c \in J[X]$, then a^2, $abc + cba \in j[X]$.

We note that if there is an element $\frac{1}{2}$ in Φ, then our definition of j-polynomial coincides with the definition mentioned in Chapter 3. In addition, $j[X]$ coincides as a set with the free special Jordan algebra SJ$[X]$.

We return again to our set $R = \{x_1, x_2, \ldots, x_k\}$, where $x_i < x_j$ for $i < j$. We recall that the set of all x_k-indecomposable words is denoted by T. We shall call an associative word α from elements of the set R *singular* if there exists a j-polynomial h_α which is homogeneous in each x_i and for which the leading term is the word α.

LEMMA 9. Every T-word α is singular (with respect to the set R).

PROOF. If the T-length of the word α equals 1, that is,

$$\alpha = \underbrace{x_k x_k \cdots x_k}_{n} x_{i_1} x_{i_2} \cdots x_{i_m}, \qquad i_r \neq k; \qquad r = 1, 2, \ldots, m,$$

then it is possible to set

$$h_\alpha = [\cdots (x_k^n \circ x_{i_1}) \circ x_{i_2} \cdots] \circ x_{i_m}.$$

Suppose the assertion of the lemma is proved for T-words for which the T-lengths are less than n, and the T-length of the T-word α equals n. Then

$$\alpha = \beta \underbrace{x_k x_k \cdots x_k}_{r} x_{i_1} x_{i_2} \cdots x_{i_m}, \qquad i_s \neq k; \qquad s = 1, 2, \ldots, m,$$

where β is a T-word satisfying the conditions of the induction assumption. Let h_β be the j-polynomial which corresponds to the word β. Then it is possible to take as h_α the j-polynomial

$$\{\cdots[(h_\beta x_k^2 x_{i_1} + x_{i_1} x_k^2 h_\beta) \circ x_{i_2}] \circ \cdots\} x_{i_m}.$$

This proves the lemma.

Now let A be an arbitrary associative Φ-algebra, and $M = \{m_i\}$ be a subset of the algebra A. We denote by $j[M]$ the set of elements of the form $h(m_1, \ldots, m_k)$, where $m_i \in M$, $h(x_1, \ldots, x_k) \in j[X]$. We shall call elements of the set $j[M]$ *j-polynomials* from elements of the set M.

THEOREM 3 (*Shirshov*). In an associative Φ-algebra A let the set $j[\{a_i\}]$ of j-polynomials from the set of generators $\{a_i\}$ satisfy some admissible polynomial identity. Then if every element of the set $j[\{a_i\}]$ is algebraic over Z, the algebra A is locally finite over Z.

PROOF. By Lemma 8 it suffices to prove that every finite subset $\bar{R} = \{a_1, \ldots, a_k\}$ of the set of generators of the algebra A generates a finite over Z subalgebra. We shall prove this assertion by means of induction on the number of generators, assuming its validity in the case when the number of elements in $\bar{R}$ equals $k - 1$. Along with the set $\bar{R}$, we consider the set $R = \{x_1, \ldots, x_k\}$ and the free associative algebra Ass$[R]$. If $f = f(x_1, \ldots, x_k)$ is an associative word from $x_1, \ldots, x_k$ or some polynomial from Ass$[R]$, then as before we denote by $\bar{f}$ the corresponding image $f(a_1, \ldots, a_k)$ of this element in the algebra A. In view of the induction assumption and the fact the element a_k is algebraic over Z, there exists a natural number m such that in the algebra A for every x_k-indecomposable word α of length m we have $\bar{\alpha} = \sum_i z_i \bar{\alpha}_i$, where $z_i \in Z$ and the α_i are words from $x_1, \ldots, x_k$ of smaller length. We denote by T_0 the set of all x_k-indecomposable words of length less than m, and by V the set of all T_0-words of T-length less than n, where n is the degree of the admissible polynomial identity. [We note that this identity can be assumed to have form (1).] In view of Lemma 9, every element of the set V is a singular word (with respect to the set R), that is, for every element $v \in V$ there exists a homogeneous j-polynomial from the generators x_i whose leading term is the word v. For each element $v \in V$ we fix a corresponding j-polynomial h_v. In view of the finiteness of the set V, the set $H(V) = \{h_v | v \in V\}$ is also finite. Moreover, under the conditions of the lemma, the element $\bar{h}_v$ is algebraic over Z for each element h_v from the set $H(V)$. We now consider an R-word α of R-length $mN(K,s,n) + 2m$, where K is the number of elements in the set T_0, and s is the maximum of the algebraic degrees over Z of the elements $\bar{h}_v$ for $h_v \in H(V)$. (Without loss of generality it is possible to assume that $s \geq 2n$.)

Everything will be proved if we establish that in the algebra A the element $\bar{\alpha}$ is representable in the form of a Z-linear combination of elements $\bar{\alpha}_i$, where the α_i are R-words having R-length less than that of α.

Let $\alpha = \alpha'\beta\alpha''$, where α' does not contain x_k, $\beta = x_k\beta' x_i$ with $i \neq k$, and $\alpha'' = x_k x_k \cdots x_k$. It is possible to assume that the R-lengths of the words α' and α'' are less than m, since otherwise everything would be proved. Then the R-length of the word β is greater than $mN(K,s,n)$. It is possible to assume further that the image of each of the x_k-indecomposable words appearing in the word β cannot be represented in the algebra A in the form of a Z-linear combination of images of R-words of lesser R-length. Each such word has R-length less than m. Consequently, β is a T_0-word, and the T-length of the word β is greater than the number $N(K,s,n)$. By Lemmas 3 and 4, as in the proof of Theorem 1, we obtain that there can be found in the word β either an n_T-partitionable subword γ or a subword of the form $(\gamma')^s$, where γ' is a T_0-word for which the T-length is less than n. We consider separately both possibilities.

1. $\beta = \beta_1(\gamma')^s\beta_2$, where the T-length of the word γ' is less than n. In this case $\gamma' \in V$, and therefore there exists a j-polynomial $h_{\gamma'} \in H(V)$ having γ' as its leading term. Because of the choice of the number s, there exist elements $z_i \in Z$ such that $\bar{h}^s_{\gamma'} = \sum_{i<s} z_i \bar{h}^i_{\gamma'}$. Hence, it follows that in the algebra A the image $(\bar{\gamma}')^s$ of the word $(\gamma')^s$ (and this means also the image $\bar{\alpha}$ of the word α) is representable in the form of a Φ-linear combination of images of lesser words (in the lexicographic sense) with the same R-length, and a Z-linear combination of images of words having smaller R-length.

2. $\beta = \beta_1\gamma_1\gamma_2\cdots\gamma_n\beta_2$, $\gamma_1\gamma_2\cdots\gamma_n > \gamma_{i_1}\gamma_{i_2}\cdots\gamma_{i_n}$ if $(i_1, i_2, \ldots, i_n) \neq (1, 2, \ldots, n)$. Let the γ_i be the leading terms of the j-polynomials h_{γ_i}. Since the set of j-polynomials from $\{a_i\}$ satisfies a relation of the form (1), the product $\bar{h}_{\gamma_1}\bar{h}_{\gamma_2}\cdots\bar{h}_{\gamma_n}$ can be expressed in the form of a linear combination of products obtained from the given one by means of permutations of the factors. Furthermore, since the leading term of a product is the product of the leading terms of the factors, it follows from this that in the algebra A the image of the word $\gamma_1\cdots\gamma_n$ is representable in the form of a linear combination of images of lesser words with the same R-length. Consequently the element $\bar{\alpha}$ can also be expressed in an analogous fashion.

Thus we arrive at the conclusion that in the algebra A the image $\bar{\alpha}$ of the word α is representable in the form of a Φ-linear combination of images of words which have the same R-length but are less than α, and a Z-linear combination of images of words of smaller R-length. Since a decreasing sequence of words of equal length terminates, then at some step we obtain a representation of the element $\bar{\alpha}$ in the form of a Z-linear combination of images of words of smaller R-length. This proves the theorem.

COROLLARY. In an associative Φ-algebra A let every j-polynomial from the generators be algebraic over Z, and in addition let the algebraic degrees over Z of the j-polynomials be bounded overall. Then the algebra A is locally finite over Z.

The proof follows from Lemma 6.

Now let there be an element $\frac{1}{2}$ in Φ. We consider the free Jordan Φ-algebra J[X] from the set of generators $X = \{x_1, x_2, \ldots, x_n, \ldots\}$. We shall call an element $f \in \mathrm{J}[X]$ *essential* if under the natural homomorphism of the algebra J[X] onto the free special Jordan algebra SJ[X] the image $\bar{f}$ of the element f is an admissible element when considered as an element of the free associative algebra Ass[X]. Now let J be some Jordan Φ-algebra. We shall say that J satisfies an *essential polynomial identity* if there exists an essential element $f(x_1, \ldots, x_n) \in \mathrm{J}[X]$ such that $f(a_1, \ldots, a_n) = 0$ for all $a_1, \ldots, a_n \in J$. If the Jordan algebra J satisfies an essential polynomial identity f, then we shall call the algebra J a *Jordan PI-algebra* for short, and shall say that J satisfies the polynomial f. Here are some examples of Jordan PI-algebras:

1. Any associative–commutative algebra is a Jordan PI-algebra since it satisfies the essential polynomial $f(x_1, x_2, x_3) = 4(x_1, x_2, x_3)$. In fact, we have

$$\overline{f(x_1, x_2, x_3)} = 4[(x_1 \odot x_2) \odot x_3 - x_1 \odot (x_2 \odot x_3)] = [x_2, [x_1, x_3]]$$

is an admissible element.

2. The Jordan algebra $\Phi_2^{(+)}$ of matrices of dimension 2×2 with elements from Φ satisfies the essential polynomial $f(x, y, z, r, s) = 128((x, y, z)^2, r, x)$.

3. A Jordan nil-algebra of index n satisfies the essential polynomial $f(x) = x^n$.

Now let J be some special Jordan algebra and A be an associative enveloping algebra for the algebra J. It is clear that if J is a Jordan PI-algebra, then J as a submodule of the algebra A satisfies some associative admissible polynomial identity. We show that the converse assertion is also valid.

LEMMA 10. Let the special Jordan algebra J considered as a submodule of the associative algebra A satisfy an associative admissible polynomial identity. Then J also satisfies some Jordan essential polynomial identity, that is, J is a Jordan PI-algebra.

PROOF. Let $f(x_1, \ldots, x_n) = 0$ be an associative admissible polynomial identity satisfied in the submodule J. Then J also satisfies the polynomial $\rho(x, y) = f(xyx, xy^2x, \ldots, xy^nx)$. If $\rho^*(x, y)$ is the polynomial obtained by applying the involution $*$ to the polynomial ρ, then J also satisfies the polynomial $\psi(x, y) = \rho(x, y) \cdot \rho^*(x, y)$. The polynomial $\psi(x, y)$ is admissible and symmetric with respect to the involution $*$. Consequently $\psi(x, y) \in \mathrm{SJ}[x, y]$ by Theorem 3.3. Now let $\psi_0(x, y) \in \mathrm{J}[x, y]$ be some inverse image of the element $\psi(x, y)$ under the canonical homomorphism of $\mathrm{J}[x, y]$ onto $\mathrm{SJ}[x, y]$. Then it is clear that $\psi_0(x, y)$ is an essential element, and that J as a Jordan algebra satisfies the identity $\psi_0(x, y) = 0$. This proves the lemma.

We note that a set of generators $\{a_\alpha\}$ of a special Jordan algebra J also generates an associative enveloping algebra A. In addition, J coincides as a subset of the algebra A with the set $j[\{a_\alpha\}]$ of all j-polynomials from $\{a_\alpha\}$.

We can now reformulate Theorem 3 and its corollary in the following manner:

THEOREM 3′ (*Shirshov*). Let every element of a special Jordan PI-algebra J over a ring Φ be algebraic over Z. Then an associative enveloping algebra A of the algebra J is locally finite over Z.

COROLLARY 1. Let every element of a special Jordan algebra J be algebraic over Z, and in addition let the algebraic degrees of the elements be bounded overall. Then an associative enveloping algebra A of J is locally finite over Z.

Setting $Z = (0)$, we also obtain

COROLLARY 2. Let every element of a special Jordan PI-algebra J be nilpotent. Then an associative enveloping algebra A of J is locally nilpotent.

From Theorem 3′ easily follows

THEOREM 4. Let J be a special Jordan PI-algebra over a ring Φ, each element of which is algebraic over Z. Then J is locally finite over Z.

PROOF. Let I be some subalgebra of the algebra J generated by a finite set of elements $R = \{b_1, \ldots, b_k\}$. By Theorem 3 the set R generates a finite over Z algebra B in an associative enveloping algebra A of the algebra J. We consider the Jordan algebra $B^{(+)}$. By the finiteness over Z, the algebra B (and consequently also $B^{(+)}$) is a finitely generated module over Φ. Hence it follows that $B^{(+)}$ is finitely generated as a Φ-algebra. It is now clear that $B^{(+)}$ is finite over Z. By Lemma 8 the subalgebra $I \subseteq B^{(+)}$ is also finite over

Z. Since the subalgebra I was arbitrary, this means that the algebra J is locally finite over Z, which proves the theorem.

From Theorem 4 are easily deduced assertions for the algebra J which are analogous to Corollaries 1 and 2 of Theorem 3′. We cite here only one very important particular case of these assertions.

COROLLARY. A special Jordan nil-algebra of bounded index is locally nilpotent.

We note that it follows from the corresponding result of Golod for associative algebras that the bounding of the index in the formulation of this corollary is essential (see Exercise 5).

The question on the local nilpotency of an arbitrary Jordan nil-algebra of bounded index still remains open.

Exercises

1. Let $J = \Phi \cdot 1 + M$ be the Jordan algebra of a symmetric bilinear form f on a vector space M which is not necessarily finite-dimensional. Prove that J is a PI-algebra.

2. Prove that the simple exceptional Jordan algebra $H(\boldsymbol{C}_3)$ is a PI-algebra.

3. Let every element of the Jordan Φ-algebra J be algebraic over Φ, and in addition let the algebraic degrees be bounded overall. Prove that J is a PI-algebra.

4. Let A be a commutative algebra with the identity $x^3 = 0$. Prove that A is a Jordan algebra, and that moreover A is locally nilpotent. *Hint.* Apply Corollary 2 of Theorem 3′ to the right multiplication algebra $R(A)$, and prove that it is locally nilpotent.

5. Let A be an associative Φ-algebra ($\frac{1}{2} \in \Phi$), $\mathfrak{N}(A)$ be its upper nil-radical, and $\mathfrak{L}(A)$ be its locally nilpotent radical. Prove that

$$\mathfrak{N}(A) = \mathfrak{N}(A^{(+)}), \qquad \mathfrak{L}(A) = \mathfrak{L}(A^{(+)}).$$

Deduce from this the existence of a special Jordan nil-algebra that is not locally nilpotent. *Hint.* Use Exercise 3.1.5.

5. ALTERNATIVE PI-ALGEBRAS

Before considering PI-algebras, we shall establish some facts we shall need about arbitrary alternative algebras.

PROPOSITION 1. Let I, J be ideals of an alternative algebra A. Then the product IJ is also an ideal of the algebra A.

PROOF. Let $i \in I, j \in J$. Then for any $a \in A$

$$(ij)a = i(ja) + (i,j,a) = i(ja) - (i,a,j) \in IJ,$$
$$a(ij) = (ai)j - (a,i,j) = (ai)j + (i,a,j) \in IJ,$$

which is what was to be proved.

Now let S and T be any subsets of an alternative algebra A. We denote by $\{STS\}$ the submodule of the Φ-module A generated by all possible elements of the form sts, where $s \in S$, $t \in T$.

PROPOSITION 2. For any ideals I, J of an alternative algebra A the set $\{IJI\}$ is also an ideal in A.

PROOF. We set $\{xyz\} = (xy)z + (zy)x = x(yz) + z(yx)$. Then for any $i, i' \in I$, $j \in J$, we have

$$\{iji'\} = (i + i')j(i + i') - iji - i'ji' \in \{IJI\},$$

whence by the Moufang identities for any $a \in A$

$$(iji)a = i[j(ia)] = i[j(ia)] + (ia)(ji) - (ia)(ji)$$
$$= \{ij(ia)\} - i(aj)i \in \{IJI\}.$$

Analogously, $a(iji) \in \{IJI\}$, which proves the lemma.

We recall that for an arbitrary algebra A by $A^{\#}$ is denoted the algebra obtained by the formal adjoining of an identity element to the algebra A. By Corollary 2 to Proposition 1.7, if the algebra A is alternative, then the algebra $A^{\#}$ is also alternative.

COROLLARY. For any ideal I of an alternative algebra A, the sets $P(I) = \{IA^{\#}I\}$ and $T(I) = \{III\}$ are ideals of the algebra A, and moreover $P(I) \supseteq T(I) \supseteq (P(I))^2$.

PROOF. We note that I is also an ideal in the algebra $A^{\#}$, and therefore by Proposition 2 the set $P(I)$ is an ideal of the algebra $A^{\#}$. But $P(I) \subseteq A$, and therefore $P(I)$ is an ideal of the algebra A. It also follows directly from Proposition 2 that $T(I)$ is an ideal of the algebra A. Furthermore, the containment $P(I) \supseteq T(I)$ is obvious. It remains to prove the containment $[P(I)]^2 \subseteq T(I)$. We shall write $x \equiv y$ if $x - y \in T(I)$. Let $i, j \in I$, $a, b \in A^{\#}$.

Then by the Moufang identities and their linearizations we have

$$\begin{aligned}(iai)(jbj) &= \{(ia)(ij)(bj)\} - (bj \cdot i)(j \cdot ia) \\ &\equiv -(bj \cdot i)(j \cdot ia) = -b\{ji(j \cdot ia)\} + ([b(j \cdot ia)]i)j \\ &\equiv ([b(j \cdot ia)]i)j = \{b(j \cdot ia)i\}j - ([i(j \cdot ia)]b)j \\ &= b[(j \cdot ia)(ij)] + i[(j \cdot ia)(bj)] - ([(iji)a]b)j \\ &\equiv b[j(iai)j] + i[j(ia \cdot b)j] \equiv 0.\end{aligned}$$

This proves the corollary.

Let $\Phi[X]$ be the free algebra from set of generators $X = \{x_\alpha\}$. We set

$$\langle x_1 \rangle = x_1, \qquad \langle x_1, \ldots, x_{n-1}, x_n \rangle = \langle x_1, \ldots, x_{n-1} \rangle \cdot x_n$$

for $n > 1$. We shall call nonassociative words of the form $\langle x_{i_1}, \ldots, x_{i_n} \rangle$ r_1-*words* from the set $\{x_\alpha\}$. If in an r_1-word $v = \langle x_{i_1}, x_{i_2}, \ldots, x_{i_n} \rangle$ the indices are ordered, $i_1 < i_2 < \cdots < i_n$, then we shall call v a *regular* r_1-*word*. Furthermore, we shall call nonassociative words of the form $\langle u_1, \ldots, u_n \rangle$, where each u_i is an r_1-word (regular r_1-word), r_2-*words* (*regular* r_2-*words*, respectively) from $\{x_\alpha\}$. If A is an arbitrary algebra and $M = \{m_i\}$ is a subset of the algebra A, then we shall call elements of the form $\langle m_{i_1}, \ldots, m_{i_n} \rangle$ r_1-*words from the set* M. Analogously we define r_2-*words from the set* M. For a multilinear word $v = v(x_1, \ldots, x_n)$ we denote by $\langle v \rangle$ the regular r_1-word with the same composition as v. If the word v is not multilinear, then we set $\langle v \rangle = 0$.

An algebra A is called *antiassociative* if the identity

$$(xy)z + x(yz) = 0$$

is valid in A.

LEMMA 11. Let A be an antiassociative and anticommutative algebra, and let $v = v(x_1, \ldots, x_n)$ be an arbitrary nonassociative word. Then the equality

$$v(a_1, \ldots, a_n) = \pm\langle v(a_1, \ldots, a_n) \rangle$$

is valid for any elements $a_1, \ldots, a_n \in A$.

PROOF. We shall first prove that $v(a_1, \ldots, a_n) = v'(a_1, \ldots, a_n)$, where $v' = v'(x_1, \ldots, x_n)$ is some not necessarily regular r_1-word of the same type as the word v. We shall prove this assertion by induction on the length $d(v)$ of the word v. The basis for the induction, $d(v) = 1$, is obvious. Now let $d(v) = n > 1$, and assume our assertion is valid for any nonassociative word u of length less than n. We have $v = v_1 v_2$ where $d(v_i) < n$. If $d(v_2) = 1$, then everything is clear, since by the induction assumption the element $\bar{v}_1 = v_1(a_1, \ldots, a_n)$ is representable in the form of an r_1-word. Now let $d(v_2) > 1$, $v_2 = v_2' v_2''$. Then for the images $\bar{v}, \bar{v}_i, \bar{v}_2', \bar{v}_2''$ of the elements v, v_i, v_2', v_2'' in the algebra A we have $\bar{v} = v(a_1, \ldots, a_n) = \bar{v}_1 \bar{v}_2 = \bar{v}_1(\bar{v}_2' \bar{v}_2'') = -(\bar{v}_1 \bar{v}_2')\bar{v}_2''$, and the proof is completed by induction on $d(v_2)$. Furthermore, let $v(a_1, \ldots, a_n) = \langle a_{i_1}, \ldots, a_i, a_j, \ldots, a_{i_m} \rangle$ and $i \geq j$. If $j = i$, then we have

$$v = \langle a_{i_1}, \ldots, a_i, a_i, \ldots, a_{i_m} \rangle = -\langle a_{i_1}, \ldots, a_i^2, \ldots, a_{i_m} \rangle = 0.$$

If $j < i$, then $v = -\langle a_{i_1}, \ldots, a_i a_j, \ldots, a_{i_m} \rangle = \langle a_{i_1}, \ldots, a_j a_i, \ldots, a_{i_m} \rangle = -\langle a_{i_1}, \ldots, a_j, a_i, \ldots, a_{i_m} \rangle$. Since the given process is monotonic (it reduces a word in the sense of the lexicographic ordering) and does not change the type of the word v, then after some finite number of steps we obtain the needed result. This proves the lemma.

PROPOSITION 3 (*Zhevlakov*). Let A be an alternative algebra and $v = v(x_1, \ldots, x_n)$ be an arbitrary nonassociative word. Then for any elements $a_1, \ldots, a_n \in A$ the element $v(a_1, \ldots, a_n)$ is representable in the form

$$v(a_1, \ldots, a_n) = \pm \langle v(a_1, \ldots, a_n) \rangle + \sum_i \alpha_i v_i(a_1, \ldots, a_n),$$

where $\alpha_i \in \Phi$ and for each i the element $v_i = v_i(x_1, \ldots, x_n)$ is a nonassociative homogeneous polynomial having the same type as v and representable in one of the forms

$$u^2, \quad u \circ u', \quad (uw)u, \quad \{uwu'\}, \tag{2}$$

where u, u', w are nonassociative words of length less than that of v.

PROOF. Without loss of generality we can assume that A is the free alternative algebra with the set of free generators $\{a_i\}$. By the corollary to Theorem 1.5 the variety of alternative algebras is homogeneous, and therefore the concepts of degree of a monomial, homogeneous polynomial, etc. make sense in A. We consider the quotient algebra $\bar{A} = A/P(A)$. For any elements $\{a, b, c\} \in A$ we have $\{abc\} \in P(A)$, $a \circ b = \{a1b\} \in P(A)$, and therefore in the quotient algebra $\bar{A}$ we obtain $\bar{a}\bar{b} = -\bar{b}\bar{a}$ and $(\bar{a}\bar{b})\bar{c} = -(\bar{c}\bar{b})\bar{a} = \bar{a}(\bar{c}\bar{b}) = -\bar{a}(\bar{b}\bar{c})$. Thus the algebra $\bar{A}$ is anticommutative and antiassociative. Hence by Lemma 11 it follows that

$$v(a_1, \ldots, a_n) = \pm \langle v(a_1, \ldots, a_n) \rangle + v_0, \tag{3}$$

where $v_0 \in P(A)$. We note that every element from the ideal $P(A)$, in particular the element v_0, is a linear combination of elements of the form (2) for some monomials u, u', w. Furthermore, since A is a free algebra of a homogeneous variety, any homogeneous component of any identity valid in A is also an identity in A. Considering now the homogeneous component of relation (3) containing the element $v(a_1, \ldots, a_n)$, we obtain the required assertion. This proves the proposition.

THEOREM 5 (*Shirshov*). Let A be an alternative algebra and $v = v(x_1, \ldots, x_n)$ be an arbitrary nonassociative word. Then for any elements $a_1, \ldots, a_n \in A$ the element $v(a_1, \ldots, a_n)$ is representable in the form of a linear combination of regular r_2-words from $a_1, \ldots, a_n$ with the same length as v.

PROOF. Again it is possible to assume that A is the free alternative algebra with set of generators $\{a_i\}$. It suffices to show that $v(a_1, \ldots, a_n)$ can be represented in the form of some linear combination of regular r_2-words. The preservation of length will follow from the fact that A is a free algebra of a homogeneous variety. Let u_1 be an arbitrary monomial from A. We shall prove by induction on the degree $d(u_1)$ of the monomial u_1 that for any regular r_2-word u_2 the element u_2u_1 is again representable in the form of a linear combination of regular r_2-words from $a_1, \ldots, a_n$. The basis for the induction, $d(u_1) = 1$, is obvious. Let $d(u_1) = m \geq 2$, and assume the assertion to be proved for words of smaller length. By Proposition 3 we have $u_1 = \alpha_0 u_0 + \sum_i \alpha_i u_{1i}$, where u_0 is a regular r_1-word and the u_{1i} are elements of the form (2). Since u_2u_0 is a regular r_2-word, it remains for us to consider the elements u_2u_{1i}. For example, let u_{1i} have the form $\{uwu'\}$ where u, u', w are monomials from A of degree less than n. Then by linearization of the right Moufang identity we have

$$u_2u_{1i} = u_2\{uwu'\} = [(u_2u)w]u' + [(u_2u')w]u.$$

Since $d(u) < n$, $d(w) < n$, $d(u') < n$, then by applying (from left to right) the induction assumption consecutively three times to each term on the right side, we obtain a representation of the element u_2u_{1i} in the form of a linear combination of regular r_2-words. The other cases are considered analogously. Thus we have proved that for any regular r_2-word u_2 and for any monomial u_1 the element u_2u_1 is a linear combination of regular r_2-words. We now note that the element $v(a_1, a_2, \ldots, a_n)$ is representable in the form $v(a_1, a_2, \ldots, a_n) = a_{i_1}R_{v_1}R_{v_2} \cdots R_{v_k}$, where a_{i_1} is a generator located in some fixed left position in $v(a_1, \ldots, a_n)$, and $v_1, v_2, \ldots, v_k$ are some monomials from A. Beginning with the regular r_2-word a_{i_1} and applying the assertion just proved k times, we obtain a representation of the element $v(a_1, \ldots, a_n)$ in the form of a linear combination of regular r_2-words. This proves the theorem.

COROLLARY. In an alternative algebra A the containment $A^{n^2} \subseteq A^{\langle n \rangle}$ is valid for any natural number n. In particular, if the algebra A is right nilpotent of index n, then A is nilpotent of index not greater than n.

PROOF. By Theorem 5 every element from A^{n^2} is a linear combination of r_2-words from generators of the algebra A of length $\geq n^2$. But any r_2-word w of length $\geq n^2$ either contains an r_1-word of length $\geq n$ or is itself an r_1-word from more than n r_1-words. Since by Proposition 1 $A^{\langle n \rangle}$ is an ideal in A, then in each of these cases it is obvious that $w \in A^{\langle n \rangle}$. This proves the corollary.

We now consider the free alternative Φ-algebra Alt$[X]$ from the set of generators $X = \{x_1, x_2, \ldots\}$. By analogy with the associative case, we shall call an element $f = f(x_1, \ldots, x_n)$ of the algebra Alt$[X]$ an *alternative j-polynomial* if f can be expressed from elements of the set X by means of the operations of addition, multiplication by elements from Φ, squaring, and the "quadratic multiplication" $xU_y = yxy$. We denote the set of all alternative j-polynomials by $J_{\mathrm{Alt}}[X]$.

Let π be the canonical homomorphism of the algebra Alt$[X]$ onto the free associative algebra Ass$[X]$. It is clear that $\pi(j_{\mathrm{Alt}}[X]) = j[X]$.

LEMMA 12. Let $f = f(x_1, \ldots, x_n) \in j_{\mathrm{Alt}}[X]$. Then

$$R_{f(x_1, \ldots, x_n)} = f^{\pi}(R_{x_1}, \ldots, R_{x_n}).$$

PROOF. Let S be the set of alternative j-polynomials for which the assertion of the lemma is satisfied. It is obvious that $S \supseteq X$ and that S is closed with respect to Φ-linear combinations. Now let $f, g \in S$. Then

$$R_{f^2(x_1, \ldots, x_n)} = R^2_{f(x_1, \ldots, x_n)} = [f^{\pi}(R_{x_1}, \ldots, R_{x_n})]^2 = (f^2)^{\pi}(R_{x_1}, \ldots, R_{x_n}),$$

and by the right Moufang identity

$$\begin{aligned} R_{fgf(x_1, \ldots, x_n)} &= R_{f(x_1, \ldots, x_n)} R_{g(x_1, \ldots, x_n)} R_{f(x_1, \ldots, x_n)} \\ &= f^{\pi}(R_{x_1}, \ldots, R_{x_n}) g^{\pi}(R_{x_1}, \ldots, R_{x_n}) f^{\pi}(R_{x_1}, \ldots, R_{x_n}) \\ &= (fgf)^{\pi}(R_{x_1}, \ldots, R_{x_n}). \end{aligned}$$

Thus the set S contains the generators and is closed with respect to the operations of addition, multiplication by elements from Φ, squaring, and the "quadratic multiplication" $xU_y = yxy$. Hence it clearly follows that $S = j_{\mathrm{Alt}}[X]$, which proves the lemma.

As in the associative case, if A is an arbitrary alternative algebra and $M = \{m_i\}$ is a subset of the algebra A, then we denote by $j_{\mathrm{Alt}}[M]$ the set of elements of the form $f(m_1, \ldots, m_k)$ where $f(x_1, \ldots, x_k) \in j_{\mathrm{Alt}}[X]$. We shall call elements of the set $j_{\mathrm{Alt}}[M]$ j-polynomials from the set M.

We call an element $f \in \mathrm{Alt}[X]$ *essential* if f^π is an admissible element in the algebra $\mathrm{Ass}[X]$. If A is some alternative Φ-algebra and M is a submodule of the algebra A, then we shall say that M satisfies an *essential polynomial identity* if there can be found an essential element $f(x_1, \ldots, x_k) \in \mathrm{Alt}[X]$ such that $f(m_1, \ldots, m_k) = 0$ for all $m_1, \ldots, m_k \in M$. We shall call an alternative algebra satisfying an essential polynomial identity an *alternative PI-algebra* for short.

We again fix some ideal Z of the ring Φ.

LEMMA 13. In an alternative Φ-algebra A with set of generators $R = \{a_1, \ldots, a_k\}$, let the submodule $M = j_{\mathrm{Alt}}[R]$ satisfy an essential polynomial identity $f = 0$ where $f \in j_{\mathrm{Alt}}[X]$. If every element of the submodule M is algebraic over Z, then for some natural number N all r_1-words from the generators a_i of length greater than or equal to N are representable in the form of a Z-linear combination of r_1-words from the generators a_i of length less than N.

PROOF. As is easy to see, it suffices for the proof of the lemma to prove the finiteness over Z of the subalgebra A^* of the algebra $R(A)$ generated by the operators of right multiplication $R_{a_1}, \ldots, R_{a_k}$. Let $d_1(x_i), \ldots, d_n(x_i)$ be arbitrary alternative j-polynomials. It is obvious that $d_t(a_1, \ldots, a_k) \in M$ for any t. Therefore by Lemma 12

$$f^\pi(d_1^\pi(R_{a_i}), \ldots, d_n^\pi(R_{a_i})) = f^\pi(R_{d_1(a_i)}, \ldots, R_{d_n(a_i)}) = R_{f(d_1(a_i), \ldots, d_n(a_i))} = 0.$$

Since $j[X] = \pi(j_{\mathrm{Alt}}[X])$, the set of j-polynomials from $\{R_{a_1}, \ldots, R_{a_k}\}$ satisfies the admissible polynomial identity $f^\pi = 0$. Since the elements of the submodule M are algebraic, for any j-polynomial $d(x_1, \ldots, x_k) \in j_{\mathrm{Alt}}[X]$ there exists a natural number m and elements $z_i \in Z$ such that

$$d^m(a_1, \ldots, a_k) = \sum_{i=1}^{m-1} z_i d^i(a_1, \ldots, a_k).$$

By Lemma 12 and the fact that $d^r(x_1, \ldots, x_k) \in j_{\mathrm{Alt}}[X]$, we have

$$(d^\pi)^m(R_{a_1}, \ldots, R_{a_k}) = (d^m)^\pi(R_{a_1}, \ldots, R_{a_k}) = R_{d^m(a_1, \ldots, a_k)}$$
$$= \sum_{i=1}^{m-1} z_i R_{d^i(a_1, \ldots, a_k)} = \sum_{i=1}^{m-1} z_i (d^\pi)^i(R_{a_1}, \ldots, R_{a_k}).$$

Thus every j-polynomial from the set $\{R_{a_1}, \ldots, R_{a_k}\}$ is algebraic over Z. By Theorem 3 the algebra A^* is finite over Z, which proves the lemma.

LEMMA 14. Every alternative PI-algebra satisfies an essential polynomial identity f, where $f \in j_{\mathrm{Alt}}[X]$.

PROOF. Let $g(x_1, \ldots, x_n) = 0$ be an essential polynomial identity satisfied in the alternative PI-algebra A. Then A also satisfies the polynomial $\rho(x, y) = g(xy, xy^2, \ldots, xy^n) \in \mathrm{Alt}[x, y]$. By Artin's theorem the algebra $\mathrm{Alt}[x, y]$ is associative, and therefore isomorphic to $\mathrm{Ass}[x, y]$. Let $\psi(x, y) = \rho(x, y)\rho^*(x, y)$. Then by a note to Theorem 3.3 $\psi(x, y)$ is a j-polynomial, that is, $\psi(x, y) \in j[\{x, y\}] = j_{\mathrm{Alt}}[\{x, y\}]$. It is clear that $\psi(x, y)$ is an essential element and that the algebra A satisfies the identity $\psi(x, y)$. This proves the lemma.

THEOREM 6 (*Shirshov*). Let A be an alternative PI-algebra over a ring Φ with set of generators $R = \{a_i\}$, and let M be the set of all r_1-words from elements of the set R. Then if every element of the submodule $j_{\mathrm{Alt}}[M]$ is algebraic over Z, the algebra A is locally finite over Z.

PROOF. In view of Lemma 8 it suffices to show that any finite subset of the set R generates a finite over Z subalgebra of the algebra A. Let B be an arbitrary subalgebra generated by a finite set $R_0 = \{a_{i_1}, \ldots, a_{i_k}\}$. From Lemmas 14 and 13 follows the existence of a natural number n such that every r_1-word from elements of the set R_0 of length $\geq n$ is representable in the form of a Z-linear combination of r_1-words of smaller length. We denote by R_1 the set of all r_1-words from elements of the set R_0 of length less than n. Again applying Lemmas 14 and 13, we obtain a natural number m for which every r_1-word from the set R_1 of R_1-length $\geq m$ is represented in the form of a Z-linear combination of r_1-words from the set R_1 of smaller R_1-length. It is now clear that every r_2-word from the set R_0 of length $\geq mn$ is a Z-linear combination of r_2-words from the set R_0 of smaller length. In view of Theorem 5, this means that every product of mn or more elements of the set R_0 is represented in the form of a Z-linear combination of products with a smaller number of factors from R_0. Consequently, the algebra B is finite over Z, which proves the theorem.

COROLLARY 1. In an alternative Φ-algebra A let every element be algebraic over Z, and in addition let the algebraic degrees be bounded by the number n. Then A is locally finite over Z.

In fact, as is easy to see, the algebra A satisfies the essential polynomial $[[x^n, y], [x^{n-1}, y], \ldots, [x, y]] \in \mathrm{Alt}[x, y]$.

COROLLARY 2. Let every element of an alternative PI-algebra A be nilpotent. Then A is locally nilpotent.

In particular, every alternative nil-algebra of bounded index is locally nilpotent.

Exercises

1. Prove that if there is an element $\frac{1}{2}$ in Φ, then $A^4 \subseteq P(A)$ for any alternative algebra A.

2. Let A be a Φ-algebra and $\frac{1}{2} \in \Phi$. Prove that if A is alternative, then $A^{(+)}$ is a special Jordan algebra. In addition, if A is finitely generated, then $A^{(+)}$ is also finitely generated.

3. (*Zhevlakov*) Let B be an alternative algebra and A be a subalgebra of the algebra B with set of generators $\{a_i\}$. Then as generators of the algebra $R^B(A)$, which is generated by the operators of right multiplication R_a by elements of the algebra A on the algebra B, can be taken the operators of right multiplication R_w, where w is a regular r_1-word from $\{a_i\}$.

4. Under the conditions of Exercise 3, prove that if the algebra A is finitely generated, then the algebra $R^B(A)$ is also finitely generated.

5. Prove that if an alternative algebra A is locally nilpotent, then the algebra $R^B(A)$ is locally nilpotent for any alternative extension algebra $B \supseteq A$.

6. Let A be a finitely generated alternative algebra. Prove that for any n there exists a number $f(n)$ such that $A^{f(n)} \subseteq A^{(n)}$.

7. Prove that the locally nilpotent radical exists in every alternative algebra.

8. Let π be the canonical homomorphism of the free alternative algebra $\mathrm{Alt}[X]$ onto the free associative algebra $\mathrm{Ass}[X]$. Prove that $j_{\mathrm{Alt}}[X] \cap \mathrm{Ker}\,\pi = (0)$. In particular, the restriction of π to the set $j_{\mathrm{Alt}}[X]$ maps $j_{\mathrm{Alt}}[X]$ one-to-one to $j[X]$.

LITERATURE

Golod [37], Jacobson [46], Zhevlakov and Shestakov [80], Kurosh [104], Medvedev [145], Procesi [170], Herstein [252], Shirshov [276, 277, 279].

CHAPTER 6

Solvability and Nilpotency of Alternative Algebras

1. THE NAGATA–HIGMAN THEOREM

In connection with Shirshov's theorem on the local nilpotency of an alternative nil-algebra of bounded index, there arises a natural question: To what extent is the localness condition in that theorem essential, that is, is not every alternative nil-algebra of bounded index nilpotent? In this section we show that in the case of associative algebras, with a sufficiently nice ring of operators, the answer to this question turns out to be yes. We shall need the following

LEMMA 1. Let A be an alternative algebra. We denote by $I_n(A)$ the set

$$I_n(A) = \left\{ \sum_i \alpha_i a_i^n \,\middle|\, \alpha_i \in \Phi,\ a_i \in A \right\}.$$

Then $(n!)^2 I_n(A)A + (n!)^2 A I_n(A) \subseteq I_n(A)$.

PROOF. We denote by $S_n(a_1, a_2, \ldots, a_n)$ the sum

$$S_n(a_1, a_2, \ldots, a_n) = \sum_{(i_1, i_2, \ldots, i_n)} v(a_{i_1}, a_{i_2}, \ldots, a_{i_n}),$$

where $v(x_1, x_2, \ldots, x_n)$ is some fixed nonassociative word of length n. By Lemma 1.4 we have

$$\begin{aligned} S_n(a_1, \ldots, a_n) = {} & v(a_1 + \cdots + a_n, \ldots, a_1 + \cdots + a_n) \\ & - \sum_{i=1}^{n} v(a_1 + \cdots + \hat{a}_i + \cdots + a_n, \ldots, a_1 + \cdots + \hat{a}_i + \cdots + a_n) \\ & + \sum_{1 \le i < j \le n} v(a_1 + \cdots + \hat{a}_i + \cdots + \hat{a}_j + \cdots + a_n, \ldots, a_1 + \cdots \\ & + \hat{a}_i + \cdots + \hat{a}_j + \cdots + a_n) \\ & - \cdots + (-1)^{n-1} \sum_{i=1}^{n} v(a_1, \ldots, a_i). \end{aligned} \tag{1}$$

In view of the fact that A is a power-associative algebra, the right-hand side of this equality does not depend on how the parentheses are arranged in the monomial v. Consequently, the sum $S_n(a_1, \ldots, a_n)$ likewise does not depend on how the parentheses are arranged in the monomial v. For example, one can assume the arrangement of the parentheses to be the standard one from the right. Furthermore, it follows from (1) that $S_n(a_1, \ldots, a_n) \in I_n(A)$ for any elements $a_1, \ldots, a_n \in A$. Now let a, b be arbitrary elements of the algebra A. Then we have

$$\begin{aligned} S_n(ab, a, \ldots, a) &= (n-1)! \sum_{i=0}^{n-1} a^i(ab)a^{n-1-i} \\ &= a(n-1)! \sum_{i=0}^{n-1} a^i b a^{n-1-i} = aS_n(b, a, \ldots, a), \end{aligned}$$

hence

$$aS_n(b, a, \ldots, a) \in I_n(A). \tag{2}$$

Linearizing this inclusion in a, for any $a_1, \ldots, a_n, b \in A$ we obtain the inclusion

$$(n-1)! \sum_{i=1}^{n} a_i S_n(b, a_1, \ldots, \hat{a}_i, \ldots, a_n) \in I_n(A). \tag{3}$$

Setting $a_1 = a$, $a_2 = a_3 = \cdots = a_n = b$ in (3), we obtain

$$(n-1)! aS_n(b, b, \ldots, b) + (n-1)(n-1)! bS_n(a, b, \ldots, b) \in I_n(A),$$

whence by (2) it follows

$$(n!)^2 ab^n \in I_n(A).$$

It is proved analogously that for any $a, b \in A$

$$(n!)^2 b^n a \in I_n(A).$$

This proves the lemma.

COROLLARY. Let A be an alternative algebra. We denote by $J_n(A)$ the set

$$J_n(A) = \{a \in A \mid (n!)^k a \in I_n(A) \text{ for some } k\}.$$

Then $J_n(A)$ is an ideal of the algebra A, and the quotient algebra $A/J_n(A)$ does not contain elements of additive order $\leq n$.

PROOF. It is clear that $J_n(A)$ is a Φ-submodule of the Φ-module A. Now let $a \in J_n(A)$, $b \in A$, and let $(n!)^k a \in I_n(A)$. By Lemma 1 $(n!)^2 b[(n!)^k a]$, $(n!)^2[(n!)^k a]b \in I_n(A)$, whence $(n!)^{k+2}ba$, $(n!)^{k+2}ab \in I_n(A)$, and $ba, ab \in J_n(A)$. By the same token we have proved that $J_n(A)$ is an ideal of the algebra A. We now note that the ideal $J_n(A)$ has the following property: if $(n!)^k a \in J_n(A)$, then $a \in J_n(A)$. Hence it follows that the quotient algebra $A/J_n(A)$ does not contain elements of additive order $\leq n$. This proves the corollary.

We can now prove the following assertion:

THEOREM 1 (*Nagata, Higman*). Let A be an arbitrary assocative algebra. Then for any natural number n

$$A^{2^n - 1} \subseteq J_n(A).$$

PROOF (*Higgins*). We shall prove the theorem by induction on n. For $n = 1$ the assertion is obvious, so assume it is true for $n - 1$. In view of the fact that $S_n(b, a, \ldots, a) \in I_n(A)$, for any $a, b \in A$ we have

$$\sum_{i=0}^{n-1} a^i b a^{n-1-i} \in J_n(A). \tag{4}$$

Now let a, b, c be arbitrary elements of the algebra A. We consider the sum

$$\sum_{i,j=1}^{n-1} a^{n-1-i} c b^j a^i b^{n-1-j} = \sum_{j=1}^{n-1} \left(\sum_{i=1}^{n-1} a^{n-1-i}(cb^j)a^i \right) b^{n-1-j}$$
$$= -\sum_{j=1}^{n-1} a^{n-1} c b^j b^{n-1-j} + k = -(n-1)a^{n-1}cb^{n-1} + k,$$

where $k \in J_n(A)$ by (4). On the other hand, we have

$$\sum_{i,j=1}^{n-1} a^{n-1-i} c b^j a^i b^{n-1-j} = \sum_{i=1}^{n-1} a^{n-1-i} c \left(\sum_{j=1}^{n-1} b^j a^i b^{n-1-j} \right)$$
$$= \sum_{i=1}^{n-1} (a^{n-1-i} c a^i) b^{n-1} + k_1 = a^{n-1} c b^{n-1} + k_2,$$

where again $k_1, k_2 \in J_n(A)$. As a result we obtain

$$n a^{n-1} c b^{n-1} \in J_n(A).$$

Because the elements a, b, c are arbitrary, it follows that

$$nI_{n-1}(A)AI_{n-1}(A) \subseteq J_n(A),$$

and furthermore

$$J_{n-1}(A)AJ_{n-1}(A) \subseteq J_n(A).$$

By the induction assumption $A^{2^{n-1}-1} \subseteq J_{n-1}(A)$, and consequently

$$A^{2^n-1} = A^{2^{n-1}-1}AA^{2^{n-1}-1} \subseteq J_n(A).$$

This proves the theorem.

COROLLARY 1. Let A be an associative nil-algebra of index n without elements of additive order $\leq n$. Then A is nilpotent of index $\leq 2^n - 1$.

For the proof it suffices to note that under the conditions of the corollary $J_n(A) = 0$.

COROLLARY 2. Let A be an arbitrary associative algebra. Then for any natural number n there exists a natural number k such that for any $a \in A^{2^n-1}$

$$(n!)^k a = \sum_i \alpha_i a_i^n, \qquad \text{where} \quad \alpha_i \in \Phi \quad \text{and} \quad a_i \in A.$$

For the proof we consider the free associative Φ-algebra Ass[X] from the set of free generators $X = \{x_1, x_2, \ldots\}$. By Theorem 1 we have

$$x_1x_2 \cdots x_{2^n-1} \in J_n(\mathrm{Ass}[X]),$$

and consequently, for some k

$$(n!)^k x_1x_2 \cdots x_{2^n-1} = \sum_i \alpha_i u_i^n, \qquad \text{where} \quad \alpha_i \in \Phi \quad \text{and} \quad u_i \in \mathrm{Ass}[X].$$

As is easy to see, the given number k is the desired one.

We note that the bound $2^n - 1$ obtained in the Nagata–Higman theorem is not exact, since in the case $n = 3$ it is known $A^6 \subseteq J_3(A)$ while $2^3 - 1 = 7$. The question of a corresponding exact bound $f(n)$ for the case of an arbitrary n remains open. Razmyslov recently showed that $f(n) \leq n^2$. On the other hand, Kuz'min has proved that $f(n) \leq n(n+1)/2$.

Exercises

1. Let A be a Jordan algebra. As in the case of alternative algebras, the subsets $I_n(A)$ and $J_n(A)$ can be considered in A. Prove that $J_n(A)$ is an ideal of the algebra A for any n.

2. Prove that for any prime number p there exists a nonnilpotent associative nil-algebra of index p over a field of characteristic p.

3. (*Higman*) Prove that the containment $A^6 \subseteq J_3(A)$ is valid for every associative algebra A.

2. DOROFEEV'S EXAMPLE

We return now to alternative algebras. It turns out that, in contrast to the associative case, alternative nil-algebras of bounded index can be nonnilpotent, that is, the Nagata–Higman theorem does not carry over to alternative algebras. This was proved by G. V. Dorofeev, who constructed an example of a nonnilpotent solvable alternative algebra over an arbitrary ring of operators. We present his example in somewhat altered form.

Let us consider two sets of symbols: $E = \{e_k\}$, $k = 1, 2, \ldots, n, \ldots$; $V = \{x, L_i, R_j\}$, $i, j = 1, 2, \ldots, n, \ldots$. If v is an arbitrary associative word from elements of V, then we shall denote by $d_R(v)$ the number of symbols R_i which appear in the composition of v. We call the word v *regular* if v has one of the following forms:

(1) $d_R(v) = 0$, $v = x$ or $v = xL_i$;
(2) $d_R(v) = 1$, $v = xR_j$ or $v = xL_iR_j$;
(3) $d_R(v) \geq 2$, $v = x\hat{L}_{i_1}R_{i_2}R_{i_3} \cdots R_{i_n}$, where $i_1 < i_2 < \cdots < i_n$ and the symbol L_{i_1} may be absent.

We denote by $T(V)$ the set of all regular words.

Now let u be an arbitrary word from V of the form

$$u = x\hat{L}_{i_1}R_{i_2}R_{i_3} \cdots R_{i_n},$$

where $d_R(u) \geq 2$ and the symbol L_{i_1} may be absent. We denote by $\bar{u}$ a regular word of the form

$$\bar{u} = x\hat{L}_{j_1}R_{j_2}R_{j_3} \cdots R_{j_n},$$

where $\{j_1, j_2, \ldots, j_n\} = \{i_1, i_2, \ldots, i_n\}$ and the symbol L_{j_1} appears in the composition of $\bar{u}$ if and only if the symbol L_{i_1} appears in u. We denote by $t(u)$ the number of inversions in the permutation $(i_1, i_2, \ldots, i_n)$. If any two of the symbols $i_1, i_2, \ldots, i_n$ are identical, then we set $\bar{u} = 0$.

Let us consider the free Φ-module A for which a basis is the set $E \cup T(V)$. We convert A into an algebra by defining multiplication on the basis according to the following rules:

1. $x \cdot y = 0$, if $x, y \in E$ or $x, y \in T(V)$.

2. (a) $x \cdot e_i = xR_i$, $xL_j \cdot e_i = xL_jR_i$;
   (b) if $u \in T(V)$ and $d_R(u) \geq 1$, then $u \cdot e_i = (-1)^{t(uR_i)}\overline{uR_i}$.
3. (a) $e_i \cdot x = xL_i$, $e_i \cdot xL_j = xR_j \cdot e_i$;
   (b) if $u \in T(V)$ and $d_R(u) \geq 1$, then $u = u'R_i$ and

$$e_j \cdot u = e_j \cdot (u'R_i) = (e_j \cdot u' + u' \cdot e_j) \cdot e_i.$$

As is easy to see, rules (1)–(3) define the products of any basis elements. We now consider the Φ-module $I = \Phi(T(V))$ generated by the set $T(V)$. It is clear that I is an ideal of the algebra A, and in addition $I^2 = (0)$ and $A^2 \subseteq I$. Consequently $(A^2)^2 = (0)$, so A is solvable and therefore a nil-algebra of bounded index. Furthermore, the algebra A is nonnilpotent, since for any n the product $(\cdots((x \cdot e_1) \cdot e_2) \cdots) \cdot e_n = xR_1R_2 \cdots R_n$ is different from zero. It remains to prove that the algebra A is alternative. We shall divide the proof of this fact into several lemmas.

We first note the following obvious equality:

$$(v \cdot e_i) \cdot e_j + (v \cdot e_j) \cdot e_i = 0, \tag{5}$$

where $v \in T(V)$ and $e_i, e_j \in E$.

LEMMA 2. For any $v \in T(V)$ and $e_i, e_k \in E$ there is the equality

$$e_i \cdot (v \cdot e_k) = (e_i \cdot v + v \cdot e_i) \cdot e_k. \tag{6}$$

PROOF. We shall prove this equality by induction on the number $d_R(v)$, that is, on the number of symbols R_i appearing in v. If $d_R(v) = 0$, then either $v = x$ or $v = xL_j$, and (6) follows at once from the rules of multiplication. Now let $d_R(v) = n \geq 1$, and assume equality (6) is valid for all words from $T(V)$ containing less than n symbols R_i. Let $v = v'R_n$. If $k \geq n$, then (6) follows from rule 3(b). Consequently, one can assume $k < n$. By the multiplication rules 2(b) and 3(b), the induction assumption, and formula (5), we have

$$\begin{aligned}
e_i \cdot (v \cdot e_k) &= e_i \cdot (v'R_n \cdot e_k) = (-1)^{t(vR_k)}e_i \cdot (\overline{v'R_k}R_n) \\
&= (-1)^{t(vR_k)}(e_i \cdot \overline{v'R_k} + \overline{v'R_k} \cdot e_i) \cdot e_n \\
&= (-1)^{t(vR_k)} \cdot (-1)^{t(v'R_k)}(e_i \cdot (v' \cdot e_k) + (v' \cdot e_k) \cdot e_i) \cdot e_n \\
&= -[(e_i \cdot v' + v' \cdot e_i) \cdot e_k] \cdot e_n + [(v' \cdot e_i) \cdot e_k] \cdot e_n \\
&= [(e_i \cdot v' + v' \cdot e_i) \cdot e_n] \cdot e_k - [(v' \cdot e_i) \cdot e_n] \cdot e_k \\
&= [e_i \cdot (v' \cdot e_n)] \cdot e_k + [(v' \cdot e_n) \cdot e_i] \cdot e_k \\
&= (e_i \cdot v'R_n + v'R_n \cdot e_i) \cdot e_k = (e_i \cdot v + v \cdot e_i) \cdot e_k.
\end{aligned}$$

This proves the lemma.

LEMMA 3. For any $v \in T(V)$ and $e_i, e_j, e_k \in E$ there are the equalities

$$[(e_i \cdot v) \cdot e_j + (e_j \cdot v) \cdot e_i] \cdot e_k = 0, \tag{7}$$

$$e_i \cdot (e_j \cdot v) = (v \cdot e_j) \cdot e_i. \tag{8}$$

PROOF. We shall first prove equality (7). If $v = x$ or $v = xL_n$, then (7) is valid. Now let $v = v'R_n$, and assume equality (7) is true for words having a smaller number of the symbols R_i. Then by (6), (5), and the induction assumption,

$$\begin{aligned}(e_i \cdot v'R_n) \cdot e_j &= [(e_i \cdot v' + v' \cdot e_i) \cdot e_n] \cdot e_j = -[(e_n \cdot v' + v' \cdot e_n) \cdot e_i] \cdot e_j \\ &= [(e_n \cdot v' + v' \cdot e_n) \cdot e_j] \cdot e_i = -[(e_j \cdot v' + v' \cdot e_j) \cdot e_n] \cdot e_i \\ &= -[e_j \cdot (v' \cdot e_n)] \cdot e_i = -(e_j \cdot v) \cdot e_i.\end{aligned}$$

Thus if $d_R(v) > 0$, then an even stronger equality than (7) is valid: $(e_i \cdot v) \cdot e_j + (e_j \cdot v) \cdot e_i = 0$. We shall now prove (8). If $d_R(v) = 0$, then (8) is valid. Now let $v = v'R_n$. By (5), (6), (7), and the induction assumption, we have

$$\begin{aligned}e_i \cdot (e_j \cdot v'R_n) &= e_i \cdot [(e_j \cdot v' + v' \cdot e_j) \cdot e_n] \\ &= [e_i \cdot (e_j \cdot v' + v' \cdot e_j) + (e_j \cdot v' + v' \cdot e_j) \cdot e_i] \cdot e_n \\ &= [(v' \cdot e_j) \cdot e_i + (e_i \cdot v' + v' \cdot e_i) \cdot e_j + (e_j \cdot v') \cdot e_i + (v' \cdot e_j) \cdot e_i] \cdot e_n \\ &= [(e_i \cdot v') \cdot e_j + (e_j \cdot v') \cdot e_i] \cdot e_n + [(v' \cdot e_n) \cdot e_j] \cdot e_i \\ &= (v \cdot e_j) \cdot e_i.\end{aligned}$$

This proves the lemma.

LEMMA 4. The algebra A is alternative.

PROOF. From the multiplication table it is clear that only associators which contain one element from $T(V)$ and two elements from E are nonzero. By equality (5) we have

$$(v, e_i, e_j) + (v, e_j, e_i) = 0.$$

Furthermore, from (8) and (5) we also have

$$(e_i, e_j, v) + (e_j, e_i, v) = 0.$$

We now consider the expression

$$\begin{aligned}(e_i, v, e_j) + (e_i, e_j, v) &= (e_i \cdot v) \cdot e_j - e_i \cdot (v \cdot e_j) - e_i \cdot (e_j \cdot v) \\ &= (e_i \cdot v) \cdot e_j - (e_i \cdot v) \cdot e_j - (v \cdot e_i) \cdot e_j - (v \cdot e_j) \cdot e_i \\ &= 0\end{aligned}$$

by (6), (8), and (5). Moreover, again by (6) we obtain

$$\begin{aligned}(v, e_i, e_j) + (e_i, v, e_j) &= (v \cdot e_i) \cdot e_j + (e_i \cdot v) \cdot e_j - e_i \cdot (v \cdot e_j) \\ &= (v \cdot e_i) \cdot e_j + (e_i \cdot v) \cdot e_j - (e_i \cdot v) \cdot e_j - (v \cdot e_i) \cdot e_j \\ &= 0.\end{aligned}$$

Finally, it is easy to see that

$$(v, e_i, e_i) = (e_i, v, e_i) = (e_i, e_i, v) = 0.$$

This proves the lemma.

Exercises

1. Let A be the solvable nonnilpotent algebra constructed in Section 2. Prove that its annihilator, $\operatorname{Ann} A = \{a \in A \mid aA = Aa = (0)\}$, is generated as a Φ-module by elements of the form xL_iR_i, $xL_iR_j + xL_jR_i$. Also prove that the quotient algebra $\bar{A} = A/\operatorname{Ann} A$ is a nonnilpotent, solvable, nil-algebra of index 3 and that $\operatorname{Ann} \bar{A} = (\bar{0})$.

2. Prove that the algebra $\bar{A}$ defined in the previous exercise has zero associative center.

3. Let T_k be the ideal of identities satisfied by the free algebra with k generators in the variety of alternative algebras of solvable index 2. Prove that the chain of T-ideals

$$T_1 \supseteq T_2 \supseteq T_3 \supseteq \cdots \supseteq T_n \supseteq \cdots$$

does not stabilize at any finite step.

3. ZHEVLAKOV'S THEOREM

In the previous section we established that the Nagata–Higman theorem does not carry over verbatim to alternative algebras. We proved an even stronger assertion, namely, the nilpotency of an alternative algebra A does not, in general, follow from its solvability. (We recall that the analogous fact also holds in the theory of Jordan algebras.) In connection with all of this there arises a natural question: Is every alternative nil-algebra of bounded index solvable? As proved by K. A. Zhevlakov, over a sufficiently nice ring of operators the answer to this question turns out to be yes.

For the proof of Zhevlakov's theorem we need a series of lemmas.

Henceforth A is an arbitrary alternative algebra.

LEMMA 5. In the algebra A there is the identity

$$(zx \circ xz, y, z) = (xzx, y, z^2). \tag{9}$$

PROOF. Applying the Moufang identities and their linearizations, we obtain

$$\begin{aligned}(zx \circ xz, y, z) &= (zx) \circ (xz, y, z) + (xz) \circ (zx, y, z)\\ &= (zx) \circ (x, yz, z) + (xz) \circ (x, zy, z)\\ &= (zx \circ x, yz, z) - x \circ (zx, yz, z) + (xz \circ x, zy, z) - x \circ (xz, zy, z)\\ &= (zx^2, yz, z) + (x^2 z, zy, z) + (xzx, y \circ z, z) - 2x \circ (x, zyz, z)\\ &= 2(x^2, zyz, z) + (xzx, y, z^2) - 2(x^2, zyz, z)\\ &= (xzx, y, z^2).\end{aligned}$$

This proves the lemma.

LEMMA 6. In the quotient algebra $\bar{A} = A/J_n(A)$ there is the identity

$$z(x^{n-1}, y, z)z = 0.$$

PROOF. We first note that in $\bar{A}$ the identity

$$0 = S_n(z, x, \ldots, x) = (n-1)! \sum_{i=0}^{n-1} x^i z x^{n-1-i}$$

is valid, whence from the properties of the ideal $J_n(A)$ we obtain

$$\sum_{i=0}^{n-1} x^{n-1-i} z x^i = 0. \tag{10}$$

Now in view of (10) we have

$$\sum_{i=0}^{n-1} (zx^i) \circ (x^{n-1-i} z) = nzx^{n-1}z + \sum_{i=0}^{n-1} x^{n-1-i} z^2 x^i = nzx^{n-1}z,$$

whence, from linearization in x of identity (9) and the Moufang identities, we have

$$\begin{aligned}nz(x^{n-1}, y, z)z = (nzx^{n-1}z, y, z) &= \left(\sum_{i=0}^{n-1} (zx^i) \circ (x^{n-1-i} z), y, z\right)\\ &= \left(\sum_{i=0}^{n-1} x^i z x^{n-1-i}, y, z^2\right) = 0.\end{aligned}$$

This proves the lemma.

LEMMA 7. In the algebra $\bar{A}$ there is the identity

$$(x^{n-1}, y^{n-1}, z^2) = 0, \qquad n \geq 2.$$

PROOF. We note first of all that in every alternative algebra the identity

$$(x^n, y, z) = \sum_{i=0}^{n-1} x^i (x, y, z) x^{n-1-i} \tag{11}$$

is valid. In fact, by Artin's theorem $(x^n, y, x) = 0$, whence

$$0 = (x_n, y, x)\Delta_x^1(z) = (x^n, y, z) + \sum_{i=0}^{n-1} (x^i z x^{n-1-i}, y, x),$$

and then by the Moufang identities, we obtain (11).

By Lemma 6 for any $k, l \geq 1$ we have

$$z^k(x^{n-1}, y, z)z^l = 0.$$

Hence by identity (11) it follows that for any $m \geq 1$

$$(x^{n-1}, y, z^m) = (x^{n-1}, y, z) \circ z^{m-1}.$$

We now have

$$0 = (x^{n-1}, y, z^n) = [(x^{n-1}, y, z) \circ z^{n-1}] = \{\cdots [(x^{n-1}, y, z) \circ z] \circ z \cdots\} \circ z.$$

Linearization of this identity in z gives

$$0 = \sum_{(i_1, i_2, \ldots, i_n)} \{\cdots [(x^{n-1}, y, z_{i_1}) \circ z_{i_2}] \circ z_{i_3} \cdots\} \circ z_{i_n}. \tag{12}$$

We set here $z_1 = z_2 = z$, $z_3 = z_4 = \cdots = z_n = y$. Since for any $k \geq 0$

$$\begin{aligned}
&\{[\cdots((x^{n-1}, y, z) \circ \underbrace{y) \circ y \cdots] \circ y}_{k}\} \circ z \\
&= ([\cdots (x^{n-1} \circ \underbrace{y) \circ y \cdots] \circ y}_{k}, y, z) \circ z \\
&= ([\cdots (x^{n-1} \circ \underbrace{y) \circ y \cdots] \circ y}_{k}, y, z^2) \\
&= \{\cdots [(x^{n-1}, y, z^2) \circ \underbrace{y] \circ y \cdots\} \circ y}_{k} \\
&= (x^{n-1}, y, z^2) \circ y^k,
\end{aligned}$$

then, after we carry out the permutations, identity (12) takes the form

$$0 = 2(n-1)!(x^{n-1}, y, z^2) \circ y^{n-2} = 2(n-1)!(x^{n-1}, y^{n-1}, z^2).$$

This proves the lemma.

LEMMA 8. $A^4 \subseteq J_2(A)$.

PROOF. For any $a, b \in A$ we have $a^2, a \circ b \in J_2(A)$ and $2aba = (a \circ b) \circ a - a^2 \circ b \in J_2(A)$, whence also $aba \in J_2(A)$. Now, as in the proof of Proposition 5.3, we obtain that the quotient algebra $\bar{A} = A/J_2(A)$ is anticommutative and antiassociative. In the algebra $\bar{A}$ we consider the associator

$$(ab, c, d) = [(ab)c]d - (ab)(cd) = [a(bc)]d + a[b(cd)]$$
$$= a[(bc)d] - a[(bc)d] = 0.$$

On the other hand, we have

$$(ab, c, d) = [(ab)c]d - (ab)(cd)$$
$$= [(ab)c]d + [(ab)c]d = 2[(ab)c]d.$$

The equality obtained proves that $\bar{A}^4 = (0)$. This proves the lemma.

Now we can prove

THEOREM 2 (*Zhevlakov*). Let A be an arbitrary alternative algebra. Then for any natural number n

$$A^{(n(n+1)/2)} \subseteq J_n(A).$$

PROOF. For $n = 1$ the assertion of the theorem is obvious. For $n = 2$ by Lemma 8 we have $A^{(3)} \subseteq A^2 \cdot A^2 \subseteq A^4 \subseteq J_2(A)$, that is, the assertion is also true. Now let $A^{(n(n-1)/2)} \subseteq J_{n-1}(A)$, $n > 2$. We consider the quotient algebra $\bar{A} = A/J_n(A)$. By the induction assumption we have

$$\bar{A}^{(n(n-1)/2)} \subseteq J_{n-1}(\bar{A}).$$

Furthermore, in view of the containment $J_{n-1}(A) \subseteq J_2(A)$, from Lemma 7 it follows that $J_{n-1}(A)$ is an associative nil-algebra of index n. By the Nagata–Higman theorem the algebra $J_{n-1}(A)$ is nilpotent of index $2^n - 1$, whence $(J_{n-1}(A))^{(n)} = (0)$. Since $(A^{(k)})^{(m)} = A^{(k+m)}$, then finally we obtain

$$\bar{A}^{(n(n+1)/2)} = (\bar{A}^{(n(n-1)/2)})^{(n)} \subseteq (J_{n-1}(A))^{(n)} = (0),$$

whence

$$A^{(n(n+1)/2)} \subseteq J_n(A).$$

This proves the theorem.

COROLLARY 1. Let A be an alternative nil-algebra of index n without elements of additive order $\leq n$. Then the algebra A is solvable of index $\leq n(n+1)/2$.

COROLLARY 2. Let A be an arbitrary alternative algebra. Then for any natural number n there exists a natural number k such that for any $a \in A^{(n(n+1)/2)}$

$$(n!)^k a = \sum_i \alpha_i a_i^n, \qquad \text{where} \quad \alpha_i \in \Phi \quad \text{and} \quad a_i \in A.$$

Thus every alternative nil-algebra of bounded index is locally nilpotent, over a sufficiently nice ring of operators even solvable, but over an arbitrary ring of operators it is possible it may not be nilpotent. In addition, as for the Nagata–Higman theorem, it is rather easy to see that the restriction on the characteristic is essential for solvability.

We note for comparison that, in the case of Jordan algebras, the situation is far from being so clear. We already remarked in Chapter 5 that the answer to the question on local nilpotency of Jordan nil-algebras of bounded index is unknown for Jordan algebras that are not special. The question on solvability of Jordan nil-algebras of bounded index is not solved even for special Jordan nil-algebras of index 3.

Exercises

Below A is an alternative Φ-algebra where $\frac{1}{2} \in \Phi$.

1. Prove that for any natural number n the set $A_n = (A^{(+)})^{(n)}$ is a subalgebra of the algebra A. *Hint*. Prove that $A_n = I_2(A_{n-1})$, then apply Lemma 1 and induction on n.

2. Prove that $A^{(2n)} \subseteq (A^{(+)})^{(n)}$. In particular, the algebra A is solvable if and only if the Jordan algebra $A^{(+)}$ is solvable. *Hint*. Apply Lemma 8 and induction on n.

3. Let I be an ideal of the algebra A, M be some submodule of the Φ-module A, and also $I \odot A^{(+)} \subseteq M$. Prove that the containment $v(I, A, A, A) \subseteq M$ is valid for any nonassociative word $v(x_1, x_2, x_3, x_4)$ of length 4. *Hint*. Consider the free alternative Φ-algebra and apply Lemma 8.

4. Prove that $A^{3n+1} \subseteq (A^{(+)})^{n+1}$. In particular, the algebra A is nilpotent if and only if the Jordan algebra $A^{(+)}$ is nilpotent. *Hint*. Apply induction on n and use Exercise 3.

LITERATURE

Dorofeev [55], Zhevlakov [66], Kuz'min [103], Nagata [157], Razmyslov [175], Higman [253].

Results on connections between solvability and nilpotency in some other varieties of algebras: Anderson [13, 14], Dorofeev [59], Zhevlakov [69], Zhevlakov and Shestakov [80], Markovichev [144], Nikitin [160], Pchelintsev [172, 174], Roomel'di [184], Shestakov [267].

CHAPTER **7**

Simple Alternative Algebras

Questions on the structure of simple algebras in this or that variety are one of the main questions in the theory of rings. The present chapter is devoted to the study of simple alternative algebras. We already know one example of a nonassociative simple alternative algebra—that is, the Cayley–Dickson algebra. As it turns out, other nonassociative simple alternative algebras do not exist. This result was proved by different authors with gradually increasing generality over a stretch of several decades: first for finite-dimensional algebras (Zorn, Schafer), next for algebras with nontrivial idempotent (Albert), for alternative division rings (Bruck, Kleinfeld, Skornyakov), for commutative alternative algebras (Zhevlakov), etc. The greatest progress was obtained by Kleinfeld, who proved that every nonassociative simple alternative algebra, which is not a nil-algebra of characteristic 3, is a Cayley–Dickson algebra. The final description of simple alternative algebras came about after the appearance of Shirshov's theorem on the local nilpotency of alternative nil-algebras with polynomial identity.

1. PRELIMINARY RESULTS

Let A be an arbitrary algebra. The following three basic central subsets can be considered in the algebra A: the *associative center* $N(A)$, *commutative center* $K(A)$, and *center* $Z(A)$. They are defined in the following manner:

$$N(A) = \{n \in A \mid (n, A, A) = (A, n, A) = (A, A, n) = 0\},$$
$$K(A) = \{k \in A \mid [k, A] = 0\},$$
$$Z(A) = N(A) \cap K(A).$$

LEMMA 1. Let A be an arbitrary algebra, $x, y, z \in A$, and $n \in N(A)$. Then in A there are the relations

$$n(x, y, z) = (nx, y, z), \tag{1}$$

$$(xn, y, z) = (x, ny, z), \tag{2}$$

$$(x, y, z)n = (x, y, zn). \tag{3}$$

In addition, if one of the elements x, y, z belongs to $N(A)$, then in A there is the relation

$$[xy, z] = x[y, z] + [x, z]y. \tag{4}$$

PROOF. We note that in every algebra there is the identity

$$(wx, y, z) + (w, x, yz) - w(x, y, z) - (w, x, y)z - (w, xy, z) = 0. \tag{5}$$

For its proof it suffices to expand all the associators. The validity of relations (1)–(3) follows from identity (5). For the proof of relation (4), it suffices to establish that in any algebra there is the identity

$$[xy, z] - x[y, z] - [x, z]y = (x, y, z) - (x, z, y) + (z, x, y). \tag{6}$$

It is proved like (5). This proves the lemma.

COROLLARY 1. The sets $N(A)$ and $Z(A)$ are always subalgebras of the algebra A. If the algebra A is alternative, then the commutative center $K(A)$ is also a subalgebra of the algebra A, and moreover $3K(A) \subseteq N(A)$.

PROOF. It follows from relations (1)–(3) that $N(A)$ is a subalgebra, and it follows from (4) that $Z(A)$ is also a subalgebra. Now let the algebra A be alternative. In this case identity (6) takes the form

$$[xy, z] - x[y, z] - [x, z]y = 3(x, y, z). \tag{7}$$

Let $k, k' \in K(A)$ and $x, y \in A$. From (7) we have

$$3(k, x, y) = 3(y, k, x) = 3(x, y, k) = [xy, k] - x[y, k] - [x, k]y = 0,$$

whence it follows that $3K(A) \subseteq N(A)$. Furthermore, again applying identity (7), we obtain

$$[kk', x] = k[k', x] + [k, x]k' + 3(k, k', x) = 0,$$

that is, $K(A)$ is a subalgebra of the algebra A.

COROLLARY 2. Let A be a commutative alternative Φ-algebra. If $\frac{1}{3} \in \Phi$, then A is associative.

Actually, in this case $A = K(A) \subseteq Z(A)$.

THEOREM 1. If A is a simple algebra, then either $Z(A) = (0)$ or $Z(A)$ is a field.

PROOF. Let $Z = Z(A) \neq (0)$. It is clear that Z is an associative commutative ring, and therefore it is sufficient for us to prove that there is an identity element in Z and that every nonzero element from Z is invertible in Z. Let $z \in Z$ with $z \neq 0$. Then the set zA is an ideal of the algebra A, and in addition, as is easy to see, $zA \neq (0)$. By the simplicity of the algebra A we have $zA = A$. This means there exists an $e \in A$ such that $ze = z$. Now let $x \in A$. Then $x = zy$ for some $y \in A$. Therefore $ex = e(zy) = (ez)y = (ze)y = zy = x$, and analogously $xe = x$, that is, e is an identity element for the algebra A. Also, there exists a $z' \in A$ such that $zz' = e$. Let x, y be arbitrary elements of the algebra A and $x = zt$. Then in view of (2) we have

$$(z', x, y) = (z', zt, y) = (z'z, t, y) = (e, t, y) = 0,$$

and analogously

$$(x, z', y) = (x, y, z') = 0,$$

that is, $z' \in N(A)$. Finally, in view of (4) we obtain

$$[z', x] = [z', zt] = z[z', t] = [zz', t] = [e, t] = 0,$$

that is, $z' \in Z$. We have proved that the element z' is an inverse for the element z in Z. This proves the theorem.

An algebra A over a field F is called *central* over F if $Z(A) = F$.

THEOREM 2. Let A be a simple central algebra over a field F, and K be any extension of the field F. Then the algebra $A_K = K \otimes_F A$ is a simple central algebra over the field K.

PROOF. Let us prove first the simplicity of the algebra A_K.

Let $U \neq (0)$ be an ideal in A_K. If u is a nonzero element from U, then we write it in the form $u = \sum_i k_i \otimes a_i$, where $k_i \in K$, $a_i \in A$, and the k_i are linearly independent over F. We shall call the number of nonzero a_i in this expression the length of the element u. We choose an element $0 \neq u \in U$ with least length. Further, we identify the algebra A with the F-subalgebra $1 \otimes A$ of the algebra A_K, and consider in the multiplication algebra $M(A_K)$ of the algebra A_K the F-subalgebra $A^* = M^{A_K}(A)$ which is generated by the operators of multiplication by elements from A. If $W \in A^*$, then $uW = \sum_i k_i \otimes a_i W \in U$. Since A is simple, there exists an element $W_0 \in A^*$ for which

$a_1 W_0 = 1$. Therefore it is possible to pass from the element u to an element $u_1 \in U$ with the same length and which has the form

$$u_1 = k_1 \otimes 1 + k_2 \otimes a_2' + \cdots + k_m \otimes a_m'.$$

For any $a \in A$ the element

$$[1 \otimes a, u_1] = k_2 \otimes [a, a_2'] + \cdots + k_m \otimes [a, a_m']$$

belongs to U. However, the length of this element is less than the length of u, therefore it must be equal to 0. Since the k_i are linearly independent over F, it then follows from the properties of the tensor product that $[a, a_i'] = 0$ for $i = 2, \ldots, m$ and any $a \in A$. Consequently $a_i' \in K(A)$. Analogously we obtain that $a_i' \in N(A)$, and finally $a_i' \in Z(A) = F$. We write $a_i' = \alpha_i \in F$. Then

$$\begin{aligned} u_1 &= k_1 \otimes 1 + k_2 \otimes \alpha_2 + \cdots + k_m \otimes \alpha_m \\ &= (k_1 + \alpha_2 k_2 + \cdots + \alpha_m k_m) \otimes 1 = k \otimes 1, \end{aligned}$$

where $k \in K$ and $k \neq 0$ by the linear independence of the k_i over F. Hence it follows that $U = A_k$, which proves the simplicity of the algebra A_K.

Now let an element $z = \sum_i k_i \otimes a_i$ be in the center of the algebra A_K. Here we again assume that the elements $k_i \in K$ are linearly independent over F. Then for any $a, b \in A$ we shall have

$$0 = [z, 1 \otimes a] = \sum_i k_1 \otimes [a_i, a],$$

$$0 = (z, 1 \otimes a, 1 \otimes b) = \sum_i k_i \otimes (a_i, a, b),$$

whence $[a_i, a] = (a_i, a, b) = 0$, and analogously $(a, a_i, b) = (a, b, a_i) = 0$, that is, the elements a_i are in $Z(A) = F$. Let $a_i = \alpha_i \in F$. Then $z = \sum_i k_i \otimes a_i = (\sum_i \alpha_i k_i) \otimes 1 \in K$. Thus we have proved that $Z(A_K) \subseteq K$. It is clear that $K \subseteq Z(A_K)$, so that finally $Z(A_K) = K$.

This proves the theorem.

PROPOSITION 1 (*Zhevlakov*). There do not exist simple locally nilpotent algebras.

PROOF. Let A be some locally nilpotent algebra. We assume that A is simple, and we take an arbitrary element $a \neq 0$ in A. Consider the ideal I_a generated by the set $Aa + aA$ in the algebra A. It is clear that $I_a \neq (0)$, and therefore $I_a = A$. This means there can be found elements x_{ij} in the algebra A such that

$$a = \sum_{i=1}^{t} aM_{x_{i1}} M_{x_{i2}} \cdots M_{x_{ik_i}}, \tag{8}$$

where each of the $M_{x_{ij}}$ equals either $R_{x_{ij}}$ or $L_{x_{ij}}$.

We take the subalgebra B generated in the algebra A by the set of elements $\{a, x_{11}, \ldots, x_{1k_1}, x_{21}, \ldots, x_{tk_t}\}$. By assumption, the algebra B is nilpotent. For example, let $B^N = (0)$. Then if in each term on the right side of equality (8) we substitute for a the expression (8) for a, and this procedure is repeated $N-1$ times, we obtain as result $a = 0$. This contradiction proves the proposition.

In this section A is subsequently always an arbitrary alternative algebra, $N = N(A)$, and $Z = Z(A)$.

We define on A the function

$$f(w, x, y, z) = (wx, y, z) - x(w, y, z) - (x, y, z)w.$$

This function is usually called the *Kleinfeld function*.

LEMMA 2. $f(w, x, y, z)$ is a skew-symmetric function of its arguments.

PROOF. It suffices to prove that for any pair of equal arguments $f(w, x, y, z) = 0$. By right alternativity $f(w, x, y, y) = 0$. Furthermore, denoting the left side of equality (5) by $g(w, x, y, z)$, we obtain

$$\begin{aligned} -f(z, w, x, y) &= g(w, x, y, z) - f(z, w, x, y) \\ &= (wx, y, z) + (w, x, yz) - (w, xy, z) - w(x, y, z) \\ &\quad - (w, x, y)z - (zw, x, y) + w(z, x, y) + (w, x, y)z \\ &= (wx, y, z) + (yz, w, x) - (xy, z, w) - (zw, x, y). \end{aligned}$$

Substituting here x, y, z, w in place of w, x, y, z, respectively, we obtain

$$f(w, x, y, z) = (wx, y, z) + (yz, w, x) - (xy, z, w) - (zw, x, y). \tag{9}$$

Hence it is clear that $f(w, x, y, z) = -f(z, w, x, y)$. By means of this equality, using the identity $f(w, x, y, y) = 0$ and its linearization, it is easy to show that $f(w, x, y, z) = 0$ for any two equal arguments. This proves the lemma.

COROLLARY. $f(w, x, y, z) = ([w, x], y, z) + ([y, z], w, x)$.

PROOF. By identity (9) it is sufficient for us to show that

$$(xy, z, w) + (zw, x, y) + (xw, z, y) + (zy, x, w) = 0.$$

This identity is, in fact, valid since it is linearization of the identity $(xy, x, y) = 0$ in both variables. This proves the corollary.

LEMMA 3. Let a, b be elements of an algebra A such that $(a, b, A) = (0)$. Then $[a, b] \in N$.

PROOF. Let x, y be arbitrary elements of the algebra A. Then $f(x, y, a, b) = (xy, a, b) - y(x, a, b) - (y, a, b)x = 0$. On the other hand, by the corollary to Lemma 2 we have $([a, b], x, y) = f(a, b, x, y) - (a, b, [x, y]) = 0$. Consequently $[a, b] \in N$. This proves the lemma.

COROLLARY 1. $[N, A] \subseteq N$.

COROLLARY 2. Let $n \in N$ and $x, y, z \in A$. Then

$$n(x, y, z) = (nx, y, z) = (xn, y, z) = (x, y, z)n. \tag{10}$$

For the proof it is sufficient to apply relations (1)–(3) and Corollary 1.

We denote by $ZN(A)$ the ideal of the algebra A generated by the set $[N, A]$. This ideal is like a measure of the difference between the associative center $N(A)$ and the center $Z(A)$ of an alternative algebra A: $ZN(A) = (0)$ if and only if $N(A) = Z(A)$.

LEMMA 4. $ZN(A) = [N, A]A^{\#} = A^{\#}[N, A]$.

PROOF. It is clear that $ZN(A) \supseteq [N, A]A^{\#}$. For the proof of the reverse containment, it suffices to show that the set $M = [N, A]A^{\#}$ is an ideal of the algebra A. By Corollary 1 to Lemma 3 we have for any $n \in N$, $y \in A^{\#}$, and $x, z \in A$

$$\begin{aligned} y[n, x] &= [n, x]y + [y, [n, x]] \in M, \\ [n, x]y \cdot z &= [n, x](yz) \in M, \\ z \cdot ([n, x])y &= (z[n, x])y \in MA^{\#} \subseteq M. \end{aligned}$$

Consequently, M is an ideal of the algebra A and $M = ZN(A)$. It is proved analogously that $ZN(A) = A^{\#}[N, A]$, which proves the lemma.

LEMMA 5 (*Slater*). $(ZN(A), A, A) \subseteq N$.

PROOF. By Lemma 4 it suffices to show that $([x, n]y, z, t) \in N$ for any $x, y, z, t \in A$ and $n \in N$. Applying relations (3), (4), and (10), we obtain by Corollary 1 to Lemma 3

$$\begin{aligned} ([x, n]y, z, t) = [x, n](y, z, t) &= [x(y, z, t), n] \\ &- x[(y, z, t), n] = [x(y, z, t), n] \in [A, N] \subseteq N. \end{aligned}$$

This proves the lemma.

LEMMA 6. Let $n \in N$ and $nA^{\#}n = (0)$. Then $(n)^2 = (0)$, where (n) is the ideal of the algebra A generated by the element n.

PROOF. We first note that $(n) = A^{\#}nA^{\#}$. In fact, it is easy to see that $(n) \supseteq A^{\#}nA^{\#}$. For the proof of the reverse containment it suffices to show that the set $A^{\#}nA^{\#}$ is an ideal of the algebra A. For any $r \in A$ and $s, t \in A^{\#}$ we have

$$r \cdot snt = rs \cdot nt - (r, s, nt) = rs \cdot nt - n(r, s, t) \in A^{\#}\, nA^{\#},$$

and analogously $snt \cdot r \in A^{\#}nA^{\#}$, which is what was to be proved. For the proof of the lemma it remains for us now to show that $rns \cdot tnv = 0$ for any $r, s, t, v \in A^{\#}$. Applying relation (10) several times, we obtain

$$\begin{aligned} rns \cdot tnv &= r \cdot (ns)(tnv) + (r, ns, tnv) \\ &= r[(ns)(tn) \cdot v] - r(ns, tn, v) + n(r, s, tnv) \\ &= r[n(st)n \cdot v] - r \cdot n(s, t, v)n + (r, s, n \cdot tnv) = 0. \end{aligned}$$

This proves the lemma.

LEMMA 7. Let a, b be elements of an algebra A such that $(a, b, A) \subseteq N$. Then for any element $x \in A$ the element $n = (a, b, x)$ belongs to N and $(n)^2 = (0)$.

PROOF. Let y, z be arbitrary elements of the algebra A. By the linearized Moufang identities and the assumptions of the lemma, we have

$$\begin{aligned} (a, y, z)(a, b, y) = &-(a, (a, b, x), z)y + ((a, b, x)a, y, z) \\ &+ (ya, (a, b, x), z) = ((a, b, ax), y, z) = 0. \end{aligned}$$

Analogously, we obtain

$$(a, b, x)(a, y, z) = 0.$$

Furthermore, we have

$$\begin{aligned} (a, b, x)y(a, b, x) &= -(a, b, y)x \cdot (a, b, x) + (a, yb, x)(a, b, x) \\ &\quad + (a, xb, y)(a, b, x) = -(a, b, y) \cdot x(a, b, x) \\ &= -(a, b, y)(a, bx, x) = 0. \end{aligned}$$

Thus $nA^{\#}n = (0)$, whence by Lemma 6, $(n)^2 = (0)$. This proves the lemma.

COROLLARY. If the algebra A is simple and not associative, then $N = Z$.

PROOF. If $N \nsubseteq Z$, then $ZN(A) \neq (0)$, and therefore by the simplicity of the algebra A we obtain $ZN(A) = A$. Then by Lemma 5 the algebra A satisfies the identity

$$((x, y, z), r, s) = 0. \tag{11}$$

Since A is not associative, there exists an element $n = (x, y, z) \neq 0$, and $n \in N$ by (11). From identity (11) and Lemma 7 it follows that $(n)^2 = (0)$, whence in view of the simplicity of A, it follows $(n) = (0)$ and $n = 0$. This contradiction proves that $ZN(A) = (0)$ and $N = Z$, which proves the corollary.

PROPOSITION 2. Let the identity $[x, y]^n = 0$ hold in the algebra A. Then the nilpotent elements of the algebra A form an ideal.

PROOF. We first consider the case when the algebra A is associative. Let I be the largest nil-ideal of the algebra A (the upper nil-radical). The quotient algebra $\bar{A} = A/I$ does not contain nonzero two-sided nil-ideals. We assume there is a nilpotent element $\bar{x}$ in the algebra $\bar{A}$ with $\bar{x}^2 = 0$. Consider the right ideal $\bar{x}\bar{A}$. For any $\bar{y} \in \bar{A}$ we have $0 = [\bar{x}, \bar{y}]^n \bar{x}\bar{y} = (\bar{x}\bar{y})^{n+1}$, that is, the ideal $\bar{x}\bar{A}$ is a right nil-ideal of index $n + 1$. By Levitzki's theorem (Corollary 2 to the theorem on height) the ideal $\bar{x}\bar{A}$ is locally nilpotent. We note that $\bar{x}\bar{A} \neq (0)$, since otherwise the element $\bar{x}$ generates a nonzero two-sided nil-ideal in the algebra $\bar{A}$, which can not be. But then, as is well known (e.g., see Exercise 4 in Section 5.3), $\mathscr{L}(\bar{A}) \neq (0)$ also, where $\mathscr{L}(\bar{A})$ is the locally nilpotent radical (Levitzki radical) of the algebra $\bar{A}$. The ideal $\mathscr{L}(\bar{A})$ is a nonzero two-sided nil-ideal of the algebra $\bar{A}$. This contradiction proves the proposition in the case when the algebra A is associative.

Now let the algebra A be alternative and x, y be two nilpotent elements of the algebra A. We consider the associative subalgebra B generated by the elements x, y. The algebra B satisfies the assumption of the proposition, and therefore by what was proved above, its nilpotent elements form an ideal. In particular, the element $x + y$ is nilpotent. Analogously one can show that if x is nilpotent and y is an arbitrary element of the algebra A, then the elements xy and yx are nilpotent. Consequently, the nilpotent elements of the algebra A form an ideal.

This proves the proposition.

LEMMA 8. Let the algebra A be commutative. Then

$$(x, y, z)^2 = 0$$

for any elements $x, y, z \in A$.

PROOF. We note first that in A there is the identity

$$4(x, (x, y, z), z) = 0. \tag{12}$$

In fact, we have

$$\begin{aligned} 4[x(x, y, z)]z - 4x[(x, y, z)z] &= [x \circ (x, y, z)] \circ z - x \circ [(x, y, z) \circ z] \\ &= (x^2, y, z^2) - (x^2, y, z^2) = 0. \end{aligned}$$

From Corollary 1 of Lemma 1 we obtain

$$9(x, y, z)^2 = 3(x, y, z) \cdot 3(x, y, z) = 0.$$

Next we have by the Moufang identities and (12)

$$\begin{aligned} 8(x, y, z)^2 &= 4(x, y, z) \circ (x, y, z) \\ &= 4(x, (x, y, z) \circ y, z) - 4y \circ (x, (x, y, z), z) \\ &= 4(x, (x, y^2, z), z) = 0. \end{aligned}$$

Finally, we obtain

$$(x, y, z)^2 = 0.$$

This proves the lemma.

THEOREM 3 (*Zhevlakov*). A simple commutative alternative algebra A is a field.

PROOF. By Proposition 2 the nilpotent elements of the algebra A form an ideal I. If A is not associative, then by Lemma 8 $I \neq (0)$, whence $I = A$ and A is a nil-algebra. The algebra A is commutative and therefore satisfies an essential polynomial identity. By Theorem 5.6, A is locally nilpotent, which in view of Proposition 1 contradicts the simplicity of A. Consequently the algebra A is associative and, as is well known, is a field.

This proves the theorem.

Exercises

1. Let A be an alternative algebra. Prove that if $k \in K(A)$, then $k^3 \in Z(A)$.

2. Let A be a commutative alternative algebra that is not associative. Prove that for any elements $a, b, c \in A$ the associator (a, b, c) generates a nil-ideal I in A of index 3. *Hint*. Prove that for any $a, b \in A$ and $x, y \in I$ the relations $(ab)^3 = a^3b^3$ and $(x + y)^3 = x^3 + y^3$ are valid.

3. Let A be a locally nilpotent algebra and I be a minimal ideal of the algebra A. Prove that $IA = AI = (0)$.

4. Prove that in an alternative algebra A the following conditions are equivalent:

(a) $ZN(A) \subseteq N(A)$;
(b) $(A[A, N], A, A) = (0)$;
(c) $(A, A, A)[A, N] = (0)$.

5. Let A be an alternative algebra and $n \in N(A)$. Then the following conditions are equivalent:

(a) $(n) \subseteq N(A)$;
(b) $(A, A, A)n = (0)$.

6. Let A be an alternative algebra, $x, y, z \in A$, and $n \in N(A)$. Prove that

$$(x, y, z)[n, z] = 0.$$

Hint. Use Corollaries 1 and 2 to Lemma 3 and relation (4).

7. (*Slater*) Prove that for any elements $n, m \in N(A)$ in an alternative algebra A the ideal $([m, n])$ generated by the commutator $[m, n]$ is contained in $N(A)$. *Hint*. Use Exercises 6 and 5.

2. ELEMENTS OF THE ASSOCIATIVE CENTER

Let $\mathfrak{M}$ be some variety of Φ-algebras. We shall denote by $N_{\mathfrak{M}}$, $K_{\mathfrak{M}}$, and $Z_{\mathfrak{M}}$ the associative center, commutative center, and center, respectively, of the $\mathfrak{M}$-free algebra $\Phi_{\mathfrak{M}}[X]$ from a countable set of generators $X = \{x_1, x_2, \ldots\}$. If P is some subset of the algebra $\Phi_{\mathfrak{M}}[X]$ and A is an arbitrary algebra from $\mathfrak{M}$, then we shall denote by $P[A]$ the collection of elements from A of the form $p(a_1, \ldots, a_n)$, where $p(x_1, \ldots, x_n) \in P$ and $a_1, \ldots, a_n \in A$. It is easy to see that $N_{\mathfrak{M}}[A] \subseteq N(A)$, $K_{\mathfrak{M}}[A] \subseteq K(A)$, and $Z_{\mathfrak{M}}[A] \subseteq Z(A)$ for any algebra A from $\mathfrak{M}$. In addition, generally speaking these containments can be strict.

PROPOSITION 3. Let $\mathfrak{M}$ be a homogeneous variety. Then the subsets $N_{\mathfrak{M}}$, $K_{\mathfrak{M}}$, $Z_{\mathfrak{M}}$ of the algebra $\Phi_{\mathfrak{M}}[X]$ are invariant under the operators of partial linearization.

PROOF. We show, for example, that for any element $n = n(x_1, \ldots, x_m) \in N_{\mathfrak{M}}$ all partial linearizations of the element n also belong to the set $N_{\mathfrak{M}}$. Let $n_i^{(k)}(x_1, \ldots, x_m; x_j) = n\,\Delta_i^k(x_j)$ be a partial linearization of the element n in x_i of degree k (see Section 1.4), and let $x_r, x_l \in X \setminus \{x_1, \ldots, x_m, x_j\}$. Since $n \in N_{\mathfrak{M}}$, then

$$(n, x_r, x_l) = (x_r, n, x_l) = (x_r, x_l, n) = 0,$$

whence by Lemma 1.2 we obtain

$$0 = (n, x_r, x_l)\,\Delta_i^k(x_j) = (n\,\Delta_i^k(x_j), x_r, x_l),$$

and analogously

$$0 = (x_r, n\,\Delta_i^k(x_j), x_l) = (x_r, x_l, n\,\Delta_i^k(x_j)).$$

This means $n\,\Delta_i^k(x_j) \in N_{\mathfrak{M}}$. The cases for the subsets $K_{\mathfrak{M}}$ and $Z_{\mathfrak{M}}$ are considered in exactly the same way. This proves the proposition.

If $\mathfrak{M} = \text{Ass}$ is the variety of associative algebras, it is easy to see $N_{\text{Ass}} = \text{Ass}[X]$ and $K_{\text{Ass}} = Z_{\text{Ass}} = (0)$. In the case of the variety Alt of alternative algebras, it turns out that $N_{\text{Alt}} \neq (0)$, $K_{\text{Alt}} \neq (0)$, and $Z_{\text{Alt}} \neq 0$. We shall consider the cases of the subalgebras K_{Alt} and Z_{Alt} later, but in this section we shall find a series of nonzero elements from N_{Alt} which we need for the proof of the theorem on simple alternative algebras.

Henceforth A is an arbitrary alternative algebra.

LEMMA 9. In the algebra A there are the identities

$$(x, y, (x, y, z)) = [x, y](x, y, z), \tag{13}$$

$$((x, y, z), x, y) = -(x, y, z)[x, y], \tag{14}$$

$$(x, y, z) \circ [x, y] = 0. \tag{15}$$

PROOF. We first note that in A there is the identity

$$(xy)(x, y, z) = y[x(x, y, z)]. \tag{16}$$

In fact, by the left Moufang identity and its linearization we have

$$\begin{aligned}(xy)[(xy)z] - (xy)[x(yz)] - y\{x[(xy)z]\} + y\{x[x(yz)]\} \\ = (xy)^2 z - [(xy)xy + yx(xy)]z + y[x^2(yz)] \\ = [(xy)^2 - (xy)^2 - yx^2y + yx^2y]z = 0.\end{aligned}$$

Now we have

$$\begin{aligned}(y, x, (x, y, z)) &= (yx)(x, y, z) - y(x(x, y, z)) \\ &= (yx)(x, y, z) - (xy)(x, y, z) = [y, x](x, y, z),\end{aligned}$$

that is, (13) is proved. Analogously we obtain identity (14). Finally, by (13) and (14) we have

$$\begin{aligned}(x, y, z) \circ [x, y] &= (x, y, z)[x, y] + [x, y](x, y, z) \\ &= -((x, y, z), x, y) + (x, y, (x, y, z)) = 0,\end{aligned}$$

that is, (15) is also proved. This proves the lemma.

LEMMA 10. Let $x, y \in A$. Then for any elements u, v from the subalgebra generated by the elements x, y there is the containment $(A, u, v) \subseteq (A, x, y)$.

PROOF. By Artin's theorem it suffices to prove that for any associative words $u_1(x, y)$, $u_2(x, y)$ there is the containment $(A, u_1, u_2) \subseteq (A, x, y)$. We shall prove this assertion by induction on the number $n = d(u_1) + d(u_2)$. The basis for the induction, $n = 2$, is obvious. Now let $n = k > 2$ and the assertion of the lemma be valid for all $n < k$. We shall write $r \equiv s$ if $r - s \in (A, x, y)$. We note that in every alternative algebra there is the identity

$$(z, x, yxy) = (z, xyx, y). \tag{17}$$

In fact, by the Moufang identities we have

$$\begin{aligned}(z, x, yxy) - (z, xyx, y) &= (zx)(yxy) - z(xyxy) \\ &\quad - [z(xyx)]y + z(xyxy) \\ &= \{[(zx)y]x\}y - \{[(zx)y]x\}y = 0.\end{aligned}$$

Linearizing (17) in x we obtain

$$(z, t, yxy) = (z, \{xyt\}, y) + (z, yty, x), \tag{18}$$

where $\{xyt\} = (xy)t + (ty)x$. We consider now two possible cases.

1. One of the words u_1, u_2 begins and ends with the same element. For example, let $u_1 = xv_1x$. If v_1 is not the empty word, then in view of (18) and by the induction assumption for any $z \in A$ we obtain

$$(z, u_1, u_2) = (z, xv_1x, u_2) = (xzx, v_1, u_2) + (\{zxv_1\}, x, u_2) \equiv 0.$$

If the word v_1 is missing, then we have

$$(z, u_1, u_2) = (z, x^2, u_2) = (z \circ x, x, u_2) \equiv 0.$$

2. At the beginning or end of the word u_2 there is the same element as at the beginning of the word u_1. For example, let $u_1 = xv_1$ and $u_2 = xv_2$. Applying the Moufang identities and their linearizations, we obtain by means of the induction assumption and case 1

$$\begin{aligned}(z, xv_1, xv_2) &= -(z, xv_1, v_2x) + (z, xv_1, x \circ v_2) \\ &= -(z, xv_1, v_2x) + (z \circ x, xv_1, v_2) + (z \circ v_2, xv_1, x) \\ &\equiv -(z, xv_1, v_2x) \\ &= (z, v_2, xv_1x) - (v_2z, xv_1, x) - ((xv_1)z, v_2, x) \equiv 0.\end{aligned}$$

This proves the lemma.

COROLLARY. Let $x, y, z \in A$. Then for any elements u_i, v_i from the subalgebra generated by the elements x, y, where $i = 1, 2$, there is the relation

$$(u_1, v_1, z) \circ [u_2, v_2] = 0. \tag{19}$$

PROOF. By Artin's theorem it suffices to prove that for any associative words $u(x, y)$, $v(x, y)$ there is the equality

$$(u_1, v_1, z) \circ [u, v] = 0. \tag{20}$$

We shall prove this equality by induction on the number $n = d(u) + d(v)$. The validity for $n = 2$, the basis of the induction, follows from Lemma 10 and equality (15). Now let $n = k > 2$ and equality (20) be valid if $n < k$. We note that by Lemma 10 we have $(u_1, v_1, z) = (x, y, z_1)$. If $d(u) > 1$, $d(v) = 1$, and $v = x$, for example, then by linearized identity (15) and the induction assumption we have

$$(u_1, v_1, z) \circ [u, v] = (x, y, z_1) \circ [u, x] = -(x, u, z_1) \circ [y, x] = 0.$$

If $d(u) > 1$ and $d(v) > 1$, then analogously we obtain

$$\begin{aligned}(u_1, v_1, z) \circ [u, v] - (x, y, z_1) \circ [u, v] \\ = -(u, y, z_1) \circ [x, v] - (x, v, z_1) \circ [u, y] - (u, v, z_1) \circ [x, y] \\ = 0.\end{aligned}$$

This proves the corollary.

THEOREM 4. Let $x, y \in A$ and B be the subalgebra of the algebra A generated by the elements x, y. Then for any $u_i, v_i \in B$, $r, s \in A$, and $w_i = [u_i, v_i]$ there are the equalities

$$(w_1 \circ w_2, r, s)w_3 = w_3(w_1 \circ w_2, r, s) = 0, \tag{21}$$

$$(w_1^2, r, s)w_2 = w_2(w_1^2, r, s) = 0, \tag{22}$$

$$(w_1^2(w_2 \circ w_3), r, s) = ((w_2 \circ w_3)w_1^2, r, s) = 0, \tag{23}$$

$$(w_1^2 w_2^2, r, s) = 0. \tag{24}$$

PROOF. We note first that for any $r \in A$ and $u \in B$

$$(w_1 \circ w_2, u, r) = (w_1^2, u, r) = 0. \tag{25}$$

In fact, by the linearized Moufang identity and identity (19) we obtain

$$(w_1 \circ w_2, u, r) = w_1 \circ (w_2, u, r) + w_2 \circ (w_1, u, r) = 0,$$

and the second equality is obtained analogously. Now from linearized identity (14), in view of (25) we obtain

$$\begin{aligned}(w_1 \circ w_2, r, s)w_3 &= (w_1 \circ w_2, r, s)[u_3, v_3] \\ &= -(w_1 \circ w_2, u_3, s)[r, v_3] - (w_1 \circ w_2, r, v_3)[u_3, s] \\ &\quad -(w_1 \circ w_2, u_3, v_3)[r, s] - ((w_1 \circ w_2, r, s), u_3, v_3) \\ &\quad -((w_1 \circ w_2, u_3, s), r, v_3) - ((w_1 \circ w_2, r, v_3), u_3, s) \\ &\quad -((w_1 \circ w_3, u_3, v_3), r, s) = -((w_1 \circ w_2, r, s), u_3, v_3),\end{aligned}$$

and furthermore, again by the linearized Moufang identity (19), and Artin's theorem,

$$\begin{aligned}((w_1 \circ w_2, r, s), u_3, v_3) &= (w_1 \circ (w_2, r, s), u_3, v_3) + (w_2 \circ (w_1, r, s), u_3, v_3) \\ &= w_1 \circ ((w_2, r, s), u_3, v_3) + (w_2, r, s) \circ (w_1, u_3, v_3) \\ &\quad + w_2 \circ ((w_1, r, s), u_3, v_3) + (w_1, r, s) \circ (w_2, u_3, v_3) = 0,\end{aligned}$$

which proves the first part of identity (21). The second part of identity (21) and identity (22) are proved analogously. In addition, from (25) we obtain

$$\begin{aligned}f(r, s, w_1^2, w_2 \circ w_3) = (rs, w_1^2, w_2 \circ w_3) &- s(r, w_1^2, w_2 \circ w_3) \\ &-(s, w_1^2, w_2 \circ w_3)r = 0.\end{aligned}$$

Consequently, in view of the left and right alternative identities, (21), and (22), we have

$$\begin{aligned}(w_1^2(w_2 \circ w_3), r, s) &= f(w_1^2, w_2 \circ w_3, r, s) + (w_2 \circ w_3)(w_1^2, r, s) + (w_2 \circ w_3, r, s)w_1^2 \\ &= w_2[w_3(w_1^2, r, s)] + w_3[w_2(w_1^2, r, s)] + [(w_2 \circ w_3, r, s)w_1]w_1 \\ &= 0,\end{aligned}$$

that is, the first part of identity (23) is proved. The second part of identity (23) and identity (24) are proved analogously. This proves the theorem.

COROLLARY 1 (*Kleinfeld*). In every alternative algebra there are the identities (Kleinfeld identities)

$$\begin{aligned}[x, y]([x, y]^2, r, s) &= 0, \\ ([x, y]^2, r, s)[x, y] &= 0, \\ ([x, y]^4, r, s) &= 0.\end{aligned}$$

Before formulating a second corollary, we define the functions

$$\begin{aligned}n_1(x, y) &= [x, y]^4, \\ n_2(x, y) &= [x, y]^2([x, y] \circ [x, yx]), \\ n_3(x, y) &= [x, y]^2[x, yx]^2.\end{aligned}$$

COROLLARY 2 (*Shestakov*). The elements $n_i(x_1, x_2)$ belong to the set N_{Alt} for $i = 1, 2, 3$. In addition, the following relation is valid:

$$n_1(x_1, x_2)x_1^2 - n_2(x_1, x_2)x_1 + n_3(x_1, x_2) = 0. \tag{26}$$

In fact, by Theorem 4 $n_i(x_1, x_2) \in N_{\mathrm{Alt}}$, and relation (26) is valid by Artin's theorem.

We note that the elements $n_i(x_1, x_2)$ are different from zero in the free associative algebra $\mathrm{Ass}[X]$, and therefore they are nonzero in the algebra $\mathrm{Alt}[X]$. In particular, $N_{\mathrm{Alt}} \neq (0)$.

Exercises

In Exercises 1–6 A is an alternative algebra, x, y, z, r, s, t are arbitrary elements from A, $v = [x, y]$, and $w = (r, s, t)$.

1. Prove that $v^{2i} (i \geq 2)$, $v^i[v^2, z](i \geq 1)$, $[v^4, z]w \in N_{\text{Alt}}[A]$. *Hint.* Apply Lemmas 3 and 5.

2. Prove that $v[w, v^2] = 0$, $[v[v^2, z], w] = 0$. *Hint.* Use identity (3.2).

3. Prove that $(v^2, r, s)^2 = 0$. *Hint.* First prove that $(v^2, (v^2, r, s), t) = 0$, then use the Moufang identities and their linearizations.

4. Let $(x, y, z) = 0$. Prove that $[[x, y]^2, x] \in N(A)$. *Hint.* Apply Lemma 3, Lemma 2 and its corollary, and also linearized identity (15).

5. Prove that $(v^2 t)(v^2, r, s) = (tv^2)(v^2, r, s) = 0$. *Hint.* Using linearized identity (16), prove that $[v(vt)](v, r, v \circ s) = 0$.

6. (*Shestakov*) Prove that $[[x, y]^2, z]^2 \in N_{\text{Alt}}[A]$. *Hint.* Prove that $[v^2, z] \circ f(v^2, z, r, s) = 0$.

7. Prove that $N_{\text{Alt}}(A) = (0)$ for any associative–commutative algebra A. *Hint.* Consider the Cayley–Dickson matrix algebra $\boldsymbol{C}(A)$.

8. Let A be an alternative algebra with an identity element over a field F, $N(A) = F$, and let there be elements a,b in A such that $[a, b]^4 \neq 0$. Prove that the commutative center $K(A)$ of the algebra A equals F.

3. KLEINFELD'S THEOREM

In this section we shall prove Kleinfeld's theorem on the structure of simple alternative algebras that are not associative.

LEMMA 11. Let A be an algebra with identity element 1 over a field F and K be some extension of the field F. Then if the algebra $\tilde{A} = K \otimes_F A$ is quadratic over K, the algebra A is quadratic over F. In addition, if $\tilde{A}$ is a composition algebra with respect to the norm $\tilde{n}(x)$, then A is also a composition algebra with respect to the restriction $n(x)$ to A of the norm $\tilde{n}(x)$.

PROOF. By assumption each element $x \in \tilde{A}$ satisfies the relation $x^2 - \tilde{t}(x)x + \tilde{n}(x) = 0$, where $\tilde{t}(x)$, $\tilde{n}(x) \in K$. We denote by $t(x)$ and $n(x)$ the restrictions of the trace $\tilde{t}(x)$ and norm $\tilde{n}(x)$ to A. For the proof that A is quadratic it suffices for us to show that $t(x), n(x) \in F$ for any $x \in A$. If $x \in F$, then $t(x) = 2x \in F$ and $n(x) = x^2 \in F$. Now let $x \notin F$, that is, 1 and x are linearly independent. In this case a basis can be selected in A of the form $\{1, x, e_1, e_2, \ldots\}$, that is, containing the elements 1 and x. As is known, this basis is also a basis for $\tilde{A}$ over K. We consider the element x^2. On the one hand, since $x^2 \in A$ then $x^2 = \alpha 1 + \beta x + \sum_i \gamma_i e_i$ where α, β, $\gamma_i \in F$. On the other hand, we have $x^2 = t(x)x - n(x) \cdot 1$. By the uniqueness of the representation for the element x^2 in the basis $\{1, x, e_1, e_2, \ldots\}$ of the algebra $\tilde{A}$ we obtain $t(x) = \beta$, $n(x) = -\alpha$, that is, $t(x), n(x) \in F$. By the same token, this proves that the algebra A is quadratic over F.

For completion of the proof of the lemma it remains for us to prove that if the form $\tilde{n}(x)$ is strictly nondegenerate on $\tilde{A}$, then the form $n(x)$ is strictly nondegenerate on A. Let $a \in A$ and $n(a+x) - n(a) - n(x) = 0$ for all $x \in A$. Then by the bilinearity of the form $\tilde{f}(x, y) = \tilde{n}(x+y) - \tilde{n}(x) - \tilde{n}(y)$ and in view of the fact that $\tilde{f} \equiv f$ on A, we obtain that $\tilde{f}(a, \tilde{A}) = 0$. Hence $a = 0$ and the form $n(x)$ is strictly nondegenerate.

This proves the lemma.

LEMMA 12. Let A be an alternative algebra with identity element 1 over an infinite field F. Assume that $n_i(x, y) \in F$ for any $x, y \in A$ and $i = 1, 2, 3$, and also $n_1(a, b) = [a, b]^4 \neq 0$ for some elements $a, b \in A$. Then the algebra A is quadratic over F.

PROOF. We first note that if $x, y \in A$ and $y^4 \neq 0$, then there exist distinct elements $\beta_i \in F$, $i = 1, 2$, such that $(x + \beta_i y)^4 \neq 0$. This follows easily from the nonfiniteness of the field F. Now let a, b be elements from A for which $[a, b]^4 \neq 0$, and x be an arbitrary element of the algebra A. In view of the symmetry of the elements a, b, we can assume that either $[x, a] \neq 0$ or $[x, b] = 0$. We select distinct elements β_1, β_2 from F such that $([b, x] + \beta_i[b, a])^4 = [b, x + \beta_i a]^4 \neq 0$. By relation (26) there exists $\alpha_i \in F$ such that

$$(x + \beta_i a)^2 + \alpha_i(x + \beta_i a) \in F, \qquad i = 1, 2.$$

Since $[a, b]^4 \neq 0$, then analogously we have $a^2 \in F + Fa$, whence

$$x^2 + \alpha_i x + \beta_i y \in F + Fa,$$

where $y = x \circ a$. Eliminating y from here and dividing by $\beta_2 - \beta_1$, we obtain

$$x^2 + \alpha x + \gamma = \theta a$$

for suitable $\alpha, \gamma, \theta \in F$. If $[x, b] = 0$, we commute this relation with b and obtain $\theta[a, b] = 0$. If $[x, a] \neq 0$, we commute this expression with x and obtain $\theta[x, a] = 0$. In both cases $\theta = 0$ and $x^2 + \alpha x + \gamma = 0$. Since the element x was arbitrary, by the same token this proves the algebra A is quadratic over F. This proves the lemma.

Now can be proved

THEOREM 5. Let A be a simple noncommutative alternative algebra and $N_{\mathrm{Alt}}[A] \subseteq Z(A)$. Then the center $Z(A)$ of the algebra A is a field, and A is either the algebra of generalized quaternions or a Cayley–Dickson algebra over its center.

PROOF. By Theorem 1, for the proof that the center $Z(A)$ is a field, it suffices for us to show that $Z(A) \neq (0)$. In view of the assumptions of our theorem and Corollary 2 to Theorem 4, for this it suffices to show there are elements a, b in A for which $[a,b]^4 \neq 0$. Suppose this is not so, that is, $[x,y]^4 = 0$ for any $x, y \in A$. By Proposition 2 the nilpotent elements of the algebra A form an ideal I, and in addition $[x,y] \in I$ for any $x, y \in A$. The algebra A is noncommutative. Therefore $I \neq (0)$, and $I = A$ because of the simplicity of the algebra A. Consequently A is a nil-algebra. Since A satisfies the essential polynomial identity $[x,y]^4 = 0$, then by Theorem 5.6 A is locally nilpotent. But this is impossible in view of Proposition 1. This contradiction proves that there exist $a, b \in A$ for which $[a,b]^4 \neq 0$. Hence $(0) \neq N_{\mathrm{Alt}}[A] \subseteq Z(A)$, and $Z = Z(A)$ is a field.

We shall now consider A as an algebra over the field Z. As is easy to see, A is simple not only as a Φ-algebra but also as a ring. In particular, A is simple over Z. Let K be some infinite extension of the field Z. By Theorem 1.6 and Theorem 2 the algebra $A_K = K \otimes_Z A$ is a central simple alternative algebra over the field K. We now note that under the conditions of the theorem there are satisfied in the algebra A the identities

$$(n_i(x,y),z,t) = [n_i(x,y),z] = 0, \qquad i = 1,2,3. \tag{27}$$

In view of Proposition 3, all partial linearizations of these identities are also satisfied in A. By Theorems 1.4 and 1.6 it follows that identities (27) are valid in the algebra A_K. Consequently, $n_i(x,y) \in Z(A_K) = K$ for any $x, y \in A_K$. In addition, there exist $a, b \in A \subseteq A_K$ such that $[a,b]^4 \neq 0$. By Lemma 12 the algebra A_K is quadratic over K, whence in view of the noncommutativity, A_K is a composition algebra by Theorem 2.4. By Lemma 11, A is also a composition algebra. It only remains to use Theorem 2.1. This proves the theorem.

COROLLARY 1 (*Kleinfeld's theorem*). Let A be a simple alternative algebra that is not associative. Then the center of the algebra A is a field, and A is a Cayley–Dickson algebra over its center.

In fact, by Theorem 3 the algebra A is noncommutative, and by the corollary to Lemma 7 $N_{\mathrm{Alt}}[A] \subseteq N(A) = Z(A)$, that is, the conditions of Theorem 5 are satisfied. It only remains to note that, in view of the nonassociativity, A cannot be the algebra of generalized quaternions.

An algebra A is called an *algebra with division* if, for any elements $a, b \in A$ with $a \neq 0$, each of the equations

$$ax = b, \qquad ya = b$$

is solvable in A. If each of these equations with $a \neq 0$ has one and only one solution, and A has an identity element, then A is called a *division ring*.

As is easy to see (see Exercise 1), every alternative algebra with division is a division ring.

COROLLARY 2 (*Bruck, Kleinfeld, Skornyakov*). Let A be an alternative algebra with division that is not associative. Then the center of the algebra A is a field, and A is a Cayley–Dickson algebra over its center.

Exercises

1. Prove that every alternative algebra with division is a division ring.

An element a of a ring R is called *completely invertible* if there exists an element a^{-1} in R, called the *complete inverse* for a, such that

$$a^{-1}(ax) = (xa)a^{-1} = x$$

for each $x \in R$.

2. Let R be an alternative division ring. Then all nonzero elements from R are completely invertible.

The converse assertion is also valid: If all nonzero elements of an arbitrary ring R with identity element are completely invertible, then R is an alternative division ring (see A. I. Mal'cev, "Algebraic Systems," p. 119).

3. Prove that in an alternative division ring the subring generated by any two elements is an associative division ring.

4. Every finite alternative division ring is a field. *Hint*. Apply Exercise 3 and Wedderburn's theorem for finite divisions rings (I. Herstein, "Noncommutative Rings," p. 100, Mir, 1972).

COROLLARY. Every Cayley–Dickson algebra over a finite field is split.

LITERATURE

Albert [6], Bruck and Kleinfeld [24], Zhevlakov [70], Kleinfeld [92, 93], Skornyakov [190], Slater [197, 199, 205], Herstein [250, 251], Zorn [258], Shestakov [270].

CHAPTER **8**

Radicals of Alternative Algebras

In the theory of rings the concept of a radical is most important. It consists of the following. Let us assume that it is necessary to describe the rings of some class of rings $\mathfrak{K}$, for example the class of alternative Φ-algebras. As a rule, the class $\mathfrak{K}$ contains very heterogeneous elements. It can contain both division rings and nilpotent rings, and so to find the general features in the structure of the rings of the class $\mathfrak{K}$ usually proves to be impossible. In this case we act in the following manner. In each ring $K \in \mathfrak{K}$ we fix some ideal $\mathscr{R}(K)$, called the radical, so that individually the radicals $\mathscr{R}(K)$ and the quotient rings $K/\mathscr{R}(K)$ have a consistent structure, and then we describe the classes of rings $\mathscr{R} = \{\mathscr{R}(K)\}$ and $\mathscr{P} = \{K/\mathscr{R}(K)\}$. Thus an arbitrary ring of the class $\mathfrak{K}$ is described as the extension of a ring from $\mathscr{P}$ by means of a ring from $\mathscr{R}$. The discovery of this method belongs to Wedderburn. In 1908 he proved that every finite-dimensional associative algebra is the extension of a direct sum of complete matrix algebras over division rings by means of a nilpotent algebra.

At the present time a series of radicals, which allow the proof of deep structure theorems, have been discovered for the class of alternative rings. The quasi-regular radical of Zhevlakov, which is the analog of the Jacobson radical in the class of associative rings, occupies a particularly important place in this series. Chapter 10 will be devoted to this radical.

1. ELEMENTS OF THE GENERAL THEORY OF RADICALS

In 1953 Kurosh and, independently, Amitsur laid the foundation for an axiomatic study of the concept of a radical. Having as our goal the proof of structure theorems on alternative and Jordan rings, we shall not delve deeply into this theory, but shall restrict ourselves to only a necessary minimum of information.

We fix some class of Φ-algebras $\mathfrak{K}$, which is closed with respect to ideals and homomorphic images. In the course of this section we shall assume that all algebras considered belong to $\mathfrak{K}$. Let $\mathscr{R}$ be some subclass of the class $\mathfrak{K}$. We shall call algebras of the class $\mathscr{R}$ $\mathscr{R}$-algebras for short. We shall call an ideal I of an algebra A an $\mathscr{R}$-ideal if I is an $\mathscr{R}$-algebra. The class $\mathscr{R}$ (for an algebra the property of belonging to the class $\mathscr{R}$) is called *radical* in the class of algebras $\mathfrak{K}$ if the following three conditions are satisfied:

(A) the homomorphic image of an $\mathscr{R}$-algebra is an $\mathscr{R}$-algebra;

(B) each algebra A from $\mathscr{R}$ contains an $\mathscr{R}$-ideal $\mathscr{R}(A)$, which contains all $\mathscr{R}$-ideals of the algebra A;

(C) the quotient algebra $A/\mathscr{R}(A)$ does not contain nonzero $\mathscr{R}$-ideals.

In this case the mapping $A \mapsto \mathscr{R}(A)$ is called a *radical* defined on the class of algebras $\mathfrak{K}$. We shall also denote it by the letter $\mathscr{R}$.

The ideal $\mathscr{R}(A)$ of the algebra A is called its *$\mathscr{R}$-radical*. Algebras coinciding with their $\mathscr{R}$-radical are called *$\mathscr{R}$-radical*, and nonzero algebras for which the radical equals zero are called *$\mathscr{R}$-semisimple*. The class $\mathscr{P}$ of all $\mathscr{R}$-semisimple algebras of the class $\mathfrak{K}$ is called the *semisimple class of the radical $\mathscr{R}$*. A radical is uniquely determined not only by its radical class, but also by its semisimple class, since is valid

PROPOSITION 1. A radical class $\mathscr{R}$ is the collection of algebras from $\mathfrak{K}$ which are not mapped homomorphically onto algebras of the class $\mathscr{P}$.

PROOF. If $A \in \mathscr{R}$, then it cannot be mapped homomorphically onto an algebra from $\mathscr{P}$ by condition (A). If $A \notin \mathscr{R}$, then $A \neq \mathscr{R}(A)$ and $A/\mathscr{R}(A)$ is a homomorphic image of the algebra A belonging to $\mathscr{P}$. This proves the proposition.

Verification of conditions (A)–(C) often turns out to be difficult. Sometimes it is much easier to establish that a class $\mathscr{R}$ is radical by using the criterion formulated in the following theorem.

THEOREM 1. The class $\mathscr{R}$ is radical if and only if the following conditions are satisfied:

(A) the homomorphic image of an $\mathscr{R}$-algebra is an $\mathscr{R}$-algebra;

(D) if every nonzero homomorphic image of an algebra A contains a nonzero $\mathscr{R}$-ideal, then A is an $\mathscr{R}$-algebra.

PROOF. If $\mathscr{R}$ is a radical class, then condition (A) is satisfied, and it is only necessary to prove condition (D). We assume there is a nonzero $\mathscr{R}$-ideal in each nonzero homomorphic image of the algebra A. We consider the radical $\mathscr{R}(A)$ in the algebra A and the quotient algebra $A/\mathscr{R}(A)$. By property (C) this homomorphic image does not contain nonzero $\mathscr{R}$-ideals, and this means it is zero. Consequently $A = \mathscr{R}(A)$, and A is an $\mathscr{R}$-algebra.

Now let us assume that the class $\mathscr{R}$ satisfies conditions (A) and (D). We shall prove that it is radical. In order to establish (B), we consider the ideal R which is the sum of all $\mathscr{R}$-ideals of the algebra A. It is necessary to prove that R is an $\mathscr{R}$-ideal. We assume that R is not an $\mathscr{R}$-algebra. Then by (D) there is a nonzero homomorphic image R/I of the algebra R which does not contain nonzero $\mathscr{R}$-ideals. This is impossible for the following reason. Since $I \neq R$, there exists an $\mathscr{R}$-ideal B of the algebra A which is not contained in I. In view of the fact that $B \subseteq R$, B is an $\mathscr{R}$-ideal of the algebra R. By the second homomorphism theorem $B + I/I \cong B/B \cap I$, and by condition (A) $B + I/I$ is a nonzero $\mathscr{R}$-ideal in R/I. The obtained contradiction proves that condition (B) is satisfied.

We shall prove that (C) is satisfied. Let A be an arbitrary algebra. By the already proved condition (B) there exists in A an $\mathscr{R}$-ideal R containing all $\mathscr{R}$-ideals of the algebra A. Let us assume that A/R contains a nonzero $\mathscr{R}$-ideal M/R. We now prove that $M \in \mathscr{R}$, and obtain with this the contradiction that $M \nsubseteq R$. Let M/N be a homomorphic image of the algebra M. If $N \supseteq R$, then M/N is a homomorphic image of the $\mathscr{R}$-algebra M/R, and therefore is an $\mathscr{R}$-algebra. If $N \nsupseteq R$, then by the second homomorphism theorem

$$N + R/N \cong R/R \cap N,$$

and by condition (A) $N + R/N$ is a nonzero $\mathscr{R}$-ideal of the algebra M/N. We see that the algebra M satisfies the premise of condition (D), and therefore $M \in \mathscr{R}$. Thus condition (C) is satisfied, and the theorem is proved.

THEOREM 2. A class $\mathscr{P}$ which does not contain the zero algebra is a semisimple class for some radical $\mathscr{R}$ if and only if $\mathscr{P}$ satisfies the conditions:

(E) every nonzero ideal of an algebra from $\mathscr{P}$ can be mapped homomorphically onto an algebra from $\mathscr{P}$;

(F) if every nonzero ideal of an algebra A can be mapped homomorphically onto an algebra from $\mathscr{P}$, then $A \in \mathscr{P}$.

PROOF. Let $\mathscr{P}$ be the semisimple class of some radical $\mathscr{R}$. Let $A \in \mathscr{P}$ and I be an ideal in A. Since A is $\mathscr{R}$-semisimple, the ideal I is not $\mathscr{R}$-radical, and by Proposition 1 it can be mapped homomorphically onto an algebra from $\mathscr{P}$. This means $\mathscr{P}$ satisfies condition (E).

If $A \notin \mathscr{P}$, then $\mathscr{R}(A) \neq (0)$, and $\mathscr{R}(A)$ is an ideal which cannot be mapped homomorphically onto an algebra from $\mathscr{P}$. This argument proves (F).

Conversely, let $\mathscr{P}$ be a class of algebras from $\mathfrak{K}$ satisfying conditions (E) and (F) and not containing the zero algebra. We denote by $\mathscr{R}$ the class of algebras which cannot be mapped homomorphically onto algebras from $\mathscr{P}$. We shall prove that $\mathscr{R}$ is a radical class and $\mathscr{P}$ is the semisimple class of the radical $\mathscr{R}$. It is obvious condition (A) is valid for the class $\mathscr{R}$. In order to prove (D) we assume that the algebra A is such that in any of its nonzero homomorphic images there is a nonzero $\mathscr{R}$-ideal. If $A \notin \mathscr{R}$, then A can be mapped homomorphically onto an algebra $A' \in \mathscr{P}$. By the assumption just made, the algebra A' possesses an $\mathscr{R}$-ideal $I \neq (0)$. According to (E) I can be mapped homomorphically onto an algebra from $\mathscr{P}$. However, this contradicts that $I \in \mathscr{R}$. Thus the class $\mathscr{R}$ is radical.

It remains to prove that $\mathscr{P}$ is the semisimple class of this radical. If A is an algebra from $\mathscr{P}$, then according to (E) no ideal of the algebra A is radical. Consequently A is an $\mathscr{R}$-semisimple algebra. If A is an $\mathscr{R}$-semisimple algebra, then no nonzero ideal of A is $\mathscr{R}$-radical. This means that every ideal $I \neq (0)$ of the algebra A can be mapped homomorphically onto an algebra from $\mathscr{P}$, and hence by (F) it follows that $A \in \mathscr{P}$. We have proved that the semisimple class of the radical $\mathscr{R}$ coincides with $\mathscr{P}$. Thus the theorem is proved.

We shall say that a class of algebras $\mathscr{R}$ is *closed with respect to extension* if from the fact that $I \in \mathscr{R}$ and $A/I \in \mathscr{R}$ it follows that $A \in \mathscr{R}$.

PROPOSITION 2. Every radical class $\mathscr{R}$ is closed with respect to extension.

PROOF. Let us assume that for some algebra A and some ideal I of A the algebras I and A/I belong to $\mathscr{R}$ but nevertheless $A \notin \mathscr{R}$. In this case $I \subseteq \mathscr{R}(A)$ and $\mathscr{R}(A) \neq A$. By the first homomorphism theorem

$$A/\mathscr{R}(A) \cong A/I/\mathscr{R}(A)/I,$$

and the desired contradiction is obtained, since on the right there is a homomorphic image of A/I, that is, a radical algebra, but on the left there is a semisimple algebra. This proves the proposition.

PROPOSITION 3. If $\mathscr{R}$ is a class of algebras satisfying condition (A) and closed with respect to extension, then the sum of any two $\mathscr{R}$-ideals of an arbitrary algebra A is again an $\mathscr{R}$-ideal of the algebra A.

PROOF. If I_1 and I_2 are $\mathscr{R}$-ideals of an algebra A, then by the second homomorphism theorem

$$I_1 + I_2/I_1 \cong I_2/I_2 \cap I_1.$$

Therefore $I_1 + I_2/I_1$ is an $\mathscr{R}$-algebra, since it is the homomorphic image of the $\mathscr{R}$-algebra I_2. From closure with respect to extension we obtain that

$I_1 + I_2$ is an $\mathscr{R}$-algebra, and consequently an $\mathscr{R}$-ideal of the algebra A. This proves the proposition.

In particular, Proposition 3 is valid if the class $\mathscr{R}$ is radical.

A radical $\mathscr{R}$ in a class of algebras $\mathfrak{K}$ is called *hereditary* if for any algebra $A \in \mathfrak{K}$ and any ideal I of A

$$\mathscr{R}(I) = I \cap \mathscr{R}(A).$$

THEOREM 3. A radical $\mathscr{R}$ in a class of algebras $\mathfrak{K}$ is hereditary if and only if the following conditions are satisfied:

(G) if $A \in \mathscr{R}$ and I is an ideal in A, then $I \in \mathscr{R}$;
(H) if $A \in \mathscr{P}$ and I is an ideal in A, then $I \in \mathscr{P}$.

The proof is broken up into two lemmas.

LEMMA 1. For any radical $\mathscr{R}$ condition (G) is equivalent to $\mathscr{R}(I) \supseteq I \cap \mathscr{R}(A)$ for any algebra $A \in \mathfrak{K}$ and any ideal I of A.

PROOF. If the radical $\mathscr{R}$ satisfies property (G) and $A \in \mathfrak{K}$, then for any ideal I of A the intersection $I \cap \mathscr{R}(A)$ is an ideal of the $\mathscr{R}$-algebra $\mathscr{R}(A)$, and by (G) we conclude that $I \cap \mathscr{R}(A)$ is an $\mathscr{R}$-ideal of the algebra I. But then it must be contained in the $\mathscr{R}$-radical of I, that is, $I \cap \mathscr{R}(A) \subseteq \mathscr{R}(I)$.

Conversely, if $\mathscr{R}(I) \supseteq I \cap \mathscr{R}(A)$ and A is an $\mathscr{R}$-algebra, then $\mathscr{R}(I) \supseteq I \cap \mathscr{R}(A) = I \cap A = I$. Consequently $\mathscr{R}(I) = I$, and $I \in \mathscr{R}$. This proves the lemma.

LEMMA 2. For any radical $\mathscr{R}$ condition (H) is equivalent to $\mathscr{R}(I) \subseteq I \cap \mathscr{R}(A)$ for any algebra $A \in \mathfrak{K}$ and any ideal I of A.

PROOF. Let us assume that (H) is valid. Let I be an ideal of the algebra A. We consider the $\mathscr{R}$-semisimple algebra $A/\mathscr{R}(A)$. By (H) the ideal $I + \mathscr{R}(A)/\mathscr{R}(A)$ of $A/\mathscr{R}(A)$ is also $\mathscr{R}$-semisimple. By the second homomorphism theorem

$$I + \mathscr{R}(A)/\mathscr{R}(A) \cong I/I \cap \mathscr{R}(A).$$

Consequently, the algebra $I/I \cap \mathscr{R}(A)$ is $\mathscr{R}$-semisimple, and therefore $\mathscr{R}(I) \subseteq I \cap \mathscr{R}(A)$.

Conversely, if the algebra A is $\mathscr{R}$-semisimple and the condition $\mathscr{R}(I) \subseteq I \cap \mathscr{R}(A)$ is satisfied for any ideal I of A, then $\mathscr{R}(I) \subseteq I \cap (0) = (0)$. This proves Lemma 2, and together with it also Theorem 3.

We devote the remainder of the section to the proof of the fact that in the class of alternative algebras condition (H) is valid for any radical.

LEMMA 3. If I is an ideal of an alternative algebra A and M is an ideal of the algebra I, then for every $a \in A$

(a) $M^2 A \subseteq M$;
(b) $(M, a, I) \subseteq M + Ma$;
(c) $M + Ma$ is an ideal in I;
(d) $M + (Ma)^2$ is an ideal in I contained in $M + Ma$;
(e) $(MA)I^2 \subseteq M$;
(f) $(Ma)^2(Ma)^2 \subseteq M$.

PROOF. Right alternativity is used for the proof of (a) and (b):

$$M^2 A \subseteq M(MA) + (M, M, A) \subseteq MI + (M, A, M) \subseteq M,$$
$$(M, a, I) = (M, I, a) \subseteq M(Ia) + (MI)a \subseteq M + Ma.$$

Assertion (c) is a corollary to (b). We shall prove (d). We have $M + (Ma)^2 \subseteq M + Ma$ by (c). Applying (c) one more time, we obtain

$$\begin{aligned}[M + (Ma)^2]I &\subseteq M + (Ma)[(Ma)I] + (Ma, Ma, I)\\ &\subseteq M + (Ma)^2 + (Ma, I, Ma) \subseteq M + (Ma)^2.\end{aligned}$$

The linearized Moufang identity is used in order to prove (e). Let $m \in M$, $x, y \in I$, $r \in A$. Then

$$(mr, x, y) + (yr, x, m) = (r, x, y)m + (r, x, m)y.$$

In view of the fact that the associators (yr, x, m) and $(r, x, y)m$ lie in M, we conclude that

$$(mr, x, y) - (r, x, m)y \in M.$$

However,

$$(mr, x, y) - (r, x, m)y = (mr, x, y) - (m, r, x)y = -(mr)(xy) + [m(rx)]y,$$

whence in view of the fact that $[m(rx)]y \in M$ it follows that $(mr)(xy) \in M$. Thus we have proved (e). Finally, by (d) and (e)

$$(Ma)^2(Ma)^2 \subseteq (M + Ma)(Ma)^2 \subseteq M + (Ma)(Ma)^2 \subseteq M + (Ma)I^2 \subseteq M,$$

which gives us (f). This proves the lemma.

THEOREM 4. If I is an ideal of an alternative algebra A, and M is an ideal of the algebra I such that the quotient algebra I/M does not contain nonzero nilpotent ideals, then M is an ideal of the entire algebra A.

PROOF. If M is not an ideal in A, then there can be found an element $a \in A$ such that either $aM \nsubseteq M$ or $Ma \nsubseteq M$. For example, let $Ma \nsubseteq M$. Then by condition (c) of Lemma 3, $M_1 = M + Ma$ is an ideal in I which strictly contains M. This means that in the quotient algebra $\bar{I} = I/M$ the image $\bar{M}_1$ of the ideal M_1 does not equal zero. But by (f) $(\bar{M}_1^2)^2 = (0)$, and therefore in view of Proposition 5.1 and the absence of nonzero nilpotent ideals in the algebra $\bar{I}$, we obtain that $\bar{M}_1^2 = (0)$ and $\bar{M}^1 = (0)$. The obtained contradiction proves that $Ma \subseteq M$. Utilizing the left analog of Lemma 3, it is proved analogously that $aM \subseteq M$. This proves the theorem.

LEMMA 4. Let $\mathscr{R}$ be some radical in a class of algebras $\mathfrak{K}$ and I be an ideal of the algebra A with $I^2 = (0)$. Then $\mathscr{R}(I)$ is an ideal in A.

PROOF. We shall assume the opposite. For some $a \in A$ let either $a\mathscr{R}(I) \nsubseteq \mathscr{R}(I)$ or $\mathscr{R}(I)a \nsubseteq \mathscr{R}(I)$. To be definite let us assume that $a\mathscr{R}(I) \nsubseteq \mathscr{R}(I)$. Then $\mathscr{R}(I) + a\mathscr{R}(I)/\mathscr{R}(I)$ is a nonzero ideal of the algebra $I/\mathscr{R}(I)$ which is $\mathscr{R}$-semisimple. We shall show that the ideal $\mathscr{R}(I) + a\mathscr{R}(I)/\mathscr{R}(I)$ is $\mathscr{R}$-radical, so that our assumption leads to a contradiction. We consider the mapping $\theta: \mathscr{R}(I) \to \mathscr{R}(I) + a\mathscr{R}(I)/\mathscr{R}(I)$ defined by the rule $\theta(y) = ay + \mathscr{R}(I)$. We shall show that this is a homomorphism. Actually, for any two elements y_1, $y_2 \in \mathscr{R}(I)$, in view of the fact that $I^2 = (0)$ we have

$$\theta(y_1 y_2) = a(y_1 y_2) + \mathscr{R}(I) = 0 + \mathscr{R}(I),$$
$$\theta(y_1)\theta(y_2) = (ay_1)(ay_2) + \mathscr{R}(I) = 0 + \mathscr{R}(I).$$

Thus θ is a homomorphism which maps $\mathscr{R}(I)$ onto $\mathscr{R}(I) + a\mathscr{R}(I)/\mathscr{R}(I)$. This means by condition (A) that the latter algebra is $\mathscr{R}$-radical, which proves the lemma.

THEOREM 5 (*Anderson–Divinsky–Sulinski*). Let $\mathscr{R}$ be an arbitrary radical defined in the class of alternative algebras, A be an alternative algebra, and I be an ideal of A. Then $\mathscr{R}(I)$ is an ideal in A.

PROOF. We denote $\mathscr{R}(I)$ by M. If M is not an ideal in A, then there exists an $a \in A$ such that either $Ma \nsubseteq M$ or $aM \nsubseteq M$. Let us suppose the first of these. Then $M + Ma$ is an ideal in I strictly containing M. To begin with, let us assume that $(Ma)^2 \subseteq M$. We consider the mapping $\theta: M \to M + Ma/M$ defined by the rule $\theta(M) = ma + M$. Let $m, n \in M$. Then it is obvious $\theta(m + n) = \theta(m) + \theta(n)$, and by part (a) of Lemma 3

$$\theta(mn) = (mn)a + M = 0 + M.$$

Furthermore, since by assumption $(Ma)^2 \subseteq M$,

$$\theta(m)\theta(n) = (ma)(na) + M = 0 + M.$$

We have proved that θ is a homomorphism of the radical algebra M onto the ideal $M + Ma/M$ of the semisimple algebra I/M. This contradicts that $M + Ma$ strictly contains M, that is, that $Ma \nsubseteq M$.

It remains to consider the case when $(Ma)^2 \nsubseteq M$. In this case by Lemma 3 $M + (Ma)^2$ is an ideal in I which strictly contains M. In view of Lemma 3, $M + (Ma)^2/M$ is an ideal with nonzero multiplication in the semisimple algebra I/M. We shall show that the algebra $M + (Ma)^2/M$ has nonzero $\mathscr{R}$-radical. In view of Lemma 4, with this we shall arrive at a contradiction to the assumption we have made, and so shall prove the theorem.

We fix an element $n \in M$ such that $(Ma)(na) \nsubseteq M$, and consider the mapping $\theta: M \to M + (Ma)^2/M$ sending an element $m \in M$ to the coset $\theta(m) = (ma)(na) + M$. By (a) we have that

$$\theta(m_1 m_2) = [(m_1 m_2)a](na) + M = 0 + M.$$

In addition, in view of (f)

$$\theta(m_1)\theta(m_2) = [(m_1 a)(na)][(m_2 a)(na)] + M = 0 + M.$$

The additivity of the mapping θ is obvious, and therefore θ is a homomorphism of the $\mathscr{R}$-radical algebra M onto the nonzero ideal $(Ma)(na) + M/M$ of the algebra $(Ma)^2 + M/M$. Consequently $\mathscr{R}(M + (Ma)^2/M) \neq (0)$, and everything is proved.

COROLLARY. Condition (H) is valid for any radical $\mathscr{R}$ in the class of alternative algebras.

PROOF. If in a semisimple alternative algebra A there could be found a nonsemisimple ideal I, then, by the theorem just proved, its radical $\mathscr{R}(I)$ would be a nonzero radical ideal of the entire algebra A, which is impossible. This proves the corollary.

Exercises

1. Let $\mathfrak{M}$ be a class of Φ-algebras with property (E), and $\bar{\mathfrak{M}}$ be the class of algebras A such that every nonzero ideal I of an algebra A can be mapped homomorphically onto a nonzero algebra from $\mathfrak{M}$. Prove that the class $\bar{\mathfrak{M}}$ satisfies conditions (E) and (F).

2. Let $\mathfrak{M}$ be the class of all simple algebras of a class $\mathfrak{K}$. Prove that for any partitioning of the class $\mathfrak{M}$ into two nonintersecting subclasses $\mathfrak{M}_1$ and $\mathfrak{M}_2$ there exists a radical in $\mathfrak{K}$ with respect to which the simple algebras from $\mathfrak{M}_1$ will be radical and the algebras from $\mathfrak{M}_2$ semisimple.

3. Let $\mathfrak{K}$ be the class of all associative rings and $\mathscr{R}$ be the class of all associative rings which cannot be mapped homomorphically onto nilpotent rings. Show that $\mathscr{R}$ is a radical class in $\mathfrak{K}$ and that condition (G) is not satisfied for $\mathscr{R}$.

4. (*Divinsky–Sulinski*) Let $\mathfrak{M}$ be a class of rings closed with respect to ideals and homomorphic images and $\mathfrak{K}$ be a class of Φ-algebras with these same properties where $\mathfrak{K} \subseteq \mathfrak{M}$. Prove that if $\mathscr{R}$ is a radical in $\mathfrak{M}$, then the $\mathscr{R}$-radical of any Φ-algebra from $\mathfrak{K}$ is a Φ-ideal, that is, invariant under multiplication by elements from Φ. Prove also that the class $\mathscr{R} \cap \mathfrak{K}$ is radical in $\mathfrak{K}$.

2. NIL-RADICALS

Let $\mathfrak{K}$ be an arbitrary class of algebras closed with respect to ideals and homomorphic images. In each algebra $A \in \mathfrak{K}$ we define the following chain of ideals, which we shall call the *Baer chain.* We set $\mathscr{B}_0(A) = (0)$, and denote by $\mathscr{B}_1(A)$ the sum of all *trivial ideals* of the algebra A(i.e., ideals with zero multiplication). Furthermore, we assume that the ideals $\mathscr{B}_\alpha(A)$ are defined for all ordinals α less than the ordinal β. If β is a limit ordinal, we set $\mathscr{B}_\beta(A) = \bigcup_{\alpha<\beta} \mathscr{B}_\alpha(A)$. If β is not a limit ordinal, then there exists an ordinal $\beta - 1$, and we define $\mathscr{B}_\beta(A)$ as the ideal such that

$$\mathscr{B}_\beta(A)/\mathscr{B}_{\beta-1}(A) = \mathscr{B}_1(A/\mathscr{B}_{\beta-1}(A)).$$

If the cardinality of the ordinal γ is greater than the cardinality of the algebra A, then $\mathscr{B}_\gamma(A) = \mathscr{B}_{\gamma+1}(A) = \cdots$. We denote $\mathscr{B}_\gamma(A)$ by $\mathscr{B}(A)$ and call it the *Baer ideal* of the algebra A.

PROPOSITION 4. The quotient algebra $A/\mathscr{B}(A)$ does not contain nonzero trivial ideals. $\mathscr{B}(A)$ is the smallest ideal with this property.

PROOF. The quotient algebra $A/\mathscr{B}(A)$ does not contain nonzero trivial ideals by the construction of the ideal $\mathscr{B}(A)$. We consider the set $\{I_\alpha\}$ of all ideals of the algebra A such that A/I_α is an algebra without nonzero trivial ideals. It is easy to see that $I = \bigcap I_\alpha$ is the smallest ideal in this set. If the containment $I \subseteq \mathscr{B}(A)$ were strict, there could be found a minimal ordinal δ such that $\mathscr{B}_\delta(A) \nsubseteq I$. It is clear that δ is not a limit ordinal, and that $\mathscr{B}_{\delta-1}(A) \subseteq I$. Then in the algebra $\bar{A} = A/\mathscr{B}_{\delta-1}(A)$ there would exist a trivial ideal $\bar{K} \neq (0)$ not contained in $\bar{I} = I/\mathscr{B}_{\delta-1}(A)$. Its complete inverse image K in the algebra A is such that $K \nsubseteq I$ but $K^2 \subseteq \mathscr{B}_{\delta-1}(A) \subseteq I$. Thus a nonzero trivial ideal $K + I/I$ has been obtained in A/I, which contradicts our assumption. This means that in fact $I = \mathscr{B}(A)$, and the proposition is proved.

An algebra is called *prime* if for any two of its ideals I and J it follows from the equality $IJ = (0)$ that either $I = (0)$ or $J = (0)$. In many classes of algebras the prime algebras can be sufficiently well characterized. Prime alternative algebras will be studied in the following chapter.

THEOREM 6. The quotient algebra $A/\mathscr{B}(A)$ is a subdirect sum of prime algebras.

PROOF. An ideal I of an algebra A is called a *prime ideal* if the algebra A/I is prime. The assertion of the theorem is equivalent to the Baer ideal $\mathscr{B}(A)$ equaling the intersection of some set of prime ideals. We shall show that it coincides with the intersection I of all prime ideals. Let P be a prime ideal. Then by Proposition 4 $P \supseteq \mathscr{B}(A)$, and consequently $I \supseteq \mathscr{B}(A)$. Let $a \notin \mathscr{B}(A)$ and (a) be the ideal generated by the element a. We have $(a)^2 \nsubseteq \mathscr{B}(A)$, and so there exists an element $a_1 \in (a)^2 \backslash \mathscr{B}(A)$. Analogously, we obtain an element $a_2 \in (a_1)^2 \backslash \mathscr{B}(A)$, etc. We construct an infinite sequence $a_0 = a$, a_1, $a_2, \ldots$, and by Zorn's lemma we select a maximal ideal Q of the algebra A such that $Q \cap \{a_i\} = \varnothing$. If the ideals C and D strictly contain the ideal Q, then for some k and m the inclusions $C \ni a_k$ and $D \ni a_m$ are valid. But then $C \cap D \supseteq (a_l)$ where $l = \max(k, m)$, and $CD \supseteq (a_l)^2 \ni a_{l+1}$. Therefore $CD \nsubseteq Q$, which proves the ideal Q is prime. Since $a = a_0 \notin Q$, then $a \notin I$. This means $\mathscr{B}(A) \subseteq I$, and the equality $\mathscr{B}(A) = I$ is proved.

An algebra without nonzero trivial ideals is called *semiprime*. Since in a semiprime algebra the Baer ideal equals zero, from Theorem 6 follows

COROLLARY. Every semiprime algebra is a subdirect sum of prime algebras.

We now explain what condition must be satisfied by a class of algebras $\mathfrak{K}$ in order that the mapping $\mathscr{B}: A \to \mathscr{B}(A)$ be radical in $\mathfrak{K}$.

PROPOSITION 5. The mapping $\mathscr{B}$ is radical in the class of algebras $\mathfrak{K}$ if and only if every nonzero ideal of a semiprime algebra from $\mathfrak{K}$ can be mapped homomorphically onto a semiprime algebra.

PROOF. The formulated necessary and sufficient condition is, in particular, condition (E) for the class $\mathscr{P} = \{A \in \mathfrak{K} \mid \mathscr{B}(A) = (0)\}$. We shall prove condition (F) for the class $\mathscr{P}$. Actually, if each nonzero ideal of some algebra A can be mapped homomorphically onto an algebra without trivial ideals, then there cannot be trivial ideals in the algebra A, and $A \in \mathscr{P}$. In view of Theorem 2, $\mathscr{P}$ is the semisimple class of some radical, for which by Proposition 1 the radical class in $\mathfrak{K}$ consists of algebras not homomorphically mapped onto algebras without trivial ideals. We conclude by Proposition 4 that $\mathscr{R} = \{A \in \mathfrak{K} \mid A = \mathscr{B}(A)\}$, and the sufficiency of the condition is proved. The necessity is also clear, because if $\mathscr{B}$ is a radical, then its semisimple class is $\mathscr{P}$, and condition (E) is valid for $\mathscr{P}$. This proves the proposition.

If the mapping $\mathscr{B}$ is radical, then this radical is called the *Baer radical* or *lower nil-radical*. In view of Proposition 5, for this map to be radical in the class of alternative algebras, it suffices to show that every ideal I of a semiprime alternative algebra A is itself a semiprime algebra. We shall prove this in the following chapter.

PROPOSITION 6. For any algebra A and ideal I of A the containment $\mathscr{B}(I) \supseteq I \cap \mathscr{B}(A)$ is valid.

PROOF. Let $\mathscr{B}(I) \not\supseteq I \cap \mathscr{B}(A)$. Then there can be found a minimal ordinal δ such that $\mathscr{B}(I) \not\supseteq I \cap \mathscr{B}_\delta(A)$. It is clear that δ is not a limit ordinal and that $I \cap \mathscr{B}_{\delta-1}(A) \subseteq \mathscr{B}(I)$. Since $I \cap \mathscr{B}_\delta(A) \not\subseteq \mathscr{B}(I)$, in the quotient algebra $A/\mathscr{B}_{\delta-1}(A)$ there can be found a trivial ideal $\bar{K}$ such that $I \cap K \not\subseteq \mathscr{B}(I)$ holds for its inverse image K. But $(I \cap K)^2 \subseteq I \cap \mathscr{B}_{\delta-1}(A) \subseteq \mathscr{B}(I)$, and in the quotient algebra $I/\mathscr{B}(I)$ there is found a nonzero trivial ideal $(I \cap K) + \mathscr{B}(I)/\mathscr{B}(I)$, which is impossible. This means $\mathscr{B}(I) \supseteq I \cap \mathscr{B}(A)$, and the proposition is proved.

Now let $\mathscr{L}$ be the class of all locally nilpotent algebras from $\mathfrak{K}$. If $\mathscr{L}$ is a radical class in $\mathfrak{K}$, then the radical it defines is called the *locally nilpotent radical* or *Levitzki radical*. Necessary and sufficient criteria that the class $\mathscr{L}$ be radical are unknown. Some sufficient criteria are known [13, 14, 59, 80, 174]. We cite the most simple of these, which is sufficient for our purposes.

THEOREM 7 (*Dorofeev*). Let $\mathfrak{M}$ be a homogeneous variety of Φ-algebras such that for each algebra $A \in \mathfrak{M}$ the elements of the chain

$$A \supseteq A^{(1)} \supseteq A^{(2)} \supseteq \cdots$$

are ideals of the algebra A. Then the class $\mathscr{L}$ of all locally nilpotent algebras is radical in $\mathfrak{M}$ if and only if all the algebras from $\mathfrak{M}$ which are solvable of index 2 belong to $\mathscr{L}$.

PROOF. The necessity of the condition follows from Proposition 2. We shall prove the sufficiency. The proof is divided into a series of lemmas.

LEMMA 5. The square of a finitely generated algebra A from $\mathfrak{M}$ is a finitely generated algebra.

PROOF. Let $a_1, a_2, \ldots, a_n$ be generators of the algebra A. We consider the free algebra F from $\mathfrak{M}$ with n free generators. By assumption the quotient algebra $F/F^{(2)}$ belongs to $\mathscr{L}$, and consequently is nilpotent. If M is the index of its nilpotency, then every monomial u from F of degree not less than M is representable in the form

$$u = \sum_i v_i' v_i'', \tag{1}$$

where v_i', v_i'' are monomials from F^2. Moreover, by homogeneity it is possible to assume that $d(v_i') + d(v_i'') = d(u)$. It is now clear that all possible elements of the form $w(a_{i_1}, a_{i_2}, \ldots, a_{i_s})$, where w is a nonassociative monomial of degree not less than 2 and not more than $M - 1$, generate the algebra A^2.

LEMMA 6. Local solvability is a radical property in $\mathfrak{M}$.

PROOF. The class of locally solvable algebras satisfies property (A). We shall show that it is closed with respect to extension. We let the ideal I and quotient algebra A/I be locally solvable, and consider in A a finitely generated subalgebra A_0. By the local solvability of A/I, we have $A_0^{(p)} \subseteq I$ for some p. The ideal I is also locally solvable, and by the previous lemma $A_0^{(p)}$ is a finitely generated algebra. Therefore $A_0^{(q)} = (0)$ for some q, which also means the algebra A is locally solvable. We denote by $\mathscr{L}(A)$ the sum of all locally solvable ideals of the algebra A. By Proposition 3 $\mathscr{L}(A)$ is a locally solvable ideal, and by closure with respect to extension there are no nonzero locally solvable ideals in the quotient algebra $A/\mathscr{L}(A)$. Consequently, the class of locally solvable algebras satisfies (B) and (C) and is radical.

LEMMA 7. Every locally solvable algebra from $\mathfrak{M}$ is locally nilpotent.

PROOF. It suffices to prove that for the free algebra F with n generators and for any $k \geq 1$ there exists a number $f(n, k)$ such that

$$F^{f(n,k)} \subseteq F^{(k)}. \tag{2}$$

We have seen that the number $f(n, 2)$ exists. Let us assume that $k \geq 3$ and a number $f(n, k-1)$ exists for any n. Then in view of Lemma 5, for some number L we have

$$(F^2)^L \subseteq (F^2)^{(k-1)} = F^{(k)}. \tag{3}$$

We consider the number $N = \max(L, f(n, 2))$. We note that expansion (1) holds for monomials of degree $\geq N$. We set $f(n, k) = N^2$ and consider a monomial u of degree $\geq N^2$. By (1) $u = \sum_i v_i' v_i''$, where $d(v_i') + d(v_i'') = d(u)$ and $\min(d(v_i'), d(v_i'')) \geq 2$. If the monomials v_i' and v_i'' have degree not less than N, then we again apply (1) etc., until u is represented in the form of a linear combination of products of monomials each of which has degree greater than 2 but less than N. But then in each product there will be not less than N such monomials, and by (3) each product will lie in $F^{(k)}$. Consequently u lies in $F^{(k)}$, and (2) is proved. This proves the lemma.

For the proof of Theorem 7 it now suffices to apply Lemmas 6 and 7.

PROPOSITION 7. Let A be an alternative algebra with n generators and $A^{(2)} = (0)$. Then $A^{n+3} = (0)$.

PROOF. Let $\tilde{R}_a$ and $\tilde{L}_a$ be the restrictions of the operators R_a and L_a to the ideal A^2. Then $\tilde{R}_{ab} = \tilde{L}_{ab} = 0$ for any $a, b \in A$, and on account of the alternative identities

$$\tilde{R}_a^2 = \tilde{L}_a^2 = 0. \tag{4}$$

In addition, by the middle Moufang identity

$$a(bc)a = (ab)(ca) = 0$$

for any $a, b, c \in A$. Therefore

$$\tilde{R}_a \tilde{L}_a = \tilde{L}_a \tilde{R}_a = 0. \tag{5}$$

Let $w = \tilde{M}_{a_1} \tilde{M}_{a_2} \cdots \tilde{M}_{a_k}$ be some monomial from operators $\tilde{R}_{a_i}$ and $\tilde{L}_{a_i}$ where $a_i \in A$. Without loss of generality the elements a_i can be considered as linear combinations of generators. From relations (4), (5) and their linearizations it follows that the element w is a skew-symmetric function of the elements $a_1, a_2, \ldots, a_k$. Therefore, if $k \geq n + 1$ then $w = 0$. It is now easy to see that $A^{n+3} = (0)$. This proves the proposition.

COROLLARY. The class of locally nilpotent algebras is radical in the variety of alternative algebras.

PROOF. Since by the corollary to Theorem 1.5 the variety of alternative algebras is homogeneous, the assertion follows from Theorem 7 and Proposition 7.

THEOREM 8 (*Shestakov*[80]). Let there be a locally nilpotent radical $\mathscr{L}$ in a variety of Φ-algebras $\mathfrak{M}$. Then for every algebra $A \in \mathfrak{M}$ the radical $\mathscr{L}(A)$ equals the intersection of all such prime ideals P_α of the algebra A for which the quotient algebra A/P_α is $\mathscr{L}$-semisimple.*

PROOF. We denote the intersection of the appropriate prime ideals P_α by J. It is clear that $\mathscr{L}(A)$ is contained in each ideal P_α, and consequently also in their intersection, that is, $\mathscr{L}(A) \subseteq J$. The theorem will be proved if for each element $x \notin \mathscr{L}(A)$ there can be found a prime ideal P such that $\mathscr{L}(A/P) = (0)$ and $x \notin P$. Then the reverse containment $J \nsubseteq \mathscr{L}(A)$ will also hold.

Let $x \notin \mathscr{L}(A)$. We consider the ideal (x) of the algebra A generated by the element x. Since $x \notin \mathscr{L}(A)$, the ideal (x) is not locally nilpotent, and this means it contains a nonnilpotent finitely generated subalgebra S. Let M be the set of ideals I of the algebra A such that the image of the algebra S under the canonical homomorphism of A onto A/I is not nilpotent. The set M is

* This theorem was first proved for associative rings by Babich [20].

nonempty, since $(0) \in M$. Furthermore, the set M is partially ordered by inclusion. Now let $T = \{I_\alpha\}$ be a linearly ordered subset of the set M. We shall show that $I = \bigcup I_\alpha$ belongs to M. We assume the contrary, that is, the image $\bar{S}$ of the subalgebra S in the quotient algebra A/I is nilpotent, or in other words $S^m \subseteq I$ for some m. We consider the set W of words from generators of the subalgebra S of length from m to $2m$. This set is finite, lies in S^m, and so also in I. But since it is finite, we have $W \subseteq I_{\alpha_0}$ for some α_0. We shall show that also $S^m \subseteq I_{\alpha_0}$. Actually, let u be a word from generators of the algebra S of length not less than m. If its length is not greater than $2m$, then $u \in W$, and consequently $u \in I_{\alpha_0}$. If $d(u) > 2m$, then $u = u_1 u_2$ where the length of one of the u_i is greater than m. Let $d(u_1) > m$. Making an induction assumption, we can assume that $u_1 \in I_{\alpha_0}$, but then also $u = u_1 u_2 \in I_{\alpha_0}$. Consequently $S^m \subseteq I_{\alpha_0}$, which contradicts the fact that $I_{\alpha_0} \in M$.

Thus $I \in M$, and the set M is inductive. By Zorn's lemma there is a maximal element P in M. It is obvious that $x \notin P$, because otherwise $(x) \subseteq P$ and $S \subseteq P$. We consider the algebra $\bar{A} = A/P$. For completion of the proof it suffices to show that $\bar{A}$ is a prime $\mathscr{L}$-semisimple algebra. We note that the subalgebra $\bar{S}$ in $\bar{A}$ is such that in any proper homomorphic image of the algebra $\bar{A}$ its image is nilpotent. For convenience we denote $\bar{A}$ by A and $\bar{S}$ by S. We shall first show that A is prime.

Let B and C be two ideals of the algebra A and $BC = (0)$. If $B \neq (0)$ and $C \neq (0)$, then the algebra S under the canonical homomorphisms of A onto A/B and A onto A/C becomes nilpotent algebras S_1 and S_2. By the homomorphism theorems

$$S_1 = S + B/B \cong S/S \cap B, \qquad S_2 = S + C/C \cong S/S \cap C.$$

We consider the algebra $S \cap B/S \cap B \cap C$. It is nilpotent since $S \cap B/S \cap B \cap C \cong S \cap B + S \cap C/S \cap C$, and the latter algebra is isomorphic to a subalgebra of the algebra S_2. In addition $(S \cap B \cap C)^2 = (0)$, and therefore by Proposition 2 the algebra $S \cap B$ is locally nilpotent. We now see that $S \cap B$ and $S/S \cap B$ are locally nilpotent algebras. Then that same Proposition 2 gives us the algebra S is locally nilpotent. However, this is impossible, since it is finitely generated and nonnilpotent. We have proved the algebra A is prime.

Finally, let B be a locally nilpotent ideal of the algebra A. If $B \neq (0)$, then the algebra $\bar{S} = S + B/B$ is nilpotent. Consequently the algebra $S/S \cap B$, which is isomorphic to $S + B/B$, is nilpotent. Thus $S \cap B$ and $S/S \cap B$ are locally nilpotent algebras. By Proposition 2 the algebra S is also locally nilpotent, but this is not so. The obtained contradiction proves the theorem.

COROLLARY. Under the conditions of Theorem 8 every $\mathscr{L}$-semisimple algebra from $\mathfrak{M}$ is a subdirect sum of prime $\mathscr{L}$-semisimple algebras.

Now let $\mathfrak{K}$ be an arbitrary class of power-associative algebras which is closed with respect to ideals and homomorphic images. It is easy to see that the class $\mathfrak{N}$ of all nil-algebras is a radical class in $\mathfrak{K}$. The corresponding radical is called the *Köthe radical* or *upper nil-radical.*

Condition (G) is satisfied for all three of the radicals $\mathscr{B}$, $\mathscr{L}$, $\mathfrak{N}$ introduced. For $\mathscr{L}$ and $\mathfrak{N}$ it is obvious, and for $\mathscr{B}$ it follows from Proposition 6 and Lemma 1. By the corollary to Theorem 5, in the class of alternative algebras condition (H) is also valid for these radicals. However, by Theorem 3, conditions (G) and (H) are equivalent to heredity. Therefore we have proved

THEOREM 9. The radicals $\mathscr{B}$, $\mathscr{L}$, $\mathfrak{N}$ are hereditary in the class of alternative algebras.

Between the radical classes introduced, the relations

$$\mathscr{B} \subseteq \mathscr{L} \subseteq \mathfrak{N}$$

are obvious. These containments are strict [37, 186]. Every radical $\mathscr{R}$ for which the radical class satisfies the relations $\mathscr{B} \subseteq \mathscr{R} \subseteq \mathfrak{N}$ is called a *nil-radical.* Clearly it is in this connection that the terms "lower nil-radical" and "upper nil-radical" come about. At one time Zhevlakov expressed the hypothesis that nil-radicals form a chain. The hypothesis proved to be incorrect, as Ryabukhin constructed an entire class of noncomparable nil-radicals in the class of associative algebras over an arbitrary field.

Exercises

1. Prove that every prime (semiprime) Φ-algebra is a prime (semiprime) ring.

2. (*Dorofeev*) Let $\mathfrak{M}$ be a homogeneous variety of algebras such that for any algebra $A \in \mathfrak{M}$ the terms of the chain

$$A \supseteq A^{(1)} \supseteq A^{(2)} \supseteq \cdots$$

are ideals in A. Then if all solvable of index 2 algebras from $\mathfrak{M}$ are nilpotent, all solvable algebras in $\mathfrak{M}$ are also nilpotent. *Hint.* Argue as in the proof of Lemma 7.

3. (*Slin'ko*) Prove that for any derivation D of an algebra A over a field of characteristic 0

(a) $\mathscr{B}(A)^D \subseteq \mathscr{B}(A)$;

(b) if $\mathscr{L}(A)$ is a locally nilpotent ideal in A and $A/\mathscr{L}(A)$ does not contain nonzero locally nilpotent ideals, then $\mathscr{L}(A)^D \subseteq \mathscr{L}(A)$. In particular, this is so if A belongs to a variety where the locally nilpotent radical exists;

(c) if A is a power-associative algebra, then $\mathfrak{N}(A)^D \subseteq \mathfrak{N}(A)$.

4. Prove that in every power-associative algebra A the Köthe radical $\mathfrak{N}(A)$ equals the intersection of all prime ideals of this algebra modulo which the quotient algebra is $\mathfrak{N}$-semisimple.

3. THE ANDRUNAKIEVICH RADICAL

An algebra A is called *subdirectly irreducible* if its *heart* $H(A)$, the intersection of all nonzero ideals of the algebra A, is different from zero. To a considerable degree, the properties of a subdirectly irreducible algebra are determined by the properties of its heart. In alternative algebras the square of an ideal is always an ideal (Proposition 5.1), so that if A is an alternative algebra then either $H(A)^2 = (0)$ or $H(A)^2 = H(A)$. Thus, subdirectly irreducible alternative algebras divide into two classes: algebras with nilpotent heart and algebras with idempotent heart. Although one cannot say much definite about the structure of the first of these, the second type permits an exhaustive description modulo associative algebras. It will be proved in this section that every nonassociative subdirectly irreducible alternative algebra with idempotent heart is a Cayley–Dickson algebra over its center. In addition, it will be shown that subdirectly irreducible alternative algebras with idempotent heart are the building blocks of a semisimple class for a certain new radical with interesting properties. In Chapter 12 this radical will play an important role in the study of alternative algebras with minimal condition.

Let X be a subset of an algebra A. The set $\mathrm{Ann}_r(X) = \{z \in A \mid Xz = (0)\}$ is called the *right annihilator* of X, the set $\mathrm{Ann}_l(X) = \{z \in A \mid zX = (0)\}$ is the *left annihilator*, and $\mathrm{Ann}(X) = \mathrm{Ann}_r(X) \cap \mathrm{Ann}_l(X)$ is the *annihilator*.

LEMMA 8. Let B be an ideal of an alternative algebra A. Then $\mathrm{Ann}_r(B)$, $\mathrm{Ann}_l(B)$, and $\mathrm{Ann}(B)$ are also ideals in A.

PROOF. By symmetry it suffices to show that $\mathrm{Ann}_r(B)$ is an ideal. Let $z \in \mathrm{Ann}_r(B)$, $a \in A$, $b \in B$. Then

$$b(za) = (bz)a - (b, z, a) = -(a, b, z) = a(bz) - (ab)z = 0,$$

that is, $za \in \mathrm{Ann}_r(B)$ and $\mathrm{Ann}_r(B)$ is a right ideal. Analogously

$$b(az) = (ba)z - (b, a, z) = (a, b, z) = 0,$$

and $az \in \mathrm{Ann}_r(B)$. Thus $\mathrm{Ann}_r(B)$ is an ideal, and the lemma is proved.

We recall that for an ideal I of an alternative algebra A by $T(I)$ and $P(I)$ are denoted the submodules $\{III\}$ and $\{IA^{\#}I\}$, respectively, of the Φ-module

A. By the corollary to Proposition 5.2, these submodules are ideals of the algebra A such that $P(I) \supseteq T(I) \supseteq P(I)^2$.

LEMMA 9. Let $P(I) = (0)$ for an ideal I of an alternative algebra A. Then for any $v \in I$ the submodule $V = vI$ is a trivial ideal of the algebra A.

PROOF. For any $i, j \in I$, $a \in A$, we have

$$i \circ j = \{i1j\} = 0,$$
$$(ia)j + (ja)i = i(aj) + j(ai) = \{iaj\} = 0.$$

Consequently,

$$\begin{aligned}(vi)a &= -(va)i + v(a \circ i) = v(a \circ i) + (ia)v \\ &= v(a \circ i) - v(ia) = v(ai) \in vI, \\ a(vi) &= -a(iv) = i(av) - (a \circ i)v \\ &= v(a \circ i) - v(ai) = v(ia) \in vI,\end{aligned}$$

that is, V is an ideal of the algebra A. By the middle Moufang identity $(vi)(vj) = -(vi)(jv) = -v(ij)v = 0$, hence $V^2 = (0)$. This proves the lemma.

LEMMA 10. Let I be an ideal of an alternative algebra A. If $I \neq (0)$ and I does not contain nonzero trivial ideals of the algebra A, then $T(I) \neq (0)$.

PROOF. Let us assume that $T(I) = (0)$. By the corollary to Proposition 5.2 we then have $P(I)^2 = (0)$. Since $P(I) \subseteq I$ and $P(I)$ is an ideal in A, then by the assumptions of the lemma $P(I) = (0)$. By Lemma 9 for any $v \in I$ it is then true $vI = (0)$, hence $I^2 = (0)$. This contradicts the assumption of the lemma. Consequently $T(I) \neq (0)$, and the lemma is proved.

An ideal $I \neq (0)$ of an algebra A is called *minimal* if, from the fact that I' is an ideal of the algebra A strictly contained in I, it follows that $I' = (0)$.

THEOREM 10 (*Zhevlakov*). If I is a minimal ideal of an alternative algebra A, then either $I^2 = (0)$ or I is a simple algebra.

PROOF. Let $I^2 \neq (0)$. Then $I^2 = I$, and we find ourselves in the situation of Lemma 10. This means in this case $I = T(I)$. Let us assume that B is an ideal of the algebra I, and $B \neq I$. We consider the set $C = (IB)I + I(BI)$ and prove that C is an ideal of the algebra A. The notation $p \equiv q$ will denote that $p - q \in C$. Let $b \in B$, $x, y \in I$, $a \in A$. Then by (7.18)

$$(a, b, xyx) = (xax, b, y) + (\{axy\}, b, x) \equiv 0.$$

In view of the equality $I = T(I)$, for any $i \in I$ we have

$$(a, b, i) \equiv 0. \tag{6}$$

Now let $i, j \in I$. Then in view of (6)

$$[(ib)j]a = (ib)(ja) + (ib, j, a) \equiv 0,$$

and furthermore

$$\begin{aligned}[i(bj)]a &= i[(bj)a] + (i, bj, a) \equiv i[(bj)a] \\ &= i(b, j, a) + i[b(ja)] \equiv 0.\end{aligned}$$

These relations mean that C is a right ideal in A. It is proved analogously that C is a left ideal. But $C \subseteq B$, and therefore C is strictly contained in I. Consequently $C = (0)$. In particular, $(IB)I = (0)$, that is, $IB \subseteq \mathrm{Ann}_l(I)$. If $\mathrm{Ann}_l(I) \cap I \neq (0)$, then by the minimality of the ideal I and Lemma 8 $\mathrm{Ann}_l(I) \supseteq I$, that is, $I^2 = (0)$. Consequently $\mathrm{Ann}_l(I) \cap I = (0)$, and analogously $\mathrm{Ann}_r(I) \cap I = (0)$. However, $IB \subseteq \mathrm{Ann}_l(I) \cap I$, and this means $IB = (0)$. But then $B \subset \mathrm{Ann}_r(I) \cap I$ and $B = (0)$. This means I is a simple algebra, which proves the theorem.

This theorem follows easily for associative algebra from the following assertion, which goes by the name of the *Andrunakievich lemma.*

LEMMA 11. Let I be an ideal of an associative algebra A, and J be an ideal of the algebra I. Then the containment $\hat{J}^3 \subseteq J$ holds for the ideal $\hat{J}$ generated by the set J.

PROOF. Since A is associative, $\hat{J} = A^{\#}JA^{\#}$. Therefore

$$\hat{J}^3 \subseteq I(A^{\#}JA^{\#})I \subseteq IJI \subseteq J.$$

This proves the lemma.

For an alternative algebra A it is as yet unknown whether there exists a number n such that $\hat{J}^n \subseteq J$. Shestakov proved [271] that for an alternative algebra A with k generators $\hat{J}^{k+3} \subseteq J$. In the general case Hentzel and Slater showed that $\hat{J}^{\omega+1} \subseteq J$, where ω is the first infinite ordinal [249].

In view of the fact that a subdirectly irreducible algebra with idempotent heart is prime, we gather together a collection of preliminary information on prime and semiprime algebras.

For an arbitrary algebra A we denote by $D(A)$ the ideal generated by the set (A, A, A) of all associators, the so called *associator ideal.*

PROPOSITION 8. $D(A) = (A, A, A)A^{\#} = A^{\#}(A, A, A)$.

PROOF. From identity (7.5) it follows that

$$A(A, A, A) \subseteq (A, A, A)A^{\#}.$$

In addition, the inclusion

$$((A, A, A)A)A \subseteq (A, A, A) + (A, A, A)A^2$$

is obvious. Therefore

$$\begin{aligned} A((A, A, A)A) &\subseteq (A(A, A, A))A + (A, A, A) \\ &\subseteq (A, A, A) + (A, A, A)A + ((A, A, A)A)A \subseteq (A, A, A)A^{\#}, \end{aligned}$$

whence it follows that the set $(A, A, A)A^{\#}$ is an ideal in A. Consequently $D(A) = (A, A, A)A^{\#}$. The second equality is also proved analogously, which proves the proposition.

We shall call every ideal contained in the associative center $N = N(A)$ a *nuclear ideal*, and the largest nuclear ideal the *associative nucleus* of the algebra A. We shall denote the latter by $U = U(A)$.

PROPOSITION 9. $U(A) = \{x \in A \mid xA^{\#} \subseteq N(A)\}$.

PROOF. It suffices to show that the set V which is the right side of the equality is an ideal. It is obvious that V is a right ideal. On the other hand, $[N, A] \subseteq N$, and therefore

$$(AV)A^{\#} = A(VA^{\#}) \subseteq AV \subseteq [V, A] + V \subseteq N.$$

This means that $AV \subseteq V$, that is, V is a left ideal. Consequently V is an ideal, and the proposition is proved.

PROPOSITION 10. $DU = UD = (0)$.

PROOF. In view of Lemma 7.1 we have $U(A, A, A) = (A, A, A)U = (0)$. Therefore

$$U((A, A, A)A^{\#}) = (U(A, A, A))A^{\#} = (0),$$
$$(A^{\#}(A, A, A))U = A^{\#}((A, A, A)U) = (0).$$

In view of Proposition 8, the proof is completed.

COROLLARY. If A is prime, then either $U = (0)$ or A is associative.

It is clear that if $U = (0)$, then the algebra A is quite distant from associative algebras, and in this case it is commonly called *purely nonassociative*. Furthermore, if A is alternative we shall call it *purely alternative*.

The concept of a *separated algebra* also proves to be useful. This is an algebra in which $D \cap U = (0)$.

Now let A be an alternative algebra. We consider the ideal $ZN(A)$ of A generated by the set $[N, A]$. We shall denote it by M. As we know (see Lemmas 7.4 and 7.5), $M = [N, A]A^{\#}$ and $(M, A, A) \subseteq N$.

As before, we shall denote the center of the algebra A by $Z = Z(A)$.

THEOREM 11. The containment $M \subseteq U$ holds in a semiprime alternative algebra A. If A is also purely alternative, then $N = Z$.

PROOF. If $M \nsubseteq U$, then $M \nsubseteq N$, and we can find elements $n \in N$, $a, b, c, d \in A$ such that $m = ([a, n]b, c, d) \neq 0$. The element $u = [a, n]b$ lies in M, and consequently $m \in (M, A, A) \subseteq N$. But then by Lemma 7.7 $(m^2) = (0)$, which contradicts the algebra A is semiprime. Consequently $M \subseteq U$.

If A is also purely alternative, it hence follows $M = (0)$. In particular, $[N, A] = (0)$, and this means that $N = Z$. This proves the theorem.

COROLLARY. In a prime alternative algebra A that is not associative, the equality $N = Z$ holds.

THEOREM 12. Every subdirectly irreducible semiprime alternative algebra is either associative or a Cayley–Dickson algebra over its center.

PROOF. Let the algebra A satisfy the conditions of the theorem and $H = H(A)$ be its heart. By Theorem 10, H is a simple algebra, and therefore it is either associative or a Cayley–Dickson algebra. In the first case it coincides with its associative center, and therefore by Lemma 10 $N(H) = T(N(H))$. In the second case this equality is also true, since by Theorem 2.5 the associative center of a Cayley–Dickson algebra is a field.

We shall show that $N(H) \subseteq N(A)$. Let $n, m \in N(H)$ and $a, b \in A$. Then by identity (7.18)

$$(nmn, a, b) = -(nan, m, b) + (n, \{mna\}, b).$$

From this equality it follows that $(N(H), A, A) \subseteq (N(H), H, A)$. Furthermore, let $h \in H$, $n, m \in N(H)$, $a \in A$. By the same identity

$$(nmn, h, a) = -(nan, h, m) + (n, h, \{mna\}) = 0,$$

and consequently $(N(H), A, A) \subseteq (N(H), H, A) = (0)$. Thus we have obtained that $N(H) \subseteq N(A)$.

Let us now assume that A is not associative. Then by the corollary to Theorem 11, we have $N(A) = Z(A)$ since the algebra A is prime. Hence it follows that if H is associative, then $H = N(H) \subseteq N(A) = Z(A)$, and H is a field. Thus if A is not associative, then in any case $N(H)$ is a field. Let e be the identity element of this field. We consider the set $E = \{ea - a \mid a \in A\}$. Since $e \in Z(A)$, then this set is an ideal. However $eE = Ee = (0)$, which by

Lemma 8 contradicts the algebra A is prime, provided e is not an identity element in A. But if e is an identity element in A, then A coincides with H and is a simple algebra. In view of the theorem on simple alternative algebras, our assertion is proved.

We turn now to the construction of a radical. Let $\mathfrak{K}$ be the class of all Φ-algebras, and $\mathfrak{B}$ be the class of all subdirectly irreducible Φ-algebras for which the heart is a simple algebra. We shall call an ideal I of an algebra A a $\mathfrak{B}$-ideal if the quotient algebra A/I belongs to $\mathfrak{B}$.

LEMMA 12. Let $A \in \mathfrak{B}$ and I be an ideal in A. Then I maps homomorphically onto an algebra from $\mathfrak{B}$. If the algebra A is alternative, then $I \in \mathfrak{B}$ and also $H(I) = H(A)$.

PROOF. Let $H(A)$ be the heart of the algebra A. The algebra $H(A)$ is simple and $I \supseteq H(A)$. By the simplicity of $H(A)$, for any ideal J of the algebra I either $J \supseteq H(A)$ or $J \cap H(A) = (0)$. By Zorn's lemma we select a maximal ideal K such that $K \cap H(A) = (0)$. It is clear that $\bar{I} = I/K$ is a subdirectly irreducible algebra with heart $\overline{H(A)}$. If the algebra A is alternative, then by Theorem 4 K is an ideal in A. However, $K \cap H(A) = (0)$, and consequently $K = (0)$, which also proves the second assertion of the lemma.

THEOREM 13. The class $\mathscr{A}$ of all algebras which are not mapped homomorphically onto algebras from the class $\mathfrak{B}$ is radical in the class of all algebras.

PROOF. Property (A) is obvious for the class $\mathscr{A}$. Property (D) follows from Lemma 12, so that the necessary and sufficient conditions for a radical formulated in Theorem 1 are satisfied.

The radical $\mathscr{A}$ is called the *Andrunakievich radical* or *antisimple radical.*

LEMMA 13. Let A be an alternative algebra, I be an ideal of A, and J be a $\mathfrak{B}$-ideal of the algebra I. Then there exists a $\mathfrak{B}$-ideal K of the algebra A such that $K \cap I = J$.

PROOF. In view of Theorem 4, J is an ideal of the algebra A. We consider $\bar{A} = A/J$ and the ideal $\bar{I} = I/J$. By assumption $\bar{I}$ is a subdirectly irreducible algebra with a simple heart. Now if $\bar{K}$ is a maximal ideal of the algebra $\bar{A}$ such that $\bar{I} \cap \bar{K} = (\bar{0})$, then the algebra $\bar{\bar{A}} = \bar{A}/\bar{K}$ belongs to $\mathfrak{B}$ since $H(\bar{\bar{A}}) \supseteq \overline{H(\bar{I})} \neq (\bar{0})$. Let K be the inverse image in A of the ideal $\bar{K}$. Then

$$\bar{\bar{A}} = \bar{A}/\bar{K} = A/J/K/J \cong A/K,$$

and K is a $\mathfrak{B}$-ideal of the algebra A. In addition, $K \cap I \subseteq J$, and consequently $K \cap I = J$. This proves the lemma.

COROLLARY. The Andrunakievich radical is hereditary in the class of alternative algebras.

PROOF. In view of the corollary to the theorem we must only prove condition (G) for the radical $\mathscr{A}$. Let $A \in \mathscr{A}$ and I be an ideal in A. If $I \notin \mathscr{A}$, then there is a proper $\mathfrak{B}$-ideal J in the algebra I. By Lemma 13 there exists a $\mathfrak{B}$-ideal K of the algebra A such that $K \cap I = J$, and this means $K \neq A$. But then $A/K \in \mathfrak{B}$, which contradicts the algebra A is radical. This proves the corollary.

THEOREM 14. The Andrunakievich radical $\mathscr{A}(A)$ of an alternative algebra A equals the intersection of all its $\mathfrak{B}$-ideals.

PROOF. We consider the intersection I of all $\mathfrak{B}$-ideals of the algebra A. By Lemma 12 every ideal of an algebra from $\mathfrak{B}$ is again an algebra from $\mathfrak{B}$, and therefore all algebras from $\mathfrak{B}$ are $\mathscr{A}$-semisimple. Hence it follows that $I \supseteq \mathscr{A}(A)$.

We shall now prove that I is an $\mathscr{A}$-radical algebra, and we shall then have $I = \mathscr{A}(A)$. Let $I \notin \mathscr{A}$. Then there exists a proper $\mathfrak{B}$-ideal J in I. By Lemma 13 there is a $\mathfrak{B}$-ideal K in A such that $K \cap I = J \neq I$, but this contradicts the definition of I. Thus I is a radical ideal, and the theorem is proved.

COROLLARY. For every alternative algebra A the quotient algebra $A/\mathscr{A}(A)$ is a subdirect sum of subdirectly irreducible associative algebras with simple heart and Cayley-Dickson algebras.

In view of Theorem 4, $\mathscr{A}$-semisimple alternative algebras are precisely the subdirect sums of subdirectly irreducible algebras with idempotent heart. Exercise 4 sheds some light on the structure of $\mathscr{A}$-radical algebras.

Exercises

1. Prove that all locally nilpotent algebras are $\mathscr{A}$-radical.

2. Prove that in the class of associative–commutative algebras the anti-simple radical $\mathscr{A}$ coincides with the quasi-regular radical $\mathscr{J}$. In the class of all associative algebras these radicals are incomparable.

3. (*Andrunakievich*) Let I be a finitely generated ideal of an associative algebra A. If $I \subseteq \mathscr{A}(A)$, then $I^2 \neq I$. *Hint*. Let $I = (a_1, \ldots, a_n)$ and $I^2 = I$. Using Zorn's lemma select a maximal ideal M in I containing $a_1, \ldots, a_{n-1}$ and not containing a_n. Prove that M is maximal in I, that is, that I/M is a simple algebra. Use for this the Andrunakievich lemma.

4. Prove that an alternative algebra is $\mathscr{A}$-radical if and only if any homomorphic image of it is a subdirect sum of subdirectly irreducible algebras with nilpotent heart.

5. Prove that every separated algebra is the subdirect sum of an associative algebra and a purely nonassociative algebra.

6. Prove that in a purely nonassociative algebra A the associator ideal $D(A)$ has a nonzero intersection with each nonzero ideal of A.

7. Prove that every purely alternative algebra is an alternative PI-algebra. *Hint.* Use Exercise 7 in Section 7.1.

LITERATURE

Amitsur [12], Anderson [13, 14], Anderson, Divinsky, Sulinski [15], Andrunakievich [16–18], Babich [20], Golod [37], Divinsky [50], Divinsky, Krempa, Sulinski [51], Divinsky and Sulinski [52], Dorofeev [59], Zhevlakov and Shestakov [80], Kleinfeld [94], Krempa [102], Kurosh [105], Nikitin [159, 161], Pchelintsev [174], Roomel'di [182], Ryabukhin [186, 188, 189], Slin'ko [208, 210, 211], Hentzel and Slater [249], Tsai [257], Shestakov [271].

CHAPTER **9**

Semiprime Alternative Algebras

In the previous chapter we have seen that for the description of the semisimple classes of many radicals it is important to know the structure of the semiprime algebras from a given class $\mathfrak{K}$. In particular, both for the nil-radicals and the antisimple radical of Andrunakievich, all the semisimple algebras are semiprime. In this chapter we shall study the structure of semiprime alternative algebras. Since every semiprime algebra is a subdirect sum of prime ones, the problem in many instances reduces to the study of prime alternative algebras. If the characteristic 3 case is excluded, these latter algebras admit a completely satisfactory description modulo associative algebras: every nonassociative prime alternative algebra is a so called "Cayley–Dickson ring," which is represented by means of itself as a subring of a definite form in a Cayley–Dickson algebra. The question on the description of prime alternative algebras of characteristic 3 is still open.

The basic results of this chapter were obtained by Kleinfeld and Slater.

1. IDEALS OF SEMIPRIME ALGEBRAS

LEMMA 1. Let A be an alternative algebra and I be an ideal of A. If $(m, I, I) = (0)$ for some element $m \in A$, then $(m, A, A) \subseteq \mathrm{Ann}_r\, T(I)$.

PROOF. By identity (7.18) for any elements $i, j \in I$, $a \in A$

$$(m, iji, a) = -(m, iai, j) + (m, i, \{jia\}) = 0,$$

whence

$$(m, T(I), A) = (0).$$

If now $b \in A$, then

$$f(a, b, m, iji) = (ab, m, iji) - b(a, m, iji) - (b, m, iji)a = 0.$$

Consequently,

$$(iji)(a, b, m) = -f(a, iji, b, m) + (a(iji), b, m) - (iji, b, m)a = 0,$$

which is what was to be proved.

THEOREM 1 (*Slater*). Let A be a semiprime alternative algebra and I be an ideal of the algebra A. Then $N(I) = I \cap N(A)$.

PROOF. By the corollary to Theorem 8.6 the algebra A is a subdirect sum of prime algebras. If $N(I) \nsubseteq N(A)$, then there exist elements $n \in N(I)$, $a, b \in A$ such that $(n, a, b) \neq 0$. But then there can be found a prime homomorphic image $\bar{A}$ of the algebra A in which $(\bar{n}, \bar{a}, \bar{b}) \neq \bar{0}$. It is clear that the image $\bar{I}$ of the ideal I is different from zero. By Lemma 8.10, then also $T(\bar{I}) \neq (\bar{0})$. The right annihilator of the ideal $T(\bar{I})$ is an ideal in $\bar{A}$, which equals zero since $\bar{A}$ is prime. We now note that $\bar{n} \in \overline{N(I)} \subseteq N(\bar{I})$, and by Lemma 1 $(\bar{n}, \bar{a}, \bar{b}) \in \mathrm{Ann}_r(T(\bar{I})) = (\bar{0})$. This contradiction proves that $N(I) \subseteq N(A)$, that is, $N(I) \subseteq I \cap N(A)$. Since the reverse inclusion is obvious, this proves the theorem.

Now let A be an alternative algebra and I be a right ideal in A. We denote by $\check{I}$ the largest ideal of the algebra A contained in I, and by $\hat{I}$ the smallest ideal of the algebra A containing I. As usual, we shall denote by $(I:A)$ the set $\{x \in A \mid Ax \subseteq I\}$.

LEMMA 2. $\check{I} = (I:A^{\#})$.

PROOF. It is clear that $(I:A^{\#}) \subseteq I$ and that each two-sided ideal contained in I lies in $(I:A^{\#})$. For the proof of the lemma it suffices to be convinced this submodule is an ideal in A. Let $s \in (I:A^{\#})$, $a \in A$, $b \in A^{\#}$. Then

$$b(sa) = (bs)a - (b, s, a) = (bs)a + (s, b, a) \in I,$$
$$b(as) = (ba)s - (b, a, s) = (ba)s + (s, a, b) \in I,$$

that is, sa and as lie in $(I:A^{\#})$. This proves the lemma.

LEMMA 3. $(I, I, A) \subseteq \check{I}$.

PROOF. Let $i, j \in I$, $a \in A$, $b \in A^{\#}$. Then by the linearized Moufang identity

$$b(i, j, a) = -j(i, b, a) + (i, j, ab) + (i, b, aj) \in I,$$

whence $(I, I, A) \subseteq (I: A^{\#}) = \check{I}$. This proves the lemma.

LEMMA 4. $\hat{I} = A^{\#} I$.

PROOF. It obviously suffices to show that the set $A^{\#} I$ is an ideal in A. Let $i \in I$, $a \in A$, $b \in A^{\#}$. Then

$$a(bi) = (ab)i - (a, b, i) = (ab)i - (i, a, b) \in A^{\#} I,$$
$$(bi)a = b(ia) + (b, i, a) = b(ia) + (i, a, b) \in A^{\#} I,$$

which is what was to be proved.

LEMMA 5. Let A be semiprime and $n \in I$ be such that $(n, I, A) = (0)$. Then $n \in N(A)$.

PROOF. Let $i \in I$, $a, b \in A$, and set $n_1 = (n, a, b)$. We have $f(a, b, n, i) = (ab, n, i) - b(a, n, i) - (b, n, i)a = 0$, whence $(a, b, n)i = -f(i, a, b, n) + (ia, b, n) - a(i, b, n) = 0$, that is,

$$n_1 I = 0. \tag{1}$$

On the other hand, $i(a, b, n) = -f(a, i, b, n) + (ai, b, n) - (i, b, n)a = (ai, b, n)$, that is,

$$in_1 = (ai, b, n). \tag{2}$$

Furthermore, it follows from (1) that $(n_1, i, a) = (n_1 i)a - n_1(ia) = 0$. Also, since the element a is arbitrary, by Lemma 7.3 we obtain $[n_1, i] \in N(A)$, which by (1) and (2) gives us

$$(ai, n, A) \subseteq N(A).$$

By Lemma 7.7 each element from (ai, n, A) generates a trivial ideal in A. Since A is semiprime $(ai, n, A) = (0)$, and we now have

$$(n, \hat{I}, A) = (0).$$

In particular, $n \in N(\hat{I})$ and by Theorem 1 $n \in N(A)$, which proves the lemma.

THEOREM 2 (*Slater*). Let A be a semiprime alternative algebra and I be a right ideal in A. Then $N(I) = I \cap N(A)$.

PROOF. As in the proof of Theorem 1, it suffices to prove the assertion in the case when the algebra A is prime. If $\check{I} = (0)$, then by Lemma 3 $(I, I, A) =$

(0), which by Lemma 5 implies $I \subseteq N(A)$. It is easy to see that in this case the assertion of the theorem is true.

If $\check{I} \neq (0)$, then by Lemma 8.10 $T(\check{I}) \neq (0)$. For any element $n \in N(I)$ we have $(n, \check{I}, \check{I}) = (0)$, whence by Lemma 1 $(n, A, A) \subseteq \mathrm{Ann}_r\, T(\check{I})$. Since the algebra A is prime, we obtain that $\mathrm{Ann}_r\, T(\check{I}) = (0)$, and consequently $n \in N(A)$. Thus the inclusion $N(I) \subseteq I \cap N(A)$ is proved. The reverse inclusion is obvious. This proves the theorem.

THEOREM 3 (*Slater*). Under the conditions of Theorem 2, $Z(I) = I \cap Z(A)$.

PROOF. Let $z \in Z(I)$. Then by Theorem 2 $z \in N(A)$. Let $i \in I$, $a, b \in A$. By identity (7.4)

$$i[z, a] = [z, ia] - [z, i]a = 0.$$

Furthermore, by Corollary 1 to Lemma 7.3 $[z, a] \in N(A)$, and therefore $(bi)[z, a] = b(i[z, a]) = 0$. We see that $[z, a] \in \mathrm{Ann}_r(\hat{I})$. In view of the fact that the element $[z, a]$ belongs to the ideal $\hat{I}$ and its right annihilator, it also belongs to their intersection, which is a trivial ideal in A. Since A is semiprime $[z, a] = 0$, which means that $z \in Z(A)$. We have proved that $Z(I) \subseteq I \cap Z(A)$. This proves the theorem, since the reverse inclusion is obvious.

Finally, we shall show that in a semiprime alternative algebra every ideal is itself a semiprime algebra.

LEMMA 6. Let A be a semiprime alternative algebra, I be an ideal of the algebra A, and V be a trivial ideal of the algebra I. Then the submodules $A^{\#}V$ and $VA^{\#}$ are also trivial ideals of the algebra I.

PROOF. We show first that $A^{\#}V$ is an ideal in I. Let $a \in A^{\#}$, $i \in I$, $v \in V$. Then

$$(av)i = a(vi) + (a, v, i) = a(vi) - (a, i, v) \in A^{\#}V,$$
$$i(av) = (ia)v - (i, a, v) = (ia)v + (a, i, v) \in A^{\#}V,$$

which is what was to be proved. Analogously, we obtain that $VA^{\#}$ is an ideal of the algebra I. Furthermore, let $u, v \in V$, $i \in I$, $a, b \in A$. We have

$$f(a, i, u, v) = ([a, i], u, v) + (a, i, [u, v]) = 0.$$

Consequently

$$i(a, u, v) = -f(a, i, u, v) + (ai, u, v) - (i, u, v)a = 0,$$

that is, $(a, u, v) \in \mathrm{Ann}_r(I)$. At the same time $(a, u, v) = (v, a, u) \in I$. Since A is semiprime $I \cap \mathrm{Ann}_r(I) = (0)$, so finally we obtain

$$(A, V, V) = (AV)V = (0). \tag{3}$$

Now from (3) by the Moufang identity we have

$$(bv, i, u) = -(iv, b, u) + (v, i, u)b + (v, b, u)i = 0,$$

that is,

$$(AV, I, V) = (0). \tag{4}$$

From (4) by Lemma 7.3 it follows that $[AV, V] \subseteq N(I)$, whence by (3) and Theorem 1 we obtain the inclusion

$$VAV \subseteq N(A). \tag{5}$$

We note further that in view of the Moufang identity and (4), (3),

$$v(i, b, u) = -i(v, b, u) + (i, bv, u) + (v, bi, u) = 0,$$

whence

$$V(I, A, V) = (0). \tag{6}$$

We now set $w = uav$. Then $w \in V$, $w^2 = 0$, and by (5) $w \in N(A)$. Let $i = ua$. Then by (6)

$$wbw = w[b(iv)] = w[(bi)v] - w(b, i, v) = 0.$$

Thus $w \in N(A)$ and $wA^{\#}w = (0)$. By Lemma 7.6 the element w generates a trivial ideal in the algebra A, and therefore $w = 0$. This means

$$VAV = (0). \tag{7}$$

Now by (4), (3), and (7) we have

$$f(i, a, bv, u) = ([i, a], bv, u) + (i, a, [bv, u]) = 0,$$

whence by (4)

$$(a, u, bv)i = -f(i, a, u, bv) + (ia, u, bv) - a(i, u, bv) = 0,$$

that is, $(a, u, bv) \in \mathrm{Ann}_l(I)$. Since $(a, u, bv) \in I$, then in view of the fact the algebra A is semiprime we obtain

$$(A, V, AV) = (0). \tag{8}$$

From (3), (7), and (8) it follows that $(A^{\#}V)^2 = (0)$. Analogously, we obtain that $(VA^{\#})^2 = (0)$, which proves the lemma.

THEOREM 4 (*Slater*). Let A be a semiprime alternative algebra. Then every ideal I of the algebra A is itself a semiprime algebra.

PROOF. Let us assume that the algebra I is not semiprime, that is, there is a trivial ideal U in I. We denote by $\mathscr{U}$ the set of all trivial ideals of the algebra I containing U. The set $\mathscr{U}$ is nonempty since $U \in \mathscr{U}$. Furthermore, as is easy to see, $\mathscr{U}$ is an inductive set partially ordered by inclusion, and therefore by Zorn's lemma $\mathscr{U}$ contains a maximal element V. By Lemma 6 $A^{\#}V \in \mathscr{U}$, and therefore $AV \subseteq V$. Analogously, $VA \subseteq V$. Consequently, V is an ideal of the algebra A, whence $V = (0)$ and $U = (0)$. The obtained contradiction proves the theorem.

COROLLARY 1. Every ideal of a prime alternative algebra is itself a prime algebra.

PROOF. Let A be a prime alternative algebra and I be an ideal of A. If the algebra I is not prime, then there exist nonzero ideals U, V in I such that $U \cdot V = (0)$. We denote by M the right annihilator $\mathrm{Ann}_r(U)$ of the ideal U in the algebra I. Then $(M \cap U)^2 = (0)$, whence $M \cap U = (0)$, since by Theorem 4 the algebra I is semiprime. We now prove that the quotient algebra $\bar{I} = I/M$ is semiprime. Let $\bar{T}$ be an ideal in $\bar{I}$ with $\bar{T}^2 = (\bar{0})$. If T is the complete inverse image of the ideal $\bar{T}$ in the algebra I, then $T^2 \subseteq M$ and $(T \cap U)^2 \subseteq T^2 \cap U \subseteq M \cap U = (0)$, whence $T \cap U = (0)$ since the algebra I is semiprime. Consequently $UT \subseteq U \cap T = (0)$, $T \subseteq M$, and $\bar{T} = (\bar{0})$, which proves the algebra $\bar{I}$ is semiprime. By Theorem 8.4 we obtain that M is an ideal of the algebra A. But $\mathrm{Ann}_l(M) \supseteq U \neq (0)$, which contradicts that A is prime. This contradiction proves that the algebra I is prime, which proves the corollary.

From Theorem 4 and Proposition 8.5 also follows

COROLLARY 2. The mapping $A \to \mathscr{B}(A)$ is radical in the class of all alternative algebras.

Thus the lower nil-radical is defined in the class of alternative algebras. We note that by Theorem 8.9 this radical is hereditary.

Exercises

1. Let I be an ideal of an alternative algebra A and e be an identity element of the algebra I. Prove that $e \in N(A)$.

In Exercises 2–6 I is a minimal right ideal of an alternative algebra A where $I^2 \neq (0)$.

2. Let $(I, I, A) = (0)$. Prove that

(a) $I = I^2 = vI$ for some $v \in I$;
(b) $I = T(I)$;
(c) $(I, \hat{I}, A) = (0)$;
(d) $I \subseteq N(A)$.

Hint. In the proof of (c) argue as in Lemma 5, and for the proof of (d) use the assertions of points (b), (c), and identity (7.18).

3. Prove that either $\hat{I} \subseteq U(A)$ or $\hat{I} \subseteq D(A)$. *Hint*. Use Lemma 3 and the previous exercise.

4. Let $\hat{I} \subseteq U(A)$. Then $I^2 = I$, $\hat{I} = \hat{I}I$, and I is a minimal right ideal of the associative algebra $\hat{I}$. In addition, if $\mathrm{Ann}_r(\hat{I}) = (0)$, then $\hat{I}$ is a simple algebra.

5. Prove that if A is associative, then $I = eA$ for some idempotent $e \in A$.

6. Let $\hat{I} \subseteq D(A)$. Then $I = \hat{I}$ is a Cayley–Dickson algebra.

7. (*Slater*) Let I be a minimal right ideal of an alternative algebra A and $I^2 \neq (0)$. Then $I^2 = I$ and I is a minimal right ideal of the algebra $\hat{I}$. In addition, $I = eA$ for some idempotent $e \in N(A)$.

8. Prove that a semiprime alternative algebra does not contain nonzero trivial right ideals.

9. (*Slater*) Let A be a semiprime alternative algebra and I be an ideal of the algebra A. Then the set of all minimal ideals (right ideals) of the algebra I coincides with the set of all minimal ideals (right ideals) of the algebra A contained in I.

10. Prove that if A is a semiprime alternative algebra, then $Z(D(A)) = N(D(A))$. *Hint*. Use Theorem 8.11.

In Exercises 11–13 A is a semiprime alternative algebra, and I is a right ideal of the algebra A.

11. Let V be a trivial ideal of the algebra I. Prove that $V \subseteq N(I)$. *Hint*. Prove that $V \cap \check{I} = (0)$.

12. Let V be a nuclear ideal of the algebra I. Prove that $VA^{\#}$ is a nuclear right ideal of the algebra A.

13. (*Slater*) Let A be semiprime and purely alternative. Then the algebra I is also semiprime and purely alternative.

2. NONDEGENERATE ALTERNATIVE ALGEBRAS

An element a of an alternative algebra A is called an *absolute zero divisor* if $aAa = (0)$. An algebra A is called *nondegenerate* (or *strongly semiprime*) if A does not contain nonzero absolute zero divisors. As is easy to see, an associative algebra A is nondegenerate if and only if A is semiprime. In the case of alternative algebras the situation is complicated. It is clear that every nondegenerate algebra is semiprime. However, the question on whether a semiprime alternative algebra can contain nonzero absolute zero divisors still remains not completely answered. In this section we shall show that semiprime alternative algebras of characteristic $\neq 3$ are nondegenerate, and study some properties of nondegenerate alternative algebras.

THEOREM 5 (*Kleinfeld*). Let A be a semiprime alternative algebra where $3A = A$. Then A is nondegenerate.

PROOF. In view of the corollary to Theorem 8.6, it again suffices for us to prove the theorem in the case when the algebra A is prime. Suppose that A is not nondegenerate, that is, A contains a nonzero absolute zero divisor a. Considering, if necessary, the element a^2 instead of a, we can assume that $aA^{\#}a = (0)$. We denote by T the set

$$T = \{t \in A \mid a(A^{\#}t) = (aA^{\#})t = t(A^{\#}a) = (0)\}.$$

We note that $T \neq (0)$, since $a \in T$. We fix some element $t \in T$ and consider the Φ-module $S = S(t)$ generated by the set (aA, t, A). Let x, y, z be arbitrary elements of the algebra A. Since $(a, A, t) = (0)$, we have $(ax, t, x) = x(a, t, x) = 0$. Linearizing this relation in x, we obtain

$$(ax, t, y) = -(ay, t, x). \tag{9}$$

Henceforth notation of the form $u \equiv v$ will denote that $u - v \in S$. We have

$$f(ax, t, y, z) = ([ax, t], y, z) + ([y, z], ax, t) \equiv 0.$$

Consequently,

$$(ax, t, y)z = (ax, t, zy) - y(ax, t, z) - f(z, y, ax, t) \equiv -y(ax, t, z),$$

that is,

$$(ax, t, y)z \equiv -y(ax, t, z). \tag{10}$$

Now by (10) and (9) we obtain

$$\begin{aligned}(ax, t, y)z &\equiv -y(ax, t, z) = y(az, t, x) \equiv -(az, t, y)x \\ &= (ay, t, z)x \equiv -z(ay, t, x) = z(ax, t, y),\end{aligned}$$

that is, $[(ax, t, y), z] \in S$. We have proved that

$$[S, A] \subseteq S. \tag{11}$$

We denote by U the submodule of the Φ-module A generated by the set $SA^{\#}$. By identity (7.7) for any $s \in S$, $x, y \in A$ we have

$$\begin{aligned}3(x, y, s) &= [xy, s] - x[y, s] - [x, s]y \\ &= [xy, s] - [y, s]x - [x, [y, s]] - [x, s]y,\end{aligned}$$

whence in view of (11) and the condition $3A = A$ we conclude that

$$(A, A, S) \subseteq U. \tag{12}$$

Now by (12) and (11) for any $s \in S$, $a \in A^{\#}$, $b \in A$ we have

$$(sa)b = s(ab) + (s, a, b) \in U,$$
$$b(sa) = (bs)a - (b, s, a) = (sb)a + [b, s]a - (b, s, a) \in U,$$

that is, U is an ideal of the algebra A. Furthermore, by the linearized Moufang identity we have

$$(ax, t, y) = -(yx, t, a) + (x, t, y)a + (x, t, a)y = (x, t, y)a,$$

whence by the right Moufang identity for any $x, y \in A$, $z \in A^{\#}$

$$[(ax, t, y)z]a = ([(x, t, y)a]z)a = (x, t, y)(aza) = 0,$$

that is, $a \in \mathrm{Ann}_r(U)$. Since the algebra A is prime and $a \neq 0$, we obtain that $U = (0)$. In an analogous manner, considering the Φ-submodule $S' = S'(t)$ generated by the set (Aa, t, A) and the ideal $U' = A^{\#}S'$, we obtain that $a \in \mathrm{Ann}_l(U')$, whence $U' = (0)$. In particular, we have proved that $S(t) = S'(t) = (0)$ for any $t \in T$, that is,

$$(aA, T, A) = (Aa, T, A) = (0). \tag{13}$$

We now note that for any $x, y \in A$, $t \in T$

$$f(x, y, a, t) = (xy, a, t) - y(x, a, t) - (y, a, t)x = 0,$$

whence in view of (13),

$$0 = (ax, y, t) = f(a, x, y, t) + x(a, y, t) + (x, y, t)a$$
$$= (x, y, t)a = -f(a, t, x, y) + (at, x, y) - t(a, x, y) = -t(a, x, y).$$

Analogously,

$$0 = (xa, y, t) = -(x, y, a)t$$
$$= f(t, x, y, a) - (tx, y, a) + x(t, y, a) = -(tx, y, a).$$

Consequently,

$$(TA, A, a) = T(a, A, A) = (0). \tag{14}$$

We shall now show that T is an ideal of the algebra A. For any $x \in A$, $y \in A^{\#}$, $t \in T$, by (13) we have

$$(ay)(tx) = -(ay, t, x) + ((ay)t)x = 0,$$

whence according to (14)

$$a(y(tx)) = (ay)(tx) - (a, y, tx) = 0.$$

Finally, by (13) and (14) we have

$$(tx)(ya) = t[x(ya)] + (t, x, ya)$$
$$= t[x(ya)] = t[(xy)a] - t(x, y, a) = (0).$$

The obtained relations show that T is a right ideal of the algebra A. Analogously it is proved that T is a left ideal in A. Since $a \in T$ and $a \in \mathrm{Ann}(T)$, we obtain that $a = 0$ because the algebra A is prime. This contradiction proves the theorem.

In order to pass from algebras with the restriction $3A = A$ to the case of algebras without elements of order 3 in the additive group, we need some additional observations.

Let R be some ring and $R^\# = R + \mathbf{Z} \cdot 1$ be the ring obtained by the formal adjoining of an identity element to R (here $\mathbf{Z}$ is the ring of integers).

PROPOSITION 1. A ring R is semiprime (nondegenerate) if and only if the ring $R^\#$ is semiprime (nondegenerate, respectively).

PROOF. Since every ideal of the ring R is an ideal in $R^\#$, it is clear that $R^\#$ semiprime implies R is semiprime. It is also clear the ring $R^\#$ nondegenerate implies the ring R is nondegenerate. For the proof of the converse assertions it suffices to note that every nilpotent element of the ring $R^\#$ lies in R. Let $r = n \cdot 1 + r' \in R^\#$ where $n \in \mathbf{Z}$, $r' \in R$, and let $r^2 = 0$. Then we have $0 = n^2 \cdot 1 + r'' \in R$ where $r'' \in R$, whence $n = 0$ and $r = r' \in R$. This proves the proposition.

We assume now that the center $Z = Z(R)$ of the ring R is different from zero. A nonempty subset S of the center Z is called a *multiplicative subset* if, together with any two elements x, y, S contains their product xy. As in the case of associative-commutative rings (see, e.g., S. Lang, "Algebra," p. 85), we can construct the *ring of quotients* of the ring R modulo S.

We consider the pairs (r, s) where $r \in R$ and $s \in S$. We define a relation $(r, s) \sim (r', s')$ between such pairs by the following condition: there exists an $s_1 \in S$ for which $s_1(s'r - sr') = 0$. It is trivially verified that this will be an equivalence relation. We denote the equivalence class containing the pair (r, s) by $s^{-1}r$, and the set of equivalence classes by $S^{-1}R$. Addition and multiplication in the set $S^{-1}R$ are introduced by the conditions

$$s^{-1}r + s_1^{-1}r_1 = (ss_1)^{-1}(s_1 r + s r_1),$$
$$(s^{-1}r)(s_1^{-1}r_1) = (ss_1)^{-1}(rr_1).$$

It is trivially verified that addition and multiplication are well defined, and that with respect to these operations the set $S^{-1}R$ is a ring.

We note that along with the ring $S^{-1}R$ we can also consider the usual associative–commutative ring of quotients $S^{-1}Z$ of the center Z modulo S.

Let s_0 be some fixed element of the set S. As is easy to see, the element $s_0^{-1}s_0$ is an identity element of the ring $S^{-1}R$, and also $s_0^{-1}s_0 = s^{-1}s$ for any $s \in S$. In addition, $s^{-1}r = (ss_1)^{-1}(rs_1)$ for any $r \in R$ and $s, s_1 \in S$, that is, we can "reduce" our fractions by elements from S. We now consider the mapping $\rho_0: R \to S^{-1}R$ for which $\rho_0(r) = s_0^{-1}(rs_0)$. It is seen at once that ρ_0 is a ring homomorphism. In addition, every element from $\rho_0(S)$ is invertible in $S^{-1}Z$ (as an inverse for $s_0^{-1}(ss_0)$ can be used $(ss_0)^{-1}s_0$). If s_1 is another element from S, then, as is easy to see, the mapping $\rho_1: R \to S^{-1}R$ for which $\rho_1(r) = s_1^{-1}(rs_1)$ coincides with the mapping ρ_0. Thus the mapping ρ_0 does not depend on the choice of the element s_0, and we shall denote it by ρ_S.

PROPOSITION 2. Let S be a multiplicative subset of the center Z of a ring R, where S does not contain zero divisors of the ring R. Then

(1) the mapping $\rho_S: R \to S^{-1}R$ is an insertion of the ring R into $S^{-1}R$;

(2) $Z(S^{-1}R) = S^{-1}Z$;

(3) the ring $S^{-1}R$ is semiprime (nondegenerate, prime) if and only if the ring R is semiprime (nondegenerate, prime, respectively);

(4) if the ring R is an algebra over Φ, then the ring $S^{-1}R$ will also be an algebra over Φ with respect to the composition $\alpha \cdot (s^{-1}r) = s^{-1}(\alpha r)$, where $\alpha \in \Phi$, $s \in S$, $r \in R$. In this connection the Φ-algebra $S^{-1}R$ satisfies a homogeneous identity $f = f(x_1, \ldots, x_n) \in \Phi[X]$ if and only if the Φ-algebra R satisfies this identity.

(5) if $S \cap I \neq \varnothing$ for some ideal I of the ring R, then the set $S_1 = S \cap I$ is a multiplicative subset of the center of the ring I. In this connection the rings of quotients $S_1^{-1}I$ and $S^{-1}R$ coincide.

PROOF. We have already noted that the mapping ρ_S is a ring homomorphism. If $\rho_S(r) = s_0^{-1}(rs_0) = 0$ for some $r \in R$, then by definition, this means there exists $s \in S$ for which $s(rs_0) = (ss_0)r = 0$, and consequently $r = 0$. By the same token (1) is proved, and we shall henceforth identify the ring R with its image $\rho_S(R)$ in the ring $S^{-1}R$.

Now let $z \in Z$, $r, t \in R$, $s, s_1, s_2 \in S$. We have

$$(s^{-1}z, s_1^{-1}r, s_2^{-1}t) = (ss_1s_2)^{-1}(z, r, t) = 0,$$
$$[s^{-1}z, s_1^{-1}r] = (ss_1)^{-1}[z, r] = 0,$$

whence it follows that $S^{-1}Z \subseteq Z(S^1R)$. Conversely, let $s^{-1}r \in Z(S^{-1}R)$. Then for any $x, y \in R$ we have

$$0 = (s^{-1}r, x, y) = s^{-1}(r, x, y) = (s^{-1}(r, x, y))s = (r, x, y),$$

that is, $r \in N(R)$. Analogously we obtain that $[r, x] = 0$ for any $x \in R$, that is, $r \in Z$ and $Z(S^{-1}R) \subseteq S^{-1}Z$. By the same token assertion (2) is proved.

We note now that for any ideal I of the ring R the set $S^{-1}I = \{s^{-1}i \mid s \in S, i \in I\}$ is an ideal of the ring $S^{-1}R$. In this connection $I \cdot J = (0)$ for some ideals I, J of the ring R if and only if $(S^{-1}I) \cdot (S^{-1}J) = (0)$. Hence it follows that the ring $S^{-1}R$ is semiprime or prime implies the ring R is semiprime or prime, respectively. For the proof of the converse assertions it obviously suffices to note that $I \cap R \neq (0)$ for any nonzero ideal I of the ring $S^{-1}R$. Let $0 \neq i \in I$, $i = s^{-1}r$, where $s \in S$, $r \in R$. Then $r \neq 0$ and $r = si \in I \cap R$, whence $I \cap R \neq (0)$. Finally, let a be an absolute zero divisor of the ring R, that is, $aRa = (0)$. Then for any $r \in R$, $s \in S$ we have $a(s^{-1}r)a = s^{-1}(ara) = 0$, that is, a is an absolute zero divisor of the ring $S^{-1}R$. Conversely, if the element $x = s^{-1}r$ is a nonzero absolute zero divisor of the ring $S^{-1}R$, then the element $sx = r$ will be a nonzero absolute zero divisor in R. Hence it follows that the ring $S^{-1}R$ is nondegenerate if and only if R is nondegenerate. This proves assertion (3).

It is clear that if R is a Φ-algebra, then the ring $S^{-1}R$ will also be a Φ-algebra with respect to the composition indicated in the proposition. If the Φ-algebra $S^{-1}R$ satisfies some identity, then the Φ-algebra R as a subalgebra of the algebra $S^{-1}R$ must also satisfy this identity. Conversely, let the algebra R satisfy a homogeneous identity $f(x_1, \ldots, x_n) \in \Phi[X]$, and let all the monomials appearing in f have type $[k_1, \ldots, k_n]$. Then for any elements $r_1, \ldots, r_n \in R$, $s_1, \ldots, s_n \in S$, we have $f(s_1^{-1}r_1, \ldots, s_n^{-1}r_n) = (s_1^{k_1}, \ldots, s_n^{k_n})^{-1} f(r_1, \ldots, r_n) = 0$, that is, the algebra $S^{-1}R$ also satisfies the identity f. By the same token (4) is also proved.

Finally, let $I \cap S = S_1 \neq \varnothing$ for some ideal I of the ring R. It is clear that S_1 is a multiplicative subset of the center of the ring I and that the inclusion $S_1^{-1}I \subseteq S^{-1}R$ is valid. We now fix some element $s_0 \in S_1$. Every element of the form $s^{-1}r$ from the ring $S^{-1}R$ can be represented in the form $s^{-1}r = (ss_0)^{-1}(rs_0)$. Since $ss_0 \in I \cap S$, $rs_0 \in I$, then by the same token it is proved that $S^{-1}R \subseteq S_1^{-1}I$. This proves the proposition.

It follows from (4) that if the ring R belongs to a homogeneous variety $\mathfrak{M}$, then the ring $S^{-1}R$ also belongs to this variety. In particular, if R is alternative then $S^{-1}R$ is also alternative.

We are now able to prove the analog of Theorem 5 for alternative algebras without elements of order 3 in the additive group.

THEOREM 5′. Let A be a semiprime alternative algebra without elements of order 3 in the additive group. Then A is nondegenerate.

PROOF. We consider the algebra A as a ring. It is easy to see that the ring A is semiprime. Then by Proposition 1 the ring $A^{\#} = A + \mathbf{Z} \cdot 1$ is also semiprime, and in addition $Z(A^{\#}) \neq (0)$. We consider in the center $Z(A^{\#})$ of the ring $A^{\#}$ the subset $S = \{3^k \cdot 1 \mid k = 1, 2, \ldots\}$. It is clear that S is a multiplicative subset of the center $Z(A^{\#})$. In addition, by the assumptions of the theorem, S does not contain zero divisors of the ring $A^{\#}$. By Proposition 2 the ring $A^{\#}$ can be isomorphically inserted into the ring of quotients $S^{-1}A^{\#}$, which also is semiprime. Furthermore, it is easy to see that $3(S^{-1}A^{\#}) = S^{-1}A^{\#}$, and therefore by Theorem 5 the ring $S^{-1}A^{\#}$ is nondegenerate. In view of Propositions 2 and 1, the ring A is also nondegenerate. Since the concept of nondegeneracy in no way depends on the ring of operators, by the same token the theorem is proved.

REMARK. We note without proof that every finitely generated semiprime algebra is nondegenerate. This follows from the following result belonging to Shestakov [271].

For any natural number m there exists a natural number $f(m)$ such that in every alternative algebra from m generators any absolute zero divisor generates a nilpotent ideal of index not greater than $f(m)$.

We turn now to the study of the properties of nondegenerate alternative algebras.

THEOREM 6 (*Kleinfeld*). Let A be a nondegenerate alternative algebra. Then either $A = Z(A)$ or $N_{\mathrm{Alt}}[A] \neq (0)$.

PROOF. We recall that we have denoted by $N_{\mathrm{Alt}}[A]$ the subalgebra of the algebra A consisting of all possible specializations in the algebra A of elements of the associative center N_{Alt} of the free alternative algebra $\mathrm{Alt}[X]$ (see Section 7.2). By the Kleinfeld identities the relations

$$([x_1, x_2]^4, x_3, x_4) = ([x_1, x_2]^2, [x_1, x_2]x_3, x_4) = 0$$

are valid in the algebra $\mathrm{Alt}[X]$. Hence it follows that $[x_1, x_2]^4 \in N_{\mathrm{Alt}}$, and by Lemma 7.3 $[[x_1, x_2]^2, [x_1, x_2]x_3] \in N_{\mathrm{Alt}}$. We assume that $N_{\mathrm{Alt}}[A] = (0)$. Then for any $x, y, z \in A$ and $v = [x, y]$ we have $v^4 = [v^2, vz] = 0$. Therefore $v[v^2, z] = [v^2, vz] = 0$ and $v^2zv^2 = v^2[z, v^2] = 0$, whence in view of the nondegeneracy of the algebra A

$$v^2 = [x, y]^2 = 0. \tag{15}$$

Linearizing this relation in y, we obtain

$$[x, y] \circ [x, z] = 0. \tag{16}$$

Now by (15) and (16) we have

$$vxv = vxv + xv^2 = [x, y]x[x, y] + x[x, y][x, y] = [x, y] \circ [x, xy],$$

whence

$$v[v, x] = [v, x]v = 0. \tag{17}$$

Furthermore, by identity (7.7) we have

$$z[v, x] = [zv, x] - [z, x]v - 3(z, v, x).$$

Taking the Jordan product of both sides of this equality with the element $[v, x]$, we obtain by (15), (16), and (7.15)

$$\begin{aligned}[v, x]z[v, x] &= [zv, x] \circ [v, x] - ([z, x]v) \circ [v, x] \\ &\quad - 3(z, v, x) \circ [v, x] = -([z, x]v) \circ [v, x].\end{aligned}$$

Then by (16), (17), and the linearized alternative identities,

$$\begin{aligned}[v, x]z[v, x] &= -([z, x]v) \circ [v, x] = (v[z, x])[v, x] - [v, x]([z, x]v) \\ &= -(v[v, x])[z, x] + v([z, x] \circ [v, x]) \\ &\quad + [z, x]([v, x]v) - ([v, x] \circ [z, x])v = 0.\end{aligned}$$

In view of the nondegeneracy of the algebra A, it follows from this that

$$[v, x] = [[x, y], x] = 0. \tag{18}$$

In particular $[[v, z], v] = 0$, and therefore by (15)

$$0 = [v^2, z] = v \circ [v, z] = 2vzv.$$

Hence, again by the nondegeneracy of A, we obtain

$$2v = 2[x, y] = 0. \tag{19}$$

We now have by (19) and (18)

$$[x^2, y] = x \circ [x, y] = [[x, y], x] + 2x[x, y] = 0.$$

Linearizing this relation in x, we obtain

$$[x \circ y, z] = 0,$$

whence in view of (19),

$$[v, z] = [[x, y], z] = [x \circ y, z] - 2[yx, z] = 0,$$

and then by (15)

$$vzv = v^2 z = 0.$$

Consequently $v = 0$, and the algebra A is commutative. By Lemma 7.8 we now have $(x, y, z)^2 = 0$, and then $(x, y, z)t(x, y, z) = (x, y, z)^2 t = 0$, whence $(x, y, z) = 0$. Thus the algebra A is associative and commutative, that is, $Z(A) = A$. This proves the theorem.

COROLLARY. Let A be a nondegenerate alternative algebra. If $A \neq (0)$, then also $N(A) \neq 0$.

In fact, we have the inclusions

$$N_{\mathrm{Alt}}[A] \subseteq N(A), \qquad Z(A) \subseteq N(A),$$

from which everything follows.

LEMMA 7. Every ideal of a nondegenerate alternative algebra is itself a nondegenerate algebra.

PROOF. Let A be a nondegenerate alternative algebra and I be an ideal of the algebra A. We assume there exists an $i \in I$ such that $iIi = (0)$. Then for any $x, y, \in A$ by the Moufang identities we obtain

$$(ixi)y(ixi) = [i(x \cdot iy)](ixi) = [i(x \cdot iy)](ix \cdot i) = i[(x \cdot iy)(ix)]i = 0.$$

Consequently $(ixi)A(ixi) = (0)$, whence by the nondegeneracy of the algebra A it follows $ixi = 0$. Since the element x was arbitrary, this means that $iAi = (0)$, whence finally we have $i = 0$. Thus the algebra I does not contain nonzero absolute zero divisors. This proves the lemma.

THEOREM 7 (*Slater*). Let A be a nondegenerate alternative algebra. If A is not associative, then $Z(A) \neq (0)$.

PROOF. We consider the associator ideal $D = D(A)$ of the algebra A. Since A is not associative, $D \neq (0)$. By Lemma 7 the ideal D is a nondegenerate algebra, and therefore by the corollary to Theorem 6 $N(D) \neq (0)$. Furthermore, in view of the nondegeneracy of D, by Theorem 1 we have $N(D) = D \cap N(A)$. We select $0 \neq n \in D \cap N(A)$. Then for any $a \in A$ by Theorem 8.11 we have

$$[n, a] \in ZN(A) \cap D \subseteq U \cap D,$$

where $U = U(A)$ is the associative nucleus of the algebra A. By Proposition 8.10 we have $(U \cap D)^2 = (0)$, whence, since the algebra A is semiprime, it follows that $U \cap D = (0)$. By the same token this proves that $[n, a] = 0$ for all $a \in A$, that is, $0 \neq n \in Z(A)$. This proves the theorem.

COROLLARY 1. Every nondegenerate alternative nil-algebra is associative.

In fact, if A is not associative, then by Theorem 7 $Z(A) \neq 0$. On the other hand, it is easy to see that no semiprime algebra can contain nonzero nilpotent elements in its center.

COROLLARY 2 (*Kleinfeld*). Let A be a semiprime alternative nil-algebra without elements of order 3 in the additive group. Then A is associative.

In concluding this section we apply the results obtained to an arbitrary semiprime alternative algebra.

Let A be some algebra and m be a natural number. We set $A_m = \{x \in A \mid mx = 0\}$, and we denote by $\bar{A} = \bar{A}(m)$ the quotient algebra A/A_m.

LEMMA 8. If the algebra A is semiprime, then the algebra $\bar{A}$ is semiprime and does not contain elements of order m in the additive group.

PROOF. Let $\bar{I}$ be an ideal of the algebra $\bar{A}$ and $\bar{I}^2 = (\bar{0})$. We consider the complete inverse image I of the ideal $\bar{I}$ in the algebra A. Then I is an ideal in A and $I^2 \subseteq A_m$. This means that $mI^2 = (0)$, whence $(mI)^2 = (0)$. Since the algebra A is semiprime, we have $mI = (0)$. Consequently $I \subseteq A_m$ and $\bar{I} = (\bar{0})$, so that the algebra $\bar{A}$ is semiprime.

We note that $(mA_{m^2})^2 = (0)$. Since the algebra A is semiprime, it follows from this that $mA_{m^2} = (0)$ and $A_{m^2} \subseteq A_m$. Now if $\bar{x} \in (\bar{A})_m$, then $m\bar{x} = \bar{0}$, and for any inverse image x of the element $\bar{x}$ in the algebra A we have $mx \in A_m$ so $m^2x = 0$. Hence $x \in A_{m^2} \subseteq A_m$, and $\bar{x} = \bar{0}$. By the same token this proves that $(\bar{A})_m = (\bar{0})$, that is, there are no elements of order m in $\bar{A}$.

This proves the lemma.

THEOREM 8 (*Slater*). Let A be a semiprime alternative algebra. Then

(1) $3a = 0$ for any absolute zero divisor a of the algebra A;
(2) if $3A \neq (0)$, then $N(A) \neq (0)$;
(3) if $3A \not\subseteq N(A)$, then $Z(A) \neq 0$.

PROOF. We consider the quotient algebra $\bar{A} = A/A_3$. By Lemma 8 and Theorem 5′ the algebra $\bar{A}$ is nondegenerate, and therefore every absolute zero divisor of the algebra A lies in A_3.

By the same token this proves (1).

Furthermore, if $3A \neq (0)$, then $\bar{A} \neq (\bar{0})$. By the corollary to Theorem 6 we have $N(\bar{A}) \neq (\bar{0})$. We select $\bar{0} \neq \bar{n} \in N(\bar{A})$, and let n be some inverse image of the element $\bar{n}$ in A. Also let $r, s \in A$, $t = (n, r, s)$. Then $\bar{t} = (\bar{n}, \bar{r}, \bar{s}) = \bar{0}$ in the algebra $\bar{A}$. This means $t \in A_3$ and $(3n, r, s) = 0$. Since the elements r, s were arbitrary, $3n \in N(A)$. If $3n = 0$, then $n \in A_3$ and $\bar{n} = \bar{0}$. Consequently $0 \neq 3n \in N(A)$ and $N(A) \neq (0)$, which proves (2).

Finally, if $3A \not\subseteq N(A)$, then the algebra $\bar{A}$ is not associative. By Theorem 7 we have $Z(\bar{A}) \neq (0)$, whence as in case (2) we obtain $Z(A) \neq (0)$.

This proves the theorem.

COROLLARY. Let A be an arbitrary alternative algebra. Then $3a \in \mathscr{B}(A)$ for any absolute zero divisor a of the algebra A.

Exercises

1. Let R be some ring with nonzero center Z, and S be a multiplicative subset of the center Z which does not contain zero divisors of the ring R. For an ideal I of the ring R we set $S^{-1}I = \{s^{-1}i \in S^{-1}R \mid s \in S, i \in I\}$. Prove that the mapping $\Psi_S: I \to S^{-1}I$ maps the set of all ideals of the ring R onto the set of all ideals of the ring of quotients $S^{-1}R$. In addition, for any ideals I, J of the ring R

$$S^{-1}(I + J) = S^{-1}I + S^{-1}J,$$
$$S^{-1}(IJ) = (S^{-1}I)(S^{-1}J),$$
$$S^{-1}(I \cap J) = (S^{-1}I) \cap (S^{-1}J).$$

2. Prove that the following isomorphism holds:

$$S^{-1}R \cong S^{-1}Z \otimes_Z R.$$

3. Let R be a prime ring. Prove that the ring $R^{\#}$ is prime if and only if for any $n \in Z$ there are no elements a in R such that $R_a = L_a = nE$.

4. Prove that in an arbitrary alternative algebra A the submodule $\mathcal{M}_1(A)$ generated by all absolute zero divisors is an ideal.

5. By analogy with the Baer chain of ideals $\{\mathcal{B}_\alpha(A)\}$ we define in an arbitrary alternative algebra A a chain of ideals $(0) \subseteq \mathcal{M}_1(A) \subseteq \mathcal{M}_2(A) \subseteq \cdots \subseteq \mathcal{M}_\alpha(A) \subseteq \cdots \subseteq \mathcal{M}(A)$, where $\mathcal{M}_1(A)$ is the ideal of the algebra A introduced in Exercise 4. Prove that the mapping $A \to \mathcal{M}(A)$ is a hereditary radical in the class of all alternative algebras. In addition, the semisimple class of the radical $\mathcal{M}$ is precisely the class of all nondegenerate alternative algebras.

6. Prove that in every alternative algebra A

$$\mathcal{M}(A) \supseteq \mathcal{B}(A) \supseteq 3\mathcal{M}(A).$$

3. PRIME ALTERNATIVE ALGEBRAS

Let R be some ring such that $Z = Z(R) \neq (0)$, and Z does not contain zero divisors of the ring R. Then the set $Z^* = Z \setminus \{0\}$ is a multiplicative subset of the center Z, and we can consider the rings of quotients $Z_1 = (Z^*)^{-1}Z$ and $R_1 = (Z^*)^{-1}R$. It is clear that in the given case the ring Z_1 will be the usual field of quotients of the associative–commutative ring Z without zero divisors. By Proposition 2 we have $Z(R_1) = Z_1$, so that the ring R_1 can in a unique manner be considered as a central algebra over the field Z_1. In addition, the ring R is isomorphically imbedded in the algebra R_1.

We now introduce the following definition. We shall call an alternative ring K, with a nonzero center Z which does not contain zero divisors of the ring K, a *Cayley–Dickson ring* if the ring of quotients $K_1 = (Z^*)^{-1}K$ is a Cayley–Dickson algebra over the field of quotients Z_1 of the center Z.

PROPOSITION 3. Every Cayley–Dickson ring is a prime nondegenerate purely alternative ring. Any nonzero ideal of a Cayley–Dickson ring is itself a Cayley–Dickson ring. Conversely, if some ideal I of a prime alternative ring R is a Cayley–Dickson ring, then the ring R is itself also a Cayley–Dickson ring.

PROOF. Let K be some Cayley–Dickson ring and $\mathbf{C} = K_1$ be the corresponding Cayley–Dickson algebra. Since every Cayley–Dickson algebra is a simple ring (in particular, prime), then in view of Proposition 2 the ring K is prime. Furthermore, since the algebra $\mathbf{C}$ is not associative the ring K also is not associative, and therefore by the corollary to Proposition 8.10, K is purely alternative. By Proposition 2, for the proof of the nondegeneracy of the ring K it suffices to establish that the Cayley–Dickson algebra $\mathbf{C}$ is nondegenerate. Let $a \in \mathbf{C}$, $a \neq 0$, and $a\mathbf{C}a = (0)$. Then in particular $a^2 = 0$, whence $t(a) = n(a) = 0$, where $t(x)$ and $n(x)$ are the trace and norm, respectively, of the element x in the Cayley–Dickson algebra $\mathbf{C}$ (see Chapter 2). Now for any $x \in \mathbf{C}$ by (2.21) we have

$$\begin{aligned} axa = (a \circ x)a - xa^2 &= t(a)xa + t(x)a^2 - t(x)t(a)a + t(xa)a \\ &= t(xa)a = 0. \end{aligned}$$

We note that by (2.8)

$$t(xa) = f(xa, 1) = -f(x, a) + f(x, 1)f(a, 1) = -f(x, a).$$

Since the form f is nondegenerate on $\mathbf{C}$, there can be found an element $x \in \mathbf{C}$ for which $t(xa) = -f(x, a) \neq 0$. But then $a = 0$. The contradiction obtained proves the nondegeneracy of the algebra $\mathbf{C}$ and the ring K.

Now let $(0) \neq I$ be an ideal of the Cayley–Dickson ring K. By Corollary 1 to Theorem 4 and Lemma 7 the ring I is prime and nondegenerate. If I is associative, then by Theorem 1 we have $I = N(I) \subseteq N(K)$, whence $I \subseteq U(K)$ and $U(K) \neq (0)$, which contradicts that the ring K is purely alternative. This means the ring I is not associative. Then by Theorems 7 and 3 we have $(0) \neq Z(I) = I \cap Z(K)$. Hence by assertion (5) of Proposition 2 it follows that $\mathbf{C} = S^{-1}I$, where $S = Z(I)\backslash\{0\}$, that is, I is a Cayley–Dickson ring.

Finally, let R be some prime alternative ring and I be an ideal of the ring R that is a Cayley–Dickson ring. Then $Z_0 = Z(I) \neq 0$, whence also $Z = Z(R) \neq (0)$. Since the ring R is prime, the set $Z^* = Z\backslash\{0\}$ does not contain zero divisors of the ring R. Consequently we can consider the ring of quotients $R_1 = (Z^*)^{-1}R$. Since $I \cap Z^* = Z_0^* = Z(I)\backslash\{0\} \neq \varnothing$, then again by assertion (5) of Proposition 2 we obtain that $(Z_0^*)^{-1}I = (Z^*)^{-1}R$. Since by assumption the ring $(Z_0^*)^{-1}I$ is a Cayley–Dickson algebra, by the same token it is proved that R is a Cayley–Dickson ring.

This proves the proposition.

COROLLARY. Every Cayley–Dickson ring is nil semisimple.

PROOF. Let $\mathfrak{N}(K)$ be the upper nil-radical of the Cayley–Dickson ring K. If $\mathfrak{N}(K) \neq (0)$, then $\mathfrak{N}(K)$ is a Cayley–Dickson ring. On the other hand, by Corollary 1 to Theorem 7 every nondegenerate alternative nil-ring is associative. The contradiction obtained proves that $\mathfrak{N}(K) = (0)$, that is, there are no nonzero nil-ideals in K. This proves the corollary.

We shall now show that except for Cayley–Dickson rings there do not exist prime nondegenerate alternative rings that are not associative.

THEOREM 9 (*Slater*). Let A be a prime nondegenerate alternative algebra that is not associative. Then A is a Cayley–Dickson ring.

PROOF. By Theorem 7 we have $Z = Z(A) \neq (0)$. In addition, since A is prime it follows easily that Z does not contain nonzero zero divisors of the ring A. Consequently we can form the corresponding ring of quotients A_1, which is an algebra over the field of quotients Z_1 of the center Z. By Proposition 2, the algebra A_1 is alternative, prime, and $Z(A_1) = Z_1$. In addition, the ring A is isomorphically imbedded in the ring A_1, and therefore the algebra A_1 is not associative. In view of Kleinfeld's theorem, for the proof of our theorem it now suffices to show that A_1 is a simple algebra.

We note that since A_1 is not associative and prime, then by the corollary to Proposition 8.10 A_1 is a purely alternative algebra, that is, $U(A_1) = (0)$. Now let I be an arbitrary nonzero ideal of the algebra A. By Corollary 1 to Theorem 4 and Lemma 7 the algebra I is prime and nondegenerate. If I is associative, then by Theorem 1 we have $I = N(I) \subseteq N(A_1)$, which contradicts the algebra A_1 is purely alternative. This means the algebra I is not associative. Then by Theorems 7 and 3 we have $(0) \neq Z(I) = I \cap Z(A_1) = I \cap Z_1$. Let $0 \neq \alpha \in I \cap Z_1$. Then we have $1 = \alpha \cdot \alpha^{-1} \in I$, whence $I = A_1$. Consequently the algebra A_1 is simple.

This proves the theorem.

COROLLARY. Let A be a prime alternative algebra that is not associative and $3A \neq (0)$. Then A is a Cayley–Dickson ring.

PROOF. In view of Theorem 5′ it suffices to prove there are no elements of order 3 in the additive group of the algebra A. Let

$$A_3 = \{x \in A \mid 3x = 0\}.$$

Then A_3 is an ideal of the algebra A, and in addition $A_3 \cdot (3A) = (0)$. Since $3A$ is also an ideal of the algebra A and by assumption $3A \neq (0)$, we obtain $A_3 = (0)$. This proves the corollary.

We shall call a prime alternative algebra *exceptional* if it is not associative and not a Cayley–Dickson ring. The question whether there exist exceptional prime alternative algebras still remains open. From what has been proved it follows that every such algebra must contain nonzero absolute zero divisors and have characteristic 3. We shall find a further series of properties of such algebras.

THEOREM 10 (*Slater*). Let there exist an exceptional prime alternative algebra. Then there exists an exceptional prime alternative algebra S with the following properties:

(1) $3S = (0)$;
(2) there are nonzero absolute zero divisors in S;
(3) $N(S) = (0)$;
(4) S is an alternative PI-algebra;
(5) S is locally nilpotent;
(6) S does not contain minimal one-sided ideals.

PROOF. Suppose we are given some exceptional prime algebra A, which, as we have seen, satisfies conditions (1) and (2). We shall now construct, in turn, new algebras satisfying an ever larger number of the enumerated properties.

First of all, we note that if B is a prime exceptional algebra, then every nonzero ideal I of the algebra B is also an exceptional prime algebra. In fact, since the algebra B is purely alternative, the algebra I is not associative, and by Proposition 3 the algebra I cannot be a Cayley–Dickson ring. Finally, by Corollary 1 to Theorem 4 the algebra I is prime.

We return to the algebra A. If $N(A) = (0)$, then property (3) is satisfied. Otherwise by Theorem 8.11 we have $Z(A) = N(A) \neq (0)$, and as usual we can form the algebra of quotients A_1. If the algebra A_1 is simple, then A_1 is a Cayley–Dickson algebra, and A is a Cayley–Dickson ring which contradicts our assumption. Consequently, the algebra A_1 is not simple. Now let I be any proper ideal of the algebra A_1. Then by what has been proved I is an exceptional prime algebra. If $N(I) \neq (0)$, then we have $(0) \neq N(I) = Z(I) = I \cap Z(A_1)$, whence, as in the proof of Theorem 10, it follows that $I = A_1$. This contradiction proves that $N(I) = (0)$, that is, the exceptional prime algebra I satisfies properties (1)–(3).

Furthermore, by the Kleinfeld identities for any $a, b \in I$ we have $[a,b]^4 \in N(I) = (0)$, that is, I satisfies the essential polynomial identity $[x,y]^4 = 0$. Consequently property (4) is also satisfied in I.

We now denote by S the set of all nilpotent elements of the algebra I. By Proposition 7.2 S is an ideal of the algebra I. Since there are nonzero absolute zero divisors in I, $S \neq (0)$, and by what was proved above, S is an exceptional prime nil-algebra.

We have $N(S) = S \cap N(I) = (0)$, that is, property (3) is satisfied in S. It is also clear that S is an alternative PI-algebra. By Theorem 5.6 the algebra S is locally nilpotent. Thus the algebra S satisfies properties (1)–(5).

Now let V be a minimal right ideal of the algebra S, and $\check{V}$ be the largest ideal of the algebra S contained in V. Since V is minimal, either $\check{V} = (0)$ or $\check{V} = V$. In the first case, by Lemma 3 the right ideal V is associative, whence by Theorem 2 $V = N(V) \subseteq N(S) = (0)$, which is not so. If $\check{V} = V$, then $\check{V}$ is a minimal two-sided ideal of the algebra S. Since the algebra S is prime, we have $V^2 \neq (0)$, and therefore by Theorem 8.10 V is a simple algebra. But the algebra V is locally nilpotent, and by Proposition 7.1 it cannot be simple. The contradiction obtained proves that there are no minimal right ideals in S, that is, the algebra S has property (6).

This proves the theorem.

We shall now prove that prime alternative algebras without nonzero locally nilpotent ideals are not exceptional.

LEMMA 9. Let A be an alternative algebra, $A \neq (0)$, and $N(A) = (0)$. Then $\mathscr{L}(A) \neq (0)$, that is, A contains nonzero locally nilpotent ideals.

PROOF. In view of the Kleinfeld identities, the algebra A satisfies the identity $[x, y]^4 = 0$. In this case, by Proposition 7.2 the set I of all nilpotent elements of the algebra A forms an ideal. By Theorem 5.6 the ideal I is locally nilpotent, that is, $I \subseteq \mathscr{L}(A)$. If $I = (0)$, then the algebra A is commutative, and by Lemma 7.7 also associative, that is, $(0) \neq A = N(A)$, which contradicts our assumption. This means $I \neq (0)$, and $\mathscr{L}(A) \neq (0)$.

This proves the lemma.

THEOREM 11. Let A be a prime alternative algebra that is not associative. If $\mathscr{L}(A) = (0)$, then A is a Cayley–Dickson ring.

PROOF. By Lemma 9 we have $N(A) \neq (0)$, and then by Theorem 8.11 $Z(A) = N(A) \neq (0)$, so that we can again consider the algebra of quotients A_1. Let I be an arbitrary nonzero ideal of the algebra A_1. Then I is a prime alternative algebra that is not associative. If $N(I) = (0)$, then by Lemma 9 $\mathscr{L}(I) \neq (0)$, and since by Theorem 8.9 the radical $\mathscr{L}$ is hereditary, then also $\mathscr{L}(A_1) \neq (0)$. It is easy to see that $(0) \neq \mathscr{L}(A_1) \cap A$ is a locally nilpotent ideal of the algebra A. Since this contradicts the assumption $\mathscr{L}(A) = (0)$, we obtain that $N(I) \neq (0)$ for any proper ideal I of the algebra A_1. But then

$(0) \neq N(I) = Z(I) = I \cap Z(A_1)$, whence, as previously, it follows that $I = A_1$. Thus the algebra A_1 is simple, and is a Cayley–Dickson algebra. This proves the theorem.

In conclusion we note that from the result of Shestakov formulated in the remark to Theorem 5′, it follows that no exceptional prime alternative algebra can be finitely generated.

Exercises

1. Let K be a Cayley–Dickson ring. If $a, b \in K$ and $(aK)b = (0)$ or $a(Kb) = (0)$, then either $a = 0$ or $b = 0$.

2. Let A be a prime alternative algebra that is not associative and I be a right ideal of the algebra A. Prove that there are no nonzero associative ideals in the algebra I. *Hint*. Use Exercise 12 from Section 1.

3. Prove that under the conditions of Exercise 2, I is a prime ideal that is not associative. *Hint*. If $PQ = (0)$ for some ideals P and Q of the algebra I, then consider in the ideal I_0, generated by the set (I, I, A) in the algebra A, the ideals P_0 and Q_0 generated by the sets (P, P, P) and (Q, Q, Q), respectively, in the algebra I. Use Corollary 1 to Theorem 4.

4. Prove that every one-sided ideal of a Cayley–Dickson ring is a Cayley–Dickson ring.

5. Prove that under the conditions of Theorem 10 every one-sided ideal of the algebra S also satisfies conditions (1)–(6).

6. Let A be an alternative algebra and a be an arbitrary absolute zero divisor in A. Prove that $a \in \mathscr{L}(A)$. *Hint*. Apply Theorem 8.8 and Theorem 11.

COROLLARY. Let $\mathscr{M}$ be the radical defined in Exercise 5 from Section 2. Then in every alternative algebra A the containment

$$\mathscr{M}(A) \subseteq \mathscr{L}(A)$$

is valid.

LITERATURE

Amitsur [12], Bruck and Kleinfeld [24], Skornyakov [190], Kleinfeld [94, 97], Zhevlakov [75], Slater [198, 203, 205], Shestakov [271].

Results on absolute zero divisors in Jordan, associative, and alternative algebras: Slin'ko [209], McCrimmon [137].

Results dealing with the description of semiprime and prime algebras in other varieties of algebras: Dorofeev [61], Kleinfeld [96], E. Kleinfeld, M. Kleinfeld, Kosier [100], Pchelintsev [172], Roomel'di [185], Thedy [231], Hentzel [246].

CHAPTER **10**

The Zhevlakov Radical

As follows from the general theory, there exist many different radicals in the class of associative algebras. Only a few, however, turn out to be useful in the structure theory. For some—as example, for the nil-radicals—the structure of the radical algebras is clear. For others, it is the organization of the semisimple algebras that is the more clear. Frequently neither the one nor the other is altogether clear.

From this point of view the quasi-regular radical of Jacobson is a fortunate exception. On the one hand, quasi-regular algebras are an adequately studied object; while every Jacobson semisimple algebra is a subdirect sum of primitive algebras, each of which is a dense ring of linear transformations of some vector space over a division ring. Thus there is essential information on the radical of an arbitrary associative algebra and on the quotient algebra modulo the radical.

Attempts to create an analogous structure theory for alternative algebras were begun long ago. In 1948 Smiley proved that in an arbitrary alternative algebra A, there exists a largest quasi-regular ideal $\mathcal{S}(A)$, modulo which the quotient algebra does not contain nonzero quasi-regular ideals. In the literature this radical became known as the *Smiley radical.*

In 1955 Kleinfeld extended the concept of primitive to alternative algebras. Since at that time it was not yet clear what to consider as representations and modules of alternative algebras, he defined primitive alternative algebras in purely ring terms—as algebras having a maximal modular one-sided ideal that does not contain nonzero two-sided ideals. For associative algebras, this condition is equivalent to the usual notion of primitive. As shown by Kleinfeld, the only primitive alternative algebras that are not associative are the Cayley–Dickson algebras.

In the same work Kleinfeld proved that in every alternative algebra A the intersection $\mathscr{K}(A)$ of all maximal modular right (left) ideals is a two-sided ideal in A, and that the quotient algebra $A/\mathscr{K}(A)$ is a subdirect sum of primitive alternative algebras. The ideal $\mathscr{K}(A)$ became known as the *Kleinfeld radical* of the algebra A. Afterwards, this name was justified when Zhevlakov proved that the mapping $\mathscr{K}: A \rightarrow \mathscr{K}(A)$ is a hereditary radical.

Until 1969, the question on the coincidence of these two radicals remained open. Zhevlakov solved this problem by proving that in an arbitrary alternative algebra the radicals of Smiley and Kleinfeld coincide. Thus there appeared for the class of alternative algebras a radical with rich information about both its radical and semisimple classes. We shall denote this radical by the letter $\mathscr{J}$, and we shall call it the *Zhevlakov radical* as well as the *quasi-regular radical.*

In the following years, in the works of Zhevlakov, Slin'ko, and Shestakov, the concept of a right (left) representation of an alternative algebra was developed; and it was proved that the Zhevlakov radical $\mathscr{J}(A)$ of an arbitrary alternative algebra A coincides with the intersection of the kernels of the right (left) irreducible representations of that algebra. It was also proved by these authors that the primitive alternative algebras are precisely those algebras having a faithful irreducible alternative module. We shall devote the chapter following this to representations of alternative algebras.

1. PRIMITIVE ALTERNATIVE ALGEBRAS

We shall call a right ideal I of an algebra A *modular* if there exists an element $e \in A$ such that $ea - a \in I$ for all $a \in A$. This element e is a left identity element of the algebra A modulo the ideal I. If the algebra A has an identity element, then every right ideal of A is obviously modular. Analogously a left ideal J of the algebra A is called *modular* if there exists an element $e \in A$ such that $ae - a \in J$ for all $a \in A$.

PROPOSITION 1. Let I be a modular right ideal of the algebra A. Then it can be imbedded in some maximal right ideal of the algebra A which will also be modular.

PROOF. It is easy to see that if e is a left identity element modulo the right ideal I, then e is a left identity element modulo any ideal of the algebra A containing I. Therefore any ideal containing I is modular.

We consider the set M of right ideals of the algebra A containing I but not containing e. By Zorn's lemma there is a maximal ideal J in M. We shall show that J is a maximal ideal of the algebra A. Let J' be a right ideal in A and $J' \supset J$. Then $e \in J'$ and for an arbitrary element $a \in A$ we have $a - ea \in I \subseteq J'$. But since $ea \in J'$, then also $a \in J'$, that is, $J' = A$. This proves the proposition.

PROPOSITION 2. Let I be a modular right ideal of the algebra A. Then the largest ideal $\check{I}$ of the algebra A contained in I coincides with the set $(I:A) = \{a \in A \mid Aa \subseteq I\}$.

PROOF. By Lemma 9.2 $\check{I} = (I:A^{\#})$. Since $(I:A^{\#}) = I \cap (I:A)$, it suffices for us to prove that $(I:A) \subseteq I$. Let e be a left identity element modulo I and $a \in (I:A)$. Then $ea \in I$ and $a - ea \in I$, and as a result $a \in I$. This proves the proposition.

We shall call an alternative algebra A (*right*) *primitive* if it has a maximal modular right ideal I which does not contain nonzero two-sided ideals of the algebra A. By Proposition 2, the relation $\check{I} = (I:A) = (0)$ is valid for the ideal I.

LEMMA 1. A primitive alternative algebra A is prime.

PROOF. Let I be a maximal modular ideal and $\check{I} = (0)$. We assume that $BC = (0)$ for ideals B and C of the algebra A. If $B \neq (0)$ then $B \nsubseteq I$, and by the maximality of I we have $I + B = A$. But then $AC = (I + B)C = IC \subseteq I$, that is, $C \subseteq (I:A)$. But $(I:A) = (0)$, and consequently $C = (0)$. Thus it is proved the algebra A is prime.

LEMMA 2. If A is an alternative algebra that is not associative and $(a, b, A) = (0)$ for some elements $a, b \in A$, then $[a, b] = 0$.

PROOF. We recall that by Theorem 8.11 $N(A) = Z(A)$ and $Z(A)$ does not contain zero divisors. Therefore $[a, b] \in Z(A)$ by Lemma 7.3. Furthermore, by the Moufang identity $(a, ab, A) = (a, b, A)a = (0)$, which by the same lemma implies the inclusion $[a, ab] \in Z(A)$. We note that $[a, ab] = a[a, b]$. Hence it follows that $0 = [a[a, b], b] = [a, b]^2$. But since there are no zero divisors in $Z(A)$, we have $[a, b] = 0$, and the lemma is proved.

LEMMA 3. Let A be a prime alternative algebra that is not associative, and I be a nonzero one-sided ideal of A. Then $\check{I} \neq (0)$.

PROOF. To be specific let I be a right ideal. If $\check{I} = (0)$, then by Lemma 9.3 $(I, I, A) = (0)$, and by the previous lemma we conclude that I is an associative and commutative subalgebra in A. By Theorem 9.3 we have $I = Z(I) \subseteq Z(A)$, whence I is a two-sided ideal. Consequently $I = \check{I} = (0)$. This contradiction proves the lemma.

COROLLARY. A Cayley–Dickson algebra $\boldsymbol{C}$ over any field is primitive and does not contain proper one-sided ideals.

PROOF. If I is a nonzero one-sided ideal in $\boldsymbol{C}$, by Lemma 3 $\check{I} \neq (0)$. But then $\check{I} = \boldsymbol{C}$ since $\boldsymbol{C}$ is a simple algebra, and consequently $I = \boldsymbol{C}$.

We see that (0) is a maximal right ideal in $\boldsymbol{C}$. In addition, in view of the presence in $\boldsymbol{C}$ of an identity element, (0) is also modular. Therefore $\boldsymbol{C}$ is primitive, and the corollary is proved.

THEOREM 1 (*Kleinfeld*). Every primitive alternative algebra is either a primitive associative algebra or a Cayley–Dickson algebra over its center.

PROOF. If A is not associative and I is a maximal modular ideal of A such that $\check{I} = (0)$, then by Lemma 3 $I = (0)$. Hence it follows there are no proper right ideals in A, and in particular A is simple. By Kleinfeld's theorem on simple alternative algebras, this proves Theorem 1.

Exercises

1. Prove that in every alternative algebra A the radical $\mathscr{L}(A)$ contains all one-sided locally nilpotent ideals. *Hint.* Use Theorem 8.8 and Lemma 3.

2. THE KLEINFELD RADICAL

PROPOSITION 3. Let $\varphi: B \to A$ be a homomorphism of the algebra B onto A, $\psi: A \to C$ be a homomorphism of the algebra A onto C, and I be a maximal modular right (left) ideal of the algebra A containing the kernel of ψ. Then $\varphi^{-1}(I)$ and $\psi(I)$ are also maximal modular right (left) ideals in the algebras B and C.

PROOF. Let e be a left identity element of the algebra A modulo I. Then every element $\tilde{e} \in \varphi^{-1}(e)$ will be a left identity element of the algebra B modulo $\varphi^{-1}(I)$. If $\varphi^{-1}(I)$ is not maximal in B, then there exists a right ideal P in B such that $P \supset \varphi^{-1}(I)$ and $B \neq P$. It is clear that $P = \varphi^{-1}(J)$ where J is a right ideal of the algebra A strictly containing I. But then $J = A$ and $P = \varphi^{-1}(A) = B$.

It is clear also that $\psi(I)$ is modular. As a left identity element modulo $\psi(I)$ it is possible to take $\psi(e)$. If $\psi(I)$ is not maximal, then either $\psi(I) = C$ or there exists a right ideal Q in C such that $Q \supset \psi(I)$ and $Q \neq C$. In the first case there can be found an element $x \in I$ such that $\psi(x) = \psi(e)$. But then $x - e \in \operatorname{Ker}\psi \subseteq I$, where $e \in I$ and $I = A$. In the second case we have $\psi^{-1}(Q) = A$, and consequently $Q = C$. This proves the proposition.

As usual, we shall call an ideal B of an alternative algebra A *primitive* if the quotient algebra A/B is a primitive alternative algebra.

THEOREM 2 (*Kleinfeld*). In every alternative algebra A the intersection of all maximal modular right (left) ideals $\mathscr{K}(A) = \bigcap I_\alpha$ is a two-sided ideal in A coinciding with the intersection of all primitive ideals of the algebra A. The quotient algebra $A/\mathscr{K}(A)$ is a subdirect sum of primitive associative algebras and Cayley–Dickson algebras.

PROOF. Let I_α be some maximal modular ideal of the algebra A, and $Q_\alpha = \check{I}_\alpha$ be the largest two-sided ideal of the algebra A contained in I_α. From Proposition 3 it follows that Q_α is a primitive ideal. This means $\mathscr{K}(A)$ contains the intersection of all the ideals Q_α, and all the more the intersection Q of all primitive ideals of the algebra A. On the other hand, for any primitive ideal Q_α the quotient algebra A/Q_α is either a Cayley–Dickson algebra or a primitive associative algebra. Thus, by the corollary to Lemma 3 and a well-known fact from the theory of associative algebras, the intersection of all the maximal modular ideals in A/Q_α equals zero. We therefore conclude by Proposition 3 that $\mathscr{K}(A)$ is contained in each primitive ideal of the algebra A, and consequently in their intersection Q. Thus $\mathscr{K}(A) = Q$. The latter assertion of the theorem follows from Theorem 1. This proves the theorem.

If an algebra A does not contain maximal modular right ideals, we shall consider that by definition $\mathscr{K}(A) = A$. We then call the ideal $\mathscr{K}(A)$ the *Kleinfeld radical* of the algebra A.

LEMMA 4. Let A be an alternative algebra, B be an ideal in A, and I be a maximal modular right ideal of the algebra B. Then I is a right ideal of the algebra A.

PROOF. Let $IA \nsubseteq I$. Then there exists an element $x \in A$ such that $Ix \nsubseteq I$. We consider the set $I_1 = I + Ix$ and show that it is right ideal in B. In fact, for any $i \in I$ and $b \in B$

$$\begin{aligned}(ix)b &= i(xb) + (i, x, b) = i(xb) - (i, b, x)\\ &= i(xb) - (ib)x + i(bx) \in I + Ix = I_1.\end{aligned}$$

This means $I_1B \subseteq I_1$, and all the other properties of a right ideal are obvious for I_1. Since $I_1 \supset I$ and I is maximal in B, then $I_1 = B$. This means the left identity element e of the algebra B modulo I is represented in the form $e = i + jx$ where $i, j \in I$. We note that $jx \notin I$ since $e \notin I$. However, in view of the modularity of the ideal I, $jx - e(jx) \in I$, and therefore we obtain

$$jx - e(jx) = jx - (i + jx)(jx) = jx - i(jx) - j(xjx) \in I.$$

Hence it follows $jx \in I$, and the contradiction sought for is obtained. This proves the lemma.

LEMMA 5. Let B be an ideal of an alternative algebra A. Then the set of maximal modular right ideals of the algebra B coincides with the set of right ideals of the form $I \cap B$, where I is a maximal modular right ideal of the algebra A not containing B.

PROOF. Let I_1 be a maximal modular right ideal of the algebra B and e be a left identity element of the algebra B modulo I_1. We consider the right ideal J generated by the element $1 - e$ in the algebra $A^\#$. We shall show that $JB \subseteq I_1$. Let $J_0 = \{j \in J \mid jB \subseteq I_1\}$. Since $b - eb \in I_1$ for any $b \in B$, then $1 - e \in J_0$. Furthermore, by Lemma 4 the ideal I_1 of the algebra B is a right ideal of the algebra A. Therefore for any $j \in J_0$, $a \in A^\#$, $b \in B$, we have

$$\begin{aligned}(ja)b &= j(ab) + (j, a, b) = j(ab) - (j, b, a)\\ &= j(ab) - (jb)a + j(ba) \in I_1,\end{aligned}$$

whence it follows that $J_0A^\# \subseteq J_0$. It is now clear that J_0 is a right ideal of the algebra $A^\#$. Since $J_0 \subseteq J$ and $1 - e \in J_0$, then $J_0 = J$, that is, $JB \subseteq I_1$. We now set $J_1 = I_1 + J \cap A$. It is clear that J_1 is a right ideal of the algebra A and that $J_1B \subseteq I_1$. If $e \in J_1$ then we would have $e^2 \in J_1B \subseteq I_1$, and then $e = (e - e^2) + e^2 \in I_1$, which is not so. This means $e \notin J_1$. For any $a \in A$ we have $a - ea \in J \cap A \subseteq J_1$, that is, J_1 is a proper modular right ideal of the algebra A. We consider a maximal modular right ideal I containing J_1. It is clear that $e \notin I$, and therefore $B \nsubseteq I$. But then $B \neq I \cap B \supseteq I_1$, whence by the maximality of I_1 we obtain that $I \cap B = I_1$.

Conversely, let I be a maximal modular right ideal of the algebra A where $B \nsubseteq I$. By the maximality of I we have $I + B = A$. If e is a left identity element of the algebra A modulo I, then $e = i + e'$ where $i \in I$, $e' \in B$. Furthermore, for any $b \in B$ we have

$$b - e'b = b - (e - i)b = (b - eb) + ib \in I \cap B.$$

Hence it follows that $I_1 = I \cap B$ is a modular right ideal of the algebra B. We consider a maximal modular right ideal I_2 of the algebra B containing I_1. By Lemma 4 I_2 is a right ideal of the algebra A. If $I_1 \neq I_2$, then $I_2 \nsubseteq I$, and therefore $I + I_2 = A$. There exist elements $i' \in I$, $i_2 \in I_2$ such that $e = i' + i_2$. Then we again have $b - i_2 b \in I \cap B = I_1 \subseteq I_2$ for any $b \in B$, whence it follows that $I_2 = B$. This contradiction proves the maximality of the ideal I_1, which proves the lemma.

COROLLARY. Under the conditions of Lemma 5, $B \subseteq \mathscr{K}(A)$ if and only if $B = \mathscr{K}(B)$.

THEOREM 3 (*Zhevlakov*). The mapping $\mathscr{K}: A \to \mathscr{K}(A)$ is a hereditary radical in the class of alternative algebras.

PROOF. If there are no maximal modular right ideals in the algebra A, then by Proposition 3 there are also no such ideals in any homomorphic image of the algebra A. Furthermore, in view of the corollary to Lemma 5, in every alternative algebra A the ideal $\mathscr{K}(A)$ is the largest $\mathscr{K}$-ideal. According to Proposition 3, in the quotient algebra $\bar{A} = A/\mathscr{K}(A)$ the intersection of all maximal modular right ideals equals zero, that is, $\mathscr{K}(\bar{A}) = (0)$. Consequently, the mapping $\mathscr{K}$ satisfies conditions (A)–(C) and is radical. The heredity of the radical $\mathscr{K}$ follows from the corollary to Lemma 5, the corollary to Theorem 8.5, and Theorem 8.3. This proves the theorem.

3. THE SMILEY RADICAL

Let A be an alternative algebra with identity element 1. An element $a \in A$ is called *invertible* if there exists an element $a^{-1} \in A$ such that $aa^{-1} = a^{-1}a = 1$. We shall prove some lemmas on the properties of invertible elements of an algebra A.

LEMMA 6. If $ab = ca = 1$, then $b = c$.

PROOF. We shall first prove that $(a,b,c) = 0$. In fact, by identity (7.16) $(a,b,c) = (ab)(a,b,c) = b(a(a,b,c)) = b(a,b,ca) = 0$. Now $c = c(ab) = (ca)b - (c,a,b) = b$. This proves the lemma.

LEMMA 7. $(x, a, a^{-1}) = f(x, y, a, a^{-1}) = 0$, where $f(x, y, z, t)$ is the Kleinfeld function.

PROOF. By identity (7.16) we have

$$(x, a, a^{-1}) = (aa^{-1})(x, a, a^{-1}) = a^{-1}(a(x, a, a^{-1})) = a^{-1}(x, a, a^{-1}a) = 0.$$

Then

$$f(x, y, a, a^{-1}) = (xy, a, a^{-1}) - y(x, a, a^{-1}) - (y, a, a^{-1})x = 0.$$

This proves the lemma.

LEMMA 8. $(a, x, y) = 0$ implies $(a^{-1}, x, y) = 0$.

PROOF. By Lemma 7

$$(a^{-1}, x, y)a = (aa^{-1}, x, y) - f(a, a^{-1}, x, y) - a^{-1}(a, x, y) = 0.$$

That same lemma now gives us

$$(a^{-1}, x, y) = (a^{-1}, x, y)(aa^{-1}) = ((a^{-1}, x, y)a)a^{-1} = 0.$$

This proves the lemma.

LEMMA 9. The following statements are equivalent:

(a) the elements a and b are invertible;
(b) the elements ab and ba are invertible.

PROOF. By Lemma 8 we have

$$(ab, b, a) = 0 \Rightarrow (ab, b^{-1}, a) = 0 \Rightarrow (ab, b^{-1}, a^{-1}) = 0.$$

Now by Lemma 7

$$(ab)(b^{-1}a^{-1}) = ((ab)b^{-1})a^{-1} - (ab, b^{-1}, a^{-1}) = 1.$$

The equalities $(b^{-1}a^{-1})(ab) = (ba)(a^{-1}b^{-1}) = (a^{-1}b^{-1})(ba) = 1$ are proved analogously.

Conversely, let ab and ba be invertible. Then by Lemma 8 $(a, b, (ab)^{-1}) = ((ba)^{-1}, b, a) = 0$, and we obtain that

$$1 = (ab)(ab)^{-1} = a(b(ab)^{-1}) + (a, b, (ab)^{-1}) = a(b(ab)^{-1}),$$
$$1 = (ba)^{-1}(ba) = ((ba)^{-1}b)a - ((ba)^{-1}, b, a) = ((ba)^{-1}b)a.$$

Hence by Lemma 6 it follows that a is an invertible element. Analogously we also invert the element b, which proves the lemma.

Now let A be an arbitrary alternative algebra. An element $a \in A$ is called *right quasi-regular* if there exists an element $b \in A$ such that $ab = a + b$, and *left quasi-regular* if there exists an element $c \in A$ such that $ca = a + c$. The element b is called a *right quasi-inverse* for the element a, and the element c is a *left quasi-inverse*. An element $a \in A$ is called *quasi-regular* if there exists an element $b \in A$ such that $ab = ba = a + b$. In this case the element b is called a *quasi-inverse* for a.

PROPOSITION 4. The following statements are equivalent:

(a) the element a of an alternative algebra A is quasi-regular with quasi-inverse element b;

(b) the element $1 - a$ of the algebra $A^{\#} = \Phi \cdot 1 + A$ is invertible with inverse element $1 - b$.

The proof is obvious.

COROLLARY. If a quasi-regular element z of an alternative algebra A lies in the center $Z(A)$ of A, then its quasi-inverse element z' also lies in $Z(A)$.

PROOF. By Proposition 4 it suffices to prove that for an invertible element $a \in A$, from $a \in Z(A)$ it follows that $a^{-1} \in Z(A)$. By Lemma 8 $a^{-1} \in N(A)$. Then $[a^{-1}, x]a = [a^{-1}a, x] = 0$. By Lemma 7

$$[a^{-1}, x] = [a^{-1}, x](aa^{-1}) = ([a^{-1}, x]a)a^{-1} = 0.$$

This means $a^{-1} \in Z(A)$, which proves the corollary.

LEMMA 10. If b is a right quasi-inverse for the element $a \in A$ and c is a left quasi-inverse for a, then the element a is quasi-regular and $b = c$.

The proof follows from Proposition 4 and Lemma 6.

COROLLARY. A quasi-regular element has a unique quasi-inverse.

An algebra is called *quasi-regular* if each of its elements is quasi-regular. An ideal of an algebra is called *quasi-regular* if it is quasi-regular as an algebra. It is easy to see that every ideal of a quasi-regular algebra is quasi-regular.

LEMMA 11. Let A be an alternative algebra with an identity element, u be an invertible element in A, and a be an element of a quasi-regular ideal B. Then $u - a$ is an invertible element.

PROOF. We see that

$$(u - a)u^{-1} = 1 - au^{-1} = 1 - b, \qquad u^{-1}(u - a) = 1 - u^{-1}a = 1 - c,$$

where $b, c \in B$. Consequently the elements $(u - a)u^{-1}$ and $u^{-1}(u - a)$ are invertible, whence by Lemma 9 we conclude that $u - a$ is invertible. This proves the lemma.

LEMMA 12. The sum of two quasi-regular ideals of an alternative algebra is a quasi-regular ideal.

PROOF. Let B, C be quasi-regular ideals of the algebra A. Then B and C are also ideals in the algebra $A^{\#} = \Phi \cdot 1 + A$. We shall show that each element of the ideal $B + C$ is quasi-regular. Let $b + c \in B + C$. Then $1 - (b + c) = (1 - b) - c = u - c$, where $u = 1 - b$ is an invertible element since b is quasi-regular. By Lemma 11 we now have $1 - (b + c)$ is an invertible element, and according to Proposition 4 this means the element $b + c$ is quasi-regular.

LEMMA 13. If B is a quasi-regular ideal of an alternative algebra A and in the quotient algebra $\bar{A} = A/B$ the image $\bar{a}$ of an element $a \in A$ is quasi-regular, then a is a quasi-regular element.

PROOF. Without loss of generality it is possible to assume that A has an identity element 1. Let $\bar{x}$ be a quasi-inverse element for $\bar{a}$. Then $(\bar{1} - \bar{a})(\bar{1} - \bar{x}) = (\bar{1} - \bar{x})(\bar{1} - \bar{a}) = \bar{1}$, and for any inverse image x of the element $\bar{x}$ we have $(1 - a)(1 - x) = 1 - b$, $(1 - x)(1 - a) = 1 - b'$, where b, $b' \in B$. By Proposition 4 the elements $(1 - a)(1 - x)$ and $(1 - x)(1 - a)$ are invertible. Consequently by Lemma 9 the element $1 - a$ is invertible, and this means that the element a is quasi-regular. This proves the lemma.

THEOREM 4 (*Smiley*). Every alternative algebra A has a largest quasi-regular ideal $\mathcal{S}(A)$ such that the quotient algebra $A/\mathcal{S}(A)$ does not have nonzero quasi-regular ideals.

PROOF. Let $\mathcal{S}(A)$ be the sum of all the quasi-regular ideals of the algebra A. By Lemma 12 this ideal is quasi-regular, and by Lemma 13 the quotient algebra $A/\mathcal{S}(A)$ does not have nonzero quasi-regular ideals. This proves the theorem.

COROLLARY. The mapping $\mathcal{S}: A \to \mathcal{S}(A)$ is a hereditary radical in the class of alternative algebras.

The radical introduced in this section goes by the name of the *Smiley radical.*

4. ZHEVLAKOV'S THEOREM ON THE COINCIDENCE OF THE KLEINFELD AND SMILEY RADICALS

PROPOSITION 5. The containment $\mathcal{S}(A) \subseteq \mathcal{K}(A)$ holds in every alternative algebra A.

PROOF. Let $\mathcal{S}(A) \nsubseteq I$ where I is a maximal modular right ideal, and let e be a left identity element modulo I. Then $I + \mathcal{S}(A) = A$, and there can be found elements $i \in I$, $s \in \mathcal{S}(A)$ such that $i + s = e$. For any $a \in A$ we have $ea - a = (i + s)a - a \in I$. But $ia \in I$, and therefore $sa - a \in I$ for any $a \in A$. In particular, if the element p, which is the quasi-inverse for s, is substituted in place of a, then we obtain $sp - p = s \in I$. But then $e = i + s \in I$, which is impossible. This means $\mathscr{S}(A)$ is contained in all maximal modular right ideals, and consequently, also in their intersection, that is, $\mathscr{S}(A) \subseteq \mathscr{K}(A)$.

PROPOSITION 6. If the radicals $\mathscr{K}$ and $\mathscr{S}$ are different, then there exists a $\mathscr{K}$-radical $\mathscr{S}$-semisimple algebra A which is a subdirect sum of Cayley–Dickson rings.

PROOF. In view of Proposition 5, the noncoincidence of the Kleinfeld and Smiley radicals means the existence of an alternative algebra B in which $\mathscr{S}(B) \subset \mathscr{K}(B)$. But then $A = \mathscr{K}(B)/\mathscr{S}(B)$ is a $\mathscr{K}$-radical $\mathscr{S}$-semisimple algebra. By Theorems 8.8 and 9.11, the algebra A is a subdirect sum of prime associative algebras without locally nilpotent ideals and Cayley–Dickson rings. Let P be the intersection of all the ideals P_α of the algebra A for which the quotient algebra A/P_α is a Cayley–Dickson ring. Then it is obvious $P \cap D = (0)$, where $D = D(A)$ is the associator ideal of the algebra A. If $P = (0)$, then everything is proved. Let $P \neq (0)$. Then we have $D(P) \subseteq D(A) \cap P = (0)$, that is, P is a nonzero associative ideal of the algebra A. But in this case $P = \mathscr{K}(P) = \mathscr{S}(P)$, and the algebra A is not $\mathscr{S}$-semisimple. This contradiction proves the proposition.

LEMMA 14. If the center $Z(A)$ of an alternative algebra A is quasi-regular and the element $x \in A$ is quadratic over the center, $x^2 = \alpha x + \beta$ where $\alpha, \beta \in Z(A)$, then x is a quasi-regular element.

PROOF. We formally adjoin an identity element 1 to the algebra A. We shall seek a quasi-inverse for x in the form $x' = \gamma x + \delta$ where $\gamma, \delta \in Z(A)$. We have

$$\begin{aligned} 0 = x + x' - xx' &= x + \gamma x + \delta - \gamma x^2 - \delta x \\ &= (1 + \gamma - \gamma\alpha - \delta)x + (\delta - \gamma\beta). \end{aligned}$$

We solve the following system of equations with respect to the unknowns γ and δ:

$$\begin{cases} 1 + \gamma - \gamma\alpha - \delta = 0, \\ \qquad\qquad \delta - \gamma\beta = 0. \end{cases}$$

Adding the equations, we have $1 + \gamma - \gamma\alpha - \gamma\beta = 0$, whence

$$\gamma(1 - \alpha - \beta) = -1.$$

But the element $\alpha + \beta$ is quasi-regular, and therefore the element $1 - \alpha - \beta$ is invertible. This means $\gamma = -(1 - \alpha - \beta)^{-1}$ and $\delta = -\beta(1 - \alpha - \beta)^{-1}$. Since γx and δ lie in the algebra A, the desired quasi-inverse element has been found.

COROLLARY. In a $\mathscr{K}$-radical alternative algebra A, every element that is quadratic over the center is quasi-regular.

PROOF. We consider the algebra $A^{\#} = \Phi \cdot 1 + A$. We note that for any $z \in Z(A)$ the set $(1 - z)A$ is a modular right ideal of the algebra A, and this means it coincides with the entire algebra, that is, $(1 - z)A = A$. There consequently exists an element $z' \in A$ such that $(1 - z)z' = -z$. Hence $z + z' = zz' = z'z$, and z' is a quasi-inverse element for z. By the corollary to Proposition 4 $z' \in Z(A)$, and the center $Z(A)$ is a quasi-regular ring. It now only remains to use Lemma 14.

LEMMA 15. Let $\boldsymbol{C}$ be some Cayley–Dickson algebra over a field F. Then for any $a, b, x \in \boldsymbol{C}$ the equality

$$t([a,b]^2x) = [a \circ x, b][a,b] - [x,b][a^2,b]$$

holds, where $t(y)$ is the trace of the element y.

PROOF. We note that by Lemma 2.8 $[a,b]^2 \in F$, and therefore $t([a,b]^2x) = [a,b]^2t(x)$. On the other hand, by equalities (2.18) and (2.21) we have

$$\begin{aligned}
&[a \circ x, b][a,b] - [x,b][a^2,b] \\
&\quad = t(a)[x,b][a,b] + t(x)[a,b]^2 - t(a)[x,b][a,b] \\
&\quad = t(x)[a,b]^2.
\end{aligned}$$

This proves the lemma.

LEMMA 16 (*Zhevlakov*). Let the algebra A be a subdirect sum of Cayley–Dickson rings. Then there is a nonzero ideal in A which is quadratic over the center.

PROOF. Let the algebra A be a subdirect sum of the Cayley–Dickson rings K_α. If the identity $[x,y]^2 = 0$ were satisfied in A, then such an identity would also be satisfied in every Cayley–Dickson ring K_α, and consequently, also in the Cayley–Dickson algebra $\boldsymbol{C}_\alpha = (Z_\alpha^*)^{-1}K_\alpha$ where $Z_\alpha = Z(K_\alpha)$. But in an arbitrary Cayley–Dickson algebra $\boldsymbol{C}(\mu, \beta, \gamma)$ for the basis elements v_1 and v_2 we have $[v_1, v_2]^2 = (2v_1v_2 - v_2)^2 = -\beta(4\mu + 1) \neq 0$. This means there can be found elements a, b in A such that $[a,b]^2 \neq 0$. Let I be the ideal in A generated by the element $[a,b]^2$. The image of the element $[a,b]^2$ under a homomorphism of the algebra A onto any Cayley–Dickson ring K_α falls into the center $Z(K_\alpha)$. Therefore $[a,b]^2 \in Z(A)$, and $I = [a,b]^2A^{\#}$. Let x be an arbitrary element of the algebra $A^{\#}$. We consider the elements $p = [a \circ x, b][a,b] - [x,b][a^2,b]$ and $q = ([a,b]^2x)^2 - p([a,b]^2x)$. By Lemma 15, under a homomorphism onto any Cayley–Dickson ring K_α, the elements p and q are carried into $Z(K_\alpha)$, and therefore p and q belong to $Z(A)$. This means $([a,b]^2x)^2 = p([a,b]^2x) + q$ where $p, q \in Z(A)$, that is, all elements of the form $[a,b]^2x$ are quadratic over the center. Consequently I is a quadratic ideal in A. This proves the lemma.

THEOREM 5 (*Zhevlakov*) The Smiley and Kleinfeld radicals concide.

PROOF. If these radicals are different, then by Proposition 6 there exists a $\mathscr{K}$-radical $\mathscr{S}$-semisimple alternative algebra A which is a subdirect sum of Cayley–Dickson rings. By Lemma 16 there is a nonzero quadratic ideal in A. However by the corollary to Lemma 14 that ideal is quasi-regular, which contradicts the $\mathscr{S}$-semisimplicity of the algebra A. This proves the theorem.

Thus there is a unique analog of the Jacobson radical in the class of alternative algebras. This radical will subsequently be called the *Zhevlakov radical* and will be denoted by the letter $\mathscr{J}$.

An important characterization of the Zhevlakov radical $\mathscr{J}(A)$ of an alternative algebra A was obtained by McCrimmon. He proved that the radical $\mathscr{J}(A)$ consists of those and only those elements z for which all elements of the form za (or az), $a \in A$, are quasi-regular. In particular, it follows from this that the Zhevlakov radical of an alternative algebra contains all the one-sided quasi-regular ideals of the algebra.

5. RADICALS OF CAYLEY–DICKSON RINGS

In the rest of this section K is a Cayley–Dickson ring, Z is its center, Z_1 is the field of quotients of the ring Z, and $\boldsymbol{C} = \boldsymbol{C}(\mu, \beta, \gamma) = (Z^*)^{-1}K$ is the Cayley–Dickson algebra over the field Z_1, which is the ring of quotients of the ring K modulo the multiplicative subset $Z^* = Z\backslash\{0\}$. We recall that the ring K is isomorphically imbedded in the algebra $\boldsymbol{C}$, and we shall simply consider K as a subring of the ring $\boldsymbol{C}$.

We now begin the preparatory work for proof of the main result of this section—the theorem of Zhevlakov on the $\mathscr{J}$-semisimplicity of finitely-generated Cayley–Dickson rings.

We fix elements a, b of a Cayley–Dickson ring K such that $[a,b]^2 \neq 0$ (the existence of such elements was shown in the proof of Lemma 16). We note that since $K \subseteq \boldsymbol{C}$, the functions $t(k)$ and $n(k)$ are defined for any element $k \in K$. Generally speaking, however, the values of these functions on elements from K may not belong to Z (but necessarily belong to Z_1).

LEMMA 17. For any $k, k_1, \ldots, k_n \in K$ the following inclusions hold:

(1) $[a,b]^2 t(k_1) \cdots t(k_n) \in K$;

(2) $[a,b]^2 n(k) \in K$;

(3) $[a,b]^4 t(k_1) \cdots t(k_n) n(k) \in K$.

PROOF. In view of Lemma 15,

$$[a,b]^2 t(k_1) = t([a,b]^2 k_1) = [a \circ k_1, b][a,b] - [k_1, b][a^2, b] \in K.$$

By the same token there is a basis for an induction on n. For $m < n$ let it already be proved that $[a,b]^2 t(k_1) \cdots t(k_m) \in K$ for any $k_1, \ldots, k_m \in K$. Then by (2.21) we have

$$[a,b]^2 t(k_1) \cdots t(k_{n-1}) t(k_n) = [a,b]^2 t(k_1) \cdots t(k_{n-2})[t(k_{n-1}k_n) + t(k_{n-1})k_n + t(k_n)k_{n-1} - k_{n-1} \circ k_n] \in K,$$

that is, (1) is proved.

Furthermore, $[a,b]^2 n(k) = [a,b]^2(-k^2 + t(k)k)$, whence (2) follows by (1).

Finally, assertion (3) follows in an obvious manner from (1) and (2).

This proves the lemma.

Let X be a set of generators of the Cayley–Dickson ring K (as a Φ-algebra). Considering the set X to be ordered, we denote by $r_1^{\text{reg}}(X)$ the set of all regular r_1-words from X.

We can consider the Cayley–Dickson algebra $\boldsymbol{C}$ as a Φ-algebra, setting $\lambda \cdot (z^{-1}k) = z^{-1}(\lambda k)$ for $\lambda \in \Phi$, $z \in Z^*$, $k \in K$. We denote by $T(K)$ the Φ-subalgebra of the center Z_1 of the algebra $\boldsymbol{C}$ generated by the set $t(K) \cup n(K)$ of all traces and norms of elements from K. We note that if Z_1 is a field of characteristic $\neq 2$, then the algebra $T(K)$ is generated by the set $t(K)$, since in view of (2.18) for any $x \in K$ the norm and trace are connected by the formula $2n(x) = (t(x))^2 - t(x^2)$.

LEMMA 18. As set of generators of the Φ-algebra $T(K)$ can be taken the set $Y = Y_1 \cup Y_2$, where $Y_1 = t(r_1^{\text{reg}}(X))$ is the set of all traces of elements from $r_1^{\text{reg}}(X)$, and $Y_2 = n(X)$ is the set of all norms of elements from X.

PROOF. We denote the Φ-algebra generated by the set Y by T_0. We need to prove that $n(k), t(k) \in T_0$ for any $k \in K$. In view of the relation $n(xy) = n(x)n(y)$, we obtain that $n(v) \in T_0$ for any nonassociative word v from elements of the set X. Since $n(x+y) = n(x) + n(y) + f(x,y)$, and by (2.8) $f(x,y) = t(x)t(y) - t(xy)$, it remains for us to prove that $t(k) \in T_0$ for any $k \in K$.

We shall prove that for any nonassociative word v from elements of the set X we have the equality

$$v = \pm\langle v \rangle + \sum_i \alpha_i v_i + \alpha, \tag{1}$$

where $\alpha, \alpha_i \in T_0$ and the v_i are regular r_1-words of length less than v. For a multilinear word v we recall that $\langle v \rangle$ denotes the regular r_1-word with the same composition as v, and if v is not multilinear then $\langle v \rangle = 0$. It is clear that the assertion we need will follow from (1).

We shall prove relation (1) by induction on the length $d(v)$ of the word v. The basis of the induction, $d(v) = 1$, is obvious. Now let $d(v) = n > 1$, and all words of lesser length admit representation in the form (1). We shall write $v \equiv v'$ if $d(v) = d(v')$ and $v - v' = \sum_i \alpha_i v_i + \alpha$, where $\alpha, \alpha_i \in T_0$ and the v_i are regular r_1-words of length less than n. We first consider the case when v is an r_1-word, $v = \langle x_{i_1}, x_{i_2}, \ldots, x_{i_n}\rangle$. If $x_{i_1} < x_{i_2} < \cdots < x_{i_n}$ then $v = \langle v \rangle$, and there is nothing to prove. Now let $x_{i_{k+1}} \leqq x_{i_k}$ for some $k < n$. If $x_{i_{k+1}} = x_{i_k}$, then in view of (2.18) we have

$$\begin{aligned} v &= \langle x_{i_1}, \ldots, x_{i_k}, x_{i_{k+1}}, \ldots, x_{i_n}\rangle = \langle x_{i_1}, \ldots, x_{i_k}^2, \ldots, x_{i_n}\rangle \\ &= t(x_{i_k})\langle x_{i_1}, \ldots, x_{i_k}, \ldots, x_{i_n}\rangle - n(x_{i_k})\langle x_{i_1}, \ldots, x_{i_{k-1}}, x_{i_{k+2}}, \ldots, x_{i_n}\rangle, \end{aligned}$$

whence by the induction assumption $v \equiv 0$. Since $\langle v \rangle = 0$ in the present case, v is represented in the form (1). If $x_{i_{k+1}} < x_{i_k}$, then by (2.21) we have

$$\begin{aligned} v = {} & \langle x_{i_1}, \ldots, x_{i_{k-1}}, x_{i_k} \circ x_{i_{k+1}}, x_{i_{k+2}}, \ldots, x_{i_n}\rangle \\ & - \langle x_{i_1}, \ldots, x_{i_{k+1}}, x_{i_k}, \ldots, x_{i_n}\rangle \\ = {} & t(x_{i_k})\langle x_{i_1}, \ldots, x_{i_{k-1}}, x_{i_{k+1}}, \ldots, x_{i_n}\rangle \\ & + t(x_{i_{k+1}})\langle x_{i_1}, \ldots, x_{i_k}, x_{i_{k+2}}, \ldots, x_{i_n}\rangle \\ & + (t(x_{i_{k+1}}x_{i_k}) - t(x_{i_{k+1}})t(x_{i_k}))\langle x_{i_1}, \ldots, x_{i_{k-1}}, x_{i_{k+2}}, \ldots, x_{i_n}\rangle \\ & - \langle x_{i_1}, \ldots, x_{i_{k+1}}, x_{i_k}, \ldots, x_{i_n}\rangle, \end{aligned}$$

whence by the induction assumption

$$v \equiv -\langle x_{i_1}, \ldots, x_{i_{k+1}}, x_{i_k}, \ldots, x_{i_n}\rangle.$$

Repeating if necessary the given process, since it is strictly monotonic after a finite number of steps, we obtain $v \equiv \pm\langle v \rangle$, that is, v is represented in the form (1).

Now let the word v be arbitrary, $v = v_1 v_2$. By the induction assumption $v_1 \equiv \pm\langle v_1 \rangle$, $v_2 \equiv \pm\langle v_2 \rangle$, and $v \equiv \pm\langle v_1 \rangle\langle v_2 \rangle$. We represent the word $\langle v_1 \rangle$ in the form $\langle v_1 \rangle = u_1 x_i$, where $x_i \in X$ and u_1 is either a regular r_1-word of length $d(v_1) - 1$ or $u_1 = 1$ (if $d(v_1) = 1$). By (2.21) we have

$$\begin{aligned} (u_1 x_i)\langle v_2 \rangle &= -(u_1\langle v_2 \rangle)x_i + u_1(\langle v_2 \rangle \circ x_i) \\ &= -(u_1\langle v_2 \rangle)x_i + t(\langle v_2 \rangle)u_1 x_i + t(x_i)u_1\langle v_2 \rangle \\ &\quad + (t(\langle v_2 \rangle x_i) - t(\langle v_2 \rangle)t(x_i))u_1. \end{aligned}$$

Since $\langle v_2 x_i \rangle$ is an r_1-word, then by what was proved above and the induction assumption $t(\langle v_2 \rangle x_i) \in T_0$. Consequently, by the induction assumption we have

$$v \equiv \pm(u_1 x_i)\langle v_2 \rangle \equiv \mp(u_1\langle v_2 \rangle)x_i \equiv \pm\langle u_1\langle v_2 \rangle\rangle x_i,$$

whence by what was proved above the word v is represented in the form (1).

This proves the lemma.

COROLLARY. If the Cayley–Dickson ring K is a finitely generated Φ-algebra, then the Φ-algebra $T(K)$ is also finitely generated.

In fact, in this case in view of the finiteness of the set X the set $r_1^{\text{reg}}(X)$ is likewise finite.

We now represent the Cayley–Dickson algebra $\boldsymbol{C} = \boldsymbol{C}(\mu, \beta, \gamma)$ in the form $\boldsymbol{C}(\mu, \beta, \gamma) = \boldsymbol{K}(\mu) + \boldsymbol{K}(\mu)^{\perp}$ (see Section 2.5). Then each element $k \in$ K can be written in the form $k = \lambda_0 + \lambda_1 v_1 + k'$, where $\lambda_0, \lambda_1 \in Z_1$, $k' \in \boldsymbol{K}(\mu)^{\perp}$, and v_1 is the basis element of the Cayley–Dickson algebra such that $v_1^2 = v_1 + \mu$.

LEMMA 19. There exists an element $\alpha \in Z^*$ such that for any $k \in K$ represented in the form $k = \lambda_0 + \lambda_1 v_1 + k'$ the inclusion $\alpha\lambda_0 \in T(K)$ holds.

PROOF. Let $v_1 = z^{-1}k$ for some elements $z \in Z^*$, $k \in K$. We shall show that $z^2(1 + 4\mu)$ can be taken as the desired element α. We have $z^2 v_1 \in K$ and $z^2 v_1^2 \in K$, whence also $z^2\mu \in K$. Now let k be an arbitrary element from $K, k = \lambda_0 + \lambda_1 v_1 + k'$ where $\lambda_0, \lambda_1 \in Z_1, k' \in \boldsymbol{K}(\mu)^{\perp}$. We have $t(k) = 2\lambda_0 + \lambda_1$. Furthermore, $kv_1 = \lambda_0 v_1 + \lambda_1(v_1 + \mu) + k'v_1$, whence $t(kv_1) = \lambda_0 + \lambda_1(1 + 2\mu)$. From the relations obtained it follows easily that $(1 + 4\mu)\lambda_0 = (1 + 2\mu)t(k) - t(kv_1)$. Consequently,

$$\begin{aligned}\alpha\lambda_0 &= z^2(1 + 4\mu)\lambda_0 = z^2(1 + 2\mu)t(k) - z^2 t(kv_1) \\ &= t((z^2 + 2z^2\mu)k) - t(k(z^2 v_1)) \in T(K).\end{aligned}$$

This proves the lemma.

We now set $T(\alpha) = \alpha[a, b]^4 T(K)$, where α is the element from Lemma 19 and a, b are the fixed elements from K for which $[a, b]^2 \neq 0$ (and consequently also $[a, b]^4 \neq 0$). By Lemma 17 $[a, b]^4 T(K) \subseteq K$, and therefore $T(\alpha) \subseteq \alpha K$. In addition, by Lemma 19 $\alpha[a, b]^4 \in T(K)$, therefore $T(\alpha) \subseteq T(K)$, and as is easy to see $T(\alpha)$ is a nonzero ideal of the Φ-algebra $T(K)$. We now denote by $K(\alpha)$ the ideal of the Φ-algebra αK generated by the set $T(\alpha)$. Then $K(\alpha) = T(\alpha) \cdot (\alpha K)^{\#}$.

LEMMA 20. $Z(K(\alpha)) = T(\alpha)$.

PROOF. It is clear that $T(\alpha) \subseteq Z(K(\alpha))$. Now let $x \in Z(K(\alpha))$, $x = \sum_i \tau_i(\alpha k_i) + \sum_i \tau_i'$ where $\tau_i, \tau_i' \in T(\alpha)$, $k_i \in K$. Since $\sum_i \tau_i' \in T(\alpha)$, it remains for us to consider the element $x' = \sum_i \tau_i(\alpha k_i)$. We note that $x' = x - \sum_i \tau_i' \in Z(K(\alpha))$.

By Corollary 1 to Theorem 9.4 and Theorem 9.3 we have the inclusions $Z(K(\alpha)) \subseteq Z(\alpha K) \subseteq Z(K) = Z \subseteq Z_1$, whence $x' \in Z_1$. We represent the elements k_i in the form $k_i = \lambda_{i0} + \lambda_{i1} v_1 + k_i'$, where λ_{i0}, $\lambda_{i1} \in Z_1$, $k_i' \in \boldsymbol{K}(\mu)^{\perp}$.

Then $x' = \sum_i \tau_i(\alpha\lambda_{i0}) + [\sum_i \tau_i(\alpha\lambda_{i1})]v_1 + \sum_i \tau_i(\alpha k_i')$. Since $x' \in Z_1$, we have $\sum_i \tau_i(\alpha\lambda_{i1}) = \sum_i \tau_i(\alpha k_i') = 0$ and $x' = \sum_i \tau_i(\alpha\lambda_{i0})$. By Lemma 19 $\alpha\lambda_{i0} \in T(K)$ for all i, whence $\tau_i(\alpha\lambda_{i0}) \in T(\alpha)$ and $x' \in T(\alpha)$.

This proves the lemma.

We recall that an associative-commutative ring Φ with identity element is called a *Jacobson ring* if each simple ideal of Φ is the intersection of some collection of maximal ideals. The class of Jacobson rings is very extensive, for every finitely generated Φ-algebra with identity element, where Φ is a Jacobson ring, will again be a Jacobson ring. (This result, which is used by us in the proof of the following theorem, is contained with a complete proof in the book of N. Burbaki, Commutative Algebra, p. 419, Mir, 1971.) In view of the fact that the ring of integers $\boldsymbol{Z}$ is a Jacobson ring, then every finitely generated associative–commutative ring will also be a Jacobson ring. We note that in Jacobson rings the Jacobson radical coincides with the lower nil-radical.

THEOREM 6 (*Zhevlakov*). Every Cayley–Dickson ring K which is finitely generated as a Φ-algebra over a Jacobson ring Φ is $\mathcal{J}$-semisimple.

PROOF. Let $\mathcal{J}(K) \neq (0)$. We consider the ideal αK in the ring K and the ideal $K(\alpha)$ in the ring αK, where α is the element from Lemma 19. By Proposition 9.3 the rings αK and $K(\alpha)$ are also Cayley–Dickson rings. By the heredity of the Jacobson radical and since the rings K and αK are prime, we have $\mathcal{J}(\alpha K) = \mathcal{J}(K) \cap \alpha K \neq (0)$ and $\mathcal{J}(K(\alpha)) = K(\alpha) \cap \mathcal{J}(\alpha K) \neq (0)$. By Proposition 9.3 we again obtain that $\mathcal{J}(K(\alpha))$ is a Cayley–Dickson ring. But then $Z(\mathcal{J}(K(\alpha)) = \mathcal{J}(K(\alpha)) \cap Z(K(\alpha)) \neq (0)$, and $Z(\mathcal{J}(K(\alpha))$ is a nonzero ideal in $Z(K(\alpha)) = T(\alpha)$. By the corollary to Proposition 4 this ideal is $\mathcal{J}$-radical, and therefore $\mathcal{J}(T(\alpha)) \neq (0)$. By the heredity of the radical $\mathcal{J}$ we now have $\mathcal{J}(T(K)) \supseteq \mathcal{J}(T(\alpha)) \neq (0)$, since $T(\alpha)$ is an ideal of the Φ-algebra $T(K)$. By the corollary to Lemma 18 the algebra $T(K)$ is a finitely generated algebra over a Jacobson ring Φ, and consequently, as noted above, is itself a Jacobson ring. Hence it follows that $\mathcal{J}(T(K))$ is a nil-ideal, and therefore $\mathfrak{N}(T(\alpha)) = \mathfrak{N}(Z(K(\alpha))) \neq (0)$. However the center $Z(K(\alpha))$ of the Cayley–Dickson ring $K(\alpha)$ does not contain nilpotent elements. This contradiction proves the theorem.

In concluding this section we explain how the radical of a Cayley–Dickson matrix algebra $\boldsymbol{C}(\Phi)$ over an arbitrary associative–commutative ring Φ is organized.

THEOREM 7. $\mathcal{J}(\boldsymbol{C}(\Phi)) = \boldsymbol{C}(\mathcal{J}(\Phi))$.

PROOF. Let x be an arbitrary element from $\boldsymbol{C}(\mathscr{J}(\Phi))$,

$$x = \sum_{i=1}^{2} \alpha_{ii} e_{ii} + \sum_{k=1}^{3} (\alpha_{12}^{(k)} e_{12}^{(k)} + \alpha_{21}^{(k)} e_{21}^{(k)}),$$

where α_{ii}, $\alpha_{ij}^{(k)} \in \mathscr{J}(\Phi)$ and e_{ii}, e_{ij} are the Cayley–Dickson matrix units. A direct verification shows that $x^2 = t(x)x - n(x)$, where $t(x) = \alpha_{11} + \alpha_{22}$, $n(x) = \alpha_{11}\alpha_{22} - \sum_{k=1}^{3} \alpha_{12}^{(k)}\alpha_{21}^{(k)}$. Consequently, the algebra $\boldsymbol{C}(\mathscr{J}(\Phi))$ is quadratic over its center $\mathscr{J}(\Phi)$. Hence it follows by Lemma 14 that the algebra $\boldsymbol{C}(\mathscr{J}(\Phi))$ is quasi-regular. Since $\boldsymbol{C}(\mathscr{J}(\Phi))$ is an ideal of the algebra $\boldsymbol{C}(\Phi)$, by the same token this proves that $\boldsymbol{C}(\mathscr{J}(\Phi)) \subseteq \mathscr{J}(\boldsymbol{C}(\Phi))$.

On the other hand, let $x \in \mathscr{J}(\boldsymbol{C}(\Phi))$,

$$x = \sum_{i=1}^{2} \alpha_{ii} e_{ii} + \sum_{k=1}^{3} (\alpha_{12}^{(k)} e_{12}^{(k)} + \alpha_{21}^{(k)} e_{21}^{(k)}).$$

We need to prove that then $\alpha_{ii}, \alpha_{ij}^{(k)} \in \mathscr{J}(\Phi)$. We shall first prove that $\alpha_{ii} \in \mathscr{J}(\Phi)$. Let α, β be arbitrary elements from Φ. We consider the element $(\alpha e_{ii})x(\beta e_{ii}) = (\alpha\alpha_{ii}\beta)e_{ii}$. Since $x \in \mathscr{J}(\boldsymbol{C}(\Phi))$, then also $(\alpha\alpha_{ii}\beta)e_{ii} \in \mathscr{J}(\boldsymbol{C}(\Phi))$. If x' is the quasi-inverse element for $(\alpha\alpha_{ii}\beta)e_{ii}$, then it is easy to see that $x' = \gamma_{ii}e_{ii}$, where γ_{ii} is the quasi-inverse element for $\alpha\alpha_{ii}\beta$ in the ring Φ. Thus the element $\alpha\alpha_{ii}\beta$ is quasi-regular for any $\alpha, \beta \in \Phi$, that is, the element α_{ii} generates a quasi-regular ideal in the ring Φ. This means $\alpha_{ii} \in \mathscr{J}(\Phi)$. Furthermore, for any $\alpha \in \Phi$ we have $x(\alpha e_{ji}^{(k)}) = \alpha_{ij}^{(k)}\alpha e_{ii} + \alpha_{jj}\alpha e_{ji}^{(k)} + \sum_{k=1}^{3} \beta_{ij}^{(k)} e_{ij}^{(k)}$ for some $\beta_{ij}^{(k)} \in \Phi$. Since $x(\alpha e_{ji}^{(k)}) \in \mathscr{J}(\boldsymbol{C}(\Phi))$, then by what was proved above $\alpha_{ij}^{(k)}\alpha \in \mathscr{J}(\Phi)$, whence $\alpha_{ij}^{(k)}\Phi \subseteq \mathscr{J}(\Phi)$ and $\alpha_{ij}^{(k)} \in \mathscr{J}(\Phi)$. By the same token this proves that $x \in \boldsymbol{C}(\mathscr{J}(\Phi))$.

This proves the theorem.

COROLLARY. There exists a $\mathscr{J}$-radical Cayley–Dickson ring.

In fact, let Φ be a quasi-regular ring without zero divisors (e.g., the ring of formal power series with zero constant terms). Then the Cayley–Dickson matrix algebra $\boldsymbol{C}(\Phi)$ is a Cayley–Dickson ring (we leave it to the reader to convince himself of this), and in addition $\mathscr{J}(\boldsymbol{C}(\Phi)) = \boldsymbol{C}(\mathscr{J}(\Phi)) = \boldsymbol{C}(\Phi)$.

Exercises

1. Let K be a Cayley–Dickson ring with center Z. Prove that $\mathscr{J}(K) = (0)$ if and only if $\mathscr{J}(Z) = (0)$. *Hint.* If $0 \neq z \in \mathscr{J}(Z)$ and I is a quadratic over Z ideal of the ring K, then the ideal zI is quadratic over the quasi-regular ring $zZ \subseteq \mathscr{J}(Z)$.

2. Let K be a Cayley–Dickson ring. Prove that the set

$$I = \{a \in K \,|\, t(ak) \in K \text{ for any } k \in K^{\#}\}$$

is the largest quadratic over the center ideal of the ring K. *Hint*. For the proof I is an ideal use relations (2.22).

3. Under the conditions of the previous exercise, prove that $[x, y]^2$, $[[x, y], z] \in I$ for any elements $x, y, z \in K$.

LITERATURE

Jacobson [46], Zhevlakov [72–75], Zhelyabin [81], Kleinfeld [95], McCrimmon [132], Smiley [215], Shestakov [269, 271, 273].

CHAPTER **11**

Representations of Alternative Algebras

The theory of representations of algebras consists of two interconnected parts—the theory of right and left representations and the theory of birepresentations. In the well-developed theory of associative algebras, the role of representations is undoubtedly more significant than that of birepresentations; while in the nonassociative case until recently there was mainly used a concept of birepresentation of algebras which goes back to Eilenberg. Different varieties of commutative and anticommutative algebras constitute the exceptions, where as a matter of fact the concepts of representation and birepresentation coincide.

In the present chapter we introduce the concepts of right representation and right module for algebras of an arbitrary variety $\mathfrak{M}$. Our basic objective is to attain, in terms of the theory of representations, a characterization for the Zhevlakov radical of an arbitrary alternative algebra, which is analogous to the well-known characterization for the Jacobson radical in the associative case. We shall study the properties of irreducible alternative right representations and modules. It turns out that each such module is either "associative from the right" or "associative from the left." This allows us to describe the primitive alternative algebras as algebras having faithful irreducible modules.

By results of the previous chapter, the characterization we need for the Zhevlakov radical is easily deduced from this.

The results of this chapter were obtained by Zhevlakov, Slin'ko, and Shestakov.

1. DEFINITIONS AND PRELIMINARY RESULTS

Let V be some Φ-module. We shall denote by $T(V)$ the tensor algebra of the Φ-module V (see S. Lang, "Algebra," p. 470):

$$T(V) = \Phi \cdot 1 \oplus V^{(1)} \oplus \cdots \oplus V^{(r)} \oplus \cdots,$$

where $V^{(r)} = V \otimes \cdots \otimes V$ (the tensor product taken r times). We recall that every homomorphism of Φ-modules $\phi: V \to W$ induces a uniquely determined homomorphism of the tensor algebras $T(\varphi): T(V) \to T(W)$ such that $T(\varphi)(1) = 1$ and $T(\varphi)(v_1 \otimes \cdots \otimes v_r) = \varphi(v_1) \otimes \cdots \otimes \varphi(v_r)$ for any $v_1, \ldots, v_r \in V$. In particular, it follows from this that every Φ-linear mapping $\varphi: V \to B$ of the module V into an associative algebra B with identity element 1 induces a unique homomorphism of the associative algebras $T(\varphi): T(V) \to B$ such that $T(\varphi)(1) = 1$ and $T(\varphi)(v_1 \otimes \cdots \otimes v_r) = \varphi(v_1) \cdots \varphi(v_r)$ for any $v_1, \ldots, v_r \in V$.

Now let $\mathfrak{M}$ be some variety of Φ-algebras and $\Phi_{\mathfrak{M}} = \Phi_{\mathfrak{M}}[X]$ be the free algebra in the variety $\mathfrak{M}$ from a countable set of generators $X = \{x_1, x_2, \ldots\}$. We consider the tensor algebra $T(\Phi_{\mathfrak{M}})$ of the module $\Phi_{\mathfrak{M}}$. We shall denote the elements of the algebra $\Phi_{\mathfrak{M}}$ by the letters f, g, $h, \ldots$, and elements of the tensor algebra $T(\Phi_{\mathfrak{M}})$ by the letters F, G, $H, \ldots$. Let $F(x_1, \ldots, x_n) \in T(\Phi_{\mathfrak{M}})$,

$$F(x_1, \ldots, x_n) = \alpha \cdot 1 + \sum_i f_1^{(i)}(x_1, \ldots, x_n) \otimes \cdots \otimes f_{n_i}^{(i)}(x_1, \ldots, x_n).$$

If A is some algebra from the variety $\mathfrak{M}$ and $a_1, \ldots, a_n \in A$, then we shall denote by $F(a_1, \ldots, a_n)$ the following element of the tensor algebra $T(A)$:

$$F(a_1, \ldots, a_n) = \alpha \cdot 1 + \sum_i f_1^{(i)}(a_1, \ldots, a_n) \otimes \cdots \otimes f_{n_i}^{(i)}(a_1, \ldots, a_n).$$

If $\varphi: \Phi_{\mathfrak{M}} \to A$ is a homomorphism of the Φ-algebras such that $\varphi(x_i) = a_i$, $i = 1, 2, \ldots, n$, and $T(\varphi): T(\Phi_{\mathfrak{M}}) \to T(A)$ is the induced homomorphism of the tensor algebras, then it is easy to see that

$$F(a_1, \ldots, a_n) = T(\varphi)(F(x_1, \ldots, x_n)).$$

Now let $\rho: A \to \mathrm{End}_\Phi(M)$ be some Φ-linear mapping of the algebra A into the algebra of endomorphisms of the Φ-module M, and $T(\rho): T(A) \to \mathrm{End}_\Phi(M)$ be the induced homomorphism of the tensor algebra $T(A)$. We shall call an element $F(x_1, \ldots, x_n)$ of the tensor algebra $T(\Phi_{\mathfrak{M}})$ an *identity of the mapping* ρ if $T(\rho)(F(a_1, \ldots, a_n)) = 0$ for any $a_1, \ldots, a_n \in A$. In other words, the element

$$F(x_1, \ldots, x_n) = \alpha \cdot 1 + \sum_i f_1^{(i)}(x_1, \ldots, x_n) \otimes \cdots \otimes f_{n_i}^{(i)}(x_1, \ldots, x_n)$$

is an identity of the mapping ρ if for any $a_1, \ldots, a_n \in A$

$$\alpha E_M + \sum_i [f_1^{(i)}(a_1, \ldots, a_n)]^\rho \cdots [f_{n_i}^{(i)}(a_1, \ldots, a_n)]^\rho = 0,$$

where E_M is the identity endomorphism of the module M. In this case we shall also say that the mapping ρ satisfies the identity $F(x_1, \ldots, x_n)$.

For example, if $F(x_1, x_2) = x_1 \otimes x_2 - x_1 x_2$, then the mapping $\rho: A \to \mathrm{End}_\Phi(M)$ satisfies the identity $F(x_1, x_2)$ if $a^\rho b^\rho - (ab)^\rho = 0$ for any $a, b \in A$.

We now consider the Φ-linear mapping R of the algebra $\Phi_{\mathfrak{M}}$ into the algebra of endomorphisms of the Φ-module $\Phi_{\mathfrak{M}}$ which puts into correspondence with the element $f \in \Phi_{\mathfrak{M}}$ the operator of right multiplication R_f. Let $T(R): T(\Phi_{\mathfrak{M}}) \to \mathrm{End}(\Phi_{\mathfrak{M}})$ be the induced homomorphism of the tensor algebra. We denote the kernel $\mathrm{Ker}\, T(R)$ of this homomorphism by $I_{\mathfrak{M}}$, and call it the *ideal of R-identities* of the variety $\mathfrak{M}$. We shall call elements of the ideal $I_{\mathfrak{M}}$ *R-identities* of the variety $\mathfrak{M}$.

If a Φ-linear mapping $\rho: A \to \mathrm{End}_\Phi(M)$ satisfies all R-identities of the variety $\mathfrak{M}$, then we shall call ρ a *right representation of the algebra A in the variety* $\mathfrak{M}$ (*right* $\mathfrak{M}$*-representation*). In this case we shall call the Φ-module M with bilinear composition $m \cdot a = ma^\rho$, for $m \in M$, $a \in A$, a *right module for the algebra A* (*A-module*) *in the variety* $\mathfrak{M}$.

The set $\mathrm{Ker}_\rho(A) = \{a \in A \mid a^\rho = 0\}$ is called the *kernel* of the representation ρ. We denote by $K_\rho(A)$ the largest ideal of the algebra A contained in $\mathrm{Ker}_\rho(A)$. A representation ρ (module M) is called *faithful* if $\mathrm{Ker}_\rho(A) = (0)$, and is called *almost faithful* if $K_\rho(A) = (0)$.

As in the associative case [46], there are also defined in a natural way the concepts of *submodule* and *irreducible module* (*representation*), and for algebras with an identity element the concept of a *unital module* (*representation*).

As before let A be an algebra of a variety $\mathfrak{M}$. We denote by $I_{\mathfrak{M}}[A]$ the ideal of the tensor algebra $T(A)$ which consists of elements of the form $F(a_1, \ldots, a_n)$, where $F(x_1, \ldots, x_n) \in I_{\mathfrak{M}}$ and $a_1, \ldots, a_n \in A$. We consider the quotient algebra $\mathscr{R}_{\mathfrak{M}}(A) = T(A)/I_{\mathfrak{M}}[A]$ and the mapping $\mathscr{R}: A \to \mathscr{R}_{\mathfrak{M}}(A)$ which sends $a \in A$ to the coset $a + I_{\mathfrak{M}}[A] \in \mathscr{R}_{\mathfrak{M}}(A)$. We shall denote the image of the element a under the mapping $\mathscr{R}$ by $\mathscr{R}_a$, and we shall denote the identity element of the algebra $\mathscr{R}_{\mathfrak{M}}(A)$ by 1.

PROPOSITION 1. The mapping $\rho: A \to \mathrm{End}_\Phi(M)$ is a right $\mathfrak{M}$-representation of the algebra A if and only if there exists a homomorphism

$\varphi:\mathscr{R}_{\mathfrak{M}}(A)\to \mathrm{End}_\Phi(M)$ such that $\varphi(1)=E_M$ and $\mathscr{R}\circ\varphi=\rho$, i.e., the following diagram is commutative:

$$\begin{array}{ccc} & \mathscr{R}_{\mathfrak{M}}(A) & \\ {}^{\mathscr{R}}\nearrow & & \searrow^{\varphi} \\ A & \xrightarrow{\rho} & \mathrm{End}_\Phi(M) \end{array}$$

PROOF. It is clear that every mapping of the form

$$A \xrightarrow{\mathscr{R}} \mathscr{R}_{\mathfrak{M}}(A) \xrightarrow{\varphi} \mathrm{End}_\Phi(M),$$

where φ is a homomorphism, is a right $\mathfrak{M}$-representation of the algebra A. Conversely, let $\rho:A\to\mathrm{End}_\Phi(M)$ be some right $\mathfrak{M}$-representation. The mapping ρ induces a homomorphism of the associative algebras $T(\rho):T(A)\to\mathrm{End}_\Phi(M)$ for which $T(\rho)(1)=E_M$. By definition of a right representation we have $I_{\mathfrak{M}}[A]\subseteq \mathrm{Ker}\, T(\rho)$, i.e., the mapping $T(\rho)$ is constant on each coset modulo $I_{\mathfrak{M}}[A]$. Now setting $\varphi(F+I_{\mathfrak{M}}[A])=T(\rho)(F)$ for any $F\in T(A)$, we obtain the desired homomorphism $\varphi:\mathscr{R}_{\mathfrak{M}}(A)\to\mathrm{End}_\Phi(M)$. This proves the proposition.

We call the algebra $\mathscr{R}_{\mathfrak{M}}(A)$ the *universal algebra for right representations of the algebra A in the variety* $\mathfrak{M}$. Proposition 1 gives us the opportunity to reduce the study of representations of a nonassociative algebra A to the study of representations of the associative algebra $\mathscr{R}_{\mathfrak{M}}(A)$. We can consider each right A-module in the variety $\mathfrak{M}$ as an associative right $\mathscr{R}_{\mathfrak{M}}(A)$-module. In addition, it is easy to see that a right $\mathfrak{M}$-representation

$$\rho:A \xrightarrow{\mathscr{R}} \mathscr{R}_{\mathfrak{M}}(A) \xrightarrow{\varphi} \mathrm{End}_\Phi(M)$$

is irreducible if and only if φ is an irreducible representation of the algebra $\mathscr{R}_{\mathfrak{M}}(A)$.

We now prove one important property of $\mathfrak{M}$-representations.

LEMMA 1. Let A be an algebra from a variety $\mathfrak{M}$ and $\rho:A\to\mathrm{End}_\Phi(M)$ be a right $\mathfrak{M}$-representation of the algebra A. In addition, let B be an ideal of the algebra A contained in $\mathrm{Ker}_\rho(A)$. Then the mapping $\bar\rho:\bar A\to\mathrm{End}_\Phi(M)$ of the algebra $\bar A=A/B$ defined by the rule $\bar\rho(a+B)=\rho(a)$ is a right $\mathfrak{M}$-representation of the algebra $\bar A$.

PROOF. It is necessary to verify that the mapping $\bar\rho$ satisfies all identities from $I_{\mathfrak{M}}$. Let $F(x_1,\ldots,x_n)\in I_{\mathfrak{M}}$,

$$F(x_1,\ldots,x_n)=\alpha\cdot 1+\sum_i f_1^{(i)}(x_1,\ldots,x_n)\otimes\cdots\otimes f_{n_i}^{(i)}(x_1,\ldots,x_n).$$

Since the mapping ρ satisfies the identity $F(x_1, \ldots, x_n)$, then for any elements $a_1, \ldots, a_n \in A$ we have

$$\begin{aligned}
T(\bar{\rho})&(F(a_1 + B, \ldots, a_n + B)) \\
&= \alpha E_M + \sum_i [f_1^{(i)}(a_1 + B, \ldots, a_n + B)]^{\bar{\rho}} \cdots [f_{n_i}^{(i)}(a_1 + B, \ldots, a_n + B)]^{\bar{\rho}} \\
&= \alpha E_M + \sum_i [f_1^{(i)}(a_1, \ldots, a_n) + B]^{\bar{\rho}} \cdots [f_{n_i}^{(i)}(a_1, \ldots, a_n) + B]^{\bar{\rho}} \\
&= \alpha E_M + \sum_i [f_1^{(i)}(a_1, \ldots, a_n)]^{\rho} \cdots [f_{n_i}^{(i)}(a_1, \ldots, a_n)]^{\rho} \\
&= T(\rho)(F(a_1, \ldots, a_n)) = 0,
\end{aligned}$$

i.e., the mapping $\bar{\rho}$ satisfies the identity $F(x_1, \ldots, x_n)$. This proves the lemma.

In order to recognize whether a mapping $\rho: A \to \mathrm{End}_\Phi(M)$ is a right $\mathfrak{M}$-representation, it is absolutely unnecessary to verify all R-identities of the variety $\mathfrak{M}$ for the mapping ρ. It is only necessary to verify those R-identities from which all the rest follow. We consider this situation from a formal point of view.

Let Ω_0 be the set of all endomorphisms of the algebra $\Phi_{\mathfrak{M}}$. Each endomorphism φ from Ω_0 induces a unique endomorphism $T(\varphi)$ of the tensor algebra $T(\Phi_{\mathfrak{M}})$. We denote by Ω the set of all endomorphisms of the algebra $T(\Phi_{\mathfrak{M}})$ of the form $T(\varphi)$, where $\varphi \in \Omega_0$. An ideal of the algebra $T(\Phi_{\mathfrak{M}})$ which is mapped into itself under all endomorphisms from Ω is called an *Ω-characteristic ideal.*

PROPOSITION 2. The ideal $I_{\mathfrak{M}}$ is an Ω-characteristic ideal of the algebra $T(\Phi_{\mathfrak{M}})$.

PROOF. Let $F(x_1, \ldots, x_n) = \alpha \cdot 1 + \sum_i f_1^{(i)}(x_1, \ldots, x_n) \otimes \cdots \otimes f_{n_i}^{(i)}(x_1, \ldots, x_n) \in I_{\mathfrak{M}}$ and $T(\varphi)$ be the endomorphism of the algebra $T(\Phi_{\mathfrak{M}})$ induced by an endomorphism φ of the algebra $\Phi_{\mathfrak{M}}$. Let $\varphi(x_i) = y_i$, $i = 1, 2, \ldots, n$. We need to show that $T(\varphi)(F(x_1, \ldots, x_n)) = F(y_1, \ldots, y_n) \in I_{\mathfrak{M}}$. Since $I_{\mathfrak{M}} = \mathrm{Ker}\, T(R)$, we have

$$0 = T(R)(F(x_1, \ldots, x_n)) = \alpha E_{\Phi_{\mathfrak{M}}} + \sum_i R_{f_1^{(i)}(x_1, \ldots, x_n)} \cdots R_{f_{n_i}^{(i)}(x_1, \ldots, x_n)}.$$

In view of the fact that $\Phi_{\mathfrak{M}}$ is a free algebra and $x_1, \ldots, x_n$ are free generators, it follows from this equality that

$$T(R)(F(y_1, \ldots, y_n)) = \alpha E_{\Phi_{\mathfrak{M}}} + \sum_i R_{f_1^{(i)}(y_1, \ldots, y_n)} \cdots R_{f_{n_i}^{(i)}(y_1, \ldots, y_n)} = 0,$$

i.e., $F(y_1, \ldots, y_n) \in \mathrm{Ker}\, T(R) = I_{\mathfrak{M}}$. This proves the proposition.

We shall call any system of generators of the ideal $I_{\mathfrak{M}}$ as an Ω-characteristic ideal a *system of defining R-identities* for the variety $\mathfrak{M}$.

PROPOSITION 3. A subset $\{F_\alpha(x_1, \ldots, x_{n_\alpha})\} \subseteq I_{\mathfrak{M}}$ is a system of defining R-identities for the variety $\mathfrak{M}$ if and only if for any algebra A from $\mathfrak{M}$ any Φ-linear mapping $\rho: A \to \mathrm{End}_\Phi(M)$ satisfying all identities from $\{F_\alpha\}$ is a right $\mathfrak{M}$-representation of the algebra A.

PROOF. Let $\{F_\alpha(x_1, \ldots, x_{n_\alpha})\}$ be a system of defining R-identities for the variety $\mathfrak{M}$, i.e., the ideal $I_{\mathfrak{M}}$ is generated as an Ω-characteristic ideal by the set $\{F_\alpha(x_1, \ldots, x_{n_\alpha})\}$. It is easy to see that $I_{\mathfrak{M}}$ is generated as an ordinary ideal by elements of the form $F_\alpha(y_1, \ldots, y_{n_\alpha})$, where $y_1, \ldots, y_{n_\alpha}$ are arbitrary elements of the algebra $\Phi_{\mathfrak{M}}$. If the mapping ρ satisfies all identities from $\{F_\alpha(x_1, \ldots, x_{n_\alpha})\}$, then it also satisfies all identities of the form $F_\alpha(y_1, \ldots, y_{n_\alpha})$, and consequently also all identities from $I_{\mathfrak{M}}$. This means ρ is a right $\mathfrak{M}$-representation of the algebra A.

Conversely, let every mapping satisfying all identities from $\{F_\alpha(x_1, \ldots, x_{n_\alpha})\}$ be a right $\mathfrak{M}$-representation. We assume that the Ω-characteristic ideal $\mathscr{T}$ generated by the set $\{F_\alpha(x_1, \ldots, x_{n_\alpha})\}$ is strictly contained in $I_{\mathfrak{M}}$. We consider the algebra $A = T(\Phi_{\mathfrak{M}})/\mathscr{T}$ and the mapping

$$\rho: \Phi_{\mathfrak{M}} \xrightarrow{T} T(\Phi_{\mathfrak{M}}) \xrightarrow{\tau} A \xrightarrow{\iota} \mathrm{End}_\Phi(M),$$

where T is the insertion, τ is the canonical homomorphism, and ι is some isomorphic representation of the associative algebra A. The mapping ρ satisfies all identities from $\{F_\alpha(x_1, \ldots, x_{n_\alpha})\}$, and so by assumption it is a right $\mathfrak{M}$-representation of the algebra $\Phi_{\mathfrak{M}}$. This is impossible, however, since the mapping ρ does not satisfy some identity from $I_{\mathfrak{M}}$. The contradiction obtained proves that $\mathscr{T} = I_{\mathfrak{M}}$.

This proves the proposition.

The question on the determination of a system of defining R-identities for a variety $\mathfrak{M}$ often proves to be very complicated, even when we know defining identities for the variety $\mathfrak{M}$. For example, a single system of defining R-identities for the variety of alternative algebras is still not known, and it is not even known if there exists such a finite system. It is not clear in general whether every finitely based variety (i.e., variety given by a finite number of defining identities) has a finite system of defining R-identities.

In the case of the variety Ass of associative algebras, it is easy to see that I_{Ass} is generated as an Ω-characteristic ideal by the single element $F(x_1, x_2) = x_1 \otimes x_2 - x_1 x_2$. A system of defining R-identities is also easily found for every variety $\mathfrak{M}$ of commutative or anticommutative algebras, when we know defining identities for the variety $\mathfrak{M}$ (see the Exercises).

We now show that every algebra from a variety $\mathfrak{M}$ has a sufficiently large supply of right $\mathfrak{M}$-representations.

Let the algebra A be a subalgebra of some algebra B from a variety $\mathfrak{M}$. We denote by $R_1^B(A)$ the subalgebra of the algebra $\mathrm{End}_\Phi(B)$ generated by the set $R^B(A)$ and the identity endomorphism E_B.

PROPOSITION 4. Let the mapping $R^B: A \to R_1^B(A)$ take $a \in A$ to the operator of right multiplication $R_a^B = R_a \in R^B(A)$, φ be an arbitrary homomorphism of the algebra $R_1^B(A)$ into the algebra $\mathrm{End}_\Phi(M)$ such that $\varphi(E_B) = E_M$, and $\rho = R^B \circ \varphi$ be the composition of the mappings R^B and φ:

$$\rho: A \xrightarrow{R^B} R_1^B(A) \xrightarrow{\varphi} \mathrm{End}_\Phi(M).$$

Then ρ is a right $\mathfrak{M}$-representation of the algebra A.

PROOF. It suffices to verify that the mapping R^B satisfies all R-identities of the variety $\mathfrak{M}$. Let $F(x_1, \ldots, x_n) = \alpha \cdot 1 + \sum_i f_1^{(i)}(x_1, \ldots, x_n) \otimes \cdots \otimes f_{n_i}^{(i)}(x_1, \ldots, x_n) \in I_{\mathfrak{M}}$. By definition of the ideal $I_{\mathfrak{M}}$, this means that the relation

$$\alpha E_{\Phi_{\mathfrak{M}}} + \sum_i R_{f_1^{(i)}(x_1, \ldots, x_n)} \cdots R_{f_{n_i}^{(i)}(x_1, \ldots, x_n)} = 0$$

is valid in the algebra $\mathrm{End}(\Phi_{\mathfrak{M}})$. Since $\Phi_{\mathfrak{M}}$ is a free algebra and $x_1, \ldots, x_n$ are free generators, from this equality it follows that for any $b_1, \ldots, b_n \in B$ the relation

$$\alpha E_B + \sum_i R_{f_1^{(i)}(b_1, \ldots, b_n)} \cdots R_{f_{n_i}^{(i)}(b_1, \ldots, b_n)} = 0$$

is valid in the algebra $\mathrm{End}_\Phi(B)$. In other words, the mapping $R: B \to \mathrm{End}_\phi(B)$, which puts into correspondence with an element $b \in B$ the operator of right multiplication R_b, satisfies the identity F. Since the mapping $R^B: A \to R_1^B(A)$ is the restriction of the mapping R to A, then it also satisfies the identity F. This proves the proposition.

We shall call the right representations of the algebra A defined in Proposition 4 *regular right $\mathfrak{M}$-representations.*

An appropriate universal algebra also exists for the class of regular right $\mathfrak{M}$-representations. For an arbitrary algebra A from $\mathfrak{M}$ we consider the algebra $C = C(A) \in \mathfrak{M}$, which is the free $\mathfrak{M}$-product (see A. I. Mal'cev, "Algebraic Systems," p. 306) of the algebra A and the singly generated free algebra $\Phi_{\mathfrak{M}}[x]$ of the variety $\mathfrak{M}$, i.e., $C = \Phi_{\mathfrak{M}}[x] *_{\mathfrak{M}} A$. We denote the subalgebra $R_1^C(A)$ by $R_{\mathfrak{M}}^*(A)$, the mapping R^C by R^*, and the identity endomorphism E_C by 1.

PROPOSITION 5. The mapping $\rho: A \to \mathrm{End}_\Phi(M)$ is a regular right $\mathfrak{M}$-representation of the algebra A if and only if there exists a homomorphism $\varphi: R_{\mathfrak{M}}^*(A) \to \mathrm{End}_\Phi(M)$ such that $\varphi(1) = E_\Phi$ and $\rho = R^* \circ \varphi$, i.e., the following diagram is commutative:

$$\begin{array}{ccc} & R_{\mathfrak{M}}^*(A) & \\ {}^{R^*}\nearrow & & \searrow^{\varphi} \\ A & \xrightarrow[\rho]{} & \mathrm{End}_\Phi(M) \end{array}$$

PROOF. It suffices to prove that for every algebra B from $\mathfrak{M}$ containing A as a subalgebra there exists a homomorphism $\psi: R_1^C(A) \to \mathrm{End}_\Phi(B)$ such that $\psi(E_C) = E_B$ and $R^B = R^C \circ \psi$. The desired homomorphism ψ will exist if, from the validity of the relation

$$\alpha E_C + \sum_i R_{a_{i1}}^C R_{a_{i2}}^C \cdots R_{a_{ik_i}}^C = 0 \tag{1}$$

in $\mathrm{End}_\Phi(C)$, it always follows that the relation

$$\alpha E_B + \sum_i R_{a_{i1}}^B R_{a_{i2}}^B \cdots R_{a_{ik_i}}^B = 0 \tag{2}$$

is valid in $\mathrm{End}_\Phi(B)$, where $\alpha \in \Phi$, $a_{ij} \in A$. We shall prove that this condition is always satisfied.

Relation (1) implies the validity in C of the equality

$$\alpha x + x \sum_i R_{a_{i1}}^C R_{a_{i2}}^C \cdots R_{a_{ik_i}}^C = 0. \tag{3}$$

Furthermore, for any $b \in B$ there exists a homomorphism of the algebra $\Phi_{\mathfrak{M}}[x]$ into B which takes x into b. By definition of the free $\mathfrak{M}$-product, there exists a homomorphism ζ of the algebra C into B which is the identity on A and which takes x into b. Applying the homomorphism ζ to equality (3), we obtain the equality

$$\alpha b + b \sum_i R_{a_{i1}}^B R_{a_{i2}}^B \cdots R_{a_{ik_i}}^B = 0,$$

whence (2) follows since b was arbitrary.

This proves the proposition.

As is easy to see, the algebra $R_{\mathfrak{M}}^*(A)$ is unique up to isomorphism. We call it the *universal algebra of right multiplications* for the algebra A in the variety $\mathfrak{M}$. We shall denote by R_a^* the image of the element $a \in A$ under the mapping R^*.

COROLLARY. The following two conditions are equivalent:

(1) every right $\mathfrak{M}$-representation of the algebra A is regular;
(2) the algebras $\mathscr{R}_{\mathfrak{M}}(A)$ and $R^*_{\mathfrak{M}}(A)$ are isomorphic.

In view of Propositions 1 and 5, the proof is obvious.

In many classical varieties of algebras all right representations are regular. For example, if $\mathfrak{M} = \text{Ass}$ is the variety of associative algebras, then it is easy to see that $\mathscr{R}^*_{\text{Ass}}(A) \cong R^*_{\text{Ass}} \cong A^{\#}$ for any algebra $A \in \text{Ass}$. All right $\mathfrak{M}$-representations are likewise regular in the case when $\mathfrak{M}$ is some variety of commutative or anticommutative algebras (see the Exercises). At the same time, in the variety Alt of alternative algebras there exist right representations which are not regular [214]. We note, too, that generally speaking the analog of Lemma 1 is not true for regular representations [214].

In concluding this section, we show that $\mathfrak{M}$-birepresentations are another of the sources for right $\mathfrak{M}$-representations.

A pair of mappings (μ, ν) of an algebra A from a variety $\mathfrak{M}$ into the algebra of endomorphisms $\text{End}_\Phi(M)$ of a module M is called a *birepresentation* of the algebra A in the variety $\mathfrak{M}$ ($\mathfrak{M}$*-birepresentation*) if the *slit null extension* $S = A + M$ of the algebra A with kernel M, i.e., the direct sum of the modules A and M with multiplication

$$(a_1 + m_1)(a_2 + m_2) = m_1 a_2^\nu + m_2 a_1^\mu + a_1 a_2,$$

belongs to the variety $\mathfrak{M}$. In this case the module M with bilinear compositions $a \cdot m = ma^\mu$, $m \cdot a = ma^\nu$ for $m \in M$, $a \in A$, is called a *bimodule* for the algebra A (A*-bimodule*) in the variety $\mathfrak{M}$. The mapping $\mu: A \to \text{End}_\Phi(M)$ is called the *left component of the birepresentation* (μ, ν), and ν is called the *right component*.

PROPOSITION 6. For any $\mathfrak{M}$-birepresentation (μ, ν) of an algebra A from a variety $\mathfrak{M}$, the right component ν of this birepresentation is a regular right $\mathfrak{M}$-representation of the algebra A.

PROOF. Let (μ, ν) be a birepresentation of the algebra A into the algebra $\text{End}_\Phi(M)$, and let $S = A + M$ be the corresponding split null extension. By assumption the algebra S belongs to the variety $\mathfrak{M}$. In addition, A is a subalgebra of the algebra S. Finally, for any $a \in A$, $m \in M$ we have $ma^\nu = m \cdot a = mR_a^S$, whence it easily follows there exists a homomorphism $\varphi: R^S(A) \to \text{End}_\Phi(M)$ for which $\nu = R^S \circ \varphi$. By the same token everything is proved.

Exercises

1. Prove that if the variety $\mathfrak{M}$ is unitally closed, then the mapping $\mathscr{R}: A \to \mathscr{R}_{\mathfrak{M}}(A)$ is injective for any algebra A from $\mathfrak{M}$.

2. Prove that if $\mathfrak{M}_1 \subseteq \mathfrak{M}_2$, then there exists an epimorphism $\varphi: \mathscr{R}_{\mathfrak{M}_2}(A) \to \mathscr{R}_{\mathfrak{M}_1}(A)$ for which $\varphi(1) = 1$ and the following diagram is commutative:

$$\begin{array}{ccc} A & \xrightarrow{E_A} & A \\ \downarrow{\scriptstyle \mathscr{R}_2} & & \downarrow{\scriptstyle \mathscr{R}_1} \\ \mathscr{R}_{\mathfrak{M}_2}(A) & \xrightarrow{\phi} & \mathscr{R}_{\mathfrak{M}_1}(A) \end{array}$$

3. Prove that for any homomorphism $\tau: A \to A'$ of algebras from the variety $\mathfrak{M}$ there exist unique homomorphisms $\mathscr{R}(\tau): \mathscr{R}_{\mathfrak{M}}(A) \to \mathscr{R}_{\mathfrak{M}}(A')$ and $R^*(\tau): R^*_{\mathfrak{M}}(A) \to R^*_{\mathfrak{M}}(A')$ of the corresponding universal algebras such that $\mathscr{R}(\tau)(1) = 1$, $R^*(\tau)(1) = 1$, and the following diagrams are commutative:

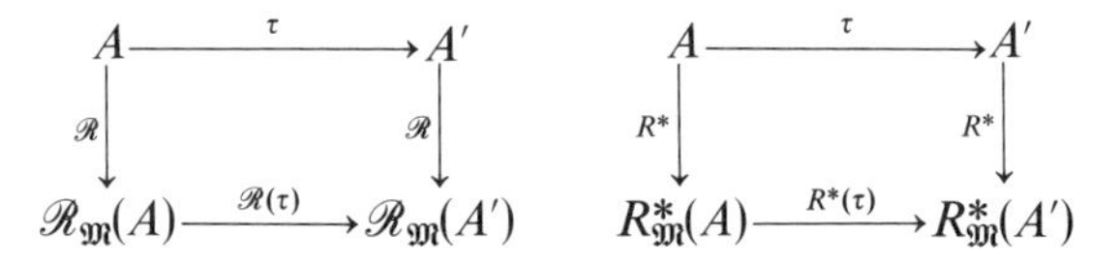

4. Let $\mathfrak{M}$ be a homogeneous variety of Φ-algebras. Prove that for any algebra A from $\mathfrak{M}$ and any extension K of the ring of operators Φ the following isomorphism holds:

$$\mathscr{R}_{\mathfrak{M}}(K \otimes_{\Phi} A) \cong K \otimes_{\Phi} \mathscr{R}_{\mathfrak{M}}(A).$$

5. An algebra A is called *right alternative* if the identity

$$[(xy)z]y = x[(yz)y]$$

is valid in the algebra $A^{\#}$. Let A be a right alternative algebra, and $\rho: A \to \mathrm{End}_{\Phi}(M)$ be a Φ-linear mapping satisfying the identities

$$F(x_1) = x_1 \otimes x_1 - x_1^2, \tag{4}$$

$$G(x_1, x_2) = x_1 \otimes x_2 \otimes x_1 - (x_1 x_2)x_1. \tag{5}$$

Prove that ρ is a regular right alternative right representation of the algebra A. *Hint.* Prove that the pair of mappings $(0, \rho)$ is a right alternative birepresentation of the algebra A.

6. Prove that it is possible to take identities (4) and (5) as a system of defining R-identities for the variety $R\mathrm{Alt}$ of right alternative algebras.

7. Prove that for any finite-dimensional right alternative algebra A the universal algebra $\mathscr{R}(A)$ for the right alternative right representations of the algebra A is also finite-dimensional.

8. Let A be an associative algebra. Prove that any associative left A-module is a right alternative right A-module.

9. Let $\mathfrak{M}$ be some homogeneous variety of commutative or anticommutative algebras. Then for any algebra A from $\mathfrak{M}$ and any $\mathfrak{M}$-representation ρ

of the algebra A, the pair of mappings (ρ, ρ) in the commutative case and $(-\rho, \rho)$ in the anticommutative case is an $\mathfrak{M}$-birepresentation of the algebra A.

10. Let $f = f(x_1, \ldots, x_n)$ be some homogeneous element of a free commutative or free anticommutative algebra F. An arbitrary partial linearization of the first degree of the element f can be written in the form

$$f_i^{(1)}(x_1, \ldots, x_n; x_j) = x_j \sum_k R_{m_{i1}^{(k)}(x_1, \ldots, x_n)} \cdots R_{m_{in_k}^{(k)}(x_1, \ldots, x_n)}$$

for suitable elements $m_{ij}^{(s)}(x_1, \ldots, x_n)$ of the free algebra F. We associate with the element f the following n elements $T_i(f)$, $i = 1, \ldots, n$, of the tensor algebra $T(F)$, setting

$$T_i(f) = \sum_k m_{i1}^{(k)}(x_1, \ldots, x_n) \otimes \cdots \otimes m_{in_k}^{(k)}(x_1, \ldots, x_n).$$

Now let some homogeneous variety $\mathfrak{M}$ of commutative or anticommutative algebras be given by a system of homogeneous identities $\{f_\alpha(x_1, \ldots, x_{n_\alpha}) | \alpha \in A\}$. Then it is possible to take the system $\{T_i(f_\alpha) | i = 1, \ldots, n_\alpha; \alpha \in A\}$ as a system of defining R-identities for the variety $\mathfrak{M}$.

2. REPRESENTATIONS OF COMPOSITION ALGEBRAS

In this section we shall study the properties of irreducible alternative modules over composition algebras.

We note first of all that it suffices to restrict ourselves to unital modules.

LEMMA 2. Let A be an alternative algebra with identity element 1. Then any irreducible alternative representation of the algebra A is unital.

PROOF. We first note that in view of the right alternative identity and the right Moufang identity, the ideal I_{Alt} of alternative R-identities contains the following identities:

$$F(x_1) = x_1 \otimes x_1 - x_1^2, \tag{4}$$

$$G(x_1, x_2) = x_1 \otimes x_2 \otimes x_1 - x_1 x_2 x_1. \tag{5}$$

Now let ρ be an irreducible representation of the algebra A, and M be the corresponding irreducible A-module. We consider the set $M_0 = \{m - m \cdot 1 | m \in M\}$. It follows from R-identity (4) and its linearization that for any $a, b \in A$ we have $(a^\rho)^2 = (a^2)^\rho$, $a^\rho \circ b^\rho = (a \circ b)^\rho$. Now for any $a \in A$ we have

$$(m - m \cdot 1) \cdot a = m \cdot a + (m \cdot a) \cdot 1 - m \cdot (a \circ 1) = (m \cdot a) \cdot 1 - m \cdot a \in M_0,$$

whence it follows that M_0 is a submodule of the module M. If M is not unital, then $M_0 \neq (0)$, and in view of the irreducibility $M_0 = M$. Furthermore,

$(m - m \cdot 1) \cdot 1 = m \cdot 1 - m \cdot 1^2 = 0$ for any $m \in M$, and therefore $M \cdot 1 = (0)$. Finally, by R-identity (5) for any a, $b \in A$ we have $(aba)^\rho = a^\rho b^\rho a^\rho$, and therefore $m \cdot a = m \cdot (1a1) = ((m \cdot 1) \cdot a) \cdot 1 = 0$ for any $m \in M$, $a \in A$. This means $M \cdot A = (0)$, which contradicts the irreducibility of M. This proves the lemma.

Let A be an alternative algebra, and M be an alternative A-module. For $a, b \in A$, $m \in M$, we set

$$(m, a, b) = (m \cdot a) \cdot b - m \cdot (ab),$$
$$\{m, a, b\} = (m \cdot a) \cdot b - m \cdot (ba).$$

Many of the assertions proved in this section are simultaneously valid, both for the composition (m, a, b) and for the composition $\{m, a, b\}$. In this connection we introduce for these trilinear compositions the common notation $[m, a, b]$. Of course all expressions $[m, a, b]$ encountered in a single formula or in the course of one fixed proof only denote some one of the compositions (m, a, b), $\{m, a, b\}$.

LEMMA 3. For any $a, b \in A$, $m \in M$, there are the relations

$$[m, a, a] = 0, \tag{6}$$
$$\{(m, a, b), a, b\} = 0, \tag{7}$$
$$(\{m, a, b\}, a, b) = 0. \tag{8}$$

PROOF. Relation (6) follows from R-identity (4). Relations (7) and (8) can be deduced from identities we know to be valid in alternative algebras (see Lemma 7.9), but we shall now prove them independently. For example, we prove (8). In view of linearized relation (6), it suffices for us to prove that $(\{m, a, b\}, b, a) = 0$. By R-identities (4) and (5) and their linearizations, for the representation ρ corresponding to the module M, we have

$$\begin{aligned}(a^\rho b^\rho &- (ba)^\rho)(b^\rho a^\rho - (ba)^\rho)\\ &= a^\rho b^\rho b^\rho a^\rho - (ba)^\rho b^\rho a^\rho - a^\rho b^\rho (ba)^\rho + (ba)^\rho (ba)^\rho\\ &= (ab^2a)^\rho - ((ba)ba + ab(ba))^\rho + (baba)^\rho = 0,\end{aligned}$$

whence follows (8). Relation (7) is proved analogously. This proves the lemma.

We shall call a module M *associative from the right* (*left*) if M has a system of generating elements $\{m_i\}$ such that $(m_i, a, b) = 0$ ($\{m_i, a, b\} = 0$, respectively) for any $m_i \in \{m_i\}$ and a, $b \in A$.

We wish to prove that any irreducible alternative module over a composition algebra is either associative from the right or associative from the left.

In this section A is subsequently always a composition algebra over a field F, and M is an irreducible alternative A-module. We note that by Lemma 2 the module M is a unital A-module.

LEMMA 4. For any $a, b \in A$, $m \in M$, there are the relations

$$[m, a, ab] = \alpha[m, a, b] \cdot a + \beta t(a)[m, a, b], \tag{9}$$

$$[m, a, ba] = \alpha'[m, a, b] \cdot a + \beta' t(a)[m, a, b], \tag{10}$$

where $t(a)$ is the trace of the element a, and α, α', β, β' are some elements of the field F.

PROOF. We first consider the case when $[,,]$ is $(,,)$. By the Moufang identity we have $(m, a, ab) = (m, a, b) \cdot a$, which proves (9). Furthermore, $(m, a, ba) = -(m, a, ab) + (m, a^2, b) = -(m, a, b) \cdot a + (m, a^2, b)$. Since A is quadratic over F, for any $a \in A$ we have

$$a^2 = t(a)a - n(a) \cdot 1, \tag{11}$$

where $t(a)$ is the trace of the element a, $n(a)$ is the norm of the element a, and $n(a), t(a) \in F$. Consequently $(m, a^2, b) = t(a)(m, a, b) - n(a)(m, 1, b) = t(a)(m, a, b)$, and $(m, a, ba) = -(m, a, b) \cdot a + t(a)(m, a, b)$, which proves (10).

Now let $[,,]$ denote the operation $\{,,\}$. Then we have $\{m, a, ab\} = (m \cdot a) \cdot (ab) - m \cdot (aba) = -[m \cdot (ab)] \cdot a + m \cdot [a \circ (ab)] - m \cdot (aba) = \{m, b, a\} \cdot a - [(m \cdot b) \cdot a] \cdot a + m \cdot (a^2 b) = -\{m, a, b\} \cdot a - (m \cdot b) \cdot a^2 + m \cdot (a^2 b) = -\{m, a, b\} \cdot a - \{m, b, a^2\} = -\{m, a, b\} \cdot a - t(a)\{m, b, a\} + n(a)\{m, b, 1\} = -\{m, a, b\} \cdot a + t(a)\{m, a, b\}$, i.e., (9) is proved. Furthermore, $\{m, a, ba\} = (m \cdot a) \cdot (ba) - m \cdot (ba^2) = -[m \cdot (ba)] \cdot a + m \cdot [a \circ (ba)] - m \cdot (ba^2) = \{m, a, b\} \cdot a - [(m \cdot a) \cdot b] \cdot a + m \cdot (aba) = \{m, a, b\} \cdot a$, which proves (10).

This proves the lemma.

COROLLARY 1. Let $[m, a, b] = 0$. Then $[m, a, ab] = [m, a, ba] = 0$.

COROLLARY 2. Let $a, v_1, v_2, v_3 \in A$, $m \in M$. If $[m, v_i, v_j] = 0$ for any $i, j \in \{1, 2, 3\}$, then

$$[m, v_i, v_j v_k] = \pm[m, v_1, v_2 v_3]; \tag{12}$$

and if $[m, a, v_i] = [m, a, v_i v_j] = 0$ for any $i, j \in \{1, 2, 3\}$, then

$$[m, a, (v_i v_j) v_k] = \pm[m, a, (v_1 v_2) v_3], \tag{13}$$

$$[m, a, v_i(v_j v_k)] = \pm[m, a, (v_1 v_2) v_3], \tag{14}$$

for any permutation (ijk) of the symbols 1, 2, 3.

PROOF. Relation (12) follows easily from linearized relations (9), (10), and the assumptions of the corollary. For the proof of relations (13) and (14), we note beforehand that the relation

$$bcb = t(bc)b + n(b)c - t(c)n(b) \cdot 1 \tag{15}$$

is valid in the algebra A. In fact, by (11) and (2.21) we have

$$\begin{aligned} bcb &= (b \circ c)b - cb^2 \\ &= [t(b)c + t(c)b - t(b)t(c) + t(bc)]b - c[t(b)b - n(b)] \\ &= t(c)b^2 - t(b)t(c)b + t(bc)b + n(b)c \\ &= t(bc)b + n(b)c - t(c)n(b) \cdot 1. \end{aligned}$$

Now from (11) and (15) follow

$$\begin{aligned} [m, a, b^2] &= t(b)[m, a, b], \\ [m, a, bcb] &= t(bc)[m, a, b] + n(b)[m,a,c]. \end{aligned}$$

In view of the assumptions of the corollary, linearizing these relations we obtain

$$\begin{aligned} [m, a, v_i \circ (v_j v_k)] &= 0, \\ [m, a, \{v_i v_j v_k\}] &= 0, \end{aligned}$$

whence by Proposition 5.3 relations (13) and (14) easily follow.
This proves the corollary.

LEMMA 5. Let $A = \boldsymbol{Q}(\mu, \beta)$ be the algebra of generalized quaternions. Then M is either associative from the right or associative from the left.

PROOF. It suffices for us to prove that there is found in M a nonzero element m for which $[m, A, A] = (0)$. In fact, then the set $m \cdot A = \{m \cdot a \mid a \in A\}$ will be a nonzero submodule in M which coincides with M by its irreducibility, and it is possible to take as a corresponding system of generating elements the system consisting of the single element m. There exist elements v_1, v_2 in the algebra $A = \boldsymbol{Q}(\mu, \beta)$ such that a basis for A over F consists of the elements $1, v_1, v_2, v_1 v_2$ (see Section 2.4). Let $m \in M, m \neq 0$. If $(m, v_1, v_2) = 0$, then by Corollary 1 to Lemma 4 $(m, A, A) = (0)$, and everything is proved. Now let $m_1 = (m, v_1, v_2) \neq 0$. Then by relation (7) we have $\{m_1, v_1, v_2\} = 0$, whence again by Corollary 1 to Lemma 4 $\{m_1, A, A\} = (0)$. This proves the lemma.

We now consider the case when $A = \boldsymbol{C}(\mu, \beta, \gamma)$ is a Cayley–Dickson algebra. In this case it is possible to select elements v_1, v_2, v_3 in A such that a basis for A over F is composed of the elements 1, v_1, v_2, v_3, $v_1 v_2$, $v_1 v_3$, $v_2 v_3$, $(v_1 v_2) v_3$ (see Section 2.4).

LEMMA 6. Let the relation $[m, v_i, v_j] = 0$ be satisfied for some $0 \neq m \in M$ and any $i, j \in \{1, 2, 3\}$. Then M is associative from the right or from the left.

PROOF. We consider two possible cases.

1. $[m, v_1, v_2 v_3] = 0$. We shall show that in this case $[m, A, A] = (0)$. By (12) and by Corollary 1 to Lemma 4 we have $[m, v_i, v_j v_k] = 0$. In view of relation (6), it now suffices for us to establish the equalities

$$[m, v_i, (v_1 v_2) v_3] = 0, \tag{16}$$

$$[m, v_i v_j, v_i v_k] = 0, \tag{17}$$

$$[m, v_i v_j, (v_1 v_2) v_3] = 0. \tag{18}$$

By (13) and Corollary 1 to Lemma 4 we have

$$[m, v_i, (v_1 v_2) v_3] = \pm [m, v_i, (v_j v_k) v_i] = 0,$$

i.e., (16) is proved. Furthermore, by linearized relation (10) we have

$$\begin{aligned} [m, v_i v_j, v_i v_k] &= -[m, v_k, v_i(v_i v_j)] = -[m, v_k, v_i^2 v_j] \\ &= -t(v_i)[m, v_k, v_i v_j] + n(v_i)[m, v_k, v_j] = 0, \end{aligned}$$

i.e., (17) is also proved. Finally, applying (17), (13), and Corollary 1 to Lemma 4 we obtain

$$[m, v_i v_j, (v_1 v_2) v_3] = \pm [m, v_i v_j, (v_i v_j) v_k] = 0,$$

which proves (18). Thus, in case 1 we have $[m, A, A] = (0)$, whence the assertion of the lemma follows as in the proof of Lemma 5.

2. $[m, v_1, v_2 v_3] = m_1 \neq 0$. We shall show that in this case $[m_1, A, A]' = (0)$, where by $[,,]'$ is denoted that one of the trilinear compositions $(,,)$, $\{,,\}$ which is different from $[,,]$. By (12) and linearized (7) and (8), we have

$$\begin{aligned} [m_1, v_i, v_j]' &= [[m, v_1, v_2 v_3], v_i, v_j]' \\ &= \pm [[m, v_i, v_j v_k], v_i, v_j]' \\ &= \mp [[m, v_i, v_j], v_i, v_j v_k]' = 0. \end{aligned}$$

Furthermore, in view of (7) and (8) we obtain

$$[m_1, v_1, v_2 v_3]' = [[m, v_1, v_2 v_3], v_1, v_2 v_3]' = 0.$$

By the already considered case 1 we now have $[m_1, A, A]' = (0)$, whence the assertion of the lemma also follows in case 2.

This proves the lemma.

Now can be proved

THEOREM 1. Every irreducible alternative module over a composition algebra is either associative from the right or associative from the left.

PROOF. From relation (6) and the fact that the module M is unital, it follows easily that in the cases $A = F$ and $A = \boldsymbol{K}(\mu)$ the module is an associative right module. In view of Lemma 5, it remains for us to consider the case when $A = \boldsymbol{C}(\mu, \beta, \gamma)$ is a Cayley–Dickson algebra. We select elements v_1, v_2, v_3 in A for which the set $\{1, v_1, v_2, v_3, v_1v_2, v_1v_3, v_2v_3, (v_1v_2)v_3\}$ forms a basis for A over F. By Lemma 6 it suffices to prove there is a nonzero element $m \in M$ such that $[m, v_i, v_j] = 0$ for any $i, j \in \{1, 2, 3\}$. Let $m \in M$, $m \neq 0$. If $(m, v_i, v_j) = 0$ for all i, j, then there is nothing to prove. It is therefore possible to assume, for example, that $m_1 = (m, v_1, v_2) \neq 0$. We note that by (7)

$$\{m_1, v_1, v_2\} = \{(m, v_1, v_2), v_1, v_2\} = 0. \tag{19}$$

If $\{m_1, v_1, v_3\} = \{m_1, v_2, v_3\} = 0$, then everything is proved. Therefore we can assume, for example, that $m_2 = \{m_1, v_1, v_3\} \neq 0$. By (8) and its linearization and in view of (19), we have

$$(m_2, v_1, v_3) = (\{m_1, v_1, v_3\}, v_1, v_3) = 0, \tag{20}$$

$$(m_2, v_1, v_2) = (\{m_1, v_1, v_3\}, v_1, v_2) = -(\{m_1, v_1, v_2\}, v_1, v_3) = 0. \tag{21}$$

If $(m_2, v_2, v_3) = 0$ also, then everything is proved. We assume that $m_3 = (m_2, v_2, v_3) \neq 0$. Applying relation (7) and its linearization, by (20) and (21) we obtain

$$\begin{aligned}
\{m_3, v_2, v_3\} &= \{(m_2, v_2, v_3), v_2, v_3\} = 0,\\
\{m_3, v_1, v_2\} &= \{(m_2, v_2, v_3), v_1, v_2\} = -\{(m_2, v_2, v_1), v_3, v_2\} = 0,\\
\{m_3, v_1, v_3\} &= \{(m_2, v_2, v_3), v_1, v_3\} = -\{(m_2, v_1, v_3), v_2, v_3\} = 0.
\end{aligned}$$

Thus in any case the module M contains a nonzero element m satisfying the condition of Lemma 6. This proves the theorem.

Exercises

1. Prove that any associative from the left, right module over an associative algebra A is an associative left A-module. (This, in some measure, explains our introduction of the term "associative from the left.")

2. Prove that the algebra of generalized quaternions $\boldsymbol{Q}(\mu, \beta)$ has exactly two nonisomorphic irreducible alternative right modules. Both of these modules are regular, and in addition one of them is associative from the right, while the other is associative from the left. *Hint.* For the construction of an associative from the left module over the algebra $\boldsymbol{Q}(\mu, \beta)$, apply the Cayley–Dickson process to the algebra $\boldsymbol{Q}(\mu, \beta)$ and consider the subspace $v\boldsymbol{Q}(\mu, \beta)$.

3. Let $\boldsymbol{C} = \boldsymbol{C}(\mu, \beta, \gamma)$ be a Cayley–Dickson algebra. We denote by Reg $\boldsymbol{C}$ and $\overline{\text{Reg}}\,\boldsymbol{C}$ the right $\boldsymbol{C}$-modules obtained by introducing on the vector space $M = \boldsymbol{C}$ the following respective actions of the algebra $\boldsymbol{C}$: if $m \in M, c \in \boldsymbol{C}$, then

$$m \cdot c = mc \text{ for the module Reg } \boldsymbol{C},$$
$$m \cdot c = m\bar{c} \text{ for the module } \overline{\text{Reg}}\,\boldsymbol{C},$$

where by ma is denoted the product of the elements m and a in the Cayley–Dickson algebra $\boldsymbol{C}$. Prove that Reg $\boldsymbol{C}$ and $\overline{\text{Reg}}\,\boldsymbol{C}$ are nonisomorphic irreducible right alternative right $\boldsymbol{C}$-modules (see Exercise 5 in Section 1). Also, Reg $\boldsymbol{C}$ is associative from the right, and $\overline{\text{Reg}}\,\boldsymbol{C}$ is associative from the left.

It is clear that the module Reg $\boldsymbol{C}$ is a regular alternative $\boldsymbol{C}$-module. Whether the module $\overline{\text{Reg}}\,\boldsymbol{C}$ is alternative is unknown. Using the description of alternative $\boldsymbol{C}$-bimodules, it is possible to prove that the module $\overline{\text{Reg}}\,\boldsymbol{C}$ is not a regular alternative $\boldsymbol{C}$-module.

4. Prove that any irreducible right alternative right $\boldsymbol{C}$-module is isomorphic either to Reg $\boldsymbol{C}$ or $\overline{\text{Reg}}\,\boldsymbol{C}$.

3. IRREDUCIBLE ALTERNATIVE MODULES

We now turn to the consideration of arbitrary irreducible alternative modules. Everywhere below A is an alternative algebra, $\mathcal{J}(A)$ is its Zhevlakov radical, ρ is an alternative right representation of the algebra A, and M is the right A-module associated with the representation ρ.

LEMMA 7. If the representation ρ is irreducible, then the quotient algebra $A/K_\rho(A)$ is prime.

PROOF. First let B, C be ideals of the algebra A such that $BC + CB \subseteq K_\rho(A)$ and $B \nsubseteq K_\rho(A)$. We consider the set $M \cdot B = \{\sum_i \alpha_i m_i \cdot b_i \,|\, \alpha_i \in \Phi, m_i \in M, b_i \in B\}$. For any $a \in A$ we have $(m_i \cdot b_i) \cdot a = -(m_i \cdot a) \cdot b_i + m_i \cdot (a \circ b_i) \in M \cdot B$, whence it follows that $M \cdot B$ is a submodule of the A-module M. Since $B \nsubseteq K_\rho(A)$, $M \cdot B \neq (0)$. Hence in view of the irreducibility of M it follows that $M \cdot B = M$. We now suppose that $C \nsubseteq K_\rho(A)$. Then there can be found elements $b \in B$, $c \in C$ such that $(M \cdot b) \cdot c \neq (0)$. We note that $(m \cdot b) \cdot c = -(m \cdot c) \cdot b + m \cdot (c \circ b) = -(m \cdot c) \cdot b$ for any $m \in M$, i.e., $(M \cdot b) \cdot c = (M \cdot c) \cdot b$. Furthermore, by the linearized R-identity (5), for any $a \in A$ and $m \in M$ we have

$$[(m \cdot b) \cdot c] \cdot a = -[(m \cdot a) \cdot c] \cdot b + m \cdot [(bc)a + (ac)b] = -[(m \cdot a) \cdot c] \cdot b,$$

whence it follows that $(M \cdot b) \cdot c = (M \cdot c) \cdot b$ is an A-submodule of the module M. In view of the irreducibility of M, we have $M = (M \cdot b) \cdot c$.

However, by the Moufang identity

$$M = (M \cdot b) \cdot c = \{[(M \cdot b) \cdot c] \cdot b\} \cdot c = [M \cdot (bcb)] \cdot c = (0),$$

which contradicts the irreducibility of M. This means if $BC + CB \subseteq K_\rho(A)$, then either $B \subseteq K_\rho(A)$ or $C \subseteq K_\rho(A)$.

Now let $BC \subseteq K_\rho(A)$. Then $(B \cap C)^2 \subseteq K_\rho(A)$, and by what has already been proved $B \cap C \subseteq K_\rho(A)$. But then also $CB \subseteq B \cap C \subseteq K_\rho(A)$. Thus $BC + CB \subseteq K_\rho(A)$, and consequently either $B \subseteq K_\rho(A)$ or $C \subseteq K_\rho(A)$.

This proves the lemma.

LEMMA 8. If the algebra A has an almost faithful irreducible module M, then either A is an associative prime algebra, or A is a Cayley–Dickson ring.

PROOF. By the previous lemma the algebra A is prime. If $3A \neq (0)$, then in view of the corollary to Theorem 9.9 everything is clear. Now let $3A = (0)$. By Theorem 9.11 it suffices for us to prove that $\mathscr{L}(A) = (0)$. Without loss of generality we can assume that $3\Phi = (0)$, whence $2 \cdot 2 = 1$ in Φ, i.e., there is a $\frac{1}{2}$ in the ring Φ. Consequently, we can consider the algebra $A^{(+)}$ with multiplication $a \odot b = \frac{1}{2}(ab + ba)$. We also consider the mapping $\mathscr{R}: A \to \mathscr{R}_{\mathrm{Alt}}(A)$ of the algebra A into the universal algebra $\mathscr{R}_{\mathrm{Alt}}(A)$. By R-identity (4), for any $a, b \in A$ we have $\mathscr{R}_a \odot \mathscr{R}_b = \frac{1}{2}\mathscr{R}_a \circ \mathscr{R}_b = \frac{1}{2}\mathscr{R}_{a \circ b} = \mathscr{R}_{a \odot b}$, whence it follows that the mapping $\mathscr{R}$ is an algebra homomorphism $\mathscr{R}: A^{(+)} \to (\mathscr{R}_{\mathrm{Alt}}(A))^{(+)}$. In addition, the image of the algebra $A^{(+)}$ will be the set $\mathscr{R}_A = \{\mathscr{R}_a \mid a \in A\}$. In particular, the set $\mathscr{R}_A$ is a subalgebra of the special Jordan algebra $(\mathscr{R}_{\mathrm{Alt}}(A))^{(+)}$, i.e., $\mathscr{R}_A$ is a special Jordan algebra. It is also clear that $\mathscr{R}_{\mathrm{Alt}}(A)$ will be an associative enveloping algebra for the Jordan algebra $\mathscr{R}_A$.

We now note that the locally nilpotent radical $\mathscr{L}(A)$ of the algebra A is a locally nilpotent ideal of the algebra $A^{(+)}$. This means its image $\mathscr{R}_{\mathscr{L}(A)}$ under the epimorphism $\mathscr{R}: A^{(+)} \to \mathscr{R}_A$ lies in the locally nilpotent radical $\mathscr{L}(\mathscr{R}_A)$ of the Jordan algebra $\mathscr{R}_A$. By Theorem 4.5 $\mathscr{L}(\mathscr{R}_A) = \mathscr{R}_A \cap \mathscr{L}(\mathscr{R}_{\mathrm{Alt}}(A))$, whence it follows that $\mathscr{R}_a \in \mathscr{L}(\mathscr{R}_{\mathrm{Alt}}(A))$ for any $a \in \mathscr{L}(A)$.

Furthermore, let ρ be an almost faithful irreducible representation of the algebra A corresponding to the A-module M. By Proposition 1 there exists a homomorphism φ of the associative algebra $\mathscr{R}_{\mathrm{Alt}}(A)$ into the algebra $\mathrm{End}_\Phi(M)$ such that $\rho = \mathscr{R} \circ \varphi$. In addition, M is an irreducible associative $\mathscr{R}_{\mathrm{Alt}}(A)$-module. It is well known that in this case $\mathscr{L}(\mathscr{R}_{\mathrm{Alt}}(A)) \subseteq \mathrm{Ker}\,\varphi$. In particular, $\mathscr{R}_a \in \mathrm{Ker}\,\varphi$ for any $a \in \mathscr{L}(A)$, whence $\mathscr{L}(A) \subseteq \mathrm{Ker}_\rho(A)$ and $\mathscr{L}(A) \subseteq K_\rho(A) = (0)$.

This proves the lemma.

LEMMA 9. Let A be an associative–commutative algebra. Then any irreducible A-module M is a right associative A-module.

PROOF. For any $m \in M$, a, b, $c \in A$, by the linearized R-identities (4) and (5) we have

$$\begin{aligned}(m, a, b) \cdot c &= [(m \cdot a) \cdot b] \cdot c - [m \cdot (ab)] \cdot c \\ &= -[(m \cdot c) \cdot b] \cdot a + 2m \cdot (abc) + (m \cdot c) \cdot (ab) - 2m \cdot (abc) \\ &= -(m \cdot c, b, a) = (m \cdot c, a, b),\end{aligned}$$

whence it follows that the subset (M, a, b) is a submodule of the A-module M. If the module M is not associative, then $(M, a, b) \neq (0)$ for some a, $b \in A$, and $(M, a, b) = M$ in view of the irreducibility of M. We note that for any $m \in M$, a, $b \in A$, the equality $(m, a, b) = \{m, a, b\}$ is valid, and therefore by (7) or (8) we have

$$((m, a, b), a, b) = 0.$$

But then $M = (M, a, b) = ((M, a, b), a, b) = (0)$. This contradiction proves the lemma.

We denote by ZN_{Alt} the ideal $ZN(\mathrm{Alt}[X])$ of the free alternative algebra $\mathrm{Alt}[X]$ from a countable set of generators $X = \{x_i\}$, and for an arbitrary alternative algebra A we denote by $ZN_{\mathrm{Alt}}[A]$ the ideal of the algebra A consisting of elements of the form $f(a_1, \ldots, a_n)$, where $f(x_1, \ldots, x_n) \in ZN_{\mathrm{Alt}}$ and $a_1, \ldots, a_n \in A$. As is easy to see, $ZN_{\mathrm{Alt}}[A] = [N_{\mathrm{Alt}}[A], A]A^{\#}$.

LEMMA 10. If the module M is irreducible and is not associative from the right, then $ZN_{\mathrm{Alt}}[A] \subseteq K_\rho(A)$. In particular, in this case

$$[N_{\mathrm{Alt}}[A], A] \subseteq K_\rho(A).$$

PROOF. Let u be an arbitrary element from ZN_{Alt}, and x_i be generators not appearing in u. By Lemma 7.5 the identity

$$((x_1 u, x_2, x_3), x_4, x_5) = 0$$

is valid in the algebra $\mathrm{Alt}[X]$. Hence it follows that for any elements $a \in ZN_{\mathrm{Alt}}[A]$ and $b_1, \ldots, b_4 \in A$,

$$a^\rho(b_1^\rho b_2^\rho - (b_1 b_2)^\rho)(b_3^\rho b_4^\rho - (b_3 b_4)^\rho) = 0. \tag{22}$$

If $ZN_{\mathrm{Alt}}[A] \nsubseteq K_\rho(A)$, then there exists an element $a \in ZN_{\mathrm{Alt}}[A]$ such that $m \cdot a \neq 0$ for some $m \in M$. We take the element $m \cdot a$ as a generating element of the A-module M. Two cases are possible:

(a) $(m \cdot a, b, c) = 0$ for any b, $c \in A$;
(b) there exist b, $c \in A$ for which $m_1 = (m \cdot a, b, c) \neq 0$.

In the second case, choosing the element m_1 as a generating element of the module M, by (22) we obtain that in any case the module M is associative from the right. This contradiction proves that $ZN_{\text{Alt}}[A] \subseteq K_\rho(A)$, which proves the lemma.

LEMMA 11. Let the algebra A have an almost faithful irreducible module M which is not associative from the right. Then $Z = Z(A) \neq (0)$, Z does not contain zero divisors of the algebra A, and the ring of quotients $A_1 = (Z^*)^{-1}A$ is either an algebra of generalized quaternions or a Cayley–Dickson algebra over the field of quotients Z_1 of the ring Z.

PROOF. By Lemma 8 the algebra A is either an associative algebra or a Cayley–Dickson ring. Since the assertion of the lemma is valid for a Cayley–Dickson ring, it remains to consider the case when A is a prime associative algebra.

We note that in view of Lemma 9 the algebra A is noncommutative. In addition, it is easy to see that the algebra A is nondegenerate. Consequently by Theorem 9.6 $N_{\text{Alt}}[A] \neq (0)$. Since by Lemma 10 $[N_{\text{Alt}}[A], A] \subseteq K_\rho(A) = (0)$, we obtain $(0) \neq N_{\text{Alt}}[A] \subseteq Z(A)$, whence $Z + Z(A) \neq (0)$. Since the algebra A is prime, the center Z does not contain zero divisors of the algebra A, and we can form the ring of quotients $A_1 = (Z^*)^{-1}A$. By Proposition 9.2 A_1 is a prime associative algebra, and the algebra A is isomorphically imbedded in A_1. Furthermore, the condition $N_{\text{Alt}}[A] \subseteq Z(A)$ is equivalent to realization in the algebra A of all identities of the form $[n(x_1, \ldots, x_m), x_{m+1}] = 0$, where $n(x_1, \ldots, x_m)$ is an arbitrary element from N_{Alt}. It is easy to see that N_{Alt} is a homogeneous subalgebra of the free algebra $\text{Alt}[X]$, and therefore by Proposition 9.2 all of these identities are also valid in the algebra A_1. This means $N_{\text{Alt}}[A_1] \subseteq Z(A_1) = Z_1$. Now let I be an arbitrary nonzero ideal of the algebra A_1. By Corollary 1 to Theorem 9.4 I will also be a prime algebra. If the algebra I were commutative, then we would have $(0) \neq I = Z(I) = I \cap Z_1 \subseteq Z_1$. This is impossible since Z_1 is a field and $A_1 \neq Z_1$ in view of the noncommutativity of the subalgebra $A \subseteq A_1$. This means I is a noncommutative prime associative algebra. It is clear that the algebra I is nondegenerate, and therefore by Theorem 9.6 we obtain $(0) \neq N_{\text{Alt}}[I] \subseteq N_{\text{Alt}}[A] \subseteq Z_1$. Hence $I \cap Z_1 \neq (0)$ and $I = A_1$. Since the ideal I is arbitrary, by the same token it is proved that the algebra A_1 is simple. Since $N_{\text{Alt}}[A_1] \subseteq Z(A_1) = Z_1$, then by Theorem 7.5 the algebra A_1 is an algebra of generalized quaternions over the field Z_1.

This proves the lemma.

Now can be proved

THEOREM 2. Let A be an arbitrary alternative algebra. Then any irreducible alternative A-module is either associative from the right or associative from the left.

PROOF. Since any irreducible A-module M is an almost faithful irreducible $\bar{A}$-module, where $\bar{A} = A/K_\rho(A)$, and since M is associative from the right (left) as an A-module if and only if M is associative from the right (left) as an $\bar{A}$-module, it suffices for us to consider almost faithful irreducible A-modules. Let M be an almost faithful irreducible alternative A-module. It is possible to assume that M is not associative from the right, since otherwise there is nothing to prove. Then by Lemma 11 $Z = Z(A) \neq (0)$, and the algebra A is isomorphically imbedded in the ring of quotients $A_1 = (Z^*)^{-1}A$, which is either an algebra of generalized quaternions or a Cayley–Dickson algebra over the field of quotients Z_1 of the ring Z.

We consider the set $N_{\mathrm{Alt}}[A]$. For any element $f = f(x_1, \ldots, x_n) \in N_{\mathrm{Alt}}$, we have

$$f \otimes x_{n+1} - fx_{n+1},\ x_{n+1} \otimes f - x_{n+1}f \in I_{\mathrm{Alt}}.$$

Hence for any elements $n \in N_{\mathrm{Alt}}[A]$, $a \in A$,

$$n^\rho a^\rho - (na)^\rho = a^\rho n^\rho - (an)^\rho = 0.$$

Since $N_{\mathrm{Alt}}[A] \subseteq Z(A)$, we finally obtain

$$n^\rho a^\rho = (na)^\rho = (an)^\rho = a^\rho n^\rho.$$

Thus $(N_{\mathrm{Alt}}[A])^\rho \subseteq \xi(M)$, where

$$\xi(M) = \{\sigma \in \mathrm{End}_\Phi(M) \mid [\sigma, a^\rho] = 0 \text{ for all } a \in A\}.$$

The set $\xi(M)$ is called the *centralizer* of the A-module M. As is easy to see, in view of the irreducibility of the module M, the set $\xi(M)$ is a division ring (with respect to the operations of the algebra $\mathrm{End}_\Phi(M)$). By Corollary 2 to Theorem 7.4, for any elements $a, b \in A$ we have $([a,b]^4)^\rho \in \xi(M)$ and $(n_2(a,b))^\rho \in \xi(M)$, where $n_2(x, y) = [x, y]^2([x, y] \circ [x, yx])$.

We now show that there is an ideal I contained in the center Z of the algebra A such that $I^\rho \subseteq \xi(M)$. First let the characteristic of the division ring $\xi(M)$ not equal 2. We select a pair of elements a, b in the algebra A for which $[a,b]^4 \neq 0$ (see the proof of Lemma 10.16). For any $z \in Z$ we have $2[a,b]^4 z = n_2(a + z, b) - n_2(a, b)$. Hence $([a,b]^4 z)^\rho \in \xi(M)$, and as the ideal I can be taken the ideal generated by the element $[a,b]^4$. If the characteristic of the division ring $\xi(M)$ equals two, then we select elements a, b in A for which $n_2(a,b) \neq 0$. Such elements always exist, since as is easy to see the composition algebra A_1 does not satisfy the identity $n_2(x, y) = 0$. We have $n_2(a + az, b) = n_2(a, b) + 5zn_2(a, b) + 10z^2 n_2(a, b) + 10z^3 n_2(a, b) + 5z^4 n_2(a, b) + z^5 n_2(a, b)$ for any $z \in Z$. Since $z^4 n_2(a, b) = n_2(a, zb)$ and $z^5 n_2(a, b) = n_2(za, b)$, then by the assumption on characteristic it follows from this that $[zn_2(a, b)]^\rho \in \xi(M)$. Hence in the given case it is possible to take the ideal generated by the element $n_2(a, b)$ as the desired ideal I.

We note that the restriction of the mapping ρ to the set I is a homomorphism of the algebra I into $\xi(M)$. Furthermore, as is easy to see, for any $i \in I \cap \mathrm{Ker}_\rho(A)$ we have $iA \subseteq K_\rho(A) = (0)$. Hence $i = 0$ since the algebra A is prime. This means $I \cap \mathrm{Ker}_\rho(A) = (0)$, and we can identify I with its image I^ρ in $\xi(M)$. Since $\xi(M)$ is a division ring, the field of quotients I_1 of the ring I is also contained in $\xi(M)$. As is easy to see, however, $I_1 = Z_1$. Consequently $Z_1 \subseteq \xi(M)$, and M can be considered as a vector space over the field Z_1. If we now set $(z^{-1}a)^\rho = z^{-1}z^\rho$, it is not difficult to verify that we obtain an alternative representation of the algebra A_1 in the same algebra $\mathrm{End}_\Phi(M)$. It is clear that M is also irreducible as an A_1-module. By Theorem 1 the module M is associative from the left as an A_1-module, and consequently also as an A-module.

This proves the theorem.

We shall now prove that the primitive alternative algebras are precisely those algebras having faithful irreducible alternative modules.

LEMMA 12. Let the alternative algebra A have an almost faithful irreducible alternative module M. Then A is primitive.

PROOF. By Theorem 2 the module M is either associative from the right or associative from the left. We suppose the first of these. Let m be a generating element of the module M such that $(m, x, y) = 0$ for any $x, y \in A$. We consider the set $(0:m) = \{x \in A \mid m \cdot x = 0\}$. As is easy to see, $(0:m)$ is a right ideal of the algebra A. Also $m \cdot A = M$, and therefore there exists an element e in A for which $m \cdot e = m$. Let a be an arbitrary element of the algebra A. Then $m \cdot (a - ea) = m \cdot a - (m \cdot e) \cdot a = m \cdot a - m \cdot a = 0$, i.e., $a - ea \in (0:m)$. By the same token this proves the right ideal $(0:m)$ is modular. Now let I be a right ideal of the algebra A which strictly contains $(0:m)$. Then $m \cdot I$ is a nonzero submodule of the module M, and consequently $M = m \cdot I$. From this there follows the existence in I of an element i such that $m \cdot i = m$. We have $m \cdot (e - i) = 0$, i.e., $e - i \in (0:m) \subseteq I$, whence $e \in I$ and $I = A$. Consequently the right ideal $(0:m)$ is maximal. Finally, we note that any two-sided ideal contained in $(0:m)$ annihilates the module M. Since the module M is almost faithful, such nonzero ideals do not exist, and the algebra A is primitive.

Now let the module M not be associative from the right. Then it is associative from the left, and by Lemma 11 the algebra A is isomorphically imbedded either in an algebra of generalized quaternions or in a Cayley–Dickson algebra. There exists a generating element m in the module M such that $\{m, x, y\} = 0$ for any $x, y \in A$. Arguing analogous to the previous case, it is easy to verify that in this case the set $(0:m)$ is a maximal modular left ideal which does not contain nonzero two-sided ideals of the algebra A, i.e., in the given case the algebra A is left primitive. Then by Theorem 10.1 A is either a Cayley–Dickson algebra or a left primitive associative algebra. In the first case, by the corollary to Lemma 10.3 the algebra A is primitive. In the second case, A is a subalgebra of an algebra of generalized quaternions, and by the corollary to Lemma 2.8 satisfies the essential polynomial identity $[[x, y]^2, z] = 0$. It is well known (e.g., see Herstein [252, p. 151]) that A is then a finite-dimensional simple algebra over its center. In particular, A is right primitive.

This proves the lemma.

COROLLARY. Any almost faithful irreducible alternative module is faithful.

PROOF. Let M be an almost faithful irreducible alternative module for an alternative algebra A. Then M is associative from the right or left and has the form $M = m \cdot A$ for some element $m \in M$. Also, the algebra A is either a Cayley–Dickson algebra $\boldsymbol{C}$ or a primitive associative algebra. Since the algebra $\boldsymbol{C}$ does not contain proper one-sided ideals, in the first case we have $(0:m) = (0)$. It follows from this that $a \in \operatorname{Ker}_\rho(\boldsymbol{C})$ if and only if $\boldsymbol{C}a = (0)$ or $a\boldsymbol{C} = (0)$, depending on whether M is associative from the right or from the left. Since there are no such elements in the algebra $\boldsymbol{C}$, in this case we obtain that $\operatorname{Ker}_\rho(\boldsymbol{C}) = (0)$, and the module M is faithful. If the algebra A is associative, then M is the usual associative right or left A-module. In each of these cases the kernel $\operatorname{Ker}_\rho(A)$ is an ideal of the algebra A, so that $\operatorname{Ker}_\rho(A) = K_\rho(A) = (0)$, and the module M is faithful. This proves the corollary.

THEOREM 3. An alternative algebra A is primitive if and only if A has a faithful irreducible regular alternative module.

PROOF. In view of Lemma 12 it suffices for us to show that any primitive alternative algebra A has a faithful irreducible regular alternative module. Let I be a maximal modular right ideal of the algebra A such that $\check{I} = (0)$. Then A and I are regular alternative right A-modules (with respect to the usual multiplication on the right by elements from A), and the quotient module $M = A/I$ is likewise. Since the right ideal I is maximal, the module M is irreducible. If $a \in A$, then $M \cdot a = (0)$ if and only if $Aa \subseteq I$. Hence by Proposition 10.2 $\operatorname{Ker}_\rho(A) = (I:A) = \check{I} = (0)$, and the module M is faithful. This proves the theorem.

We can now prove the basic result of this chapter.

THEOREM 4 (*Zhevlakov, Slin'ko, Shestakov*). Let A be an alternative algebra and $\mathcal{J}(A)$ be its Zhevlakov radical. In addition, let P_1 be the set of all irreducible alternative right representations of the algebra A, and P_2 be the set of all regular representations from P_1. Then

$$\mathcal{J}(A) = \bigcap_{\rho \in P_1} \operatorname{Ker}_\rho(A) = \bigcap_{\rho \in P_2} \operatorname{Ker}_\rho(A).$$

PROOF. Let $\rho \in P_1$. Then by Lemma 12 and its corollary $\operatorname{Ker}_\rho(A) = K_\rho(A)$, and the quotient algebra $A/\operatorname{Ker}_\rho(A)$ is primitive. This means $\operatorname{Ker}_\rho(A)$ is a primitive ideal of the algebra A. By Theorems 10.5 and 10.2 the radical $\mathcal{J}(A)$ equals the intersection of all primitive ideals of the algebra A. Thus.

$$\mathcal{J}(A) \subseteq \bigcap_{\rho \in P_1} \operatorname{Ker}_\rho(A).$$

Now let I be a primitive ideal of the algebra A. The algebra $\bar{A} = A/I$ is primitive, and by Theorem 3 has a faithful irreducible regular module M. Let $\varphi: A \to \bar{A}$ be the canonical homomorphism and $\rho: \bar{A} \to \operatorname{End}_\Phi(M)$ be the representation of the algebra $\bar{A}$ corresponding to the module M. Then it is easy to see (see Exercise 3 in Section 1) that the mapping $\phi \circ \rho: A \to \operatorname{End}_\Phi(M)$ is an irreducible regular alternative representation of the algebra A. In addition, $\operatorname{Ker}_{\varphi \circ \rho}(A) = I$. We have proved that any primitive ideal of the algebra A is the kernel of some irreducible regular representation. Consequently

$$\bigcap_{\rho \in P_2} \operatorname{Ker}_\rho(A) \subseteq \mathcal{J}(A).$$

Finally, we obtain

$$\mathcal{J}(A) = \bigcap_{\rho \in P_1} \operatorname{Ker}_\rho(A) = \bigcap_{\rho \in P_2} \operatorname{Ker}_\rho(A) \subseteq \mathcal{J}(A),$$

from which follows the assertion of the theorem.

This proves the theorem.

COROLLARY. For an alternative algebra A the following inclusions are equivalent:

(1) $a \in \mathcal{J}(A)$;
(2) $R_a^* \in \mathcal{J}(R_{\mathrm{Alt}}^*(A))$;
(3) $\mathcal{R}_a \in \mathcal{J}(\mathcal{R}_{\mathrm{Alt}}(A))$.

For the proof it suffices to recall that an alternative A-module M (regular A-module M) is irreducible if and only if M is an associative irreducible $\mathcal{R}_{\mathrm{Alt}}(A)$-module ($R_{\mathrm{Alt}}^*(A)$-module, respectively).

Exercises

We call a regular representation ρ of an algebra A of the form

$$\rho : A \xrightarrow{R} R_1^A(A) \xrightarrow{\phi} \mathrm{End}_\Phi(M)$$

completely regular.

1. Prove that for any alternative algebra A

$$\mathscr{J}(A) = \bigcap_{\rho \in P_3} \mathrm{Ker}_\rho(A),$$

where P_3 is the set of all completely regular representations of the algebra A.

2. Prove that the inclusions $a \in \mathscr{J}(A)$ and $R_a \in \mathscr{J}(R(A))$ are equivalent.

LITERATURE

Jacobson [44, 46], Zhevlakov [76], McCrimmon [118], Slin'ko and Shestakov [214], Shestakov [273], Eilenberg [282].

Results on representations of Jordan algebras: Jacobson [44, 47], Osborn [165].

CHAPTER 12

Alternative Algebras with Finiteness Conditions

One of the classical divisions of the theory of algebras is the structure theory of algebras with various finiteness conditions. The structure theorems of Wedderburn describing the structure of finite-dimensional associative algebras over an arbitrary field were the first major success of this theory and the starting point for all subsequent investigations. Later there were observed two tendencies. One of these rejected associativity and led to the creation by Zorn and Schafer of a structure theory for finite-dimensional alternative algebras. In the confines of the other there was a transfer to infinite-dimensional associative algebras and rings with various conditions on the structure of ideals. Thus Artin extended the theorems of Wedderburn to rings with maximal and minimal conditions for right (left) ideals. Next Noether and Hopkins showed that the minimal condition alone was sufficient for the validity of the theorems proved by Artin. These results have long since become classical and are recorded in a series of monographs.

Both theories were subsequently combined by Zhevlakov into one single theory—the theory of alternative algebras with minimal condition, which includes as special cases the theories of Zorn–Schafer and Artin–Noether–

Hopkins. In this chapter we obtain the results of Zhevlakov by applying the general theory developed in previous chapters.

1. ALTERNATIVE ALGEBRAS WITH MINIMAL CONDITION ON TWO-SIDED IDEALS

Let A be an alternative algebra, and X, Y be subalgebras of A. We introduce the notation $(X, Y)_1 = X \cdot Y$ and $(X, Y)_{k+1} = (X, Y)_k \cdot Y$.

LEMMA 1. Let B and I be ideals of an alternative algebra A which satisfies the minimal condition for ideals contained in B. Then there exists a number k such that

$$(B, I)_k \subseteq BI^2.$$

PROOF. We consider the chain of ideals

$$B \supseteq (B, I)_1 \supseteq \cdots \supseteq (B, I)_m \supseteq \cdots.$$

Beginning with some term, which we denote by B', all of the terms of this chain are equal to each other, that is,

$$(B, I)_n = (B, I)_{n+1} = \cdots = B'.$$

Let $B' \not\subseteq BI^2$. Then the set of ideals $\{J\}$ satisfying the three conditions

$$\text{(a)} \quad J \subseteq B, \qquad \text{(b)} \quad JI = J, \qquad \text{(c)} \quad J \not\subseteq BI^2$$

is nonempty since $B' \in \{J\}$. By the minimal condition there exists in this set a minimal ideal C. Since $CI = C$, there can be found an element $r \in I$ such that $Cr \not\subseteq BI^2$. We pick such an element r and consider the set $D = Cr + BI^2$. We shall prove that D is an ideal in A.

We shall write $x \equiv y$ if $x - y \in BI^2$. Let $c \in C$, $x \in I$, $a \in A$. Then

$$\begin{aligned}[(cx)r]a &= -[(ca)r]x + c[(xr)a + (ar)x] \equiv -[(ca)r]x \\ &= [(ca)x]r - (ca)(xr + rx) \equiv [(ca)x]r \in Cr,\end{aligned}$$

that is, $[(CI)r]a \subseteq Cr + BI^2$. But $CI = C$, and therefore D is a right ideal in A.

Furthermore, for any $c \in C$, x, y, $\in I$,

$$\begin{aligned} y(cx)y &\equiv -y[(cy)x] = -(ycy)x + (y, cy, x) \\ &= -y[c(yx)] + (y, c, xy) \equiv 0,\end{aligned}$$

and therefore

$$\begin{aligned} xcx &\equiv 0, \\ (xc)y + (yc)x = x(cy) + y(cx) &\equiv 0.\end{aligned}$$

Hence for all $c \in C$, $x, y \in I$, $a \in A$, we have

$$\begin{aligned} a\{[(cy)x]r\} &\equiv -a\{[(cy)r]x\} \\ &= a[(xr)(cy)] - \{[a(cy)]r\}x - [(ax)r](cy) \\ &\equiv -a\{y[c(xr)]\} + \{[a(cy)]x\}r + y\{c[(ax)r]\} \\ &\equiv \{[a(cy)]x\}r \in Cr. \end{aligned}$$

Since $(CI)I = I$, then $a(Cr) \subseteq Cr + BI^2$, and D is a left ideal in A.

Thus D is an ideal in A. At the same time $Dr \subseteq BI^2$. Consequently $D' = D \cap C \neq C$, and therefore the ideal D' is not contained in the set $\{J\}$. Of the three conditions characterizing the set $\{J\}$, conditions (a) and (c) are satisfied for D'. Consequently D' does not satisfy condition (b). Thus we have a chain of ideals

$$D' \supset (D', I)_1 \supseteq \cdots \supseteq (D', I)_m \supseteq \cdots,$$

which is strictly decreasing at the first step.

Beginning at someplace in this chain there occur only equalities

$$(D', I)_k = (D', I)_{k+1} = \cdots.$$

Conditions (a) and (b) are satisfied for the ideal $(D', I)_k$, but $(D', I)_k$ is strictly contained in C, and therefore condition (c) does not hold for this ideal, that is, $(D', I)_k \subseteq BI^2$. In particular, this means that for any $x_1, x_2, \ldots, x_k \in I$ there is the containment

$$CR_r R_{x_1} \cdots R_{x_k} \subseteq BI^2.$$

But it follows from this that

$$CR_{x_1} \cdots R_{x_k} R_r \subseteq BI^2,$$

or equivalently

$$Cr = (C, I)_k r \subseteq BI^2.$$

The contradiction obtained proves the validity of the containment

$$(B, I)_n \subseteq BI^2,$$

which proves the lemma.

COROLLARY 1. Under the conditions of the lemma, for any natural number k there exists a number $f(k)$ such that

$$(B, I)_{f(k)} \subseteq BI^{(k)}.$$

PROOF. The assertion of Lemma 1 gives a basis for induction. We suppose that for any pair of ideals $\tilde{B}$ and $\tilde{I}$ of the algebra A, which satisfy the conditions of the lemma, there exists a number $\tilde{f}(k-1)$ depending on this pair such that

$$(\tilde{B}, \tilde{I})_{\tilde{f}(k-1)} \subseteq \tilde{B}\tilde{I}^{(k-1)}.$$

In particular, for some number $f(k-1)$

$$(B, I)_{f(k-1)} \subseteq BI^{(k-1)}.$$

Applying the induction assumption p times, we obtain that there exists a number $g(p)$ such that

$$(B, I)_{g(p)} \subseteq (B, I^{(k-1)})_p.$$

If we now apply Lemma 1 to the pair of ideals B and $I^{(k-1)}$, we obtain that for some number t the containments

$$(B, I)_{g(t)} \subseteq (B, I^{(k-1)})_t \subseteq BI^{(k)}$$

are valid, and this means that the number $g(t)$ can be taken as $f(k)$. This proves the corollary.

COROLLARY 2 (*Slater*). If an alternative algebra A satisfies the minimal condition for ideals contained in some ideal B of A, then for any number k there exists a number $h(k)$ such that

$$B^{h(k)} \subseteq B^{(k)}.$$

In particular, any solvable ideal B of an alternative algebra A with minimal condition for ideals contained in B is nilpotent.

PROOF. By the previous corollary there exists a number $f(k)$ such that

$$B^{\langle f(k)+1\rangle} \subseteq (B, B)_{f(k)} \subseteq BB^{(k)} \subseteq B^{(k)},$$

and since by the corollary to Theorem 5.5

$$B^{n^2} \subseteq B^{\langle n\rangle}$$

for any number n, then the number $(f(k)+1)^2$ can be taken as $h(k)$. This proves the corollary.

LEMMA 2. Let B be an Andrunakievich-radical ideal of an alternative algebra A which satisfies the minimal condition for ideals contained in B, and let I be a minimal ideal of the algebra A. Then $BI = IB = (0)$.

PROOF. It suffices to prove the lemma for the case when $I \subseteq B$, since otherwise the assertion of the lemma is trivial.

By the heredity of the Andrunakievich radical (corollary to Lemma 8.13), the ideal I cannot be a simple algebra, and therefore by Zhevlakov's theorem

(8.10) $I^2 = (0)$. In this case $C = \mathrm{Ann}_r(I) \neq (0)$. Let $B \not\subseteq C$, or equivalently $B_1 = C \cap B \neq B$.

We consider the quotient algebra $\bar{A} = A/B_1$. In the algebra $\bar{A}$ the image $\bar{B}$ of the ideal B is different from zero. Therefore there is contained in $\bar{B}$ a minimal ideal $\bar{D}$ of the algebra $\bar{A}$. The same argument as above shows that it is trivial, $\bar{D}^2 = (\bar{0})$. Consequently its complete inverse image D satisfies the relations

$$B \supseteq D \supset B_1, \qquad D^2 \subseteq B_1.$$

The ideal ID is contained in I and different from zero. Therefore $ID = I$. However by Lemma 1 there exists a number k such that

$$I = (I, D)_k \subseteq ID^2 \subseteq IB_1 \subseteq IC = (0).$$

This contradiction proves that $IB = (0)$. Analogously $BI = (0)$, which proves the lemma.

We recall that for an ideal I of an alternative algebra A by $P(I)$ is denoted the submodule generated by the collection of elements of the form iai, where $i \in I$, $a \in A^{\#}$. According to the corollary to Proposition 5.2, $P(I)$ is an ideal of the algebra A.

LEMMA 3. Let B be an ideal of an alternative algebra A which satisfies the minimal condition for ideals contained in B. Then the containment

$$B^m \subseteq P(B)$$

holds for some number m.

PROOF. We consider the chain of ideals

$$B \supseteq B^{\langle 2\rangle} \supseteq \cdots \supseteq B^{\langle n\rangle} \supseteq \cdots.$$

Beginning with some term of this chain there occur only equalities, $B^{\langle n\rangle} = B^{\langle n+1\rangle} = \cdots$. We assume that $B^{\langle n\rangle} \not\subseteq P(B)$. Then the set $\{J\}$ of ideals J of the algebra A having the properties

$$\text{(a)}\quad J \subseteq B, \qquad \text{(b)}\quad JB = J, \qquad \text{(c)}\quad J \not\subseteq P(B),$$

is nonempty. Let the ideal C be a minimal element in this set. By (b) and (c) for some $a \in B$ we have $Ca \not\subseteq P(B)$. We consider the set $D = Ca + P(B)$ and prove it is an ideal in A. We shall write $x \equiv y$ if $x - y \in P(B)$. It is easy to see that for any $x, y \in B, c \in C, r \in A$, we have the congruences

$$(cx)y \equiv -(cy)x, \qquad [(cx)r]y \equiv -[(cy)r]x.$$

Therefore

$$\begin{aligned}[(cx)a]r &\equiv -[(ca)x]r = [(ca)r]x - (ca)(x \circ r)\\ &\equiv [(cx)r]a + [c(x \circ r)]a \in Ca.\end{aligned}$$

By (b) it follows from this that D is a right ideal of the algebra A. Furthermore, using the same notation,

$$\begin{aligned}r[(cx)a] &\equiv -r[a(cx)] \equiv r[x(ca)] = -x[r(ca)] + (x \circ r)(ca)\\ &\equiv (ca)(rx) - a[c(x \circ r)] \equiv -[c(rx)]a + [c(x \circ r)]a\\ &= [c(xr)]a \in Ca.\end{aligned}$$

Since $CB = C$, it follows from this that D is a left ideal. Thus D is an ideal in A. For this ideal D the relation $Da \subseteq P(B)$ is valid. Therefore $D' = D \cap C \neq C$. Consequently there exists a number k such that

$$(D', B)_k \subseteq P(B).$$

This means, however, that for any $b_1, \ldots, b_k \in B$

$$CR_aR_{b_1} \cdots R_{b_k} \subseteq P(B).$$

But it follows from this that

$$CR_{b_1} \cdots R_{b_k}R_a \subseteq P(B),$$

that is, $(C, B)_k a \subseteq P(B)$. In view of condition (b), this means that $Ca \subseteq P(B)$. The contradiction obtained proves that $B^{\langle n \rangle} \subseteq P(B)$. By the corollary to Theorem 5.5 we have $B^{n^2} \subseteq P(B)$, which is what was to be proved.

REMARK. In the case of algebras over a ring Φ containing an element $\frac{1}{2}$ Lemma 3 is unnecessary, since by Lemma 6.8 $B^4 \subseteq P(B)$ even in the absence of minimal condition. However, in the general case one cannot get along without it.

THEOREM 1 (*Zhevlakov*). Let B be an ideal of an alternative algebra A which satisfies the minimal condition for ideals contained in B, and $B \subseteq \mathscr{A}(A)$. Then B is a nilpotent algebra.

PROOF. We consider the descending chain

$$B \supseteq B^{(1)} \supseteq \cdots \supseteq B^{(n)} \supseteq \cdots.$$

Let k be a number such that

$$B^{(k)} = B^{(k+1)} = \cdots = B'.$$

If $B' = (0)$, then the ideal B is nilpotent by Corollary 2 to Lemma 1. Suppose that $B' \neq (0)$. We consider the annihilator $\mathrm{Ann}_l(B')$ and the quotient algebra modulo it, $\bar{A} = A/\mathrm{Ann}_l(B')$. Since $B' = B'^2$, $B' \nsubseteq \mathrm{Ann}_l(B')$ and B' has a nonzero image $\bar{B}'$ in $\bar{A}$. Let $\bar{C}$ be a minimal ideal of the algebra $\bar{A}$ contained in $\bar{B}'$, and C be its complete inverse image in the algebra A. Then $CB' \neq (0)$ since $C \supset \mathrm{Ann}_l(B')$, but by Lemma 2 $CB' \subseteq \mathrm{Ann}_l(B')$, and therefore $(CB')B' = (0)$. However, by Lemma 3 there exists a number t such that

$$CB'^t \subseteq C \cdot P(B') \subseteq (CB')B' = (0).$$

Since $CB' \neq (0)$ and $B' = B'^2 = \cdots = B'^t$, we have obtained a contradiction.
This proves the theorem.

COROLLARY 1 (*Slater*). If B is a locally nilpotent ideal of an alternative algebra A that satisfies the minimal condition for ideals contained in B, then this ideal is nilpotent.

Proof. Since there do not exist simple locally nilpotent algebras (Proposition 7.1), the locally nilpotent radical is contained in the antisimple radical, which also proves the corollary.

COROLLARY 2. Let A be an alternative algebra with minimal condition for ideals. Then its Andrunakievich radical $\mathscr{A}(A)$ is nilpotent, and the quotient algebra $A/\mathscr{A}(A)$ is a subdirect sum of a finite number of algebras, each of which is either a Cayley–Dickson algebra over its center or an associative subdirectly irreducible algebra with simple heart.

Proof. The Andrunakievich radical $\mathscr{A}(A)$ is nilpotent by Theorem 1. By Theorem 8.14 it is the intersection of all $\mathfrak{B}$-ideals of the algebra A. Using the minimal condition, it is possible to choose a finite number of $\mathfrak{B}$-ideals whose intersection gives $\mathscr{A}(A)$. For the completion of the proof it now only remains to refer to Theorem 8.12, which describes the structure of subdirectly irreducible alternative algebras with simple heart.

2. ALTERNATIVE ARTINIAN ALGEBRAS

An algebra is called (*right*) *Artinian* if it satisfies the minimal condition for right ideals.

THEOREM 2 (*Zhevlakov*). Let B be a quasi-regular ideal of an alternative algebra A which satisfies the minimal condition for right ideals contained in B. Then B is nilpotent.

PROOF. If $B \subseteq \mathscr{A}(A)$, then B is nilpotent by Theorem 1. Let $B \not\subseteq \mathscr{A}(A)$. Then in the $\mathscr{A}$-semisimple quotient algebra $\bar{A} = A/\mathscr{A}(A)$, the image $\bar{B}$ of the ideal B does not equal zero. Let $\bar{C}$ be a minimal right ideal of the algebra $\bar{A}$ contained in $\bar{B}$. Since the algebra $\bar{A}$ is semiprime, we have $\overline{CA} \neq (\bar{0})$. Therefore $\bar{C}$ is an irreducible right alternative $\bar{A}$-module. The Zhevlakov radical $\mathscr{J}(\bar{A})$ lies in the kernel of each irreducible $\bar{A}$-module, and consequently $\bar{C} \cdot \mathscr{J}(\bar{A}) = (\bar{0})$. Since $C \subseteq \mathscr{J}(\bar{A})$, we have a contradiction to the semiprimeness of the algebra $\bar{A}$. This proves the theorem.

THEOREM 3 (*Zhevlakov*). Every semiprime alternative Artinian algebra A decomposes into a direct sum of a finite number of its minimal ideals $A = A_1 \oplus \cdots \oplus A_n$, each of which is cither a complete matrix algebra over some division ring, or a Cayley–Dickson algebra over its center. This decomposition is unique up to a permutation of the summands.

PROOF. By Theorem 2 the algebra A is $\mathscr{J}$-semisimple. Therefore the intersection of all primitive ideals in the algebra A equals zero. By the minimal condition, it is possible to find a finite set $B_1, \ldots, B_n$ of primitive ideals such that $\bigcap_{i=1}^n B_i = (0)$ and $C_j = \bigcap_{i \neq j} B_i \neq (0)$ for any $j = 1, 2, \ldots, n$. As is well known, a primitive associative Artinian algebra is simple. A Cayley–Dickson algebra is also simple, and therefore by Theorem 10.1 the quotient algebras A/B_i are simple, and the ideals B_i are maximal. Therefore $B_i + C_i = A$, whence, since $B_i \cap C_i = (0)$, we obtain $A = B_i \oplus C_i$. We set $D_r = B_1 \cap \cdots \cap B_r$. We shall show by induction that

$$A = C_1 \oplus \cdots \oplus C_k \oplus D_k.$$

As we just proved, this is true for $k = 1$. Let this be true for $k = m$, and consider the ideal D_m. By the second homomorphism theorem we have

$$D_m/D_m \cap B_{m+j} \cong D_m + B_{m+j}/B_{m+j} = A/B_{m+j},$$

where $j = 1, 2, \ldots, n - m$. Consequently, the ideals $D_m \cap B_{m+j}$ are maximal in D_m. In addition,

$$\bigcap_{j=1}^{n-m} (D_m \cap B_{m+j}) = (0),$$

and no proper subset has a zero intersection. Consequently, as earlier we obtain

$$D_m = D_{m+1} \oplus \bigcap_{j=2}^{n-m} (D_m \cap B_{m+j}) = D_{m+1} \oplus C_{m+1}.$$

Therefore $A = C_1 \oplus \cdots \oplus C_{m+1} \oplus D_{m+1}$, and the assertion is proved for $k = m + 1$. This means the required decomposition into a direct sum holds for all k. In particular, for $k = n$ we obtain

$$A = C_1 \oplus \cdots \oplus C_n.$$

Since $C_i \cong A/B_i$, C_i is a simple Artinian algebra, that is, either a complete matrix algebra over some division ring or a Cayley–Dickson algebra. It is easy to see that the ideals C_i are minimal, and that there are no other minimal ideals in the algebra A. From this follows the uniqueness of the decomposition.

This proves the theorem.

Theorems 2 and 3 together give an idea of the structure of an arbitrary alternative Artinian algebra.

COROLLARY. The radical $\mathcal{J}(A)$ of an alternative Artinian algebra A is the largest nilpotent ideal of A. The quotient algebra $A/\mathcal{J}(A)$ is isomorphic to a finite direct sum of complete matrix algebras over division rings and Cayley–Dickson algebras.

We now begin a study of nil-subrings of alternative Artinian algebras.

PROPOSITION 1. Let B be a nil-subring of a ring A of all linear transformations of a finite-dimensional vector space V over a division ring Δ. Then B is nilpotent, and there can be chosen a basis for V in which all transformations from B are strictly triangular.

PROOF. We consider an arbitrary element $b \in B$ and the chain of subspaces

$$V \supset Vb \supset \cdots \supset Vb^n = (0).$$

It is clear that $n \leq \dim_\Delta V$, that is, B is a nil-ring of bounded index. Consequently B is locally nilpotent.

Let U be a subspace in V with basis $\{e_1, \ldots, e_k\}$. We suppose that $UB = U$. Then there exist elements $b_{ij} \in B$ such that

$$e_i = \sum_{j=1}^{k} e_j b_{ji}, \qquad i = 1, \ldots, k.$$

Iteration of these relations leads to a contradiction to the nilpotency of the subring B_0 generated by the elements b_{ij}.

It is now obvious that the chain of subspaces

$$V \supset VB \supset VB^2 \supset \cdots$$

is strictly decreasing, and therefore $B^n = (0)$ for $n = \dim_\Delta V$. Choosing a basis for VB^{n-1}, extending it to a basis of VB^{n-2}, etc., we obtain a basis of the space V in which all transformations from B are strictly triangular. This proves the proposition.

PROPOSITION 2. Let B be a nil-subring of a Cayley–Dickson algebra $\mathbf{C}$ over a field F. Then the F-subalgebra D generated by the set B is nilpotent.

PROOF. Since $\mathbf{C}$ is quadratic over F, the index of nilpotency of each element $b \in B$ equals two, $b^2 = 0$. By Corollary 2 to Theorem 5.6 the ring B is locally nilpotent. Since any element from D is a finite F-linear combination of elements from B, the algebra D is also locally nilpotent. However it is finite-dimensional, and consequently nilpotent. This proves the proposition.

THEOREM 4 (*Zhevlakov*). Every nil-subring of an alternative Artinian algebra is nilpotent.

PROOF. Let A be an alternative Artinian algebra, R be its radical, and B be a nil-subring in A. We consider the quotient algebra $\bar{A} = A/R$. By Theorem 3 it is decomposable into a direct sum of its minimal ideals $\bar{A} = \bar{A}_1 \oplus \cdots \oplus \bar{A}_s$. Let $\bar{B}$ be the image of B in $\bar{A}$ and $\bar{B}_i$ be the projection of $\bar{B}$ on $\bar{A}_i$. If $\bar{A}_i$ is associative, then $\bar{A}_i$ is isomorphic to a matrix algebra $\Delta_{n_i}^{(i)}$ over some division ring $\Delta^{(i)}$. Also, by Proposition 1 $\bar{B}_i$ is contained in a subalgebra $\bar{C}_i$, which under some isomorphism of the algebra $\bar{A}_i$ onto $\Delta_{n_i}^{(i)}$ becomes the algebra of strictly triangular matrices. If $\bar{A}_i$ is a Cayley–Dickson algebra, then we denote by $\bar{C}_i$ the nilpotent subalgebra which the subring $\bar{B}_i$ generates in $\bar{A}_i$. Let

$$\bar{C} = \bar{C}_1 \oplus \cdots \oplus \bar{C}_s$$

and C be the complete inverse image of $\bar{C}$ in A. It is clear that C is a nil-subring and $C \supset B$, so that it suffices to prove C is nilpotent.

If $R = (0)$, then C is nilpotent by Propositions 1 and 2 and also Theorem 3. This gives us a basis for induction on the index of nilpotency of the radical. We suppose that the theorem is true when the index of nilpotency of the radical does not exceed $k - 1$. Let $R^k = (0)$ and $R^{k-1} \neq (0)$.

Under the canonical homomorphism of A onto $\tilde{A} = A/R^{k-1}$ the image $\tilde{R}$ of the radical R will obviously be the radical of the algebra $\tilde{A}$. Since $\tilde{R}^{k-1} = (0)$, the image $\tilde{C}$ in $\tilde{A}$ of the subring C is nilpotent. In other words, there exists a natural number n such that

$$C^n \subseteq R^{k-1} \cap C.$$

We denote $R^{k-1} \cap C$ by P. Since right nilpotency and nilpotency are equivalent, for the nilpotency of the subring C it suffices to show the existence of a natural number m such that

$$(P, C)_m = (0). \tag{1}$$

We denote by A_i the complete inverse image in A of the minimal ideal $\bar{A}_i$, and by C_i the complete inverse image of $\bar{C}_i$. We note that if $x \in A_i$, $y \in A_j$, and $i \neq j$, then $xy \in R$, and for any $p \in P$ we have the equality

$$pR_xR_y = -pR_yR_x.$$

Therefore the existence of numbers m_i such that

$$(P, C_i)_{m_i} = (0)$$

implies the existence of a number m such that

$$(P, C)_m = (0).$$

It is sufficient to set $m = m_1 + \cdots + m_s$.

We shall prove the existence of the number m_1. We consider two cases: (1) $\bar{A}_1$ is a matrix algebra over a division ring Δ; (2) $\bar{A}_1$ is a Cayley–Dickson algebra over a field F.

First let $\bar{A}_1 \cong \Delta_n$. Then $\bar{C}_1$ can be identified with the algebra of strictly triangular matrices. Any element $x \in \bar{C}_1$ is representable in the form

$$x = \sum_{i<j} \alpha_{ij}e_{ij}, \qquad \alpha_{ij} \in \Delta.$$

We denote by T_{ij} the complete inverse image of the set $\Delta \cdot e_{ij}$. We call elements belonging to the same set T_{ij} elements of the *same-type*.

Let $\bar{A}_1$ be a Cayley–Dickson algebra over a field F. We select a basis $\{e_1, \ldots, e_l\}$ for $\bar{C}_1$ in the following manner. If t is the index of nilpotency of the algebra $\bar{C}_1$, then we first choose a basis for $\bar{C}_1^{t-1}$, we extend it to a basis of $\bar{C}_1^{t-2}$, etc. We denote by T_i the complete inverse image of the set $F \cdot e_i$. We call elements belonging to the same T_i elements of the *same-type*.

We note that in each of the cases the number of types is finite. In addition, for any i and j we have the containments

$$T_{ij}^2 \subseteq R, \qquad T_i^2 \subseteq R. \tag{2}$$

LEMMA 4. If the elements $c_1, c_2, c_3 \in C_1$ are the same-type, then

$$PR_{c_1}R_{c_2}R_{c_3} = (0).$$

PROOF. Without loss of generality it is possible to assume that $c_i = d_ic$, where c is the inverse image of the element e_{ij} or e_i and d_i is the inverse image of the element $\alpha_i \cdot 1$, where $\alpha_i \in \Delta$ or $\alpha_i \in F$, respectively. It is obvious the elements d_i are such that

$$[c, d_i] \in R, \qquad (c, d_i, d_j) \in R. \tag{3}$$

Let $r \in R^{k-1}$. We consider the element $(rc_1)c_2$. By the middle Moufang identity, (2), and (3), we have

$$(rc_1)c_2 = [r(cd_1)]c_2 = -[c_2(cd_1)]r + (rc)(d_1c_2) + (c_2c)(d_1r)$$
$$= (rc)(d_1c_2) = (rc)(cd_4),$$

where d_4 is the inverse image of the element $\alpha_1\alpha_2 \cdot 1$. We now denote rc by r_1 and consider $[r_1(cd_4)]c_3$. Analogous to the preceding we obtain

$$[r_1(cd_4)]c_3 = (r_1c)(cd_5).$$

But by (2) $r_1c = (rc)c = rc^2 = 0$, and consequently,

$$rR_{c_1}R_{c_2}R_{c_3} = 0,$$

which also proves the lemma.

We now define the concept of weight for elements from C_1. We call the largest number v such that the coset $c + R$ lies in $\bar{C}_1^v$ but not in $\bar{C}_1^{v+1}$ the *weight* $\rho(c)$ of the element $c \in C_1 \setminus R$. Elements of the radical R are assigned infinite weight.

The weight function has the following property: for any $c, c' \in C_1$ there is the inequality

$$\rho(cc') \geqq \rho(c) + \rho(c'). \tag{4}$$

We shall call the number $\rho(w) = \sum_{i=1}^m \rho(c_i)$ the weight of the word $w = R_{c_1} \cdots R_{c_m}$.

Same-type elements have one and the same weight, and therefore we shall speak about the weight of a type. We order the types so that the greater the weight the greater the type. We then introduce a partial order on elements of the set $T = \bigcup T_{ij}$ ($T = \bigcup T_i$, respectively). We shall say that $c > c'$ if the type of the element c is greater than the type of the element c'. We shall consider same-type elements incomparable.

We call a word $w = R_{c_1} \cdots R_{c_m}$ *regular* if (1) $c_i \in T$, $i = 1, \ldots, m$; (2) either $\rho(w) = \infty$, or for each $t \in \{1, \ldots, m-1\}$ $c_t < c_{t+1}$ or c_t and c_{t+1} are incomparable.

LEMMA 5. Any word $w = R_{c_1}R_{c_2} \cdots R_{c_m}$, $c_i \in C_1$, is representable in the form of a sum of regular words

$$w = \sum w_i$$

such that $\rho(w_i) \geqq \rho(w)$.

PROOF. It is obviously possible to assume that $c_1, c_2, \ldots, c_m \in T$. Let $c_t > c_{t+1}$. Then we represent the word w in the following manner:

$$w = -R_{c_1} \cdots R_{c_{t+1}} R_{c_t} \cdots R_{c_m} + R_{c_1} \cdots R_{c_t \circ c_{t+1}} \cdots R_{c_m}.$$

In the first word c_t and c_{t+1} are already situated in the necessary order, and by (4) the weight of the second word is not less than the weight of the first. The second word also has fewer operators. The proof of the lemma is concluded by an obvious induction.

We now complete the proof of the theorem. Let t be the index of nilpotency of the subalgebra $\bar{C}_1$ and n be the number of types. Then for any regular word w of weight greater than $2(t-1)n$ we have

$$Pw = (0).$$

By Lemma 5 this means that for $m_1 = 2(t-1)n$

$$(P, C_1)m_1 = (0).$$

Thus the number m_1 is found, and as we have seen, this proves the theorem.

Exercises

1. (*Zhevlakov*) Let A be an alternative Artinian algebra, C be a right ideal of A, and B be a subring. If for any element $b \in B$ there exists a natural number $n = n(b)$ such that $b^n \in C$, then also $B^N \subseteq C$ for some natural number N.

A subset S of an algebra A is called a *weakly closed subsystem* if for any $a, b \in S$ there exists an element $\gamma(a, b) \in \Phi$ such that $ab + \gamma(a, b)ba \in S$. The subalgebra S^* generated by a weakly closed subsystem S is called the *enveloping algebra* for S.

2. If S is a weakly closed nil-subsystem of a Cayley–Dickson algebra $\boldsymbol{C}$ and $\gamma(a, b) \neq 1$ for any $a, b \in S$, then the enveloping algebra S^* of this subsystem is nilpotent. Find an example of a weakly closed nil-subsystem in $\boldsymbol{C}$ for which the enveloping algebra is not nilpotent.

3. (*Markovichev*) Let A be an alternative Artinian algebra and S be a weakly closed nil-subsystem of A such that $\gamma(a, b) \neq 1$ for any $a, b \in S$. Then the enveloping algebra S^* of this subsystem is nilpotent.

4. (*Markovichev*) The radical of an alternative Artinian algebra coincides with the intersection of all its maximal nilpotent subalgebras.

3. ALTERNATIVE ALGEBRAS WITH MAXIMAL CONDITION

An alternative algebra is called (*right*) *Noetherian* if it satisfies the maximal condition for right ideals. It is a well-known theorem of Levitzki that in an associative Noetherian algebra any one-sided nil-ideal is nilpotent [252, p. 40]. In this section we shall prove the analogous theorem for alternative algebras, with a single small restriction on characteristic.

The basic technical difficulties are absorbed in

LEMMA 6. Let A be an alternative algebra, B and I be ideals of A, $B^2 \subseteq I$, and the ideal I be nilpotent. If the algebra A satisfies the maximal condition for two-sided ideals contained in B, then B is a nilpotent ideal.

PROOF. We shall prove by induction on k the existence of a number $f(k)$ such that

$$B^{f(k)} \subseteq I^k.$$

The lemma will thus be proved, since $I^n = (0)$ for some n and, consequently, $B^{f(n)} = (0)$. As a basis for induction we have $f(1) = 2$. Suppose that the number $f(k)$ exists. We shall prove the existence of the number $f(k+1)$ by assuming it does not exist and reducing this assumption to a contradiction.

We introduce some notation. Let P be some ideal in A. We set

$$E_0(P) = \{x \in B \mid Px \subseteq I^{k+1}\}, \qquad E_i(P) = E_0((P, B)_i).$$

It is easy to see the $E_i(P)$ are ideals in A. We denote by $E(P)$ the ideal at which the ascending chain

$$E_0(P) \subseteq E_1(P) \subseteq \cdots \subseteq E_i(P) \subseteq \cdots$$

stabilizes. Since $B^{f(k)} \subseteq I^k$, the assumption we made about the lack of a number $f(k+1)$ implies

$$E(I^k) \neq B.$$

Let $E(C)$ be a maximal element of the set of ideals $\{E(C_\alpha) \mid C_\alpha \subseteq I^k, E(C_\alpha) \neq B\}$. Considering in place of C, if necessary, the ideal $(C, B)_i$ for a suitable i, we can assume that

$$E_0(C) = E_1(C) = \cdots = E(C).$$

In addition, taking in place of C the ideal $C + I^{k+1}$, we can also assume $C \supseteq I^{k+1}$.

Let $b \in B \setminus E(C)$. Then $(CB)b \not\subseteq I^{k+1}$. We consider the set $D = (CB)b + I^{k+1}$. We shall prove that D is an ideal in A.

We shall write $u \equiv v$ if $u - v \in I^{k+1}$.

By the right alternative and Moufang identities, for any $c \in C$, $x, y \in B$, $a \in A$, the relations

$$(cx)y \equiv -(cy)x, \tag{5}$$

$$[(cx)y]a \equiv -[(ca)y]x, \tag{6}$$

are valid. Therefore

$$[(cx)b]a \equiv -[(ca)b]x \equiv [(ca)x]b = (c_1x)b,$$

where $c_1 \in C$. This proves D is a right ideal. Furthermore,

$$\begin{aligned} a[(cx)b] &\equiv -a[(cb)x] = a[(xb)c] - [(ac)b]x - [(ax)b]c \\ &\equiv -[(ac)b]x \equiv [(ac)x]b = (c_2x)b, \end{aligned}$$

where $c_2 \in C$. This means D is a left ideal in A, that is, D is an ideal.

Since $D \subseteq C$, then $E(D) \supseteq E(C)$. However $b \in E(D)$, and therefore $E(D) \neq E(C)$. Consequently, $E(D) = B$. This means that

$$(D, B)_q \subseteq I^{k+1}$$

for some q. But then by (5)

$$(C, B)_q b \subseteq I^{k+1},$$

which contradicts the choice of the element b.

This proves the lemma.

THEOREM 5 (*Zhevlakov*). Let A be an alternative algebra with the maximal condition for two-sided ideals. Then the Baer radical $\mathscr{B}(A)$ of A is nilpotent.

PROOF. Let B be a maximal solvable ideal of the algebra A. Since the quotient algebra A/B does not have nonzero trivial ideals, then $B = \mathscr{B}(A)$. We consider the chain

$$B \supset B^{(1)} \supset \cdots \supset B^{(n)} = (0).$$

It is clear that $B^{(n-1)}$ is a nilpotent ideal. By Lemma 6 we can now conclude that the ideal $B^{(n-2)}$ is also nilpotent. Applying Lemma 6 again several times, we obtain that the ideal B is nilpotent.

This proves the theorem.

THEOREM 6 (*Zhevlakov*). Let A be an alternative Noetherian algebra over a ring Φ containing an element $\frac{1}{3}$. Then any one-sided nil-ideal of A is nilpotent.

PROOF. Let B be a one-sided nil-ideal in A. We shall show that $B \subseteq \mathscr{B}(A)$ and use Theorem 5. The quotient algebra $A/\mathscr{B}(A)$ is a subdirect sum of prime alternative Noetherian algebras. Moreover, by the restriction on characteristic, there are among them no exceptional prime algebras of characteristic 3. The rest cannot have nonzero one-sided nil-ideals. Consequently $B \subseteq \mathscr{B}(A)$, and the theorem is proved.

Exercises

1. Prove that in an associative Noetherian ring an nil-subring is nilpotent. *Hint*. Use Theorems 4 and 5, and also a theorem of Goldie [252].

2. Let A be an alternative Noetherian algebra over a ring Φ containing $\frac{1}{3}$. Then any nil-subring of A is solvable. Whether it is nilpotent is unknown.

LITERATURE

Andrunakievich [16–18], Zhevlakov [67, 68, 71, 77, 78], Markovichev [143], Oslowski [166], Slater [200, 202, 204], Herstein [252], Zorn [258, 259], Schafer [261].

Results on algebras with finiteness conditions in other varieties of algebras: Pchelintsev [174], Roomel'di [183], Thedy [234]. See also the literature to Chapter 15.

CHAPTER **13**

Free Alternative Algebras

Free algebras play an important role in the theory of any variety of algebras. The present chapter is devoted to free alternative algebras. We first consider the varieties generated by free finitely generated alternative algebras. Next we show that free alternative algebras can contain nonzero quasi-regular and nilpotent ideals, and we obtain a series of characterizations for the Zhevlakov radical of a free alternative algebra. The third theme of this chapter is that of centers and subalgebras of a special form in free alternative algebras.

1. IDENTITIES IN FINITELY GENERATED ALTERNATIVE ALGEBRAS

Let $\mathfrak{M}$ be some variety of Φ-algebras. We denote by $\mathfrak{M}_n$ the variety generated by the $\mathfrak{M}$-free algebra $\Phi_{\mathfrak{M}}[x_1, \ldots, x_n]$ from n free generators. The ideal of identities for this variety coincides with the ideal of identities for the algebra $\Phi_{\mathfrak{M}}[x_1, \ldots, x_n]$. As is easy to see, the following containments are valid:

$$\mathfrak{M}_1 \subseteq \mathfrak{M}_2 \subseteq \cdots \subseteq \mathfrak{M}_n \subseteq \mathfrak{M}_{n+1} \subseteq \cdots \subseteq \mathfrak{M}.$$

It is also clear that $\mathfrak{M} = \bigcup_n \mathfrak{M}_n$. The smallest value n for which $\mathfrak{M}_n = \mathfrak{M}$ is called the *base rank* for the variety $\mathfrak{M}$ and is denoted by $r_b(\mathfrak{M})$. In other words,

$r_b(\mathfrak{M})$ is the smallest number k such that each algebra from $\mathfrak{M}$ satisfies all identities realized in the $\mathfrak{M}$-free algebra from k free generators (or $\aleph_0$ if no such number exists).

By Lemma 3.6 a free associative algebra from a countable set of generators is imbedded in a free 2-generated associative algebra. Hence it obviously follows that $r_b(\text{Ass}) = 2$, where Ass is the variety of associative algebras. Analogously, it follows from Shirshov's theorem (Theorem 3.7) that $r_b(S\,\text{Jord}) = 2$, where S Jord is the variety generated by all special Jordan algebras. Shirshov also proved that $r_b(\text{Lie}) = 2$ for the variety Lie of Lie algebras. In 1958 he formulated the question on the determination of the exact value of the base rank for the variety Alt of alternative rings. From Artin's theorem it follows that $r_b(\text{Alt}) \geqq 3$. In 1963 Dorofeev proved that $r_b(\text{Alt}) \geqq 4$, and in 1977 Shestakov proved that $r_b(\text{Alt}) = \aleph_0$. Thus any finitely generated alternative ring satisfies some identity which is not realized in the free alternative ring from a countable set of generators. This section is devoted to the proof of this result.

Let S_n be the symmetric group of nth degree. For any partitioning $n = n_1 + n_2 + \cdots + n_k$ of the number n, we consider in S_n the subgroups S_{n_1}, $S_{n_2}, \ldots, S_{n_k}$, where S_{n_j} is the group of all permutations of the set from the n_j elements $\{n_1 + \cdots + n_{j-1} + 1, \ldots, n_1 + \cdots + n_{j-1} + n_j\}, j = 1, 2, \ldots, k$. It is clear that the set $S^n_{n_1, n_2, \ldots, n_k} = S_{n_1}S_{n_2} \cdots S_{n_k}$ is a subgroup of the group S_n. We denote by $V^n_{n_1, n_2, \ldots n_k}$ some complete system of representatives for the left cosets of the group S_n modulo the subgroup $S^n_{n_1, n_2, \ldots, n_k}$.

If $f = f(x_1, \ldots, x_n)$ is some nonassociative polynomial and $\sigma \in S_n$, then we set

$$f^\sigma = [f(x_1, \ldots, x_n)]^\sigma = f(x_{\sigma(1)}, \ldots, x_{\sigma(n)}).$$

In an arbitrary alternative algebra we define by induction a countable collection of skew-symmetric functions. We set $f_1(x_1, x_2, x_3) = (x_1, x_2, x_3)$, and if the skew-symmetric function $f_{n-1}(x_1, \ldots, x_{2n-1})$ is already defined, then

$$f_n(x_1, \ldots, x_{2n+1}) = \sum_{\sigma \in V^{2n+1}_{3, 2n-2}} (-1)^{\operatorname{sgn}\sigma}[f_{n-1}((x_1, x_2, x_3), x_4, \ldots, x_{2n+1})]^\sigma.$$

We note that in view of the skew-symmetry of the functions $f_{n-1}(x_1, \ldots, x_{2n-1})$ and (x_1, x_2, x_3), the value of the sum on the right side does not depend on the choice of the set $V^{2n+1}_{3, 2n-2}$. In analogous situations we shall subsequently not specifically state this fact. It is obvious from the definition that the function f_n is a skew-symmetric function of its arguments.

LEMMA 1. Any alternative algebra satisfies the identity

$$f_n(xyx, y, x_1, \ldots, x_{2n-1}) = f_n(x, yxy, x_1, \ldots, x_{2n-1}). \tag{1}$$

PROOF. We shall first prove that in any alternative algebra the identity

$$(r, s, xyx) = x(r, s, y)x + \{xy(r, s, x)\} \tag{2}$$

is valid, where $\{xyz\} = (xy)z + (zy)x = x(yz) + z(yx)$. Let $f(x, y, z, t)$ be the Kleinfeld function. In view of the Moufang identity $(r, s, x^2) = x \circ (r, s, x)$ and its linearization, we obtain

$$\begin{aligned}(r, s, xyx) &= -(r, s, x^2y) + (r, s, (xy) \circ x)\\ &= (x^2y, s, r) + x \circ (r, s, xy) + (xy) \circ (r, s, x)\\ &= f(x^2, y, s, r) + y[x \circ (x, s, r)] + (y, s, r)x^2\\ &\quad + x \circ (r, s, xy) + (xy) \circ (r, s, x).\end{aligned}$$

Furthermore, we have

$$\begin{aligned}f(x^2, y, s, r) &= ([x^2, y], s, r) + (x^2, y, [s, r])\\ &= (x \circ [x, y], s, r) + x \circ (x, y, [s, r])\\ &= x \circ ([x, y], s, r) + [x, y] \circ (x, s, r) + x \circ (x, y, [s, r])\\ &= x \circ f(x, y, s, r) + [x, y] \circ (x, s, r)\\ &= x \circ (xy, s, r) - x \circ (y(x, s, r)) - x \circ ((y, s, r)x)\\ &\quad + [x, y] \circ (x, s, r).\end{aligned}$$

Substituting the obtained expression for $f(x^2, y, s, r)$ into the previous equality, we obtain (2).

We shall prove equality (1) by induction on n. The validity of the basis for the induction, $n = 1$, follows from identity (7.17). Suppose it is already proved that

$$f_{n-1}(xyx, y, x_1, \ldots, x_{2n-3}) = f_{n-1}(x, yxy, x_1, \ldots, x_{2n-3}). \tag{3}$$

Linearizing this identity in y, we obtain

$$\begin{aligned}&f_{n-1}(xyx, z, x_1, \ldots, x_{2n-3}) + f_{n-1}(xzx, y, x_1, \ldots, x_{2n-1})\\ &\quad = f_{n-1}(x, \{zxy\}, x_1, \ldots, x_{2n-3}).\end{aligned} \tag{4}$$

Now by definition of the function f_n,

$$\begin{aligned}&f_n(xyx, y, x_1, \ldots, x_{2n-1})\\ &\quad = \sum_{\sigma \in V^{2n-1}_{1,2n-2}} (-1)^{\operatorname{sgn}\sigma}[f_{n-1}((x_1, xyx, y), x_2, \ldots, x_{2n-1})]^\sigma\\ &\qquad - \sum_{\sigma \in V^{2n-1}_{2,2n-3}} (-1)^{\operatorname{sgn}\sigma}[f_{n-1}((x_1, x_2, y), xyx, x_3, \ldots, x_{2n-1})]^\sigma\\ &\qquad + \sum_{\sigma \in V^{2n-1}_{2,2n-3}} (-1)^{\operatorname{sgn}\sigma}[f_{n-1}((x_1, x_2, xyx), y, x_3, \ldots, x_{2n-1})]^\sigma\\ &\qquad + \sum_{\sigma \in V^{2n-1}_{3,2n-4}} (-1)^{\operatorname{sgn}\sigma}[f_{n-1}((x_1, x_2, x_3)xyx, y, x_4, \ldots, x_{2n-1})]^\sigma\\ &\quad = S_1(x, y) - S_2(x, y) + S_3(x, y) + S_4(x, y),\end{aligned}$$

where in this connection we consider that $x^\sigma = x$, $y^\sigma = y$. By (7.17) and (3) we have

$$\begin{aligned} &f_{n-1}((x_1, xyx, y), x_2, \ldots, x_{2n-1}) \\ &\quad = -f_{n-1}((x_1, yxy, x), x_2, \ldots, x_{2n-1}), \\ &f_{n-1}((x_1, x_2, x_3), xyx, y, x_4, \ldots, x_{2n-1}) \\ &\quad = -f_{n-1}((x_1, x_2, x_3), yxy, x, x_4, \ldots, x_{2n-1}), \end{aligned}$$

whence

$$S_1(x, y) + S_4(x, y) = -S_1(y, x) - S_4(y, x).$$

Furthermore, according to (4) and (2) we obtain

$$\begin{aligned} &-f_{n-1}((x_1, x_2, y), xyx, x_3, \ldots, x_{2n-1}) \\ &\qquad + f_{n-1}((x_1, x_2, xyx), y, x_3, \ldots, x_{2n-1}) \\ &\quad = -f_{n-1}(x(x_1, x_2, y)x, y, x_3, \ldots, x_{2n-1}) \\ &\qquad - f_{n-1}(\{yx(x_1, x_2, y)\}, x, x_3, \ldots, x_{2n-1}) \\ &\qquad + f_{n-1}(x(x_1, x_2, y)x, y, x_3, \ldots, x_{2n-1}) \\ &\qquad + f_{n-1}(\{xy(x_1, x_2, x)\}, y, x_3, \ldots, x_{2n-1}) \\ &\quad = -f_{n-1}(\{yx(x_1, x_2, y)\}, x, x_3, \ldots, x_{2n-1}) \\ &\qquad + f_{n-1}(\{xy(x_1, x_2, x)\}, y, x_3, \ldots, x_{2n-1}), \end{aligned}$$

whence

$$-S_2(x, y) + S_3(x, y) = S_2(y, x) - S_3(y, x).$$

Thus

$$\begin{aligned} f_n(xyx, y, x_1, \ldots, x_{2n-1}) &= -f_n(yxy, x, x_1, \ldots, x_{2n-1}) \\ &= f_n(x, yxy, x_1, \ldots, x_{2n-1}). \end{aligned}$$

This proves the lemma.

PROPOSITION 1 (*Shestakov*). Let A be an alternative algebra, M be some Φ-module, and let the mapping f for a fixed natural number n put into correspondence with elements $a_1, \ldots, a_n \in A^\#$ some submodule $f(a_1, \ldots, a_n) \subseteq M$. Also let f satisfy the conditions

(a) $f(a_{\sigma(1)}, \ldots, a_{\sigma(n)}) = f(a_1, \ldots, a_n)$ for any permutation $\sigma \in S_n$;
(b) $f(a, a, a_1, \ldots, a_{n-2}) = (0)$;
(c) $f(1, a_1, a_2, \ldots, a_{n-1}) = (0)$;
(d) $f(\alpha a + \beta b, a_1, \ldots, a_{n-1}) \subseteq f(a, a_1, \ldots, a_{n-1}) + f(b, a_1, \ldots, a_{n-1})$, $\alpha, \beta \in \Phi$;
(e) $f(aba, a_1, \ldots, a_{n-1}) \subseteq f(a, a_1, \ldots, a_{n-1}) + f(b, a_1, \ldots, a_{n-1})$.

Then for any elements $a_1, \ldots, a_n \in A$ the containment

$$f(a_1, \ldots, a_n) \subseteq \sum_i f(u_1^{(i)}, \ldots, u_n^{(i)}) \tag{5}$$

holds, where the $u_j^{(i)}$ are all possible regular r_1-words from some fixed ordered set X of generators of the algebra A. In addition, if the mapping f satisfies the condition

(f) $f(a, ba, a_1, \ldots, a_{n-2}) \subseteq f(a, b, a_1, \ldots, a_{n-2})$,

then for any elements $a_1, \ldots, a_n \in A$

$$f(a_1, \ldots, a_n) \subseteq \sum_i f(x_1^{(i)}, \ldots, x_{n-1}^{(i)}, u^{(i)}), \tag{6}$$

where $x_j^{(i)} \in X$ and the $u^{(i)}$ are regular r_1-words which do not contain the elements $x_1^{(i)}, \ldots, x_{n-1}^{(i)}$.

PROOF. We denote by M_0 the sum of all submodules of the form $f(u_1, \ldots, u_n)$, where $u_1, \ldots, u_n$ are regular r_1-words. In view of condition (d), for the proof of containment (5) it suffices to show that $f(v_1, \ldots, v_n) \subseteq M_0$ for any nonassociative words $v_1, \ldots, v_n$ from generators of the algebra A. We shall prove this assertion by induction on the number $m(v_1, \ldots, v_n) = \sum_{i=1}^n d(v_i) + k(v_1, \ldots, v_n)$, where $k(v_1, \ldots, v_n)$ is the number of words among $v_1, \ldots, v_n$ that are not regular r_1-words. The basis for the induction, $m(v_1, \ldots, v_n) = n$, is obvious. Now let $m(v_1, \ldots, v_n) > n$. If all v_i are regular r_1-words, then there is nothing to prove. Let the word v_1, for example, not be a regular r_1-word. Then according to Proposition 5.3, we have

$$v_1 = \pm\langle v_1\rangle + \sum_i \alpha_i a_i b_i a_i + \sum_j \beta_j \{a'_j b'_j a''_j\},$$

where $\alpha_i, \beta_j \in \Phi$, a_i, a'_j, a''_j are nonassociative words of length less than v_1, and b_i, b'_j either equal 1 or are also nonassociative words of length less than v_1. We now note that by condition (d) the following linearized form of condition (e) is valid:

(e′) $f(\{abc\}, a_1, \ldots, a_{n-1}) \subseteq f(a, a_1, \ldots, a_{n-1}) + f(b, a_1, \ldots, a_{n-1}) + f(c, a_1, \ldots, a_{n-1})$.

According to (d), (e), and (e′), we obtain

$$\begin{aligned} f(v_1, v_2, \ldots, v_n) \subseteq{}& f(\langle v_1\rangle, v_2, \ldots, v_n) + \sum_i f(a_i, v_2, \ldots, v_n) \\ &+ \sum_i f(b_i, v_2, \ldots, v_n) + \sum_j f(a'_j, v_2, \ldots, v_n) \\ &+ \sum_j f(a''_j, v_2, \ldots, v_n) + \sum_j f(b'_j, v_2, \ldots, v_n), \end{aligned}$$

whence by the induction assumption and in view of condition (c) we have

$$f(v_1, v_2, \ldots, v_n) \subseteq M_0.$$

By the same token, containment (5) is proved.

Now let the mapping f satisfy condition (f). We denote the right side of containment (6) by M_1. It suffices for us to show that $f(v_1, \ldots, v_n) \subseteq M_1$ for any nonassociative words $v_1, \ldots, v_n$ from a set of generators X of the algebra A. We shall first prove that $f(x_1, \ldots, x_{n-1}, v) \subseteq M_1$ for any elements $x_1, \ldots, x_{n-1} \in X$ and for any word v. We carry out an induction on $d(v)$. The basis for the induction, $d(v) = 1$, is obvious. Let $d(v) > 1$. By what was proved above, without loss of generality it is possible to assume that v is a regular r_1-word. If v does not contain the elements $x_1, \ldots, x_{n-1}$, then everything is proved. For example, let x_1 appear in v, and let $v' = v\Delta_1^1$ be the word obtained by deletion of the element x_1 in the word v. By Proposition 5.3 and in view of conditions (d), (e), (e′) and the induction assumption, we then obtain

$$f(x_1, \ldots, x_{n-1}, v) \subseteq f(x_1, \ldots, x_{n-1}, v'x_1) + M_1,$$

whence in view of condition (f) and the induction assumption it follows that

$$f(x_1, \ldots, x_{n-1}, v) \subseteq M_1.$$

In the general case we carry out an induction on the number $m = \sum_{i=1}^{n-1} d(v_i)$. The basis for induction when $m = n - 1$ is a case already considered. For example, let $d(v_1) > 1$, $v_1 = v_1'v_1''$. By (a), (d), and (f) we obtain

$$\begin{aligned} f(v_1, v_2, v_3, \ldots, v_n) &= f(v_1'v_1'', v_n, v_3, \ldots, v_2) \\ &= f(v_1'(v_1'' + v_n) - v_1'v_n, (v_1'' + v_n) - v_1'', v_3, \ldots, v_2) \\ &\subseteq f(v_1', v_1'', v_3, \ldots, v_2) + f(v_1', v_n, v_3, \ldots, v_2) \\ &\quad + f(v_1'v_n, v_1'', v_3, \ldots, v_2) = f(v_1', v_2, v_3, \ldots, v_1'') \\ &\quad + f(v_1', v_2, v_3, \ldots, v_n) + f(v_1'', v_2, v_3, \ldots, v_1'v_n), \end{aligned}$$

whence by the induction assumption

$$f(v_1, v_2, \ldots, v_n) \subseteq M_1.$$

This proves the proposition.

COROLLARY. Under the conditions of Proposition 1, if the algebra A has a set of generators from m elements, then $f(a_1, \ldots, a_n) = (0)$ for $n \geqq 2^m$ and any elements $a_1, \ldots, a_n \in A$. If the mapping f also satisfies condition (f), then $f(a_1, \ldots, a_n) = (0)$ for $n \geqq m + 2$.

PROOF. As is easy to see, in the given case in the algebra A there exist in all not more than $2^m - 1$ different regular r_1-words from the generators.

This means that for $n \geqq 2^m$ among any regular r_1-words $u_1, \ldots, u_n$ there are found two that are identical, and by condition (b) we obtain $f(u_1, \ldots, u_n) = (0)$. Hence by (5) also $f(a_1, \ldots, a_n) = (0)$ for any elements $a_1, \ldots, a_n \in A$. The second assertion of the corollary follows analogously from (6).

Let $X_m = \{x_1, x_2, \ldots, x_m\}$ and $\mathrm{Alt}[X_m]$ be the free alternative algebra from set of generators X_m.

THEOREM 1 (*Shestakov*). The algebra $\mathrm{Alt}[X_m]$ satisfies the identity

$$f_{2^m-1}(y_1, \ldots, y_{2^m+1}) = 0.$$

PROOF. We consider the mapping f which puts into correspondence with arbitrary elements $a_1, \ldots, a_{2^m}$ of the algebra $\mathrm{Alt}[X_m]$ the following Φ-module of $\mathrm{Alt}[X_m]$:

$$f(a_1, \ldots, a_{2^m}) = \{f_{2^m-1}(a_1, \ldots, a_{2^m}, y) \mid y \in \mathrm{Alt}[X_m]\}.$$

In view of the skew-symmetry and multilinearity of the function f_{2^m-1}, the mapping f satisfies conditions (a), (b), and (d) of Proposition 1. In addition, it easily follows from the definition of the functions f_n that $f_n(1, b_1, \ldots, b_{2n}) = 0$ for any n and any elements $b_1, \ldots, b_{2n} \in \mathrm{Alt}[X_m]$. Consequently, the mapping f satisfies condition (c) of Proposition 1. Finally, linearizing relation (1) in y we obtain the identity

$$\begin{aligned} &f_n(xyx, z, x_1, \ldots, x_{2n-1}) \\ &\quad = f_n(x, \{yxz\}, x_1, \ldots, x_{2n-1}) + f_n(y, xzx, x_1, \ldots, x_{2n-1}), \end{aligned}$$

whence in view of the skew-symmetry of the function f_{2^m-1} it follows that the mapping f also satisfies condition (e). By the corollary to Proposition 1 we obtain that $f(a_1, \ldots, a_{2^m}) = (0)$ for any elements $a_1, \ldots, a_{2^m} \in \mathrm{Alt}[X_m]$. But this is equivalent to the validity in the algebra $\mathrm{Alt}[X_m]$ of the identity

$$f_{2^m-1}(y_1, \ldots, y_{2^m+1}) = 0.$$

This proves the theorem.

LEMMA 2. In any alternative algebra that is solvable of index 2, the identity

$$f_n(x_1, \ldots, x_{2n+1}) = \varphi(n) \sum_{\sigma \in V_{3,\,2n-2}^{2n+1}} (-1)^{\operatorname{sgn} \sigma}[(x_1, x_2, x_3)R_{x_4} \cdots R_{x_{2n+1}}]^{\sigma}$$

is valid, where $\varphi(n) = (n-1)!(2n-3)!!$

PROOF. We note first of all that in view of the relation $R_x \circ R_y = R_{x \circ y}$, the right side of the equality to be proved is skew-symmetric in the variables

$x_4, \ldots, x_{2n+1}$, and therefore the value of the sum on the right side does not depend on the choice of the set $V_{3,2n-2}^{2n+1}$. We shall prove the lemma by induction on n. For $n = 1$ the assertion of the lemma is obvious, so that there is a basis for the induction. Suppose it is already proved that in an alternative algebra that is solvable of index 2, the relation

$$f_{n-1}(x_1, \ldots, x_{2n-1}) = \varphi(n-1) \sum_{\sigma \in V_{3,2n-4}^{2n-1}} (-1)^{\operatorname{sgn} \sigma}[(x_1, x_2, x_3)R_{x_4} \cdots R_{x_{2n-1}}]^\sigma$$

is valid. Then by the definition of the function f_n, for any $x_1, \ldots, x_{2n+1} \in A$ we have

$$\begin{aligned} f_n(x_1, \ldots, x_{2n+1}) &= \sum_{\sigma \in V_{3,2n-2}^{2n+1}} (-1)^{\operatorname{sgn} \sigma}[f_{n-1}((x_1, x_2, x_3), x_4, \ldots, x_{2n+1})]^\sigma \\ &= \varphi(n-1) \sum_{\sigma \in V_{3,2n-2}^{2n+1}} (-1)^{\operatorname{sgn} \sigma} \sum_{\sigma_1 \in V_{3,2n-5}^{2n-2}} (-1)^{\operatorname{sgn} \sigma_1}[(x_4, x_5, x_6)R_{x_7} \cdots \\ &\quad R_{x_{2n+1}} R_{(x_1,x_2,x_3)}]^{\sigma\sigma_1} \\ &\quad + \varphi(n-1) \sum_{\sigma \in V_{3,2n-2}^{2n+1}} (-1)^{\operatorname{sgn} \sigma} \sum_{\sigma_1 \in V_{2,2n-4}^{2n-2}} (-1)^{\operatorname{sgn} \sigma_1}[((x_1, x_2, x_3), \\ &\quad x_4, x_5)R_{x_6} \cdots R_{x_{2n+1}}]^{\sigma\sigma_1}; \end{aligned}$$

where also we have made use of the fact that $(x_1, x_2, x_3)^\sigma = (x_1, x_2, x_3)^{\sigma\sigma_1}$. As is easy to see, each summand in the first sum equals zero, and therefore it only remains for us to consider the second sum. Since $yR_{xz} = 0$ and $y(R_x \circ R_z) = yR_{x \circ z} = 0$ for any $y \in A^2$, we obtain

$$\begin{aligned} &\sum_{\sigma \in V_{3,2n-2}^{2n+1}} (-1)^{\operatorname{sgn} \sigma} \sum_{\sigma_1 \in V_{2,2n-4}^{2n-2}} (-1)^{\operatorname{sgn} \sigma_1}[((x_1, x_2, x_3), x_4, x_5)R_{x_6} \cdots R_{x_{2n+1}}]^{\sigma\sigma_1} \\ &\quad = \sum_{\sigma \in V_{3,2n-2}^{2n+1}} (-1)^{\operatorname{sgn} \sigma} \sum_{\sigma_1 \in V_{2,2n-4}^{2n-2}} (-1)^{\operatorname{sgn} \sigma_1}[(x_1, x_2, x_3)R_{x_4}R_{x_5}R_{x_6} \cdots \\ &\quad\quad R_{x_{2n+1}}]^{\sigma\sigma_1} \\ &\quad = \sum_{\sigma \in V_{3,2n-2}^{2n+1}} (-1)^{\operatorname{sgn} \sigma} \frac{(2n-2)!}{2!(2n-4)!}[(x_1, x_2, x_3)R_{x_4} \cdots R_{x_{2n+1}}]^\sigma \\ &\quad = \frac{(2n-2)!}{2!(2n-4)!} \sum_{\sigma \in V_{3,2n-2}^{2n+1}} (-1)^{\operatorname{sgn} \sigma}[(x_1, x_2, x_3)R_{x_4} \cdots R_{x_{2n+1}}]^\sigma. \end{aligned}$$

Since

$$\varphi(n-1) \cdot \frac{(2n-2)!}{2!(2n-4)!} = (n-2)!(2n-5)!!(n-1)(2n-3) = (n-1)!(2n-3)!! = \varphi(n),$$

from this it follows the lemma is valid.

COROLLARY. Let there be no elements of order $\leqq 2n-1$ in the additive group of the ring of operators Φ. Then in the free alternative Φ-algebra $\mathrm{Alt}[X_{2n+1}]$ from generators $x_1, x_2, \ldots, x_{2n+1}$ the element $f_n(x_1, x_2, \ldots, x_{2n+1})$ is different from zero.

PROOF. Let A be the solvable of index 2 algebra over a ring Φ from Dorofeev's example (Section 6.2). By definition of multiplication in the algebra A and in view of Lemma 3, we have for the elements $x, e_1, \ldots, e_{2n}$ of the algebra A

$$f_n(x, e_1, \ldots, e_{2n}) = (n-1)!(2n-3)!!\frac{(2n)!}{2!(2n-2)!}(x, e_1, e_2)R_{e_3}\cdots R_{e_{2n}}$$

$$= n!(2n-1)!!xR_1R_2\cdots R_{2n} \neq 0.$$

Hence follows the assertion of the corollary.

We denote by $\mathrm{Alt}_m(\Phi)$ the variety of alternative Φ-algebras generated by the free alternative Φ-algebra $\mathrm{Alt}[X_m]$ from m generators, and by Alt_m the variety of alternative rings generated by the free alternative ring from m generators. It is clear that $\mathrm{Alt}_m = \mathrm{Alt}_m(\boldsymbol{Z})$, where $\boldsymbol{Z}$ is the ring of integers. As is easy to see, there is the chain of containments

$$\mathrm{Alt}_1(\Phi) \subseteq \mathrm{Alt}_2(\Phi) \subseteq \cdots \subseteq \mathrm{Alt}_n(\Phi) \subseteq \cdots. \qquad (*)$$

The first three containments in this chain are strict. The strictness of the containments $\mathrm{Alt}_1(\Phi) \subset \mathrm{Alt}_2(\Phi) \subset \mathrm{Alt}_3(\Phi)$ follows from Artin's theorem, and the strictness of the containment $\mathrm{Alt}_3(\Phi) \subset \mathrm{Alt}_4(\Phi)$ was established in works of Dorofeev and Humm and Kleinfeld. Shirshov raised the question of the chain (*) stabilizing at some finite step for $\Phi = \boldsymbol{Z}$. The answer to this questions turns out to be no.

THEOREM 2 (*Shestakov*). Let there be no elements of finite order in the additive group of the ring Φ. Then for any natural number m the containment $\mathrm{Alt}_m(\Phi) \subset \mathrm{Alt}_{2m+1}(\Phi)$ is strict. In particular, in this case the chain of varieties (*) does not stabilize at any finite step.

The proof follows from Theorem 1 and the corollary to Lemma 2.

Setting $\Phi = \boldsymbol{Z}$ we obtain

COROLLARY. $r_b(\mathrm{Alt}) = \aleph_0$.

The question remains open on whether there can be equality at some places in the chain (*).

It would also be interesting to find some system of defining identities for the varieties $\mathrm{Alt}_m, m \geqq 3$.

We now consider how the concepts of solvability and nilpotency are related in the varieties $\mathrm{Alt}_m(\Phi)$. We know (see Section 6.2) that these concepts are not equivalent in the variety of all alternative Φ-algebras. For any ring of operators Φ there exists a solvable alternative Φ-algebra that is not nilpotent. It turns out that in the varieties $\mathrm{Alt}_m(\Phi)$ the situation is substantially changed.

THEOREM 3 (*Shestakov*). Let Φ be a field of characteristic 0. Then any solvable algebra from the variety $\mathrm{Alt}_m(\Phi)$ is nilpotent.

PROOF. We shall first prove that any solvable of index 2 algebra A from the variety $\mathrm{Alt}_m(\Phi)$ is nilpotent. By Theorem 1 and Lemma 2, the identity

$$\sum_{\sigma \in V_{3,\,2m-2}^{2m+1}} (-1)^{\operatorname{sgn}\sigma}[(x_1, x_2, x_3)R_{x_4} \cdots R_{x_{2m+1}}]^\sigma = 0$$

is satisfied in the algebra A. Setting here $x_{2m+1} = y_1 y_2$, we obtain the identity

$$\sum_{\sigma \in V_{2,\,2m-2}^{2m}} (-1)^{\operatorname{sgn}\sigma}[(y_1 y_2, x_1, x_2)R_{x_3} \cdots R_{x_{2m}}]^\sigma = 0,$$

where in this connection we consider that $y_1^\sigma = y_1$, $y_2^\sigma = y_2$. Since $yR_{xz} = 0$ and $y(R_x \circ R_z) = yR_{x \circ z} = 0$ for any $y \in A^2$, our identity reduces to the form

$$2^m(2^m - 1)y_1 y_2 R_{x_1} \cdots R_{x_{2m}} = 0.$$

Consequently the algebra A is right nilpotent of index $2^m + 2$. In view of the corollary to Theorem 5.5, we obtain that the algebra A is nilpotent of index $(2^m + 2)^2$. Now let F be the free algebra in the variety $\mathrm{Alt}_m(\Phi)$ from a countable set of generators. It follows from what has been proved that $F^{(2^m+2)^2} \subseteq F^{(2)}$. Suppose it is already proved that for a natural number $n \geqq 2$ there exists a number $f(m, n)$ such that $F^{f(m,n)} \subseteq F^{(n)}$. Then for any algebra $A \in \mathrm{Alt}_m(\Phi)$ we have $A^{f(m,n)} \subseteq A^{(n)}$. In particular, $(F^2)^{f(n,m)} \subseteq (F^2)^{(n)} = F^{(n+1)}$. We note that by Corollary 2 to Theorem 1.3 the variety $\mathrm{Alt}_m(\Phi)$ is homogeneous. Repeating now, verbatim, the arguments carried out for the proof of Lemma 8.8 (see also Exercise 2 in Section 8.2), we obtain that the number $[f(m, n)]^2$ can be taken as $f(m, n + 1)$. By induction we conclude that the number $f(m, n)$ exists for any natural number n, from which follows the equivalence of solvability and nilpotency in the variety $\mathrm{Alt}_m(\Phi)$.

This proves the theorem.

COROLLARY. Let A be a finitely generated alternative algebra over a field of characteristic 0. Then any solvable subalgebra of the algebra A is nilpotent.

We note here that the analogous result is not true for Jordan algebras. By Theorem 3.7 the solvable nonnilpotent special Jordan algebra from the Zhevlakov example (Section 4.1) is imbeddable in a 2-generated Jordan algebra.

Exercises

1. Let $\mathfrak{M}$ be the variety of associative algebras over a field of characteristic 2 defined by the identity $x^2 = 0$. Prove that $r_b(\mathfrak{M}) = \aleph_0$.

2. Let $\frac{1}{2} \in \Phi$. We denote by $\text{Jord}(^2)$ the variety of Jordan algebras over Φ that are solvable of index 2. Prove that $r_b(\text{Jord}(^2)) = \aleph_0$.

3. (*Shestakov*) Let J be a finite-dimensional Jordan algebra over a field of characteristic 0. Prove that in the variety $\text{Var}(J)$ generated by the algebra J every solvable algebra is nilpotent. *Hint*. It suffices to prove that for any algebra $A \in \text{Var}(J)$ the equality $(A^3)^3 = (0)$ implies the nilpotency of A.

4. Prove that for the variety Jord of all Jordan algebras $r_b(\text{Jord}) \geqq 3$.

2. RADICALS AND NILPOTENT ELEMENTS OF FREE ALTERNATIVE ALGEBRAS

In this section we shall show that, unlike free associative or free nonassociative algebras, free alternative algebras over an integral domain can contain nonzero divisors of zero and nilpotent elements. The set of all nilpotent elements of a free alternative algebra forms a two-sided ideal in the algebra which coincides with the Zhevlakov radical. We deduce a series of characterizations of this radical, and give a certain description of the quotient algebra modulo the radical.

Everywhere below Φ is an arbitrary integral domain with identity element 1, $X = \{x_1, x_2, \ldots\}$, and $X_n = \{x_1, \ldots, x_n\}$.

We recall that the ideal of the algebra A generated by all its associators is denoted by $D(A)$.

LEMMA 3. Let $\mathfrak{M}$ be an arbitrary homogeneous subvariety of the variety Alt of all alternative Φ-algebras, and let $\Phi_{\mathfrak{M}}[X]$ be the free algebra in the variety $\mathfrak{M}$ from a countable set of generators X. Then $\mathcal{J}(\Phi_{\mathfrak{M}}[X])$ is a nil-ideal, i.e., $\mathcal{J}(\Phi_{\mathfrak{M}}[X]) = \mathfrak{N}(\Phi_{\mathfrak{M}}[X])$.

PROOF. Let $f \in \mathcal{J}(\Phi_{\mathfrak{M}}[X])$, $f = f(x_1, \ldots, x_n)$. We set $f' = f \cdot x_{n+1}$. Then f' is homogeneous of degree 1 in x_{n+1} and $f' \in \mathcal{J}(\Phi_{\mathfrak{M}}[X])$. Let g be a quasi-inverse element for f' and $g = \sum_{i=0}^{m} g_i$, where the g_i are homogeneous of degree i in x_{n+1}. We have

$$f' + \sum_{i=0}^{m} g_i = \sum_{i=0}^{m} f'g_i,$$

whence in view of the homogeneity of the variety $\mathfrak{M}$

$$g_0 = 0, \qquad f'g_m = 0,$$

and furthermore

$$f' + g_1 = 0, \qquad g_i = f'g_{i-1}, \qquad i = 2, \ldots, m.$$

Consequently, $g_i = -(f')^i$, whence $(f')^{m+1} = 0$. Now setting $x_{n+1} = f$ in the identity $(f')^{m+1} = 0$, we obtain $f^{2(m+1)} = 0$. This proves the lemma.

COROLLARY 1. Let Ass$[X]$ be the free associative Φ-algebra. Then $\mathscr{J}(\text{Ass}[X]) = (0)$.

In fact, the algebra Ass$[X]$ does not contain nonzero nilpotent elements.

COROLLARY 2 (*Zhevlakov*). The Zhevlakov radical $\mathscr{J}(\text{Alt}[X])$ of the free alternative Φ-algebra Alt$[X]$ is locally nilpotent, i.e., $\mathscr{J}(\text{Alt}[X]) = \mathscr{L}(\text{Alt}[X])$.

PROOF. We assume that $\mathscr{J}(\text{Alt}[X]) \nsubseteq \mathscr{L}(\text{Alt}[X])$. Then the image $\bar{\mathscr{J}}$ of the radical $\mathscr{J}(\text{Alt}[X])$ in the quotient algebra $\bar{A} = \text{Alt}[X]/\mathscr{L}(\text{Alt}[X])$ is different from zero. By Theorems 8.8 and 9.11 the algebra $\bar{A}$ is a subdirect sum of prime associative algebras without locally nilpotent ideals and Cayley–Dickson rings. Since $\bar{\mathscr{J}} \neq (0)$, there exists a subdirect summand A_α in which under the homomorphism of the algebra $\bar{A}$ the image of the ideal $\bar{\mathscr{J}}$ is different from zero. By Corollary 1 $\mathscr{J}(\text{Alt}[X]) \subseteq D(\text{Alt}[X])$, and therefore $\bar{\mathscr{J}} \subseteq D(\bar{A})$. Hence the ideal $\bar{\mathscr{J}}$ maps to zero under any homomorphism of the algebra $\bar{A}$ onto an associative algebra. It remains to consider the case when the appropriate summand A_α is a Cayley–Dickson ring. In view of Lemma 3, $\mathscr{J}(\text{Alt}[X])$ is a nil-ideal, and therefore $\bar{\mathscr{J}}$ is a nil-ideal of the algebra $\bar{A}$. Under the homomorphism of the algebra $\bar{A}$ onto A_α, the ideal $\bar{\mathscr{J}}$ must pass to a nil-ideal of the algebra A_α. But by the corollary to Proposition 9.3 the Cayley–Dickson ring A_α is nil-semisimple, and therefore under any homomorphism of the algebra $\bar{A}$ onto A_α the ideal $\bar{\mathscr{J}}$ maps to zero. The obtained contradiction proves that $\bar{\mathscr{J}} = (0)$, i.e., $\mathscr{J}(\text{Alt}[X]) = \mathscr{L}(\text{Alt}[X])$. This proves the corollary.

We denote by $\boldsymbol{C}$ the Cayley–Dickson matrix algebra $\boldsymbol{C}(\Phi)$ over the base ring of operators Φ, and by $T(\boldsymbol{C})$ the subset of the algebra Alt$[X]$ which consists of those identities of the algebra $\boldsymbol{C}$ for which all partial linearizations are also identities in $\boldsymbol{C}$. If Φ is infinite, then by Theorem 1.5 the set $T(\boldsymbol{C})$ coincides with the ideal of all identities of the algebra $\boldsymbol{C}$. In the general case, $T(\boldsymbol{C})$ is equal to the ideal of identities of the Cayley–Dickson matrix Φ-algebra $\boldsymbol{C}(\Phi_1)$, where Φ_1 is an arbitrary infinite associative–commutative Φ-algebra without zero divisors. For example, $\Phi_1 = \Phi[t]$ is the ring of polynomials in the variable t. Thus $T(\boldsymbol{C})$ is a completely characteristic ideal of the algebra $\mathrm{Alt}[X]$. We note that by Theorems 1.4 and 1.6 for any associative–commutative Φ-algebra K the algebra $\boldsymbol{C}(K) = K \otimes_\Phi \boldsymbol{C}$ satisfies all identities from $T(\boldsymbol{C})$.

THEOREM 4 (*Shestakov, Slater*). $\mathscr{J}(\mathrm{Alt}[X]) = T(\boldsymbol{C}) \cap D(\mathrm{Alt}[X])$.

PROOF. Let $f = f(x_1, \ldots, x_n) \in \mathscr{J}(\mathrm{Alt}[X])$. By Corollary 1 to Lemma 3 $\mathscr{J}(\mathrm{Ass}[X]) = (0)$, and therefore $\mathscr{J}(\mathrm{Alt}[X]) \subseteq D(\mathrm{Alt}[X])$. For the proof that $f \in T(\boldsymbol{C})$, it suffices for us to show that the identity f is valid in the algebra $\boldsymbol{C}_1 = \boldsymbol{C}(\Phi[t])$. Let $f \not\equiv 0$ in $\boldsymbol{C}_1$, i.e., there exist elements $c_1, c_2, \ldots, c_n \in \boldsymbol{C}_1$ such that $f(c_1, c_2, \ldots, c_n) \neq 0$. Since the Φ-algebra $\boldsymbol{C}_1$ is finitely generated, there exists a homomorphism φ of the algebra $\mathrm{Alt}[X]$ onto $\boldsymbol{C}_1$ such that $\varphi(x_i) = c_i$ for $i = 1, 2, \ldots, n$. Since by Corollary 2 to Lemma 3 $\mathscr{J}(\mathrm{Alt}[X]) = \mathscr{L}(\mathrm{Alt}[X])$, we then have

$$\mathscr{L}(\boldsymbol{C}_1) \supseteq \varphi(\mathscr{L}(\mathrm{Alt}[X]) = \varphi(\mathscr{J}(\mathrm{Alt}[X])) \ni \varphi(f) = f(c_1, \ldots, c_n) \neq 0,$$

that is, $\mathscr{L}(\boldsymbol{C}_1) \neq (0)$. But as is easy to see, the algebra $\boldsymbol{C}_1$ is a Cayley–Dickson ring, and by the corollary to Proposition 9.3 every Cayley–Dickson ring is nil-semisimple. In particular, $\mathscr{L}(\boldsymbol{C}_1) = (0)$. The obtained contradiction proves that $f \equiv 0$ in $\boldsymbol{C}_1$, that is, $f \in T(\boldsymbol{C})$ and $\mathscr{J}(\mathrm{Alt}[X]) \subseteq T(\boldsymbol{C}) \cap D(\mathrm{Alt}[X])$.

Now let $f = f(x_1, \ldots, x_n) \in T(\boldsymbol{C}) \cap D(\mathrm{Alt}[X])$. If $f \notin \mathscr{J}(\mathrm{Alt}[X])$, then there exists a homomorphism φ of the algebra $\mathrm{Alt}[X]$ onto some primitive algebra A under which

$$\varphi(f(x_1, \ldots, x_n)) = f(\varphi(x_1), \ldots, \varphi(x_n)) \neq 0.$$

Since $D(A) \supseteq \varphi(D(\mathrm{Alt}[X]) \ni \varphi(f) \neq 0$, the algebra A is not associative. Consequently, by Theorem 10.1 A is a Cayley–Dickson algebra over some field F. We recall that A is also an algebra over Φ, and therefore the center F of the algebra A is an algebra over Φ. If F_1 is some extension of the field F, then the field F_1 and the F_1-algebra $A_1 = F_1 \otimes_F A$ can in an obvious way also be considered as algebras over Φ. Now extending the field F to its algebraic closure, if necessary, by the corollary to Theorem 2.6 and Theorem 2.7 we obtain a Cayley–Dickson matrix algebra $\boldsymbol{C}(F_1)$ over the field F_1 in which $f(c_1, \ldots, c_n) \neq 0$ for some elements $c_1, \ldots, c_n$. Let $c_i = \sum_{j=1}^{8} \alpha_{ij} e_j$, where $\alpha_{ij} \in F_1$, $i = 1, 2, \ldots, n$; $j = 1, 2, \ldots, 8$; and $e_1, \ldots, e_8$ are the Cayley–Dickson matrix units. We consider the ring of polynomials $\Phi[t_{ij}]$, $i = 1, 2, \ldots, n$; $j = 1, 2, \ldots, 8$, and the Cayley–Dickson matrix algebra $\boldsymbol{C}(\Phi[t_{ij}])$. There exists a homomorphism of Φ-algebras $\psi : \Phi[t_{ij}] \to F_1$ under which $\psi(t_{ij}) = \alpha_{ij}$. Let $e'_1, \ldots, e'_8$ be the matrix units of the Cayley–Dickson algebra $\boldsymbol{C}(\Phi[t_{ij}])$. Setting

$$\bar{\psi}\left(\sum_{i=1}^{8} \lambda_i e'_i\right) = \sum_{i=1}^{8} \bar{\psi}(\lambda_i) e_i,$$

we obtain an extension of the homomorphism ψ to a homomorphism of Φ-algebras $\bar{\psi} : \boldsymbol{C}(\Phi[t_{ij}]) \to \boldsymbol{C}(F_1)$. Let $y_i = \sum_{j=1}^{8} t_{ij} e'_j$. Then $\bar{\psi}(y_i) = c_i$ and $\bar{\psi}(f(y_1, \ldots, y_n)) = f(c_1, \ldots, c_n) \neq 0$. Hence, it follows that also $f(y_1, \ldots, y_n) \neq 0$. We now recall that, as was noted earlier, the algebra $\boldsymbol{C}(\Phi[t_{ij}]) = \Phi[t_{ij}] \otimes_\Phi \boldsymbol{C}$ satisfies all identities from $T(\boldsymbol{C})$. In particular, the algebra $\boldsymbol{C}(\Phi[t_{ij}])$ satisfies the identity f. The obtained contradiction proves that $f \in \mathscr{J}(\mathrm{Alt}[X])$, that is, $\mathscr{J}(\mathrm{Alt}[X]) \supseteq T(\boldsymbol{C}) \cap D(\mathrm{Alt}[X])$.

This proves the theorem.

Theorem 4 gives us the following criterion for determining whether an element f of the algebra $\mathrm{Alt}[X]$ is a member of the radical $\mathscr{J}(\mathrm{Alt}[X])$.

COROLLARY 1. An element f of the algebra $\mathrm{Alt}[X]$ belongs to the radical $\mathscr{J}(\mathrm{Alt}[X])$ if and only if f together with all its partial linearizations vanish in the free associative algebra $\mathrm{Ass}[X]$ and in the Cayley–Dickson algebra $\boldsymbol{C}$.

COROLLARY 2. Let $f \in \mathscr{J}(\mathrm{Alt}[X])$. Then all partial linearizations of the element f also belong to $\mathscr{J}(\mathrm{Alt}[X])$.

In particular, $\mathscr{J}(\mathrm{Alt}[X])$ is a homogeneous ideal of the algebra $\mathrm{Alt}[X]$, that is, if $f \in \mathscr{J}(\mathrm{Alt}[X])$, then each homogeneous component of the element f also belongs to $\mathscr{J}(\mathrm{Alt}[X])$.

The criterion formulated in Corollary 1 allows us to establish that under certain restrictions on the ring Φ the radical $\mathscr{J}(\mathrm{Alt}[X])$ is different from zero.

COROLLARY 3. Let there be no elements of order less than 8 in the additive group of the ring of operators Φ. Then $\mathscr{J}(\mathrm{Alt}[X]) \neq (0)$.

PROOF. Let $f_n(x_1, \ldots, x_{2n+1})$ be the skew-symmetric function defined in Section 1. It is clear that $f_n(x_1, \ldots, x_{2n+1}) \in D(\mathrm{Alt}[X])$ for all n. On the other hand, since a Cayley–Dickson algebra $\boldsymbol{C}$ is an 8-dimensional module over Φ, then any skew-symmetric multilinear function with $\geqq 9$ arguments equals zero on $\boldsymbol{C}$. In particular, $f_n(x_1, \ldots, x_{2n+1}) \in T(\boldsymbol{C})$ for $n \geqq 4$. By the corollary to Lemma 2 we now obtain that $0 \neq f_4(x_1, \ldots, x_9) \in T(\boldsymbol{C}) \cap D(\text{Alt}[X]) = \mathscr{J}(\text{Alt}[X])$, that is, $\mathscr{J}(\text{Alt}[X]) \neq (0)$. This proves the corollary.

In particular, in a free alternative ring from nine or more generators there are no nonzero nilpotent elements.

REMARK. In fact, $\mathscr{J}(\text{Alt}[X]) \neq (0)$ for any ring of operators Φ. For example, in view of the corollary to Lemma 2.8 it is easy to see that the elements $k = ([x, y]^2, r, s)$ and $k_1 = ([x, y] \circ [z, t], r, s)$ belong to $\mathscr{J}(\text{Alt}[X])$. At the same time, for any ring Φ there exist alternative Φ-algebras in which $k \neq 0$, $k_1 \neq 0$. An appropriate example for the element k_1 was constructed in 1963 by Dorofeev, and for the element k in 1967 by Humm and Kleinfeld. Consequently the elements k and k_1 are different from zero in $\text{Alt}[X]$. Since the element k only depends on four variables, there are already nonzero nilpotent elements contained in free alternative rings from four (or more) generators. We note, too, that as shown by Shestakov the element k is an absolute zero divisor in $\text{Alt}[X]$.

LEMMA 4. Let $K[X] = \text{Alt}[X]/T(\boldsymbol{C})$. Then the algebra $K[X]$ does not contain zero divisors.

PROOF. Let $f_1, f_2 \in \text{Alt}[X]$ and $f_1 \cdot f_2 \in T(\boldsymbol{C})$. It is necessary for us to show that then either $f_1 \in T(\boldsymbol{C})$ or $f_2 \in T(\boldsymbol{C})$. Let Φ' be the field of quotients of the ring Φ. By Theorem 2.8 there exist an infinite extension F of the field Φ' and a Cayley–Dickson algebra A over F which is a division ring. Let F_1 be the algebraic closure of the field F. Then the algebra $A_1 = F_1 \otimes_F A$ is a Cayley–Dickson matrix algebra over F_1, whence $A_1 = \boldsymbol{C}(F_1) = F_1 \otimes_\Phi \boldsymbol{C}$. Any element from $T(\boldsymbol{C})$ is an identity of the algebra A_1, and consequently, also of the algebra A. In particular, the identity $f_1(x_1, \ldots, x_n) \cdot f_2(x_1, \ldots, x_n)$ is valid in the algebra A. Since there are no zero divisors in A, for any elements $a_1, \ldots, a_n \in A$ either $f_1(a_1, \ldots, a_n) = 0$ or $f_2(a_1, \ldots, a_n) = 0$. Now let I_1 and I_2 be the ideals of the algebra $\text{Alt}[X]$ generated by the elements f_1 and f_2, respectively. Then all identities of the form $f \cdot g = 0$, where $f \in I_1$, $g \in I_2$, are valid in the algebra A. Moreover, since the field F is infinite, all these identities remain valid in the algebra $F' \otimes_F A$ for any extension F' of the field F. Let $F' = F[t_{ij}]$, $i = 1, \ldots, n$; $j = 1, \ldots, 8$. We then consider in the algebra $A' = F' \otimes_F A$ the elements

$$y_i = \sum_{j=1}^{8} t_{ij}(1 \otimes e_j), \qquad i = 1, 2, \ldots, n,$$

where $e_1, e_2, \ldots, e_8$ is some basis for the algebra A over F. Let $\bar{I}_1$ and $\bar{I}_2$ be the ideals of the algebra A' generated by the elements $f_1(y_1, \ldots, y_n)$ and

$f_2(y_1, \ldots, y_n)$, respectively. Then by what was proved above we have $\bar{I}_1 \cdot \bar{I}_2 = (0)$. But the algebra A' is a Cayley–Dickson ring, and therefore A' is prime. Thus either $\bar{I}_1 = (0)$ or $\bar{I}_2 = (0)$, and consequently, either $f_1(y_1, \ldots, y_n) = 0$ or $f_2(y_1, \ldots, y_n) = 0$. We suppose that $f_1(y_1, \ldots, y_n) = 0$, and let $a_i = \sum_{j=1}^{8} \alpha_{ij} e_j$, $i = 1, \ldots, n$, be arbitrary elements of the algebra A. Considering the specializations $t_{ij} \to \alpha_{ij}$, as in the proof of Theorem 4 we obtain a homomorphism $\varphi: A' \to A$ for which $\varphi(y_i) = a_i$. But then

$$f_1(a_1, \ldots, a_n) = \varphi(f_1(y_1, \ldots, y_n)) = 0,$$

and since the elements a_i are arbitrary it follows that the algebra A satisfies the identity f_1. Since the field F is infinite, the algebra $A_1 = \boldsymbol{C}(F_1)$, where F_1 is the algebraic closure of the field F, also satisfies the identity f_1. What is more, since F_1 is infinite all partial linearizations of the element f_1 also vanish in the algebra A_1. Since $\boldsymbol{C} \subseteq A_1$, it follows from this that $f_1 \in T(\boldsymbol{C})$. This proves the lemma.

COROLLARY 1. Let $K[X_n] = \mathrm{Alt}[X_n]/(\mathrm{Alt}[X_n] \cap T(\boldsymbol{C}))$. Then the algebra $\mathrm{K}[X_n]$ does not contain zero divisors.

We shall call an algebra S with set of generators Y a *free Cayley–Dickson ring* if S is a Cayley–Dickson ring and for any algebra A which is a Cayley–Dickson ring any mapping of the set Y into A can be extended uniquely to a homomorphism of the algebra S into A.

COROLLARY 2. Let $\mathrm{card}(Y) \geqq 3$. Then the algebra $K[Y]$ is a free Cayley–Dickson ring from set of generators Y.

PROOF. Since the algebra $\mathrm{K}[Y]$ does not have zero divisors and is not associative, by Theorem 9.9 $\mathrm{K}[Y]$ is a Cayley–Dickson ring. In addition, as is easy to see, $\mathrm{K}[Y]$ is a free algebra from set of generators Y in the variety of Φ-algebras $\mathrm{Var}(\boldsymbol{C})$ generated by the Cayley–Dickson matrix algebra $\boldsymbol{C}$. For completion of the proof it suffices now to show that any algebra A which is a Cayley–Dickson ring satisfies all identities from $T(\boldsymbol{C})$. Let Z be the center of the algebra A and Z_1 be the field of quotients of the ring Z. Then Z and Z_1 are algebras over Φ, and the Cayley–Dickson algebra $A_1 = (Z^*)^{-1}A$ over the field Z_1 is also an algebra over Φ. Then reasoning as in the second half of the proof of Theorem 4, we obtain that the Φ-algebra A_1 satisfies all identities from $T(\boldsymbol{C})$. Since $A \subseteq A_1$, by the same token all is proved.

LEMMA 5. The algebra $\mathrm{K}[X]$ does not contain nonzero quasi-regular elements.

PROOF. Since $T(\boldsymbol{C})$ is a homogeneous ideal of the algebra $\mathrm{Alt}[X]$, the concepts of degree of a polynomial, homogeneous polynomial, etc., make sense in the algebra $\mathrm{K}[X]$. Let f be a quasi-regular element of the algebra

K$[X]$ and g be a quasi-inverse element for f. Then

$$f + g = fg. \tag{7}$$

If $f \neq 0$, then also $g \neq 0$, and we have

$$f = \sum_{k=1}^{n} f_k, \qquad g = \sum_{k=1}^{m} g_k; \qquad f_n \neq 0, \quad g_m \neq 0,$$

where by f_k and g_k are denoted the sums of all monomials of degree k appearing in the polynomials f and g, respectively. Comparing terms of the same degree in equality (7), we obtain $f_n g_m = 0$, whence by the absence of zero divisors in the algebra K$[X]$ either $f_n = 0$ or $g_m = 0$. The obtained contradiction proves that $f = 0$. This proves the lemma.

COROLLARY. $\mathscr{J}(\mathrm{K}[X]) = (0)$.

We shall now prove the following result.

THEOREM 5 (*Shestakov, Slater*). The Zhevlakov radical $\mathscr{J}(\mathrm{Alt}[X])$ of a free alternative algebra $\mathrm{Alt}[X]$ coincides with the collection of all nilpotent elements of the algebra $\mathrm{Alt}[X]$. The quotient algebra $\overline{\mathrm{Alt}}[X] = \mathrm{Alt}[X]/\mathscr{J}(\mathrm{Alt}[X])$ is isomorphic to a subdirect sum of the free associative algebra $\mathrm{Ass}[X]$ and the free Cayley–Dickson ring K$[X]$.

PROOF. We shall first prove the second half of the theorem. By Theorem 4 the algebra $\overline{\mathrm{Alt}}[X]$ is isomorphic to a subdirect sum of the algebras $\mathrm{Alt}[X]/D(\mathrm{Alt}[X])$ and $\mathrm{K}[X] = \mathrm{Alt}[X]/T(\boldsymbol{C})$. It is clear that

$$\mathrm{Alt}[X]/D(\mathrm{Alt}[X]) \cong \mathrm{Ass}[X].$$

On the other hand, by Corollary 2 to Lemma 4 K$[X]$ is a free Cayley-Dickson ring.

Since by Lemma 3 $\mathscr{J}(\mathrm{Alt}[X])$ is a nil-ideal, for the completion of the proof it suffices to show there are no nilpotent elements in the algebra $\overline{\mathrm{Alt}}[X]$. But this follows from the already proved second half of the theorem, since there are no nilpotent elements in the algebras $\mathrm{Ass}[X]$ and K$[X]$. This proves the theorem.

COROLLARY.

$$\mathscr{J}(\mathrm{Alt}[X_n]) = \mathscr{J}(\mathrm{Alt}[X]) \cap \mathrm{Alt}[X_n] = T(\boldsymbol{C}) \cap D(\mathrm{Alt}[X_n]).$$

PROOF. First of all we note that $\mathscr{J}(\mathrm{Alt}[X]) \cap \mathrm{Alt}[X_n]$ is a nil-ideal of the algebra $\mathrm{Alt}[X_n]$, and therefore $\mathscr{J}(\mathrm{Alt}[X]) \cap \mathrm{Alt}[X_n] \subseteq \mathscr{J}(\mathrm{Alt}[X_n])$. For the proof of the containment $\mathscr{J}(\mathrm{Alt}[X_n]) \subseteq \mathscr{J}(\mathrm{Alt}[X])$, it suffices to show that the quotient algebra $\overline{\mathrm{Alt}}[X] = \mathrm{Alt}[X]/\mathscr{J}(\mathrm{Alt}[X])$ does not contain nonzero quasi-regular elements. By Theorem 5 the algebra $\overline{\mathrm{Alt}}[X]$

is isomorphic to a subdirect sum of the algebras $\mathrm{Ass}[X]$ and $\mathrm{K}[X]$, each of which does not contain nonzero quasi-regular elements (the proof of Lemma 5 is also suitable verbatim for the algebra $\mathrm{Ass}[X]$). Consequently, this property is also satisfied by the algebra $\overline{\mathrm{Alt}}[X]$, and we have $\mathscr{J}(\mathrm{Alt}[X_n]) \subseteq \mathscr{J}(\mathrm{Alt}[X])$. This proves the first of the required equalities. The second equality follows from Theorem 5 and the equality $D(\mathrm{Alt}[X_n]) = D(\mathrm{Alt}[X]) \cap \mathrm{Alt}[X_n]$. This proves the corollary.

From the just proved corollary and Corollaries 1 and 2 to Lemma 4, it follows that results analogous to Theorems 4 and 5 are also valid for the finitely generated free alternative algebra $\mathrm{Alt}[X_n]$ (for $n \leqq 2$ they are trivial). In particular, the remark made after Corollary 3 to Theorem 4 shows that $\mathscr{J}(\mathrm{Alt}[X_n]) \neq (0)$ for $n \geqq 4$. On the other hand, by Artin's theorem $\mathrm{Alt}[X_n] \cong \mathrm{Ass}[X_n]$ for $n \leqq 2$, and therefore $\mathscr{J}(\mathrm{Alt}[X_n]) = (0)$ for $n \leqq 2$. The question remains open on the validity of the equality $\mathscr{J}(\mathrm{Alt}[X_3]) = (0)$. By what was proved above, this question is equivalent to the question on the presence of nonzero nilpotent elements in the algebra $\mathrm{Alt}[X_3]$. We note that from Exercise 6 in Section 7.1 it follows that the algebra $\mathrm{Alt}[X_3]$ contains nonzero divisors of zero. What is more, as we shall see in the following section, the algebra $\mathrm{Alt}[X_3]$ is not prime.

In concluding this section we obtain yet one more characterization of the Zhevlakov radical $\mathscr{J}(\mathrm{Alt}[X_n])$ of a free finitely generated alternative algebra $\mathrm{Alt}[X_n]$.

We need the following.

LEMMA 6. Any element from the right multiplication algebra $R(\mathrm{Alt}[X])$ of the algebra $\mathrm{Alt}[X]$ is a linear combination of elements of the form

$$R_{u_1}R_{u_2}\cdots R_{u_n}R_x, \tag{8}$$

where x is an arbitrary element of the algebra $\mathrm{Alt}[X]$ and $u_1, u_2, \ldots, u_n$ are mutually distinct regular r_1-words from the set X. In addition, the operators $R_{u_1}, R_{u_2}, \ldots, R_{u_n}$ can be missing.

PROOF. Let I_n be the submodule of the Φ-module $R(\mathrm{Alt}[X])$ generated by the set of elements of the form $R_{a_1}R_{a_2}\cdots R_{a_n}$, where $a_i \in \mathrm{Alt}[X]$. Then $R(\mathrm{Alt}[X]) = \sum_{n=1}^{\infty} I_n$. We denote by J_n the submodule of the Φ-module I_n generated by all elements of the form (8) that are in I_n. It is necessary for us to prove that $R(\mathrm{Alt}[X]) = \sum_{n=1}^{\infty} J_n$. Since $I_1 = J_1$, it suffices for this to establish that for any $n > 0$

$$I_{n+1} \subseteq J_{n+1} + \sum_{k=1}^{n} I_k. \tag{9}$$

We consider the quotient module $R_{n+1} = R(\mathrm{Alt}[X])/\sum_{k=1}^{n} I_k$, and the canonical homomorphism $\Pi: R(\mathrm{Alt}[X]) \to R_{n+1}$. For any elements $a_1, \ldots, a_n \in \mathrm{Alt}[X]$ we set

$$f(a_1, \ldots, a_n) = \{\Pi(R_{a_1} \cdots R_{a_n} R_x) \mid x \in \mathrm{Alt}[X]\}.$$

It is easy to see that $f(a_1, \ldots, a_n)$ is a submodule of the Φ-module $\Pi(I_{n+1})$, and also

$$\Pi(I_{n+1}) = \sum_{a_1, \ldots, a_n \in \mathrm{Alt}[X]} f(a_1, \ldots, a_n).$$

Consequently, for the proof of containment (9) it suffices for us to establish that $f(a_1, \ldots, a_n) \subseteq \Pi(J_{n+1})$ for any $a_1, \ldots, a_n \in \mathrm{Alt}[X]$. We shall show that the mapping f satisfies conditions (a)–(e) of Proposition 1. In view of the right alternative identity, we have

$$R_x^2 = R_{x^2}, \qquad R_x \circ R_y = R_{x \circ y},$$

whence it follows that f satisfies conditions (a) and (b). Conditions (c) and (d) are obvious. Finally, from relation (7.18) we obtain

$$R_{yxy} R_t \equiv R_x R_{yty} + R_y R_{\{xyt\}} \qquad (\mathrm{mod}\, I_1),$$

whence it follows that f also satisfies condition (e). By Proposition 1 we now obtain that $f(a_1, \ldots, a_n) \subseteq \Pi(J_{n+1})$. This proves the lemma.

COROLLARY. Any element from the algebra $R(\mathrm{Alt}[X_m])$ is a linear combination of elements of the form (8), where $n < 2^m$.

Actually, in this case by the corollary to Proposition 1 $\Pi(I_n) = (0)$ for $n > 2^m$, and therefore $R(\mathrm{Alt}[X_m]) = \sum_{k=1}^{2^m} J_k$.

THEOREM 6 (*Shestakov*). In the algebra $\mathrm{Alt}[X_m]$ any element $f \in \mathscr{J}(\mathrm{Alt}[X_m])$ generates a right nil-ideal of bounded index.

PROOF. We denote by J the right ideal generated by the element f. Then $J = \Phi f + J_1$, where $J_1 = \{fW \mid W \in R(\mathrm{Alt}[X_m])\}$. Since $J^2 \subseteq J_1$, it suffices for us to show that the right ideal J_1 is a nil-ideal of bounded index. Let $u_1, \ldots, u_k$ be some ordered collection of distinct regular r_1-words from $\mathrm{Alt}[X_m]$. We associate to each such collection a generator x_j different from $x_1, \ldots, x_m$, so that to different collections are corresponded different generators, and we set

$$F(u_1, u_2, \ldots, u_k) = f R_{u_1} R_{u_2} \cdots R_{u_k} R_{x_j},$$

$$F = \sum_{k=0}^{2^m - 1} \sum_i F(u_1^{(i)}, u_2^{(i)}, \ldots, u_k^{(i)}),$$

where $u_1^{(i)}, \ldots, u_k^{(i)}$ are arbitrary mutually distinct regular r_1-words from the elements $x_1, \ldots, x_m$. By the corollary to Lemma 6, each element of the ideal J_1 is the image of an element F under a suitable endomorphism of the algebra Alt[X]. Since by the corollary to Theorem 5 $\mathscr{J}(\mathrm{Alt}[X_m]) \subseteq \mathscr{J}(\mathrm{Alt}[X]$, then $f \in \mathscr{J}(\mathrm{Alt}[X])$. Hence it follows that also $F \in \mathscr{J}(\mathrm{Alt}[X])$, and therefore the element F is nilpotent. Let $F^n = 0$. Then J_1 is a nil-algebra of index n, and J is a nil-algebra of index $2n$. This proves the theorem.

COROLLARY 1. The Zhevlakov radical $\mathscr{J}(\mathrm{Alt}[X_m])$ of the algebra Alt[X_m] equals the sum of all right nil-ideals of bounded index of the algebra Alt[X_m].

In the case when Φ is a field of characteristic 0, a stronger assertion is valid.

COROLLARY 2. The Zhevlakov radical of a free finitely generated alternative algebra over a field of characteristic 0 equals the sum of all solvable right ideals of this algebra.

Actually, by Zhevlakov's theorem (Theorem 6.2) every alternative nil-algebra of bounded index over a field of characteristic 0 is solvable.

We turn now to two-sided ideals.

Everywhere below A is an alternative algebra, I is a right ideal of the algebra A, and $\hat{I} = A^{\#}I$ is the two-sided ideal of the algebra A generated by the set I. As in Chapter 12, for subalgebras B, C of the algebra A we set $(B, C)_1 = BC$, $(B, C)_{k+1} = (B, C)_k \cdot C$. If $B + I = C + I$, then we shall write $B \equiv C$. Analogously, if $a, b \in A$ and $a - b \in I$, then we shall write $a \equiv b$. We note that if $a \equiv b$, then $ax \equiv bx$ for any $x \in A$.

LEMMA 7. Let $v(x_1, \ldots, x_n)$ be an arbitrary multilinear monomial of degree n, and $v(I)$ be the submodule of the Φ-module A generated by the valuations $v(y_1, \ldots, y_n)$ where $y_1, \ldots, y_n \in I$. Then for any subalgebra C of the algebra A

$$C \cdot v(I) \equiv (C, I)_n.$$

PROOF. We carry out an induction on the degree n of the monomial v. The basis for the induction, $n = 1$, is clear. Now let $n > 1$. Then $v = v_1 v_2$ where $d(v_1) = n_1$, $d(v_2) = n_2$, and $n_i < n$. By the induction assumption we have

$$\begin{aligned} C \cdot v(I) &= C \cdot (v_1(I) \cdot v_2(I)) \subseteq (Cv_1(I))v_2(I) + (C, v_1(I), v_2(I)) \\ &\equiv ((C, I)_{n_1}, I)_{n_2} + (v_1(I), C, v_2(I)) \equiv (C, I)_n. \end{aligned}$$

The same reasoning carried out in reverse order shows that $(C, I)_n \subseteq C \cdot v(I) + I$. This proves the lemma.

COROLLARY. $A \cdot I^{2^n} \equiv A \cdot I^{(n)}$.

LEMMA 8. $\hat{I}^{\langle n \rangle} \subseteq AI^n + I$.

PROOF. We again carry out an induction on n. The basis for the induction, $n = 1$, is obvious. Suppose it is already proved that

$$\hat{I}^{\langle n-1 \rangle} \subseteq AI^{n-1} + I.$$

Then we have

$$\hat{I}^{\langle n \rangle} = \hat{I}^{\langle n-1 \rangle}\hat{I} \subseteq (AI^{n-1} + I)(A^{\#}I) \subseteq I + (AI^{n-1})(A^{\#}I).$$

We consider

$$\begin{aligned}(AI^{n-1})(A^{\#}I) &\subseteq (AI^{n-1} \cdot A^{\#})I + (AI^{n-1}, A^{\#}, I) \\ &\subseteq (AI^{n-1} \cdot A^{\#})I + (I, AI^{n-1}, A^{\#}) \equiv (AI^{n-1} \cdot A^{\#})I.\end{aligned}$$

By Lemma 7 we obtain

$$AI^{n-1} \equiv (A, I)_{n-1} = (A, I)_{n-2} \cdot I \equiv (AI^{n-2})I,$$

whence

$$\begin{aligned}AI^{n-1} \cdot A^{\#} &\equiv (AI^{n-2} \cdot I)A^{\#} \subseteq AI^{n-2} \cdot IA^{\#} + (AI^{n-2}, I, A^{\#}) \\ &\subseteq AI^{n-2} \cdot I + (I, A^{\#}, AI^{n-2}) \equiv AI^{n-1}.\end{aligned}$$

Consequently,

$$(AI^{n-1})(A^{\#}I) \subseteq (AI^{n-1} \cdot A^{\#})I + I \subseteq AI^{n-1} \cdot I + I \subseteq AI^n + I.$$

This proves the lemma.

PROPOSITION 2. Let the right ideal I of an alternative algebra A be solvable of index n. Then the two-sided ideal $\hat{I}$ is solvable of index $\leqq 3n$.

PROOF. By Lemma 8 and the corollary to Lemma 7 we obtain

$$\hat{I}^{\langle 2^n \rangle} \subseteq AI^{2^n} + I \subseteq AI^{(n)} + I = I.$$

Now by the corollary to Theorem 5.5 we have

$$\hat{I}^{(2^n)^2} = \hat{I}^{2^{2n}} \subseteq I,$$

whence

$$\hat{I}^{(2n)} \subseteq \hat{I}^{2^{2n}} \subseteq I,$$

and furthermore

$$\hat{I}^{(3n)} = (\hat{I}^{(2n)})^{(n)} \subseteq I^{(n)} = (0).$$

This proves the proposition.

From Corollary 2 to Theorem 6, Proposition 2, and the corollary to Theorem 3 follows

THEOREM 7 (*Shestakov*). The Zhevlakov radical of a free finitely generated alternative algebra over a field of characteristic 0 equals the sum of all two-sided nilpotent ideals of this algebra.

COROLLARY. In a free finitely generated alternative algebra over a field of characteristic 0, every nilpotent element generates a nilpotent ideal.

In concluding this section we formulate some still open questions.

1. Find a basis for the identities of a Cayley–Dickson matrix algebra **C**.
2. Describe $\mathscr{J}(\text{Alt}[X])$ as a completely characteristic ideal.
3. Is the ideal $\mathscr{J}(\text{Alt}[X_n])$ nilpotent?

3. CENTERS OF ALTERNATIVE ALGEBRAS

One of the important directions in the study of free alternative algebras is the study of the centers of these algebras. We already know a series of nonzero elements from the associative center N_{Alt} of a free alternative algebra $\text{Alt}[X]$ (see Section 7.2). In this section we shall prove that the center Z_{Alt} of the algebra $\text{Alt}[X]$ is also different from zero. We shall also continue the investigation of the question on coincidence of the centers $N(A)$ and $Z(A)$ in an alternative algebra A, which was carried out earlier for simple and prime algebras. We shall prove that if $\frac{1}{2} \in \Phi$, then $N(A) = Z(A)$ in any purely alternative algebra A. Finally, we shall prove that various centers of a free algebra of an arbitrary homogeneous variety are completely characteristic subsets (i.e., are invariant under all endomorphisms of this algebra).

Everywhere below Φ is an arbitrary associative–commutative ring with identity element 1, A is an alternative Φ-algebra, $N = N(A)$, $Z = Z(A)$, $D = D(A)$, and $U = U(A)$. We recall that the ideal of the algebra A generated by the set $[N, A]$ is denoted by $ZN(A)$. By Lemma 7.4

$$ZN(A) = [N, A]A^{\#} = A^{\#}[N, A].$$

LEMMA 9. Let $n \in N$, $a, b, c, d \in A$. Then the element $[n, a](b, c, d)$ is a skew-symmetric function of the elements a, b, c, d.

PROOF. In view of the skew-symmetry of the associator, it suffices for us to prove that for any $a, b, c \in A$

$$[n, a](a, b, c) = 0.$$

According to Corollaries 1 and 2 to Lemma 7.3 and relation (7.4), we have

$$\begin{aligned}[n,a](a,b,c) &= (a,[n,a]b,c) = (a,[n,ab],c) - (a,a[n,b],c) \\ &= -(a,[n,b],c)a = 0.\end{aligned}$$

This proves the lemma.

LEMMA 10 (*Dorofeev, Shelipov*). Let $n \in N$, a, b, c, r, s, t, $u \in A$. Then the equality

$$2(a,b,c)(r,s,t)[n,u] = 0$$

is valid in A.

PROOF. We set $d = (a,b,c)$, $d_1 = (r,s,t)$. We note first that the equality

$$d[n,u] = [n,du] = [n,u]d \tag{10}$$

is valid in A. In fact, in view of (7.4) and (7.10) we have

$$d[n,u] = [n,du] - [n,d]u = [n,du].$$

On the other hand, since $[n,u] \in N$ by (7.10)

$$d[n,u] = [n,u]d.$$

Now according to (10) we obtain

$$d_1 d[n,u] = d_1[n,du] = [n,du]d_1 = d[n,u]d_1 = dd_1[n,u].$$

By Lemma 9 the product $x = (a,b,c)(r,s,t)[n,u]$ is skew-symmetric in the variables r, s, t, u. By what has been proved $x = (r,s,t)(a,b,c)[n,u]$, and therefore the indicated product x is also skew-symmetric in the variables a, b, c, u. As a result, we obtain that x is skew-symmetric in all the variables a, b, c, r, s, t, u. In particular, we have

$$\begin{aligned}(r,s,t)(a,b,c)[n,u] &= -(a,s,t)(r,b,c)[n,u] \\ &= (a,b,t)(r,s,c)[n,u] = -(a,b,c)(r,s,t)[n,u].\end{aligned}$$

Comparing this relation with the one proved above, we obtain the assertion of the lemma.

THEOREM 8 (*Dorofeev*). In any alternative algebra A there is the containment $2D \cdot ZN(A) \subseteq U$.

PROOF. By Proposition 5.1 the set $2D \cdot ZN(A)$ is an ideal of the algebra A, and therefore it suffices for us to prove that $2D \cdot ZN(A) \subseteq N$. An arbitrary element of the ideal $2D \cdot ZN(A)$ is a linear combination of elements of the form $2(x,y,z)u \cdot v[t,n]$ where $n \in N$, u, $v \in A^{\#}$, x, y, z, $t \in A$. Since

$[t,n] \in N$, we have

$$2(x,y,z)u \cdot v[t,n] = 2[(x,y,z)u]v \cdot [t,n]$$
$$= 2((x,y,z),u,v)[t,n] + 2(x,y,z)(uv) \cdot [t,n].$$

The first of the summands obtained equals zero by Lemma 7.5, and for the second by Lemmas 9 and 10 and in view of (10) we have

$$2((x,y,z)w \cdot [t,n], r, s) = 2((x,y,z)w, r, s)[t,n]$$
$$= -2(t,r,s)[(x,y,z)w, n] = -2(t,r,s)(x,y,z)[w,n] = 0.$$

Consequently, $2D \cdot ZN(A) \subseteq N$. This proves the theorem.

COROLLARY 1. If A is a separated alternative algebra, then

$$2ZN(A) \subseteq N(A).$$

PROOF. We recall that an algebra A is called separated if $D(A) \cap U(A) = (0)$. Under the conditions of the corollary, it follows from Theorem 8 that $2D \cdot ZN(A) \subseteq D \cap U = (0)$. In particular, for any $a, b, c, d \in A$, $n \in N$ we have

$$2(a,b,[n,c]d) = 2(a,b,d)[n,c] = 0,$$

whence it follows that $2ZN(A) \subseteq N$. This proves the corollary.

COROLLARY 2 (*Dorofeev, Shelipov*). If A is a purely alternative algebra, then $2N \subseteq Z$. In particular, if $\frac{1}{2} \in \Phi$ then under the conditions of the corollary $N = Z$.

Actually, a purely alternative algebra is separated, and therefore $2ZN(A) \subseteq N$. Then $2ZN(A)$ is a nuclear ideal of the algebra A, that is, $2ZN(A) \subseteq U(A) = (0)$. But this is equivalent to the containment $2N \subseteq Z$.

THEOREM 9 (*Dorofeev*). In any alternative algebra A there is the containment $2(N \cap D^2)^2 \subseteq Z$.

PROOF. Let $a, b \in N \cap D^2$, $c \in A$. Then according to (7.4)

$$2[ab,c] = 2a[b,c] + 2[a,c]b \in 2D^2 \cdot ZN(A) + 2ZN(A) \cdot D^2, \tag{11}$$

and we shall prove that $2D^2 \cdot ZN(A) = 2ZN(A) \cdot D^2 = (0)$. In fact, for any $x, y, z, t \in A$, $s, u \in A^{\#}$, $n \in N$, $d \in D$, according to the corollary to Lemma 7.3 and Theorem 8 we obtain

$$2(s[t,n])((x,y,z)u \cdot d) = 2s[([t,n](x,y,z)u \cdot d]$$
$$= 2s((x,y,z)[t,n]u \cdot d) \in A^{\#}[(2D \cdot ZN(A))D]$$
$$\subseteq U \cdot D = (0),$$

whence it follows that $2ZN(A) \cdot D^2 = (0)$. Furthermore, for any $c, d \in D$ we have

$$2(cd) \cdot [t,n]s = 2(cd \cdot [t,n])s$$
$$= 2(c \cdot d[t,n])s \in [D \cdot (2D \cdot ZN(A))]A^{\#} \subseteq D \cdot U = (0),$$

whence $2D^2 \cdot ZN(A) = (0)$. Now from (11) we obtain $2[ab,c] = 0$, that is, $2(N \cap D^2)^2 \subseteq Z$, which is what was to be proved.

COROLLARY 1 (*Dorofeev, Shelipov*). For any $x, y, z, t \in A$ the element $[(x,y,z),t]^8$ belongs to Z.

PROOF. We recall that by Kleinfeld's identities $[x,y]^4 \in N$ for any $x, y \in A$. Consequently $[(x,y,z),t]^8 \in (N \cap D^2)^2$. If Φ contains $\frac{1}{2}$, the assertion of the corollary now follows directly from Theorem 9. We shall prove that this assertion is true in the general case as well. Let $d = (z,y,z)$. Then for any $t, r \in A$ we have

$$[[d,t]^8, r] = [d,t]^4 \circ [[d,t]^4, r]$$
$$= 2[d,t]^4[[d,t]^4, r] + [[d,t]^4, [[d,t]^4, r]].$$

Since $2[d,t]^4[[d,t]^4,r] \in 2D^2 \cdot ZN(A) = (0)$, it suffices for us to prove that $[[d,t]^4, n] = 0$ for any $n \in N$. In view of identity (3.2) we have

$$[d,t] = (xy,z,t) + (yz,x,t) + (zx,y,t),$$

whence by (7.10) it follows that $[[d,t],n] = 0$. It is clear that then $[[d,t]^4,n] = 0$ also. This proves the corollary.

COROLLARY 2. $Z_{\mathrm{Alt}} \neq (0)$.

Actually, as is easy to see, the element $f = [(x_1,x_2,x_3),x_4]^8$ is not an identity in a Cayley–Dickson algebra. This means $f \neq 0$ in the algebra $\mathrm{Alt}[X]$. At the same time, by Corollary 1 $f \in Z_{\mathrm{Alt}}$.

REMARK. Since $K_{\mathrm{Alt}} \supseteq Z_{\mathrm{Alt}}$, from Corollary 2 it follows that $K_{\mathrm{Alt}} \neq (0)$. In fact, there are contained in K_{Alt} nonzero elements of a more simple form than considered above. For example, Shestakov proved [270] that if $\frac{1}{2} \in \Phi$ then $(x_1,x_2,x_3)^4 \in K_{\mathrm{Alt}}$. We note that the corresponding identity $[(x,y,z)^4,t] = 0$ is in a definite sense the "dual" to Kleinfeld's identity $([x,y]^4,z,t) = 0$. By Corollary 1 to Lemma 7.1 we have the inclusion $3(x_1,x_2,x_3)^4 \in Z_{\mathrm{Alt}}$. It is unknown whether the inclusion $(x_1,x_2,x_3)^4 \in Z_{\mathrm{Alt}}$ is true. More generally, it is still unknown whether the containment $Z_{\mathrm{Alt}} \subseteq K_{\mathrm{Alt}}$ is strict. If $Z_{\mathrm{Alt}} \neq K_{\mathrm{Alt}}$, then it is easy to see that there are found nonzero elements of order 3 in the additive group of the algebra $\mathrm{Alt}[X]$. In this connection, we note that the question on the presence of torsion in the additive group of the algebra Alt$[X]$ is also open. A more general formulation of this question is: Is the algebra Alt$[X]$ a free Φ-module? If this is so, then it would be interesting to find some concrete basis for this Φ-module.

We now consider the ideals $ZN_{\text{Alt}} = ZN(\text{Alt}[X])$ and $U_{\text{Alt}} = U(\text{Alt}[X])$ in the algebra Alt$[X]$. We have $0 \neq [[x_1, x_2]^4, x_3] \in ZN_{\text{Alt}}$, and therefore $ZN_{\text{Alt}} \neq (0)$. We shall now show that also $U_{\text{Alt}} \neq (0)$.

LEMMA 11 (*Slater*). Let A be an arbitrary alternative algebra. Then for any elements $m, n \in N(A)$ the element $[m, n]$ belongs to $U(A)$.

PROOF. Let $a, b \in A$, $c \in A^{\#}$. Then according to Lemma 9 we have

$$([m, n]c, a, b) = [m, n](c, a, b) = -[m, c](n, a, b) = 0.$$

This means $[m, n]A^{\#} \subseteq N(A)$, whence by Proposition 8.9 $[m, n] \in U(A)$. This proves the lemma.

COROLLARY 1. $U_{\text{Alt}} \neq (0)$.

Actually, we have $0 \neq [[x_1, x_2]^4, [x_3, x_4]^4] \in U_{\text{Alt}}$.

COROLLARY 2. A free alternative algebra from three or more generators is not prime.

PROOF. We consider a free alternative algebra Alt$[Y]$, where card$(Y) \geqq 3$. It is then obvious $D(\text{Alt}[Y]) \neq (0)$. On the other hand, for any three different elements $y_1, y_2, y_3 \in Y$ we have $0 \neq [[y_1, y_2]^4, [y_1, y_3]^4] \in U(\text{Alt}[Y])$, that is, $U(\text{Alt}[Y]) \neq (0)$. Since $D(\text{Alt}[Y]) \cdot U(\text{Alt}[Y]) = (0)$, by the same token everything is proved.

We note here the following interesting open question: Is the algebra Alt$[X]$ separated, i.e., is it true that $U_{\text{Alt}} \cap D(\text{Alt}[X]) = (0)$? If this is so, then by Corollary 1 to Theorem 8 we would have the containment $2ZN_{\text{Alt}} \subseteq N_{\text{Alt}}$, on the validity of which there is also nothing yet known.

We shall now prove the following

PROPOSITION 3. The subalgebras N_{Alt}, Z_{Alt} and ideals U_{Alt}, ZN_{Alt} of the algebra Alt$[X]$ are completely characteristic (i.e., they are invariant under all endomorphisms of the algebra Alt$[X]$).

PROOF. We shall prove, for example, that the subalgebra N_{Alt} is completely characteristic. Let $n = n(x_1, \ldots, x_m) \in N_{\text{Alt}}$. It is necessary for us to show that $n(a_1, \ldots, a_m) \in N_{\text{Alt}}$ for any elements $a_1, \ldots, a_m \in \text{Alt}[X]$. Let

x_i, x_j be free generators which do not appear in the listing of the elements $a_1, \ldots, a_m$ and which are different from $x_1, \ldots, x_m$. We consider an endomorphism φ of the algebra Alt$[X]$ under which $\varphi(x_k) = a_k$ for $k = 1, \ldots, m$, $\varphi(x_i) = x_i$, $\varphi(x_j) = x_j$. Since $n \in N_{\text{Alt}}$, we have $(n, x_i, x_j) = 0$. Applying the endomorphism φ to this equality we obtain

$$0 = \varphi((n, x_i, x_j)) = (\varphi(n), \varphi(x_i), \varphi(x_j)) = (n(a_1, \ldots, a_m), x_i, x_j),$$

whence $n(a_1, \ldots, a_m) \in N_{\text{Alt}}$. The remaining cases are considered analogously. For the consideration of U_{Alt} it is necessary to use Proposition 8.9. This proves the proposition.

In connection with this proposition there arises the natural question of determining a system of generating elements for the completely characteristic subalgebras N_{Alt} and Z_{Alt}, and also for the T-ideals ZN_{Alt} and U_{Alt}. It is not clear as yet whether there exists a finite system of generating elements in even one of the cases considered.

We note that in the proof of Proposition 3 the infiniteness of the set X was used substantially. For a finitely generated free algebra Alt$[X_n]$ an analogous assertion is proved, but not so simply. We devote the remaining part of this section to that proof.

Let $f = f(x_1, \ldots, x_n) \in \Phi[X]$ be an arbitrary nonassociative polynomial. For a collection of natural numbers

$$[r] = [r_{11}, \ldots, r_{1s_1}; \ldots; r_{n1}, \ldots, r_{ns_n}] \qquad (s_j \geqq 0, \quad j = 1, 2, \ldots, n),$$

we denote by $f^{[r]} = f^{[r]}(x_1, \ldots, x_n; x_{11}, \ldots, x_{1s_1}; \ldots; x_{n1}, \ldots, x_{ns_n})$ the element

$$f\,\Delta_1^{r_{11}}(x_{11}) \cdots \Delta_1^{r_{1s_1}}(x_{1s_1}) \cdots \Delta_n^{r_{n1}}(x_{n1}) \cdots \Delta_n^{r_{ns_n}}(x_{ns_n}), \tag{12}$$

where $x_{ij} \in X \backslash \{x_1, \ldots, x_n; x_{11}, \ldots, x_{1s_1}; \ldots; x_{i1}, \ldots, x_{i(j-1)}\}$. It is easy to be convinced of the validity of the relations

$$f\,\Delta_i^r(x_k)\,\Delta_j^s(x_l) = f\,\Delta_j^s(x_l)\,\Delta_i^r(x_k), \tag{13}$$

$$f\,\Delta_i^r(x_k)\,\Delta_k^s(x_l) = f\,\Delta_i^{r-s}(x_k)\,\Delta_i^s(x_l), \tag{14}$$

where $x_i \neq x_j$; $x_i, x_j \in \{x_1, \ldots, x_n\}$; $x_k \neq x_l$; $x_k, x_l \in X \backslash \{x_1, \ldots, x_n\}$. From these relations it follows that any element from the set $f\,\Delta$ (see Section 1.4) has the form $f^{[r]}$ for a suitable collection $[r]$. If $\{i, \ldots, j\}$ is some subset of the set $\{1, \ldots, n\}$, then we denote by $f\,\Delta_{\{i, \ldots, j\}}$ the collection of all elements of the form (12) for which $s_k = 0$ when $k \notin \{i, \ldots, j\}$, and by $f\,\Delta_{\{i, \ldots, j\}}^{\{k_i, \ldots, k_j\}}$ the collection of elements $f^{[r]}$ from the set $f\,\Delta_{\{i, \ldots, j\}}$ for which $r_{i1} + \cdots + r_{is_i} = k_i, \ldots, r_{j1} + \cdots + r_{js_j} = k_j$.

LEMMA 12. Let $f = f(x_1, \ldots, x_n)$ be an arbitrary nonassociative polynomial. Then for any elements $x, y_1, \ldots, y_n \in \Phi[X]$ the equality

$$f(y_1, \ldots, y_n)\,\Delta_i^k(x) = \sum_{\langle [m],[r]\rangle = k} f^{[r]}(y_1, \ldots, y_n; y_1^{(m_{11})}, \ldots, y_1^{(m_{1s_1})}, \ldots; y_n^{(m_{n1})}, \ldots, y_n^{(m_{ns_n})})$$

is valid, where $y_j^{(t)} = y_j\,\Delta_i^t(x)$, $[m] = [m_{11}, \ldots, m_{1s_1}; \ldots; m_{n1}, \ldots, m_{ns_n}]$, $m_{j1} > m_{j2} > \cdots > m_{js_j} > 0$ for $s_j > 0$, $[r] = [r_{11}, \ldots, r_{1s_1}; \ldots; r_{n1}, \ldots, r_{ns_n}]$, $\langle [m], [r]\rangle = \sum_{i,j} m_{ij} r_{ij}$.

PROOF. It suffices to consider the case when the element f is a monomial. Let the degree of f in the variable x_i equal t_i. We consider the multilinear monomial $v(x_{11}, \ldots, x_{1t_1}; \ldots; x_{n1}, \ldots, x_{nt_n})$ for which

$$f(x_1, \ldots, x_n) = v(x_1, \ldots, x_1; \ldots; x_n, \ldots, x_n).$$

By Lemma 1.2 we have

$$f(y_1, \ldots, y_n)\,\Delta_i^k(x) = \sum_{\substack{m_{11} + \cdots + m_{1t_1} \\ + \cdots + m_{n1} + \cdots + m_{nt_n} = k}} v(y_1^{(m_{11})}, \ldots, y_1^{(m_{1t_1})}; \ldots; y_n^{(m_{n1})}, \ldots, y_n^{(m_{nt_n})}).$$

In this sum we group together the summands which depend on the same collection of arguments. In some collection $[m]$ let there be s_j different nonzero numbers $m'_{j1} > m'_{j2} > \cdots > m'_{js_j} > 0$ among the numbers m_{j1}, $m_{j2}, \ldots, m_{jt_j}$, and let the number m'_{jl} occur r_{jl} times in the collection $m_{j1}, \ldots, m_{jt_j}$. Then it is obvious

$$m_{j1} + m_{j2} + \cdots + m_{jt_j} = r_{j1}m'_{j1} + r_{j2}m'_{j2} + \cdots + r_{js_j}m'_{js_j}.$$

The sum of all those summands in which for each $j = 1, 2, \ldots, n$ there appear $t_j - s_j$ of the elements y_j, r_{j1} of the elements $y_j^{(m'_{j1})}, \ldots, r_{js_j}$ of the elements $y_j^{(m'_{js_j})}$, equals exactly

$$f^{[r]}(y_1, \ldots, y_n; y_1^{(m'_{11})}, \ldots, y_1^{(m'_{1s_1})}; \ldots; y_n^{(m'_{n1})}, \ldots, y_n^{(m'_{ns_n})}).$$

Now summing the obtained expressions for all the different possible collections of arguments, we obtain the assertion of the lemma.

We now assume that f is linear in x_1. For an arbitrary algebra A we set

$$N_f^{(1)}(A) = \{a \in A \mid g(a, A, \ldots, A) = 0 \text{ for any } g \in f\,\Delta_{\{2,\ldots,n\}}\}.$$

Let $\Phi_{\mathfrak{M}} = \Phi_{\mathfrak{M}}[Y]$ be the free algebra of a homogeneous variety $\mathfrak{M}$ from an arbitrary (possibly finite) set of generators $Y = \{y_1, y_2, \ldots\}$.

LEMMA 13. If $n = n(y_1, \ldots, y_m) \in N_f^{(1)}(\Phi_{\mathfrak{M}})$, then $n\,\Delta_y^k(z) \in N_f^{(1)}(\Phi_{\mathfrak{M}})$ for any natural number k and any $y \in Y$, $z \in \Phi_{\mathfrak{M}}$.

PROOF. We carry out an induction on k. The basis for the induction, $k = 0$, is obvious. Now we assume that the lemma is true for all $k' < k$, and consider an arbitrary element $g \in f\,\Delta_{\{2,\ldots,n\}}$, $g = g(x_1, x_2, \ldots, x_n; \ldots, x_t)$. For any elements $z_i \in \Phi_{\mathfrak{M}}$, by assumption we have $h = g(n, z_2, \ldots, z_n; \ldots, z_t) = 0$. But then $h\,\Delta_y^k(z) = 0$ by Proposition 1.3, and therefore according to Lemma 12 we shall have

$$0 = g(n^{(k)}, z_2, \ldots, z_t) + \sum_s g_s(n^{(k_s)}, z_2, \ldots, z_t; z_2^{(1)}, \ldots, z_t^{(1)}, \ldots),$$

where $k_s < k$, $g_s \in g\,\Delta_{\{2,\ldots,t\}} \subseteq f\,\Delta_{\{2,\ldots,n\}}$, $u^{(m)} = u\,\Delta_y^m(z)$. By the induction assumption we have

$$g_s(n^{(k_s)}, z_2, \ldots, z_t; z_2^{(1)}, \ldots, z_t^{(1)}, \ldots) = 0,$$

whence it follows that $g(n^{(k)}, z_2, \ldots, z_t) = 0$. Consequently $n\,\Delta_y^k(z) \in N_f^{(1)}(\Phi_{\mathfrak{M}})$. This proves the lemma.

LEMMA 14. Let I be a submodule of the Φ-module $\Phi_{\mathfrak{M}}$ such that $i\,\Delta(z) \subseteq I$ for any $i \in I$, $z \in \Phi_{\mathfrak{M}}$ (see Section 1.4). Then I is a completely characteristic subset of the algebra $\Phi_{\mathfrak{M}}$.

PROOF. Let $i \in I$, $i = f(y_1, \ldots, y_n)$, where $f = f(x_1, \ldots, x_n)$ is some nonassociative polynomial. We shall prove by induction on $n - r$ that for any $g = g(x_1, \ldots, x_r, x_{r+1}, \ldots, x_t) \in f\,\Delta_{\{r+1,\ldots,n\}}$ and for any $z_i \in \Phi_{\mathfrak{M}}$ the inclusion $g(y_1, \ldots, y_r, z_{r+1}, \ldots, z_t) \in I$ is valid. The inclusion $f(y_1, \ldots, y_n) \in I$ serves as a basis for the induction when $n - r = 0$. Having made the appropriate induction assumption, we consider an arbitrary element $h \in f\,\Delta_{\{r,r+1,\ldots,n\}}$. Let

$$h = h(x_1, \ldots, x_r, x_{r+1}, \ldots, x_m) \in f\,\Delta_{\{r,r+1,\ldots,n\}}^{\{k',l,\ldots,s\}}.$$

We make a second induction assumption: if $k' < k$; $s, \ldots, l$ are arbitrary natural numbers, and

$$g(x_1, \ldots, x_r, x_{r+1}, \ldots, x_p) \in f\,\Delta_{\{r,r+1,\ldots,n\}}^{\{k',l,\ldots,s\}},$$

then $g(y_1, \ldots, y_r, z_{r+1}, \ldots, z_p) \in I$ for any $z_i \in \Phi_{\mathfrak{M}}$. The first induction assumption serves as the basis for this induction assumption, $k = 0$. Now let $k > 0$. In view of relations (13) and (14) we can assume that

$$h(x_1, \ldots, x_r, \ldots, x_{m-1}, x_m) = h_1(x_1, \ldots, x_r, \ldots, x_{m-1})\,\Delta_r^q(x_m),$$

where $h_1 \in f\,\Delta_{\{r,\ldots,n\}}^{\{k-q,\ldots,s\}}$. By the second induction assumption we have $v = h_1(y_1, \ldots, y_r, z_{r+1}, \ldots, z_{m-1}) \in I$. But then also $v\,\Delta_{y_r}^q(z_m) \in I$, whence by

Lemma 12

$$h(y_1, \ldots, y_r, z_{r+1}, \ldots, z_{m-1}, z_m) + \sum_s h'_s(y_1, \ldots, y_r, z_{r+1}, \ldots, z_m; z^{(1)}_{r+1}, \ldots, z^{(1)}_{m-1}, \ldots) \in I,$$

where $h'_s \in h_1 \Delta^{\{q', \ldots, l', \ldots, t'\}}_{\{r, \ldots, n, \ldots, m-1\}}$, $q' < q$, $z^{(j)}_i = z_i \Delta^j_{y_r}(z_m)$. In view of relations (13) and (14), we have $h'_s \in f\Delta^{\{k', \ldots, l\}}_{\{r, \ldots, n\}}$ where $k' < k$. Hence by the induction assumption

$$h'_s(y_1, \ldots, y_r, z_{r+1}, \ldots, z_m; z^{(1)}_{r+1}, \ldots, z^{(1)}_{m-1}, \ldots) \in I,$$

and furthermore $h(y_1, \ldots, y_r, z_{r+1}, \ldots, z_m) \in I$. Since h was arbitrary, by the same token it is proved that for any $g = g(x_1, \ldots, x_r, x_{r+1}, \ldots, x_t) \in f\Delta_{\{r, r+1, \ldots, n\}}$ and for any $z_i \in \Phi_{\mathfrak{M}}$ the inclusion $g(y_1, \ldots, y_r, z_{r+1}, \ldots, z_t) \in I$ is valid. Now let h have degree n_r in x_r. Without loss of generality it is possible to assume that h is homogeneous in x_r. We have

$$h(y_1, \ldots, y_r, z_{r+1}, \ldots, z_m) \Delta^{n_r}_{y_r}(z_r) \in I,$$

whence by Lemma 12 we obtain

$$h(y_1, \ldots, y_{r-1}, z_r, \ldots, z_m) + \sum_s h''_s(y_1, \ldots, y_{r-1}, y_r, z_{r+1}, \ldots, z_m; z_r, z^{(1)}_{r+1}, \ldots) \in I,$$

where $h''_s \in f\Delta_{\{r, r+1, \ldots, n\}}$, $z^{(j)}_i = z_i \Delta^j_{y_r}(z_r)$. By what was proved above we have

$$h''_s(y_1, \ldots, y_r, z_{r+1}, \ldots, z_m; z_r, z^{(1)}_{r+1}, \ldots) \in I,$$

whence also $h(y_1, \ldots, y_{r-1}, z_r, \ldots, z_m) \in I$. Since h was arbitrary, by the first induction assumption it hence follows that $f(z_1, \ldots, z_n) \in I$ for any $z_i \in \Phi_{\mathfrak{M}}$. This proves the lemma.

We note that in the process of the proof of Lemma 14 we have proved the following more general assertion.

COROLLARY. Under the conditions of Lemma 14, let $f = f(x_1, \ldots, x_n)$ be a nonassociative polynomial such that for some integer r ($0 \leqq r \leqq n$, $r \leqq \operatorname{card}(Y)$), for any $g \in f\Delta_{\{r+1, \ldots, n\}}$, $g = g(x_1, \ldots, x_r, x_{r+1}, \ldots, x_n; \ldots, x_m)$, and for any $z_i \in \Phi_{\mathfrak{M}}$, the inclusion

$$g(y_1, \ldots, y_r, z_{r+1}, \ldots, z_m) \in I$$

is valid. Then $f(z_1, \ldots, z_n) \in I$ for any $z_i \in \Phi_{\mathfrak{M}}$.

From Lemmas 13 and 14 follows

THEOREM 10 (*Shestakov*). Let $\Phi_{\mathfrak{M}}[Y]$ be the free algebra of a homogeneous variety $\mathfrak{M}$ from an arbitrary set of generators Y, and $f(x_1, \ldots, x_n)$

be a nonassociative polynomial which is linear in x_1. Then the submodule $N_f^{(1)}(\Phi_{\mathfrak{M}}[Y])$ is a completely characteristic submodule of the algebra $\Phi_{\mathfrak{M}}[Y]$.

COROLLARY. Under the conditions of Theorem 10, the subalgebras $N(\Phi_{\mathfrak{M}}[Y])$ and $Z(\Phi_{\mathfrak{M}}[Y])$ are completely characteristic subalgebras of the algebra $\Phi_{\mathfrak{M}}[Y]$.

From Theorem 10 it also follows that in a finitely generated alternative algebra $\mathrm{Alt}[X_n]$ the ideals $ZN(\mathrm{Alt}[X_n])$ and $U(\mathrm{Alt}[X_n])$ are completely characteristic.

Exercises

1. Let A be an alternative algebra, $x, y, z, r, s \in A$, $n \in N(A)$. Prove that

$$2(x, y, z)[r, n][s, n] = 0.$$

2. (*Dorofeev*) Prove that if $n \in N(A)$, then

$$2[A, n][A, n]A \subseteq N(A).$$

3. Prove that $D(\mathrm{Alt}[X])$ is an alternative PI-algebra. *Hint*. The identity $[[x, y]^8, t] = 0$ is satisfied in $D(\mathrm{Alt}[X])$.

4. ALTERNATIVE ALGEBRAS FROM THREE GENERATORS

In this section we consider the free 3-generated alternative algebra $\mathrm{Alt}[X_3] = \mathrm{Alt}[x_1, x_2, x_3]$. We shall prove that in the associative center $N(\mathrm{Alt}[X_3])$ and in the center $Z(\mathrm{Alt}[X_3])$ there are contained nonzero elements of a substantially more simple form than in the corresponding centers of the algebra $\mathrm{Alt}[X]$. On the basis of the identities obtained, we next prove that in a free alternative algebra from three or more generators there is contained a subalgebra which has the structure of a Cayley–Dickson algebra over its centers.

As a preliminary we prove two lemmas.

LEMMA 15. In any alternative algebra A the identities

$$([x, y]^2, x, z) = 0, \tag{15}$$

$$[(x, y, z)^2, x] = 0, \tag{16}$$

$$((x, y, z)^2, x, y) = 0, \tag{17}$$

are valid.

PROOF. Identity (15) is valid by relation (7.25). Furthermore, we note that in the algebra A the identity

$$[(x^2, y, z), (x, y, z)] = 0 \tag{18}$$

is valid. In fact, by the Moufang identities we have

$$\begin{aligned}
&(x^2, y, z)(x, y, z) \\
&\quad = -(x^2, y, (x, y, z))z + (x^2, (x, y, z)y, z) + (x^2, zy, (x, y, z)) \\
&\quad = -(x, y, x \circ (x, y, z))z + (x, x \circ (x, y, yz), z) + (x, zy, x \circ (x, y, z)) \\
&\quad = -(x, y, (x^2, y, z))z + (x, (x^2, y, z)y, z) + (x, zy, (x^2, y, z)) \\
&\quad = (x, y, z)(x^2, y, z).
\end{aligned}$$

Consequently

$$[(x, y, z)^2, x] = [(x, y, z), (x, y, z) \circ x] = [(x, y, z), (x^2, y, z)] = 0,$$

that is, (16) is proved. Now let $w = (x, y, z)$. According to (7.14) and (7.15) we obtain

$$\begin{aligned}
(w^2, x, y) &= w \circ (w, x, y) = -w \circ (w[x, y]) = -w^2[x, y] - w[x, y]w \\
&= -w([x, y] \circ w) = 0.
\end{aligned}$$

This proves the lemma.

LEMMA 16 (*Bruck, Kleinfeld*). Let A be an alternative algebra with set of generators S. Then an element $z \in A$ belongs to the center $Z(A)$ if and only if $[z, S] = (z, S, S) = (0)$.

PROOF. On one hand the assertion is obvious, if $z \in Z(A)$ then $[s, S] = (z, S, S) = (0)$. Now let $[z, S] = (z, S, S) = (0)$. We consider the set $P = \{p \in A \mid (z, p, S) = (0)\}$. It is easy to see that P is a submodule of the Φ-module A. Furthermore, for any $p, p' \in P$, $s \in S$ we have

$$\begin{aligned}
(pp', s, z) &= p'(p, s, z) + (p', s, z)p + f(p, p', s, z) = f(p, p', s, z) \\
&= ([p, p'], s, z) + (p, p', [s, z]) = ([p, p'], s, z),
\end{aligned}$$

whence it follows that $(z, PP, S) = (0)$, that is, P is a subalgebra of the algebra A. Since $P \supseteq S$, $P = A$, that is, $(z, A, S) = (0)$. We now consider the set $Q = \{q \in A \mid (z, A, q) = (0)\}$. Then Q is a submodule of the Φ-module A and $Q \supseteq S$. For any $q, q' \in Q$, $a \in A$ we have

$$\begin{aligned}
(qq', a, z) &= q'(q, a, z) + (q', a, z)q + f(q, q', a, z) \\
&= f(q, q', a, z) = f(q', a, q, z) \\
&= (q'a, q, z) - a(q', q, z) - (a, q, z)q' = 0,
\end{aligned}$$

that is, $QQ \subseteq Q$ and Q is a subalgebra of the algebra A. Since $Q \supseteq S$, $Q = A$, and we obtain that $(z, A, A) = (0)$, that is, $z \in N(A)$. Finally, let

$$R = \{r \in A | [z, r] = 0\}.$$

For any $r, r' \in R$, by (7.4) we have

$$[rr', z] = r[r', z] + [r, z]r' = 0,$$

that is, $R \cdot R \subseteq R$. It is clear that R is a subalgebra of A, and since $R \supseteq S$, $R = A$. Thus $z \in Z(A)$. This proves the lemma.

THEOREM 11 (*Dorofeev*). In any alternative 3-generated algebra the identities

$$([x, y]^2, r, s) = 0, \tag{19}$$

$$([x, y] \circ [z, t], r, s) = 0, \tag{20}$$

are valid.

PROOF. It suffices to prove that identities (19) and (20) are valid in the free algebra $\mathrm{Alt}[x_1, x_2, x_3]$. Let $N = N(\mathrm{Alt}[x_1, x_2, x_3])$. We shall first prove that $[x_1, x_2]^2 \in N$ for generators x_1, x_2. For arbitrary elements $r, s \in \mathrm{Alt}[x_1, x_2, x_3]$ we denote by $f(r, s)$ the ideal of the algebra $\mathrm{Alt}[x_1, x_2, x_3]$ generated by the element $([x_1, x_2]^2, r, s)$. Then the mapping f satisfies conditions (a)–(f) of Proposition 1. Conditions (a)–(d) are obvious, (e) is true by identity (2), and (f) is true by the Moufang identities. By Proposition 1 $f(r, s) \subseteq \sum_{i<j} f(x_i, x_j) + \sum_{i=1}^{3} f(x_i, x_j x_k)$. In addition, it is easy to see that

$$f(x_i, x_j x_k) + \sum_{i<j} f(x_i, x_j) = f(x_1, x_2 x_3) + \sum_{i<j} f(x_i, x_j),$$

and therefore $f(r, s) \subseteq \sum_{i<j} f(x_i, x_j) + f(x_1, x_2 x_3)$. In view of (15), we have $f(x_i, x_j) = f(x_1, x_2 x_3) = (0)$, whence also $f(r, s) = (0)$. Thus $[x_1, x_2]^2 \in N$. By Theorem 10 the subalgebra N is completely characteristic, and therefore $[x, y]^2 \in N$ for all $x, y \in \mathrm{Alt}[x_1, x_2, x_3]$. By the same token identity (19) is proved.

We shall now prove that $[x_1, x_2] \circ [x_3, r] \in N$ for any $r \in \mathrm{Alt}[x_1, x_2, x_3]$. As in the previous case, it suffices for this to prove that

$$([x_1, x_2] \circ [x_3, r], x_i, x_j) = 0, \tag{21}$$

$$([x_1, x_2] \circ [x_3, r], x_1, x_2 x_3) = 0. \tag{22}$$

Linearizing (19) in y we obtain the following identity in $\mathrm{Alt}[x_1, x_2, x_3]$:

$$([x, y] \circ [x, z], r, s) = 0. \tag{23}$$

In addition, in any alternative algebra there is the identity

$$([x, y] \circ z, x, y) = 0. \tag{24}$$

Actually, by (7.15) and Artin's theorem we have

$$([x, y] \circ z, x, y) = [x, y] \circ (z, x, y) + z \circ ([x, y], x, y) = 0.$$

We return to relation (21). By (24) it remains for us to consider the case when either $x_i = x_3$ or $x_j = x_3$. For example, let $x_i = x_3$. Then by linearized (24) and by (23) we have

$$([x_1, x_2] \circ [x_3, r], x_3, x_j) = -([x_1, x_2] \circ [x_3, x_j], x_3, r) = 0,$$

that is, (21) is proved. Furthermore, applying again linearized identity (24), in view of (23) we obtain

$$\begin{aligned}([x_1, x_2] \circ [x_3, r], x_1, x_2x_3) &= -([x_1, x_2x_3] \circ [x_3, r], x_1, x_2)\\ &= ([x_1, x_2x_3] \circ [x_1, r], x_3, x_2) + ([x_1, x_2x_3] \circ [x_3, x_2], x_1, r)\\ &\quad + ([x_1, x_2x_3] \circ [x_1, x_2], x_3, r) = ([x_1, x_2x_3] \circ [x_3, x_2], x_1, r).\end{aligned}$$

We note that in any algebra there is the identity

$$[x, yz] + [y, zx] + [z, xy] = -(x, y, z) - (y, z, x) - (z, x, y). \tag{25}$$

Consequently, according to (25), (7.15), and (23), we have

$$\begin{aligned}[x_1, x_2x_3] \circ [x_3, x_2] &= -[x_2, x_3x_1] \circ [x_3, x_2] - [x_3, x_1x_2] \circ [x_3, x_2]\\ &\quad - 3(x_1, x_2, x_3) \circ [x_3, x_2]\\ &= [x_2, x_3x_1] \circ [x_2, x_3] - [x_3, x_1x_2] \circ [x_3, x_2] \in N,\end{aligned}$$

whence follows (22). Thus $[x_1, x_2] \circ [x_3, r] \in N$ for any $r \in \mathrm{Alt}[x_1, x_2, x_3]$. Then by the corollary to Lemma 14, $[x, y] \circ [z, t] \in N$ for any $x, y, z, t \in \mathrm{Alt}[x_1, x_2, x_3]$, which proves identity (20).

This proves the theorem.

COROLLARY. For any elements $x, y, z, t \in \mathrm{Alt}[X_3]$

$$[x, y]^2, [x, y] \circ [z, t] \in N(\mathrm{Alt}[X_3]).$$

We now find some elements from the center $Z(\mathrm{Alt}[X_3])$ of the algebra $\mathrm{Alt}[X_3]$.

THEOREM 12 (*Shestakov*). For any elements x, y, z, r, s of the algebra $\mathrm{Alt}[x_1, x_2, x_3]$

$$(x, y, z)^2, (x, y, z) \circ [r, s] \in Z(\mathrm{Alt}[x_1, x_2, x_3]).$$

PROOF. Let $Z = Z(\mathrm{Alt}[x_1, x_2, x_3])$. In view of Lemma 16, for a proof that $f = f(x_1, x_2, x_3) \in Z$ it suffices to show that $[f, x_i] = (f, x_i, x_j) = 0$ for any generators x_i, x_j. Hence from relations (16), (17) it follows that

$(x_1, x_2, x_3)^2 \in Z$. Then by Theorem 10 also $(x, y, z)^2 \in Z$ for any

$$x, y, z \in \mathrm{Alt}[x_1, x_2, x_3].$$

By the same token the first of the inclusions of Theorem 12 is proved.

For the proof of the second inclusion, by the corollary to Lemma 14 (or Theorem 10) it suffices to show that $(x_1, x_2, x_3) \circ [r, s] \in Z$ for any $r, s \in \mathrm{Alt}[x_1, x_2, x_3]$. Without loss of generality the elements r, s can be considered monomials from x_1, x_2, x_3, and we shall prove the given assertion by induction on the number $n = d(r) + d(s)$. Equality (7.15) serves as a basis for the induction when $n = 2$. As we have already noted, it suffices for us to show that for any generators x_i, x_j

$$((x_1, x_2, x_3) \circ [r, s], x_i, x_j) = 0, \tag{26}$$

$$[(x_1, x_2, x_3) \circ [r, s], x_i] = 0. \tag{27}$$

We shall first prove equality (26). If $d(r) > 1$, $d(s) > 1$, then by linearized identity (24) and the induction assumption we obtain

$$((x_1, x_2, x_3) \circ [r, s], x_i, x_j) = -((x_1, x_2, x_3) \circ [x_i, s], r, x_j)$$
$$- ((x_1, x_2, x_3) \circ [r, x_j], x_i, s) - ((x_1, x_2, x_3) \circ [x_i, x_j], r, s) = 0.$$

Now let $r = x_1$, for example. In view of the fact that

$$((x_1, x_2, x_3) \circ [x_i, s], x_i, x_j) = -((x_1, x_2, x_3) \circ [x_i, x_j], x_i, s) = 0, \tag{28}$$

it remains for us to consider the element $f = ((x_1, x_2, x_3) \circ [x_1, s], x_2, x_3)$. Repeating the previous argument, we see that

$$f \equiv -((x_1, x_2, x_3) \circ [x_2, s], x_1, x_3) = -((x_1, x_2, x_3) \circ [x_3, s], x_2, x_1). \tag{29}$$

Furthermore, from the alternative identities it follows that the monomial s can be represented in the form $s = \sum_{i=1}^{3} (\alpha_i x_i s_i + \beta_i s'_i x_i)$, where $\alpha_i, \beta_i \in \Phi$, s_i, s'_i are monomials from x_1, x_2, x_3, and $d(s_i) = d(s'_i) = d(s) - 1$. We leave the proof of this fact to the reader. We consider the element $f_i = ((x_1, x_2, x_3) \circ [x_1, x_i s_i], x_2, x_3)$. By applying relation (29), if necessary, we can assume that $f_i = \pm((x_1, x_2, x_3) \circ [x_j, x_i s_i], x_i, x_k)$, where $\{i, j, k\} = \{1, 2, 3\}$, $x_j \neq s_i$. By (25) we have

$$(x_1, x_2, x_3) \circ [x_j, x_i s_i] = -(x_1, x_2, x_3) \circ [x_i, s_i x_j]$$
$$- (x_1, x_2, x_3) \circ [s_i, x_j x_i] - 3(x_1, x_2, x_3) \circ (x_j, x_i, s_i).$$

From linearization in z of the inclusion $(x, y, z)^2 \in Z$, it follows that $(x_1, x_2, x_3) \circ (x_j, x_i, s_i) \in Z$. Furthermore, according to (28) we have

$$((x_1, x_2, x_3) \circ [x_i, s_i x_j], x_i, x_k) = 0.$$

Finally, we note that either $d(s_i) > 1$ or $s_i \in \{x_i, x_k\}$, and therefore by what was proved above or in view of (28)

$$((x_1, x_2, x_3) \circ [s_i, x_j x_i], x_i, x_k) = 0.$$

Consequently $f_i = 0$ for $i = 1, 2, 3$. It is proved analogously that

$$f'_i = ((x_1, x_2, x_3) \circ [x_1, s'_i x_i], x_2, x_3) = 0$$

for $i = 1, 2, 3$. This means $f = 0$, and relation (26) is proved.

Furthermore, according to identity (20) and linearized (7.15)

$$\begin{aligned}
&[(x_1, x_2, x_3) \circ [r, s], x_1] \\
&\quad = (x_1, x_2, x_3) \circ [[r, s], x_1] + [r, s] \circ [(x_1, x_2, x_3), x_1] \\
&\quad = (x_1, x_2, x_3) \circ [[r, s], x_1] + [r, s] \circ (x_1, x_2, [x_1, x_3]) \\
&\quad = (x_1, x_2, x_3) \circ [[r, s], x_1] \\
&\qquad + (x_1, x_2, [r, s] \circ [x_1, x_3]) - [x_1, x_3] \circ (x_1, x_2, [r, s]) \\
&\quad = (x_1, x_2, x_3) \circ [[r, s], x_1] + (x_1, x_2, [r, s]) \circ [x_3, x_1] = 0,
\end{aligned}$$

which proves (27).

This proves the theorem.

Now let Alt$[Y]$ be the free alternative Φ-algebra from set of generators Y, where card$(Y) \geqq 3$ and there are no elements of order 2 in Φ. We select arbitrary distinct elements $a, b, c \in Y$, and set $u = [a, b]$, $v = (a, b, c)$, $w = (u, v, a)$. We denote by $A_1 = \Phi[u, v]$ and $A = \Phi[u, v, w]$ the subalgebras of the algebra Alt$[Y]$ generated by the sets $\{u, v\}$ and $\{u, v, w\}$, respectively.

LEMMA 17. $u^2, v^2, w^2 \in Z(A)$.

PROOF. In view of Theorem 12, it only remains to verify that $u^2 \in Z(A)$. By Theorem 11 $u^2 \in N(A)$. In addition, we have

$$\begin{aligned}
&[u^2, v] = [u, u \circ v] = [u, [a, b] \circ (a, b, c)] = 0, \\
&[u^2, w] = [u, u \circ w] = [u, u \circ (u, v, a)] = [u, (u^2, v, a)] = 0,
\end{aligned}$$

whence by Lemma 16 it follows that $u^2 \in Z(A)$. This proves the lemma.

LEMMA 18. The elements $(u^i v^j) w^k$, where i, j, k are arbitrary non-negative integers ($u^0 = v^0 = w^0 = 1$), are linearly independent over Φ in the algebra Alt$[Y]^\#$.

PROOF. For some $\alpha_{ijk} \in \Phi$ let

$$\sum_{i,j,k} \alpha_{ijk} (u^i v^j) w^k = 0. \tag{30}$$

In the first place it is clear that $\alpha_{000} = 0$. We now note that each element $(u^i v^j)w^k$ for $i + j + k > 0$ lies in the homogeneous component of type

$$[i + j + 3k, i + j + 2k, j + k]$$

of the free algebra $\mathrm{Alt}[a, b, c]$. As is easy to see, these components are different for various collections (i, j, k). Consequently, by the homogeneity of the variety of alternative algebras, it follows from (30) that

$$\alpha_{ijk}(u^i v^j)w^k = 0 \tag{31}$$

for any i, j, k. Relation (31) is valid in the free algebra, and therefore it is also true in any alternative algebra. But in the Cayley–Dickson matrix algebra $\boldsymbol{C}(\Phi)$ it is easy to find elements $\bar{a}, \bar{b}, \bar{c}$ such that

$$([\bar{a}, \bar{b}]^i(\bar{a}, \bar{b}, \bar{c})^j)([\bar{a}, \bar{b}], (\bar{a}, \bar{b}, \bar{c}), \bar{a})^k \neq 0$$

for all i, j, k. Since $\boldsymbol{C}(\Phi)$ is a free Φ-module, it follows from this that $\alpha_{ijk} = 0$ for all i, j, k. This proves the lemma.

COROLLARY. The subalgebra $\tilde{\Phi} = \Phi[u^2, v^2, w^2]$ generated by the elements u^2, v^2, w^2 is isomorphic to the algebra of polynomials $\Phi[t_1, t_2, t_3]$ from independent variables t_1, t_2, t_3.

LEMMA 19. The center $Z(A_1)$ of the algebra A_1 equals $\Phi[u^2, v^2]$. In addition, the algebra A_1 is an algebra of generalized quaternions over its center.†

PROOF. By Lemma 17 $\Phi[u^2, v^2] \subseteq Z(A_1)$. Conversely, let $z \in Z(A_1)$. In view of (7.15) we have $vu = -uv$, whence it follows that the element z can be represented in the form

$$z = \alpha + \beta u + \gamma v + \delta uv,$$

where $\alpha, \beta, \gamma, \delta \in \Phi[u^2, v^2]$. Furthermore, we have

$$0 = [z, u] = -2\gamma uv - 2\delta u^2 v.$$

Let $\gamma = \sum_{i,j} \gamma_{ij} u^{2^i} v^{2^j}$, $\delta = \sum_{k,l} \delta_{kl} u^{2^k} v^{2^l}$, where $\gamma_{ij}, \delta_{kl} \in \Phi$. Then we obtain

$$\sum_{i,j} 2\gamma_{ij} u^{2^i+1} v^{2^j+1} + \sum_{k,l} 2\delta_{kl} u^{2^k+2} u^{2^l+1} = 0.$$

† The notion of the Cayley–Dickson process is extended in a natural way to the case of algebras over an arbitrary ring of operators Φ. If there are no elements of order 2 in Φ, then by an algebra of generalized quaternions over Φ is understood an algebra of the form $((\Phi, \alpha), \beta)$ which is obtained from Φ by a double application of the Cayley–Dickson process with parameters $\alpha, \beta \in \Phi$, $\alpha\beta \neq 0$. A Cayley–Dickson algebra over Φ is also defined in a natural way.

It is easy to see that in these sums there are no common summands of the form $u^i v^j$. Therefore by Lemma 18

$$2\gamma_{ij} = 2\delta_{kl} = 0$$

for any i, j, k, l, whence $\gamma = \delta = 0$. From the equality $[z, v] = 0$ it follows analogously that $\beta = 0$. Therefore $z = \alpha \in \Phi[u^2, v^2]$, that is, $Z(A_1) = \Phi[u^2, v^2]$.

For the proof of the lemma it now suffices to show that the elements 1, u, v, uv of the algebra $\mathrm{Alt}[a, b, c]^{\#}$ are linearly independent over $\Phi[u^2, v^2]$. Let

$$h = \alpha + \beta u + \gamma v + \delta uv = 0,$$

where $\alpha, \beta, \gamma, \delta \in \Phi[u^2, v^2]$. Then $h \in Z(A_1)$, whence by what was proved above it follows that $\beta = \gamma = \delta = 0$. But then also $\alpha = h = 0$. This proves the lemma.

It is possible in the usual manner to define an involution $h \to \bar{h}$ in the algebra A_1. If $h = \alpha + \beta u + \gamma v + \delta uv$, then we set $\bar{h} = \alpha - \beta u - \gamma v - \delta uv$.

LEMMA 20. Let a_1, a_2 be arbitrary elements of the algebra A_1. Then the following relations are true:

$$wa_1 = \bar{a}_1 w, \tag{32}$$

$$a_1(a_2 w) = (a_2 a_1)w, \tag{33}$$

$$(a_2 w)a_1 = (a_2 \bar{a}_1)w, \tag{34}$$

$$(a_1 w)(a_2 w) = w^2(\bar{a}_2 a_1). \tag{35}$$

PROOF. Since $u^2, v^2 \in Z(A)$, then

$$w \circ u = (u, v, a) \circ u = (u^2, v, a) = 0,$$

and analogously $w \circ v = 0$. Furthermore,

$$\begin{aligned} w \circ (uv) &= (u, v, a) \circ (uv) = (u \circ uv, v, a) - u \circ (uv, v, a) \\ &= (u(u \circ v), v, a) - u \circ (u, v, av) = -(u^2, v, av) = 0. \end{aligned}$$

Relation (32) follows from these obtained equalities. We now note that $a_1 + \bar{a}_1 \in Z(A)$ for any $a_1 \in A_1$. Consequently

$$(a_2 w)a_1 = -(a_2 a_1)w + a_2(w \circ a_1) = -(a_2 a_1)w + a_2 w(a_1 + \bar{a}_1) = (a_2 \bar{a}_1)w,$$

which proves (34). Analogously $a_1(wa_2) = w(\bar{a}_1 a_2)$, whence by (32) follows (33). Finally, in view of (32) we have

$$(a_1 w)(a_2 w) = (w\bar{a}_1)(a_2 w) = w(\bar{a}_1 a_2)w = w^2(\bar{a}_2 a_1).$$

This proves the lemma.

From Lemma 20 it follows that any element $x \in A$ can be represented in the form

$$x = \alpha + \beta u + \gamma v + \delta w + \lambda uv + \mu uw + \nu vw + \sigma(uv)w,$$

where $\alpha, \beta, \gamma, \delta, \lambda, \mu, \nu, \sigma \in \tilde{\Phi} = \Phi[u^2, v^2, w^2]$. As in the case of the algebra A_1, with the help of Lemma 18 it is easy to prove that $Z(A) = \tilde{\Phi}$ and that the elements $1, u, v, w, uv, uw, vw, (uv)w$ of the algebra $\mathrm{Alt}[a, b, c]^{\#}$ are linearly independent over $\tilde{\Phi}$. In addition, relations (32)–(35) show that the algebra A is obtained from the algebra $\tilde{A}_1 = \tilde{\Phi} \otimes_{\Phi[u^2, v^2]} A_1$ by means of the Cayley–Dickson process with parameter w^2, that is, $A = (\tilde{A}_1, w^2)$. It is clear that $\tilde{A}_1$ is an algebra of generalized quaternions with center $\tilde{\Phi}$, and therefore A is a Cayley–Dickson algebra with center $\tilde{\Phi}$.

Thus is proved the following

THEOREM 13 (*Dorofeev*). The algebra $\Phi[u, v, w]$ is a Cayley–Dickson algebra over its center $\Phi[u^2, v^2, w^2]$.

We note that if Φ does not have zero divisors, then the algebra $\Phi[u, v, w]$ also does not have zero divisors (see Lemma 2.13 and the corollary to Lemma 18).

Exercises

1. (*Dorofeev*) Prove that in any 3-generated alternative algebra A the containment

$$ZN(A) \subseteq U(A)$$

is valid. *Hint*. For a fixed $n \in N$ we denote by $F(x, y, z, t)$ the ideal of the algebra A generated by the element $([x, n]y, z, t)$. Then the mapping F satisfies conditions (a)–(f) of Proposition 1.

2. (*Shestakov*) Prove that $(x, y, z) \circ (r, s, t) \in Z(\mathrm{Alt}[X_3])$ for any $x, y, z, r, s, t \in \mathrm{Alt}[X_3]$.

LITERATURE

Dorofeev [56, 57, 60], Zhevlakov [75], Slater [197, 206], Humm and Kleinfeld [256], Shelipov [265], Shestakov [269–272], Shirshov [278, 279].

Free algebras of other varieties are studied in works of Pchelintsev [171–173], Roomel'di [185], and Shestakov [268].

CHAPTER **14**

Radicals of Jordan Algebras

We have already earlier touched upon the question discussed in this chapter. The locally nilpotent radical in Jordan algebras was studied in Section 4.5. In addition, we can apply a series of general results on radicals from Chapter 8 to radicals in Jordan algebras.

In this chapter still two more radicals are introduced and studied in the class of Jordan algebras, and questions on the heredity of radicals are also investigated.

As always, for the Jordan Φ-algebras considered we assume that $\frac{1}{2} \in \Phi$.

1. THE McCRIMMON RADICAL

An element z of a Jordan algebra J is called an *absolute zero divisor* if $JU_z = (0)$, that is, if U_z is the zero operator. If the only absolute zero divisor in J is 0, the algebra J is called *nondegenerate*. We shall denote by $\mathscr{Z}(J)$ the ideal of the Jordan algebra J generated by all its absolute zero divisors. It is obvious that the algebra J is nondegenerate if and only if $\mathscr{Z}(J) = (0)$.

Let I be a trivial ideal of the algebra J. Then for any elements $x \in J$ and $z \in I$ we have

$$xU_z = 2(xz)z - xz^2 = 0,$$

that is, each element of the ideal I is an absolute zero divisor in J, and in particular $I \subseteq \mathscr{Z}(J)$. Consequently, nondegeneracy of an algebra J implies J is semiprime. Whether the converse is true is unknown. It is also unknown whether the ideal $\mathscr{Z}(J)$ is locally nilpotent. It is only known that $\mathscr{Z}(J)$ is a nil-ideal.

The class of nondegenerate Jordan algebras is thus a broad subclass of the class of semiprime Jordan algebras. The naturalness of this class is confirmed by the fact that it is the semisimple class of a certain radical, to whose construction we now proceed.

LEMMA 1. Let J be a Jordan algebra and $J^{\#}$ be the algebra obtained by the formal adjoining to J of an identity element 1. Then if z is an absolute zero divisor in J and $x \in J^{\#}$, zU_x is also an absolute zero divisor in J.

PROOF. Since $J^{\#}$ is a Jordan algebra, Macdonald's identity is valid in $J^{\#}$. Therefore, for any element $a \in J$

$$aU(zU_x) = aU_xU_zU_x = 0$$

since $aU_x \in J$. This proves the lemma.

LEMMA 2. The ideal $\mathscr{Z}(J)$ is generated as a Φ-module by the absolute zero divisors.

PROOF. It suffices to note that for any absolute zero divisor z and any element $x \in J$

$$2zx = zU_{1+x} - zU_x - z,$$

that is, the element zx is a sum of three absolute zero divisors, and consequently it is in the submodule generated by the absolute zero divisors. This means that the submodule indicated is an ideal, which proves the lemma.

For an arbitrary Jordan algebra J, we set $\mathscr{M}_1(J) = \mathscr{Z}(J)$ by definition, and assume the ideal $\mathscr{M}_\alpha(J)$ to be already defined for all ordinals α such that $\alpha < \beta$. If β is a limit ordinal, we set

$$\mathscr{M}_\beta(J) = \bigcup_{\alpha<\beta} \mathscr{M}_\alpha(J).$$

If the ordinal β is not a limit ordinal, we define $\mathscr{M}_\beta(J)$ as the ideal such that

$$\mathscr{M}_\beta(J)/\mathscr{M}_{\beta-1}(J) = \mathscr{Z}(J/\mathscr{M}_{\beta-1}(J)).$$

The chain of ideals

$$\mathscr{M}_1(J) \subseteq \mathscr{M}_2(J) \subseteq \cdots \subseteq \mathscr{M}_\alpha(J) \subseteq \cdots$$

stabilizes at some ordinal γ. We set

$$\mathscr{M}(J) = \mathscr{M}_\gamma(J).$$

PROPOSITION 1. The quotient algebra $J/\mathcal{M}(J)$ is nondegenerate, and $\mathcal{M}(J)$ is the smallest ideal with this property.

The proof of this proposition is analogous to the proof of Proposition 8.4, and so we shall not carry it out here.

COROLLARY. An algebra J does not map homomorphically onto a nondegenerate algebra if and only if $J = \mathcal{M}(J)$.

We shall denote by $\mathcal{M}$ the class of Jordan algebras J which coincide with their ideal $\mathcal{M}(J)$.

THEOREM 1. $\mathcal{M}$ is a radical class.

PROOF. By the corollary to Proposition 1 property (A) for the class $\mathcal{M}$ is obvious. We shall prove property (D). For this it suffices to show the following.

LEMMA 3. Let I be an ideal of a Jordan algebra J and x be an absolute zero divisor of the algebra I. Then J also has an absolute zero divisor contained in I.

PROOF. If $JU_x = (0)$, then x is an absolute zero divisor in J and everything is proved. If $aU_x \neq 0$ for some $a \in J$, then

$$JU(aU_x) = JU_xU_aU_x = (0)$$

since $JU_xU_a \subseteq I$, that is, aU_x is an absolute zero divisor in J. This proves the lemma.

Now we complete the proof of Theorem 1. If the algebra J is such that there is a nonzero $\mathcal{M}$-ideal in any homomorphic image of J, then by Lemma 3 there is an absolute zero divisor in any homomorphic image of J, that is, the algebra J does not map homomorphically onto a nondegenerate algebra. But this means that $J \in \mathcal{M}$, which proves condition (D) for the class $\mathcal{M}$, and consequently also the entire theorem.

The radical $\mathcal{M}$ is called the *McCrimmon radical.*

THEOREM 2 (*McCrimmon*). In any Jordan algebra J the inclusions

$$\mathcal{B}(J) \subseteq \mathcal{M}(J) \subseteq \mathfrak{N}(J)$$

are valid.

PROOF. The first inclusion follows from Proposition 1, and Proposition 8.4, and the fact that any nondegenerate algebra is semiprime. For the proof of the second inclusion, it suffices to show that $\mathcal{Z}(J)$ is a nil-ideal, since in this case it can be proved by an obvious induction that all of the ideals $\mathcal{M}_\beta(J)$

are nil-ideals. Since every element from $\mathscr{L}(J)$ has the form $z_1 + \cdots + z_n$, where the z_i are absolute zero divisors, the assertion can be proved by induction on n. The basis for the induction, $z_1^3 = z_1 U_{z_1} = 0$, is obvious. For the proof of the inductive step it suffices to show that for any nilpotent element $y \in J$ and any absolute zero divisor z the element $y + z$ is also nilpotent. We consider the subalgebra J_0 in J generated by the elements y and z. By Shirshov's theorem J_0 is special and so has an associative enveloping algebra A_0. Computing in the algebra A_0, we have

$$(y + z)^k = y^k + \sum_{i=0}^{k-1} y^i z y^{k-i-1} + \sum_{i=0}^{k-2} y^i z^2 y^{k-i-2}.$$

Therefore if the index of nilpotency of the element y is m, then $(y + z)^{2m+1} = 0$. This proves the theorem.

It is pertinent to note here that the fact that $\mathscr{B}$ is a radical class in the class of Jordan algebras has up till now neither been proved nor disproved.

We now go on to the study of the McCrimmon radical of special algebras.

THEOREM 3 (*Slin'ko*). If a special Jordan algebra J is generated by a finite number of elements, each of which is an absolute zero divisor, then an associative enveloping algebra A for J is nilpotent.

PROOF. We carry out an induction on the number of generators of the algebra J, and assume that the theorem is valid if the number of generators equals $k - 1$. Let the algebra J be generated by the set $\bar{R} = \{a_1, \ldots, a_k\}$, the elements of which are absolute zero divisors. This set also generates the algebra A. Translated into associative terms, the fact that a_i is an absolute zero divisor of J means that

$$a_i j(a_1, \ldots, a_k) a_i = 0$$

for any Jordan polynomial $j(x_1, \ldots, x_k)$.

Let $R = \{x_1, \ldots, x_k\}$. If $f = f(x_1, \ldots, x_k)$ is some polynomial from Ass$[R]$, then as usual we shall denote by $\bar{f}$ the image $f(a_1, \ldots, a_k)$ of the element f under the homomorphism sending x_i to a_i.

Let l be the maximal length of nonzero words from elements of the set $\bar{R} \backslash \{a_k\}$. Since $a_k^3 = 0$, then for any x_k-indecomposable word α of length greater than $l + 2$ we have $\bar{\alpha} = 0$. If an x_k-indecomposable word α has degree >2 in x_k, then also $\bar{\alpha} = 0$. We denote by T_0 the set of all x_k-indecomposable words of length $\leqq l + 2$ and of degree $\leqq 2$ in x_k.

We consider a word α of length $(k + 2)(l + 2)$, and assume that $\bar{\alpha} \neq 0$. Then $\alpha = \beta_1 \gamma \beta_2$, where β_1 does not contain x_k and has length $\leqq l$, β_2 contains only x_k and has length $\leqq 2$, and γ is a T_0-word with T_0-length $\geqq k + 1$.

Consequently

$$\gamma = t_1 t_2 \cdots t_k x_k \gamma',$$

where $t_i \in T_0$. We note that the "remainder" of each x_k-indecomposable word t_i has length $\geqq 2$, otherwise $\overline{\gamma} = 0$. This means

$$\gamma = t'_1 x_{i_1} t'_2 x_{i_2} \cdots t'_k x_{i_k} x_k \gamma',$$

where $t'_1, \ldots, t'_k \in T_0$, and the elements $x_{i_1}, \ldots, x_{i_k}$ are different from x_k so that some two of them are the same. Between any two elements of the set $\{x_{i_1}, \ldots, x_{i_k}\}$ there is situated a T-word, which by Lemma 5.9 is singular. In particular, there is a singular word situated between two identical elements of this set.

We have proved that any word α of length $(k + 2)(l + 2)$ and such that $\overline{\alpha} \neq 0$ contains a subword of the form $x_i \delta x_i$, where δ is a singular word. Let j_δ be a homogeneous Jordan polynomial the leading word of which is δ. In view of the fact that $\overline{x_i j_\delta x_i} = 0$, we have

$$\overline{x_i \delta x_i} = \sum_n \varphi_n \overline{x_i \delta_n x_i},$$

where the words δ_n have the same-type as δ but are lexicographically less than δ. Consequently if $\overline{\alpha} \neq 0$, then $\overline{\alpha}$ is representable in the form of a linear combination of images of words of the same-type that are lexicographically smaller. We have obtained a contradiction to the fact there is a finite number of words of a fixed type in the alphabet R. Consequently $\overline{\alpha} = 0$ for any word α of length $(k + 2)(l + 2)$, that is, A is nilpotent of index not greater than $(k + 2)(l + 2)$.

This proves the theorem.

THEOREM 4 (*Slin'ko–Skosyrskiy*) The inclusion

$$\mathscr{M}(J) \subseteq \mathscr{L}(J)$$

holds in any special Jordan algebra J.

PROOF. Since the ideal $\mathscr{Z}(J)$ is the Φ-module generated by all absolute zero divisors, Theorem 3 implies its local nilpotency in any special algebra J. We consider the quotient algebra $\overline{J} = J/\mathscr{L}(J)$. By the corollary to Theorem 4.5 it is special, and consequently, $\mathscr{Z}(\overline{J}) = (\overline{0})$. But then $\mathscr{M}(\overline{J}) = (\overline{0})$ also. This means that $\mathscr{M}(J) \subseteq \mathscr{L}(J)$, which is what was to be proved.

In the general case it is not known whether the inclusion $\mathscr{M}(J) \subseteq \mathscr{L}(J)$ is valid. This question is closely connected with the Shirshov problem on Jordan nil-algebras of bounded index (see Exercise 1).

Exercises

1. Prove that any Jordan nil-algebra J of bounded index is $\mathcal{M}$-radical. *Hint*. If the identity $x^n = 0$ is valid in the algebra J, then for any $x \in J$ the element x^{n-1} is an absolute zero divisor.

Throughout the following three exercises A is an associative algebra with involution $*$, and $H = H(A, *)$ is the Jordan algebra of symmetric with respect to $*$ elements of the algebra A.

2. Prove that $\mathcal{B}(A) \cap H \subseteq \mathcal{B}(H)$. *Hint*. Prove by induction on α that $\mathcal{B}_\alpha(A) \cap H \subseteq \mathcal{B}(H)$ for any ordinal α.

3. Prove that if $a \in H$ and $aHa \subseteq \mathcal{B}(A)$, then $a \in \mathcal{B}(A)$.

4. (*Erickson–Montgomery*) Prove that $\mathcal{B}(H) = \mathcal{M}(H) = \mathcal{B}(A) \cap H$.

2. INVERTIBLE ELEMENTS, ISOTOPY, AND HOMOTOPY

Let J be a Jordan algebra with identity element 1. An element $x \in J$ is called *invertible* if there exists an element $y \in J$ such that

$$xy = 1, \qquad x^2 y = x.$$

In this case the element y is called an *inverse element* for the element x.

At first glance it may seem that in the stated definition x and y are disparate; however, this is not so.

LEMMA 4. The following conditions are equivalent for elements $x, y \in J$:

(a) x is invertible with inverse y;
(b) y is invertible with inverse x;
(c) $yU_x = x$, $y^2 U_x = 1$.

PROOF. The proof will be carried out according to the scheme (a) ⇒ (c) ⇒ (b). This will prove the equivalence of conditions (a), (b), and (c), since the implications (a) ⇒ (b) and (b) ⇒ (a) are equivalent.

(a) ⇒ (c). First of all we note that $yU_x = 2(yx)x - yx^2 = 2x - x = x$. Furthermore, by (3.27) the relations

$$[R_{x^2}, R_y] = [R_{y^2}, R_x] = 0 \tag{1}$$

are valid for the elements x and y, whence $y^2 U_x = 2(y^2 x)x - y^2 x^2 = y^2 x^2 = y(yx^2) = yx = 1$. This proves (c).

(c) ⇒ (b). By Macdonald's identity (3.48) we have

$$U(y^2 U_x) = U_x U(y^2) U_x = E,$$

and consequently the operator U_x is invertible. We now note that $[R_x, U_x] = 0$, and therefore

$$(1 - yx)U_x = x^2 - (yU_x)x = x^2 - x^2 = 0,$$
$$(y - y^2x)U_x = yU_x - (y^2U_x)x = x - x = 0.$$

By the invertibility of the operator U_x we have $yx = 1$, $y^2x = y$, that is, (b) is proved.

The relations from (c) in Lemma 4 are sometimes taken as the definition of invertibility.

LEMMA 5. The following conditions are equivalent for an element $x \in J$: (a) x is invertible, (b) $1 \in JU_x$, (c) U_x is an invertible operator.

PROOF. (a) $\Rightarrow$ (b) follows from Lemma 4. The implication (b) $\Rightarrow$ (c) was also essentially proved. If $1 = zU_x$, then by Macdonald's identity $E = U_1 = U_xU_zU_x$, whence it follows the operator U_x is invertible.

(c) $\Rightarrow$ (a). We consider the element $y = xU_x^{-1}$. Then $(1 - yx)U_x = x^2 - yR_xU_x = x^2 - yU_xR_x = x^2 - x^2 = 0$, and $(x - yx^2)U_x = x^3 - yR_{x^2}U_x = x^3 - yU_xR_{x^2} = x^3 - x^3 = 0$, which by the invertibility of the operator U_x gives us $yx = 1$, $yx^2 = x$.

The equivalence of conditions (a), (b), and (c) is proved.

LEMMA 6. The elements $x, z \in J$ are invertible if and only if $\{xzx\}$ is an invertible element.

PROOF. By Macdonald's indentity $U(\{xzx\}) = U_xU_zU_x$. The linear transformation $U_xU_zU_x$ is invertible if and only if the operators U_x and U_z are invertible. Therefore the validity of Lemma 6 follows from Lemma 5.

Let the element $x \in J$ be invertible with inverse element y. Then $xU_y = y$, and by Macdonald's identity $U_y = U_yU_xU_y$. But since the operator U_y is invertible, it can be cancelled. We obtain

$$U_xU_y = U_yU_x = E. \tag{2}$$

The operator R_x commutes with the operator U_x, and consequently it also commutes with the inverse element for U_x. Therefore

$$[R_x, U_y] = 0. \tag{3}$$

By (1) this is equivalent to

$$[R_x, R_y^2] = 0. \tag{4}$$

If in relation (3.25) we now substitute in place of y the element x, and in place of z and t the element y, we shall have $R_xR_y^2 + R_y^2R_x + R_y = R_xR_{y^2} + R_y + R_y$.

This by (4) is equivalent to the relation $2R_xR_y^2 - R_xR_{y^2} = R_y$, or what is the same

$$R_y = R_xU_y. \tag{5}$$

This relation together with (3) implies

$$[R_x, R_y] = 0. \tag{6}$$

Let H be the subalgebra of J generated by the elements x and y. As follows from Proposition 3.2, the algebra $R^J(H)$ is generated by the operators R_x, R_{x^2}, R_y, R_{y^2}. Since $R_{x^2} = 2R_x^2 - U_x$, the algebra $R^J(H)$ is also generated by the elements R_x, U_x, R_y, U_y. However, these operators commute by (2), (3), and (6). Consequently, the algebra $R^J(H)$ is commutative, which is equivalent to the strong associativity of the subalgebra H in J.

We now note that an invertible element $x \in J$ has a unique inverse, since $yU_x = x$ and $y'U_x = x$ imply $y = y'$ by the invertibility of the operator U_x. We shall denote this unique inverse element by x^{-1}. We have proved the following theorem.

THEOREM 5. For any invertible element x of a Jordan algebra J the subalgebra generated by the elements x and x^{-1} is strongly associative.

If we set $x^{-n} = (x^{-1})^n$, then powers of the element x will be multiplied according to the usual formula

$$x^kx^m = x^{k+m}.$$

We note that these relations would be invalid if we were restricted in the definition of invertibility to the single relation $xy = 1$. Another advantage of our definition is the validity of the following proposition.

PROPOSITION 2. An element a of an associative algebra A with identity element 1 is invertible if and only if it is invertible in the Jordan algebra $A^{(+)}$. In addition, the inverse elements in A and $A^{(+)}$ for the element a coincide.

PROOF. If $ab = 1 = ba$, then $a \odot b = 1$ and $a^2 \odot b = a$, that is, if a is invertible in A, then it is also invertible in $A^{(+)}$.

If a is invertible in $A^{(+)}$ with inverse b, then $1 = b^2U_a = ab^2a = -abab + 2ab(a \odot b) = -(bU_a)b + 2ab = -ab + 2ab = ab$. Analogously $ba = 1$, and a is invertible in A with inverse b.

This proves the proposition.

A Jordan algebra in which each nonzero element is invertible is called a *Jordan division algebra*. A large number of examples of Jordan division algebras can be obtained by means of Proposition 2. By that proposition, if A is an associative division algebra, then $A^{(+)}$ is a Jordan division algebra. In particular, if $\boldsymbol{Q}$ is the division ring of quaternions over the field of real numbers, then $\boldsymbol{Q}^{(+)}$ is a Jordan division algebra. We note that in $\boldsymbol{Q}^{(+)}$ some quaternions are zero divisors in the usual sense, for example $i \odot j = 0$. In particular, this means that the operator R_x may not be invertible even if the element x is invertible.

There exist examples due to Albert of exceptional Jordan division algebras. For any algebra J from these ones determined by Albert the center $F = Z(J)$ is a field and $\bar{F} \bigotimes_F J \cong H(\boldsymbol{C}_3)$, where $\bar{F}$ is the algebraic closure of the field F and $\boldsymbol{C}$ is the Cayley–Dickson algebra over $\bar{F}$. In particular, the dimension of the algebra J over the center F equals 27. There exists the hypothesis that this number is equal to the dimension of any exceptional division algebra over its center.

We now consider the concept of homotopy.

Let J be a Jordan algebra and $a \in J$. We define on the additive Φ-module of the algebra J a new operation of multiplication by the formula

$$x \cdot_a y = \{xay\}.$$

We denote the algebra obtained by $J^{(a)}$, and we call it the *a-homotope* of the algebra J. By Corollary 2 to Shirshov's theorem the identity

$$\{\{\{xax\}ay\}ax\} = \{\{xax\}a\{yax\}\}$$

is valid in any Jordan algebra. But this means that $((x \cdot_a x) \cdot_a y) \cdot_a x = (x \cdot_a x) \cdot_a (y \cdot_a x)$, that is, $J^{(a)}$ is a Jordan algebra.

We now suppose that J is special and that A is an associative enveloping algebra for J. We consider an associated algebra $A^{(a)}$, which is represented by the Φ-module of the algebra A itself with multiplication $x \times_a y = xay$. In view of the fact that $x \cdot_a y = \{xay\} = \frac{1}{2}(xay + yax) = \frac{1}{2}(x \times_a y + y \times_a x)$, we have the equality

$$(A^{(+)})^{(a)} = (A^{(a)})^{(+)}.$$

Consequently, the homotope $J^{(a)}$ of the algebra J is a subalgebra in $(A^{(a)})^{(+)}$, and therefore is a special algebra.

If the algebra J contains an identity element 1 and a is an invertible element in J, then in this case the a-homotope $J^{(a)}$ is called an *a-isotope*. By Theorem 5 $x \cdot_a a^{-1} = \{xaa^{-1}\} = (xa)a^{-1} + (a^{-1}a)x - (xa^{-1})a = x$. Consequently the element a^{-1} is an identity element in the isotope $J^{(a)}$.

LEMMA 7. For any $a, b \in J$

$$(J^{(a)})^{(b)} = J^{(\{aba\})}.$$

Isotopy is an equivalence relation.

PROOF. In order to prove that the b-homotope of an a-homotope of an algebra J is the $\{aba\}$-homotope of J, it is necessary to prove the identity

$$\{x\{aba\}y\} = \{\{xab\}ay\} + \{\{yab\}ax\} - \{\{xay\}ab\}. \tag{7}$$

This identity is linearization of the identity

$$\{x\{aba\}x\} = 2\{\{xab\}ax\} - \{\{xax\}ab\}, \tag{8}$$

which, since it is from three variables and linear in b, is valid by Corollary 2 to Shirshov's theorem.

The transitivity of the isotopy relation follows from the formula just proved and Lemma 6, by which the element $\{aba\}$ is invertible if a and b are invertible. In addition,

$$(J^{(a)})^{(a^{-2})} = J^{(\{aa^{-2}a\})} = J,$$

and so this relation is symmetric, and consequently, it is an equivalence relation.

COROLLARY. An isotope $J^{(a)}$ of a Jordan algebra J is special if and only if the algebra J is special.

Let D be an algebra with involution $d \to \bar{d}$, and $A = \operatorname{diag}\{a_1, \ldots, a_n\}$ be a diagonal invertible matrix with symmetric elements along the diagonal which are in the associative center $N(D)$ of the algebra D. We consider the algebra of matrices D_n. As is easy to see, the mapping

$$S_A : X \to A^{-1}\bar{X}^t A$$

is an involution D_n. We denote by $H(D_n, S_A)$ the algebra of symmetric with respect to S_A elements of the algebra D_n, and shall denote by $H(D_n)$ the algebra $H(D_n, S_E)$ where E is the identity matrix.

THEOREM 6. Let $H(D_n)$ be a Jordan algebra. Then the mapping $\varphi : X \to XA$ is an isomorphism of the isotope $H(D_n)^{(A)}$ of the algebra $H(D_n)$ onto the algebra $H(D_n, S_A)$.

PROOF. It is easy to see that $A \in N(D_n)$. Therefore if $X \in H(D_n)$, then $XA \in H(D_n, S_A)$ since

$$(XA)^{S_A} = A^{-1}(\overline{XA})^t A = A^{-1}A\bar{X}^t A = XA.$$

Furthermore, $\varphi(X \odot_A Y) = \frac{1}{2}(XAY + YAX)A = \frac{1}{2}[(XA)(YA) + (YA)(XA)] = XA \odot YA = \varphi(X) \odot \varphi(Y)$, that is, φ is a homomorphism. Since the matrix A is invertible, φ is an isomorphism, and the theorem is proved.

COROLLARY. Let $\boldsymbol{C}$ be a Cayley–Dickson algebra over a field F and $A = \operatorname{diag}\{a_1, a_2, a_3\}$, where $a_i \in F$. Then $H(\boldsymbol{C}_3, S_A)$ is an exceptional Jordan algebra.

Exercises

1. Let the element z of the Jordan algebra J with identity element 1 be nilpotent. Prove that in the algebra J there is found an invertible element u lying in the subalgebra generated by 1 and z and such that $u^2 = 1 - z$.

2. (*McCrimmon*) Let J be a Jordan algebra with identity element 1 over a field F of characteristic $\neq 2$, each element of which has the form $\alpha \cdot 1 + z$ where $\alpha \in F$ and z is a nilpotent element. Prove that the set of nilpotent elements of the algebra J is an ideal.

Oborn [163] proved a somewhat more general fact. Namely, if in a Jordan algebra J with identity element, each element is either invertible or nilpotent, then the set of nilpotent elements of this algebra forms an ideal.

3. THE QUASI-REGULAR RADICAL

In the class of associative algebras and the class of alternative algebras the quasi-regular radical is the unique radical for which both the radical and semisimple classes are satisfactorily described. It is natural to expect that an appropriate analog of this radical will also have nice properties in the class of Jordan algebras. The quasi-regular radical was constructed in the class of Jordan algebras by McCrimmon in 1969 [124], and in fact has a series of nice properties. Nevertheless, nothing is as yet known on the semisimple class of this radical. Recently Osborn [165] established that the quasi-regular radical of a Jordan algebra may not coincide with the intersection of the kernels of the irreducible representations of this algebra. However, hopes for a nice description of the semisimple class of the quasi-regular radical remain just the same.

An element a of a Jordan algebra J is called *quasi-regular* if there exists an element $b \in J$ such that in the algebra $J^{\#} = \Phi \cdot 1 + J$ obtained by the formal adjoining of an identity element to J the element $1 - a$ is invertible with inverse element $1 - b$. The element b is called a *quasi-inverse* for the element a. Since an invertible element of a Jordan algebra has only one inverse element, the quasi-inverse of a quasi-regular element is also unique.

THEOREM 7. The following conditions are equivalent for elements a and b of a Jordan algebra J:

(a) a is quasi-regular with quasi-inverse b;
(b) b is quasi-regular with quasi-inverse a;

(c) the relations are valid

$$\begin{cases} a + b - ab = 0, \\ (a, a, b) = 0; \end{cases}$$

(d) the relation $a + b - ab = 0$ is valid, and the subalgebra generated by the elements a and b is strongly associative.

PROOF. The equivalence (a) $\Leftrightarrow$ (b) follows from Lemma 4.

(a) $\Leftrightarrow$ (c). We write out the relations $(1-a)(1-b)=1$ and $(1-a)^2(1-b)= 1-a$ to obtain the relations

$$a + b - ab = 0,$$
$$-a - b + a^2 + 2ab - a^2b = 0.$$

Transforming the second relation twice by means of the first, we obtain $0 = -a - b + a^2 + 2ab - a^2b = a^2 + ab - a^2b = a(a + b) - a^2b = a(ab) - a^2b = -(a, a, b)$. Thus the equivalence of conditions (a) and (c) is proved.

(a) $\Rightarrow$ (d). Since strong associativity of the subalgebra A generated by the elements a and b is equivalent to commutativity of the algebra $R^J(A)$, it suffices to prove that some system of generators of the algebra $R^J(A)$ consists of pairwise commuting elements. By Proposition 3.2 and the fact that $ab = a + b$, the set $\{R_a, R_b, R_{a^2}, R_{b^2}\}$ generates $R^J(A)$. By Theorem 5 the subalgebra generated by the elements $1 - a$ and $1 - b$ is strongly associative, and therefore

$$[R_a, R_b] = [R_{1-a}, R_{1-b}] = 0,$$
$$[R_a, R_{b^2}] = -[R_{1-a}, R_{(1-b)^2}] - 2[R_a, R_b] = 0.$$

Analogously we have

$$[R_{a^2}, R_b] = 0.$$

Furthermore,

$$[R_{a^2}, R_{b^2}] = [R_{(1-a)^2}, R_{(1-b)^2}] + 2[R_a, R_{b^2}] + 2[R_{a^2}, R_b] - 4[R_a, R_b] = 0.$$

We have proved that the elements of the set $\{R_a, R_b, R_{a^2}, R_{b^2}\}$ pairwise commute. Consequently, the algebra $R^J(A)$ is commutative, and this means that the subalgebra A is strongly associative.

The implication (d) $\Rightarrow$ (c) is obvious.

This proves the theorem.

COROLLARY. If there is an identity element 1 in the algebra J, then an element a is quasi-regular with quasi-inverse element b if and only if $1 - a$ is invertible with inverse $1 - b$.

This follows, for example, from (c) of Theorem 7.

An algebra is called *quasi-regular* if each of its elements is quasi-regular. An ideal is called *quasi-regular* if it is quasi-regular as an algebra.

PROPOSITION 3. Let I be an ideal of an algebra J and a be a quasi-regular element with quasi-inverse element b. If $a \in I$, then $b \in I$ also. Each ideal of a quasi-regular algebra is quasi-regular.

PROOF. This first assertion is obvious in view of Theorem 7(c). The second assertion follows from the first.

LEMMA 8. Let the element a belong to a quasi-regular ideal I of the algebra J and u be an invertible element in $J^{\#}$. Then the element $u - a$ is also invertible.

PROOF. By Lemma 6 the element $u - a$ is invertible if and only if $\{(u - a)1(u - a)\} = (u - a)^2$ is invertible. The invertibility of the element $(u - a)^2$ is equivalent to the invertibility of the element $(u - a)^2 U_{u^{-1}}$. We consider this element. We have

$$(u - a)^2 U_{u^{-1}} = (u^2 - 2au + a^2) U_{u^{-1}} = 1 - a',$$

where the element $a' = (2au - a^2) U_{u^{-1}}$ belongs to I. Therefore the element $(u - a)^2 U_{u^{-1}}$ is invertible, and together with it also the element $u - a$. This proves the lemma.

LEMMA 9. The sum of two quasi-regular ideals is a quasi-regular ideal.

PROOF. Let K and L be quasi-regular ideals of the algebra J. We consider an arbitrary element $a = k + l$ of the ideal $K + L$, and prove that it is quasi-regular. Actually $1 - a = (1 - k) - l$, the element l belongs to the quasi-regular ideal L, and the element $1 - k$ is invertible. By Lemma 8 the element $1 - a$ is invertible also, but this means the element a is quasi-regular, which proves the lemma.

LEMMA 10. Let $\varphi: J \to J/I$ be some homomorphism of the algebra J, $a \in J$, and $\bar{a}$ be the image of the element a under the homomorphism φ. If a is quasi-regular, then $\bar{a}$ is also. If $\bar{a}$ is quasi-regular and I is a quasi-regular ideal, then a is quasi-regular.

PROOF. If a is quasi-regular with quasi-inverse b, then $\bar{a}$ will obviously be quasi-regular with quasi-inverse $\bar{b} = \varphi(b)$.

Now let $\bar{a}$ have quasi-inverse element $\bar{c}$ and I be a quasi-regular ideal in J. The set I will also be an ideal in $J^{\#}$. In the quotient algebra $\bar{J}^{\#} = J^{\#}/I$ we have the relation $(\bar{1} - \bar{c})^2 U_{\bar{1} - \bar{a}} = \bar{1}$, and passing to the inverse image we obtain $(1-c)^2 U_{1-a} = 1 - x$ where $x \in I$. Since I is quasi-regular, $1 - x$ is an invertible element, and by Lemma 6 the element $1 - a$ is also invertible. Consequently a is quasi-regular, and the lemma is proved.

Now let J be an arbitrary Jordan algebra. We denote by $\mathscr{J}(J)$ the sum of all quasi-regular ideals of the algebra J. In view of Lemma 9, this ideal will be quasi-regular. In addition, by Lemma 10 the quotient algebra $J/\mathscr{J}(J)$ does not contain nonzero quasi-regular ideals. Thus we have proved the following theorem:

THEOREM 8 (*McCrimmon*). The class $\mathscr{J}$ of quasi-regular Jordan algebras is radical.

We now turn to the properties of this radical.

PROPOSITION 4. $\mathscr{J}(J) = J \cap \mathscr{J}(J^{\#})$.

PROOF. The containment $\mathscr{J}(J) \subseteq J \cap \mathscr{J}(J^{\#})$ is true since $\mathscr{J}(J)$ is a radical ideal of the algebra $J^{\#}$. The reverse containment is also valid by the corollary to Theorem 7. This proves the proposition.

PROPOSITION 5. If for some $n \geqq 1$ the element a^n is quasi-regular, then a is also quasi-regular.

PROOF. The element a is quasi-regular in J if and only if $1 - a$ is invertible in $J^{\#}$, and this, in turn, by Lemma 5, is equivalent to the invertibility of the operator U_{1-a}. Analogously a^n is quasi-invertible if and only if U_{1-a^n} is invertible. However, by Shirshov's theorem

$$U_{1-a^n} = U_{1-a} U_{1+a+\cdots+a^{n-1}} = U_{1+a+\cdots+a^{n-1}} U_{1-a},$$

whence it follows that invertibility of the operator U_{1-a^n} implies invertibility of the operator U_{1-a}. By what was said above, this means that quasi-regularity of the element a^n implies quasi-regularity of the element a.

This proves the proposition.

If an element a is nilpotent, then $a^n = 0$ for some n. Since 0 is a quasi-regular element, by what was just proved a is quasi-regular.

We note that the converse assertion to Proposition 5 is not true. Since $(-1) + \frac{1}{2} = (-1) \cdot \frac{1}{2}$, -1 is a quasi-regular element, but $(-1)^2 = 1$ is not a quasi-regular element.

An element a of a Jordan algebra J is called *von Neumann regular* if $a = xU_a$ for some $x \in J$.

THEOREM 9 (*McCrimmon*). If a is a von Neumann regular element of a Jordan algebra J and $a \in \mathscr{J}(J)$, then $a = 0$.

PROOF. We suppose first that $a \in \mathscr{J}(J)$ and $a = uU_a$, where u is an invertible element from $J^\#$. Then by Lemma 8 the element $v = u - aU_u$ is invertible, and consequently, by Lemma 5, the operator U_v must also be invertible. We consider the element aU_v. We have

$$\begin{aligned} aU_v &= a(U_u - 2U(u, aU_u) + U(aU_u)) = aU_u - 2\{ua\{uau\}\} + aU(aU_u) \\ &= aU_u - 2\{u\{aua\}u\} + aU(aU_u) = aU_u - 2aU_u + aU_u = 0, \end{aligned}$$

since by Macdonald's identity

$$aU(aU_u) = aU_uU_aU_u = uU_aU_uU_aU_u = uU(uU_a)U_u = uU_aU_u = aU_u.$$

Since $aU_v = 0$ and the element v is invertible, then $a = 0$. In this case the theorem is proved.

Now let $a = xU_a$, where x is an arbitrary element from $J^\#$. We have

$$a = xU_a = xU(xU_a) = xU_aU_xU_a = x'U_a,$$

where $x' = xU_aU_x \in \mathscr{J}(J)$. Therefore it is possible to assume that $x \in \mathscr{J}(J)$. Furthermore, by identity (3.47) we obtain

$$a^2 = (xU_a)^2 = a^2U_xU_a = yU_a,$$

where $y = a^2U_x \in \mathscr{J}(J)$. We now consider the invertible element $u = 1 - (y - x)$, and we apply to it the operator U_a. We obtain

$$uU_a = (1 + x - y)U_a = a^2 + a - a^2 = a,$$

that is, $uU_a = a$. But this case was analyzed before. Hence it follows that $a = 0$.

This proves the theorem.

COROLLARY. The radical $\mathscr{J}(J)$ does not contain idempotents.

THEOREM 10. If Φ is a field, then any algebraic over Φ element of the algebra J contained in $\mathscr{J}(J)$ is nilpotent.

PROOF. Let a be an algebraic element and $a \in \mathscr{J}(J)$. Then $J_0 = \Phi[a]$ is a finite-dimensional associative subalgebra in J. We consider the chain of subspaces

$$J_0 \supseteq aJ_0a \supseteq a^2J_0a^2 \supseteq \cdots.$$

In view of the finite-dimensionality of the subalgebra J_0, there can be found a number n such that $a^nJ_0a^n = \cdots = a^{2n+1}J_0a^{2n+1}$. But then for some $x \in J_0$ we shall have

$$a^{2n+1} = a^{2n+1}xa^{2n+1} = xU_{a^{2n+1}},$$

whence by Theorem 9 $a^{2n+1} = 0$. This proves the theorem.

COROLLARY. In any finite-dimensional Jordan algebra J over a field Φ the radical $\mathscr{J}(J)$ is a nil-ideal.

It is well-known that the Jacobson radical of an associative algebra A consists precisely of the *properly quasi-regular elements*, that is, such elements $z \in A$ for which all elements az (or all za) are quasi-regular. We reformulate the condition of proper quasi-regularity in other terms.

PROPOSITION 6 (*McCrimmon*). An element az of an associative algebra A is quasi-regular if and only if the element z is quasi-regular in the homotope $A^{(a)}$. An element z is properly quasi-regular if and only if it is quasi-regular in each homotope of the algebra A.

We leave the proof of this proposition to the reader.

We shall call an element of a Jordan algebra J *properly quasi-regular* if it is quasi-regular in every homotope of the algebra J. McCrimmon proved that the quasi-regular radical $\mathscr{J}(J)$ of a Jordan algebra J is precisely the set PQI(J) of all its properly quasi-regular elements. We devote the remainder of this section to the proof of this characterization of the quasi-regular radical.

We define yet one more operator in the multiplication algebra of a Jordan algebra:

$$xV_{y,z} = \{xyz\} = x(R_yR_z - R_zR_y + R_{yz}).$$

Operators in the multiplication algebra of a homotope $J^{(a)}$ will be denoted $R_x^{(a)}$, $U_x^{(a)}$, $V_{x,y}^{(a)}$. The operation of squaring in the homotope $J^{(a)}$ is denoted $x^{2(a)}$, and the identity element of the homotope $J^{(a)}$ is denoted by $1^{(a)}$. It follows from the definitions that

$$x^{2(a)} = aU_x, \qquad R_x^{(a)} = V_{a,x}. \tag{9}$$

In addition, by identities (7) and (8)

$$V_{x,y}^{(a)} = V_{xU_a,y}, \qquad U_x^{(a)} = U_aU_x. \tag{10}$$

We note that by Lemma 5 invertibility of the element $1 - x$ in the Jordan algebra $J^{\#}$ is equivalent to invertibility of the operator

$$U_{1-x} = E - 2R_x + U_x. \tag{11}$$

By (9) and (10) the operator $U_{1^{(a)}-x}^{(a)}$ acts on $J^{(a)}$ like $E - 2V_{a,x} + U_aU_x$. Therefore we introduce in the multiplication algebra of an arbitrary Jordan algebra J the operators

$$T_{x,y} = E - 2V_{x,y} + U_xU_y.$$

For the proof of properties of the operator $T_{x,y}$ we need the *Koecher principle for Jordan algebras*. We reformulate it in a form convenient for us.

KOECHER PRINCIPLE. If in any Jordan algebra J with identity element, a nonassociative polynomial $f = f(x_1, \ldots, x_n)$ vanishes under the substitution in f of any invertible elements of the algebra J, then f is an identity in all Jordan algebras.

PROOF. In view of Proposition 1.9 it is only necessary to note that elements of the form $1+x$, where x is nilpotent, are invertible by Proposition 5.

LEMMA 11. The operators $T_{x,y}$ satisfy the relations

$$
\begin{aligned}
&\text{(a)}\quad T_{x,0} = T_{0,x} = E, \qquad T_{\alpha x,y} = T_{x,\alpha y}, \qquad \alpha \in \Phi;\\
&\text{(b)}\quad T_{x,y} \cdot T_{-x,y} = T_{x,y} \cdot T_{x,-y} = T_{x,xU_y} = T_{yU_x,y};\\
&\text{(c)}\quad T_{x,y}U_zT_{y,x} = U(zT_{y,x});\\
&\text{(d)}\quad T_{x,y} \cdot T_{y,x} = E - 2R_w + U_w,
\end{aligned}
\tag{12}
$$

where $w = 2xy - y^2U_x$.

PROOF. Relation (a) is obvious. Relations (b) and (d) can be proved by means of Corollary 2 to Shirshov's theorem. However, (c) cannot be proved that way. We employ the Koecher principle.

We note that for invertible x and y

$$T_{x,y} = U_xU_{x^{-1}-y} = U_{y^{-1}-x}U_y. \tag{13}$$

In fact, by Theorem 5 and relations (3.31) and (3.31′) we have

$$
\begin{aligned}
U_xU_{x^{-1},y} &= 2R_xR_y - U_{x,y} = V_{x,y},\\
U_{y^{-1},x}U_y &= 2R_xR_y - U_{y,x} = V_{x,y}.
\end{aligned}
$$

In addition, by (2) $U_xU_{x^{-1}} = U_{x^{-1}}U_x = E$. Therefore $U_xU_{x^{-1}-y} = U_xU_{x^{-1}} - 2U_xU_{x^{-1},y} + U_xU_y = T_{x,y}$. The second equality is also proved analogously.

Since in the isotope $J^{(x)}$ the element x^{-1} is an identity element, relation (13) can by (10) be written in the following manner:

$$T_{x,y} = U^{(x)}_{1^{(x)}-y} = U_{y^{-1}}U^{(y)}_{1^{(y)}-x}U_y. \tag{14}$$

We now have

$$
\begin{aligned}
T_{x,y}T_{x,-y} &= U^{(x)}_{1^{(x)}-y}U^{(x)}_{1^{(x)}+y} = U^{(x)}_{1^{(x)}-y^{2(x)}}\\
&= U^{(x)}_{1^{(x)}-xU_y} = T_{x,xU_y}.
\end{aligned}
$$

Analogously

$$
\begin{aligned}
T_{x,y}T_{-x,y} &= U_{y^{-1}}U^{(y)}_{1^{(y)}-x}U^{(y)}_{1^{(y)}+x}U_y = U_{y^{-1}}U^{(y)}_{1^{(y)}-x^{2(y)}}U_y\\
&= U_{y^{-1}}U^{(y)}_{1^{(y)}-yU_x}U_y = T_{yU_x,y}.
\end{aligned}
$$

The equality $T_{x,y}T_{-x,y} = T_{x,y}T_{x,-y}$ follows from the fact that $T_{-x,y} = T_{x,-y}$. Thus (b) is proved.

We shall prove (c). By (13) and Macdonald's identity, for invertible elements x, y, z we have

$$T_{x,y}U_zT_{y,x} = U_xU_{x^{-1}-y}U_zU_{x^{-1}-y}U_x = U(zU_{x^{-1}-y}U_x) = U(zT_{y,x}).$$

By the Koecher principle (c) is proved.

If we set $z = 1$, relation (d) follows from (c):

$$T_{x,y}T_{y,x} = U(1T_{y,x}) = U(1 - 2xy + y^2U_x) = E - 2R_w + U_w$$

This proves the lemma.

LEMMA 12. The following conditions are equivalent for elements x, y of a Jordan algebra J:

(a) x is quasi-regular in $J^{(y)}$;
(b) $T_{y,x}$ is invertible on J;
(c) $T_{y,x}$ is surjective on J;
(d) $2x - yU_x$ lies in the image of $T_{y,x}$;
(e) $2xy - y^2U_x$ lies in the image of $T_{y,x}$;
(f) $2xy - y^2U_x$ is quasi-regular in J.

If z is the quasi-inverse element for x in $J^{(y)}$, then the operators $T_{y,x}$ and $T_{y,z}$ are mutual inverses.

PROOF. If x is quasi-regular in $J^{(y)}$ with quasi-inverse element z, the operators $U^{(y)}_{1^{(y)}-x}$ and $U^{(y)}_{1^{(y)}-z}$ are mutual inverses on $(J^{(y)})^{\#}$. Their restrictions $T_{y,x}$ and $T_{y,z}$ on the ideal $J = J^{(y)}$ are also mutual inverses. Thus (a) implies (b), and in addition the final assertion of the lemma is proved.

The implications (b) $\Rightarrow$ (c), (c) $\Rightarrow$ (d), and (c) $\Rightarrow$ (e) are obviously correct. Furthermore

$$1^{(y)}U^{(y)}_{1^{(y)}-x} = (1^{(y)} - x)^{2(y)} = 1^{(y)} - 2x + yU_x.$$

It is now clear by (14) that the presence of the element $2x - yU_x$ in the image of the operator $T_{y,x}$ implies the presence of the identity element $1^{(y)}$ in the image of the operator $U^{(y)}_{1^{(y)}-x}$, which implies by Lemma 5 the invertibility of the element $1^{(y)} - x$ in the algebra $(J^{(y)})^{\#}$, and this in turn means the quasi-regularity of the element x in $J^{(y)}$. We have proved that (d) $\Rightarrow$ (a).

Now let $w = 2xy - y^2U_x$ lie in the image of $T_{y,x}$, that is, $zT_{y,x} = w$ for some $z \in J$. Since $1T_{y,x} = 1 - w$, then $(1 + z)T_{y,x} = 1$. By (12c) we have

$$E = U((1 + z)T_{y,x}) = T_{x,y}U_{1+z}T_{y,x},$$

whence it follows the operator $T_{y,x}$ is surjective. In view of the already proved equivalence (b) $\Leftrightarrow$ (c), the operator $T_{y,x}$ is invertible on J. But then the product $T_{x,y}U_{1+z} = T_{y,x}^{-1}$ is invertible, and consequently, the operator U_{1+z} is surjective on J. This implies the invertibility of the operator U_{1+z}, which means also the operator $T_{x,y} = T_{y,x}^{-1}U_{1+z}^{-1}$ is invertible. Now by (12d) the operator $U_{1-w} = E - 2R_w + U_w$ is invertible on J, and consequently, the element w is quasi-regular. Thus (e) ⇒ (f). From relation (12d) it follows, too, that (f) ⇒ (c). This proves the lemma.

In the process of the proof of this lemma we have established that invertibility of the operator $T_{x,y}$ implies invertibility of the operator $T_{y,x}$. This means that the following proposition is valid.

PROPOSITION 7 (*Symmetry Principle*). An element x is quasi-regular in $J^{(y)}$ if and only if y is quasi-regular in $J^{(x)}$.

Also valid is

PROPOSITION 8 (*Shift Principle*). The element xU_z is quasi-regular in $J^{(y)}$ if and only if x is quasi-regular in $J^{(yU_z)}$.

PROOF. We need Macdonald's identity

$$yU(xU_z) = yU_zU_xU_z, \tag{15}$$

and its linearization in x

$$yU(xU_z, wU_z) = yU_zU_{x,w}U_z. \tag{16}$$

We suppose that x is quasi-regular in $J^{(yU_z)}$ with quasi-inverse w. This means by (c) of Theorem 7 that

$$x + w = \{x\{zyz\}w\},$$
$$\{\{x\{zyz\}x\}\{zyz\}w\} = \{x\{zyz\}\{x\{zyz\}w\}\},$$

or equivalently

$$x + w = yU_zU_{x,w}, \tag{17}$$

$$yU_zU(yU_zU_{x,w}) = yU_zU(x, yU_zU_{x,w}). \tag{18}$$

Now applying the operator U_z to both sides of equality (17), by (16) we shall have

$$xU_z + wU_z = yU_zU_{x,w}U_z = yU(xU_z, wU_z). \tag{19}$$

Applying U_z to both sides of equality (18), by (15) and (16) we obtain

$$yU(yU(xU_z), wU_z) = yU(xU_z, yU(xU_z, wU_z)). \tag{20}$$

Equalities (19) and (20) can be written in the following manner:

$$xU_z + wU_z = (xU_z) \cdot_y (wU_z),$$
$$(xU_z)^{2(y)} \cdot_y (wU_z) = (xU_z) \cdot_y [(xU_z) \cdot_y (wU_z)].$$

But by Theorem 7(c) this means that xU_z is quasi-regular in the homotope $J^{(y)}$ with quasi-inverse wU_z.

Conversely, if xU_z is quasi-regular in $J^{(y)}$, then by the symmetry principle y is quasi-regular in $J^{(xU_z)}$, whence by what was proved above yU_z is quasi-regular in $J^{(x)}$. Applying the symmetry principle once again, we obtain that x is quasi-rgular in $J^{(yU_z)}$.

This proves the proposition.

In view of Lemma 12 and the symmetry principle, the set $PQI(J)$ of properly quasi-regular elements of the algebra J can be described in the three following ways:

$$\begin{aligned} \mathrm{PQI}(J) &= \{z \mid T_{x,z} \text{ is invertible for all } x \in J\} \\ &= \{z \mid T_{z,x} \text{ is invertible for all } x \in J\} \\ &= \{z \mid \mathscr{J}(J^{(z)}) = J^{(z)}\}. \end{aligned} \tag{21}$$

The following lemma justifies our terminology.

LEMMA 13. Any properly quasi-regular element is quasi-regular.

PROOF. Let $z \in \mathrm{PQI}(J)$. If there is an identity element in J, then everything is obvious since z is quasi-regular in $J^{(1)}$, that is, in J. In the general case it is easy to see that the element z^2 is quasi-regular, since the operator

$$U_{1-z^2} = E - 2R_{z^2} + U_{z^2} = E - 2V_{z,z} + U_z U_z = T_{z,z}$$

is invertible on J. By Proposition 5 the element z is also quasi-regular. This proves the lemma.

It will be convenient for us now to consider only algebras with an identity element. For this we need

LEMMA 14. The equality

$$\mathrm{PQI}(J) = \mathrm{PQI}(J^{\#}) \cap J$$

holds for any Jordan algebra J.

PROOF. Let $z \in \mathrm{PQI}(J^{\#}) \cap J$. Then for any $x \in J$ this element is quasi-regular in $(J^{\#})^{(x)}$. Since J is an ideal in $(J^{\#})^{(x)}$, by Proposition 3 the quasi-inverse element w for z in $(J^{\#})^{(x)}$ lies in J. This means that z is quasi-regular in $J^{(x)}$ for any $x \in J$, that is, $z \in \mathrm{PQI}(J)$.

Now let $z \in \mathrm{PQI}(J)$. We shall show that for any $x' \in J^{\#}$ the operator $T_{x',z}$ is invertible on $J^{\#}$. By relation (12b) we have

$$T_{-x',z} T_{x',z} = T_{zU_{x'},z},$$

whence by the invertibility on J of the last operator it follows $T_{x',z}$ is surjective on J. In particular, the element $2z - x'U_z$ lies in the image of $T_{x',z}$, whence by Lemma 12 z is quasi-regular in $(J^{\#})^{(x')}$, that is, $z \in \mathrm{PQI}(J^{\#})$.

LEMMA 15. $\mathrm{PQI}(J) \subseteq \mathrm{PQI}(J^{(x)})$ for any homotope $J^{(x)}$.

The proof is obvious, since by Lemma 7 any homotope of a homotope is again a homotope, $J^{(x)(y)} = J^{(yU_x)}$.

LEMMA 16. If J is a Jordan algebra with identity element and $z \in \mathrm{PQI}(J)$, then the element $u - z$ is invertible for any invertible element u.

PROOF. The element u is the identity element in the isotope $J^{(v)}$, where $v = u^{-1}$. Since z is quasi-regular in $J^{(v)}$, then $u - z = 1^{(v)} - z$ is invertible in $J^{(v)}$. This is equivalent to the fact that the operator $U^{(v)}_{u-z} = U_v U_{u-z}$ is invertible. By the invertibility of v, the operator U_v is invertible. Consequently the operator U_{u-z} is invertible, and this is equivalent to the invertibility of $u - z$ in J.

COROLLARY. Let z be a properly quasi-regular element of an arbitrary Jordan algebra J. Then the element $u - z$ is invertible in $(J^{(x)})^{\#}$ for any invertible element u of the algebra $(J^{(x)})^{\#}$.

PROOF. By Lemma 15 $z \in \mathrm{PQI}(J^{(x)})$, and by Lemma 14 $z \in \mathrm{PQI}((J^{(x)})^{\#})$. It now only remains to apply Lemma 16.

THEOREM 11 (*McCrimmon*). $\mathcal{J}(J) = \mathrm{PQI}(J)$ for any Jordan algebra J.

PROOF. If $z \in \mathcal{J}(J)$, then for any $x \in J$ the element $2zx - x^2U_z$ is quasi-regular in J, and by Lemma 12 this means that z is quasi-regular in $J^{(x)}$, that is, $z \in \mathrm{PQI}(J)$. Thus $\mathcal{J}(J) \subseteq \mathrm{PQI}(J)$.

It remains to prove that $\mathrm{PQI}(J)$ is an ideal. That the set $\mathrm{PQI}(J)$ is closed with respect to scalar multiplication follows from (12a) and (21). We shall prove that $\mathrm{PQI}(J)$ is closed with respect to addition. Let $x, y \in \mathrm{PQI}(J)$. Then in the algebra $(J^{(z)})^{\#}$ the element

$$1^{(z)} - (x + y) = (1^{(z)} - x) - y$$

is invertible by the corollary to Lemma 16, and this means that $x + y$ is quasi-regular in $J^{(z)}$. Since the element z is arbitrary, we have $x + y \in \mathrm{PQI}(J)$.

Furthermore, if $x \in \mathrm{PQI}(J)$ and $y \in J$, then for any $z \in J$ the element x is quasi-regular in $J^{(zU_y)}$, whence by the shift principle we have the quasi-regularity of the element xU_y in any homotope $J^{(z)}$, that is, $xU_y \in \mathrm{PQI}(J)$. By Proposition 4 and Lemma 14, we can assume there is an identity element in the algebra J. But then

$$xy = \tfrac{1}{2}(xU_{y+1} - xU_y - x) \in \mathrm{PQI}(J),$$

and we have proved that the set $\mathrm{PQI}(J)$ is an ideal. This proves the theorem.

Exercises

1. An element a of an associative algebra A is quasi-regular if and only if it is quasi-regular in $A^{(+)}$. It is also true that

$$\mathcal{J}(A) = \mathcal{J}(A^{(+)}).$$

Hint. Use Proposition 2 and Exercise 3.1.5.

2. (*McCrimmon*) Let A be an associative algebra with involution$*$ and $H = H(A, *)$. Prove that if an element $a \in H$ is quasi-regular with quasi-inverse b, then $b \in H$. Show that the element $a \in A$ is quasi-regular if and only if the elements $a + a^* - aa^*$ and $a + a^* - a^*a$ are both quasi-regular. Establish the equality

$$\mathcal{J}(H) = H \cap \mathcal{J}(A).$$

3. (*Lover*) Let J be a Jordan algebra and $x \in J$. Then

$$\mathcal{J}(J^{(x)}) = \{z \mid zU_x \in \mathcal{J}(J)\}.$$

4. HEREDITY OF RADICALS

We shall begin with the following theorem.

THEOREM 12 (*Slin'ko*). Let J be a Jordan algebra, I be an ideal in J, M be an ideal in I, and the quotient algebra I/M not contain nonzero nilpotent ideals. Then M is an ideal in J.

Under the conditions of Theorem 12 we shall first prove a series of lemmas.

LEMMA 17. $M^2J \subseteq M$.

PROOF. We shall prove that $M + M^2J/M$ is a trivial ideal in the algebra I/M. For this it suffices to show that $(M^2J)I \subseteq M$. Since M^2 is the Φ-module spanned by the squares of elements from M, it suffices to prove that $(m^2a)i \in M$ for any $m \in M$, $a \in J$, $i \in I$. (In the course of the proof of Theorem 12 we shall subsequently use this method without stipulation.) We now write the basic Jordan identity in the form $(x^2, y, x) = 0$. Linearizing it we obtain

$$(xy, z, t) + (xt, z, y) + (yt, z, x) = 0. \tag{22}$$

By (22) we now have

$$(m^2a)i = m^2(ai) + (m^2, a, i) = m^2(ai) - 2(mi, a, m) \in M.$$

In view of the absence of trivial ideals in I/M, $M^2J \subseteq M$. This proves the lemma.

LEMMA 18. $M + xM$ is a subalgebra in J for any element $x \in J$.

PROOF. Let m, $n \in M$. By the previous lemma $(mn)x \in M$. Therefore relation (3.22) implies

$$(xm)(xn) = \tfrac{1}{2}\{-x^2(mn) + [(xm)x]n + [(xn)x]m + x[(mn)x]\} \in M + xM.$$

The lemma is proved, and at the same time it is established that under the conditions of Theorem 12 for any $x \in J$

$$(xm)(xn) \equiv \tfrac{1}{2}[(mn)x]x \qquad (\mathrm{mod}\ M). \tag{23}$$

LEMMA 19. For any $x \in J$, $C_x = M + xM + (xM)I$ is an ideal in I such that $C_x^2 \subseteq M + xM$.

PROOF. In view of Lemma 18, it suffices to show that $((xM)I)I \subseteq M + xM$. Let $m \in M$, $i, j \in I$. Then by (3.22) we have

$$\begin{aligned}[(xm)i]j = &-[(xj)i]m - x[(mj)i] + (xm)(ij) \\ &+ (xi)(mj) + (xj)(mi) = m' + xm'' + (xm)(ij),\end{aligned}$$

where m', $m'' \in M$. We transform $(xm)(ij)$ according to (22):

$$\begin{aligned}(xm)(ij) &= x[m(ij)] + (x, m, ij) \\ &= x[m(ij)] - (i, m, xj) - (j, m, xi) \in M + xM.\end{aligned}$$

This proves the lemma.

LEMMA 20. For any x, $y \in J$ the following containments hold:

(a) $((IM)M)x \subseteq M$;
(b) $(IM^2)x \subseteq M$;
(c) $(((IM^2)M)x)x \subseteq M$;
(d) $(M^3y)x \subseteq M$;
(e) $(((M^2M^2)y)x)x \subseteq M$;
(f) $((M^3x)x)x \subseteq M$.

PROOF. Let $i \in I$, m, $n \in M$.

(a) By identity (3.22) we obtain

$$[(im)n]x = -[(ix)n]m - i[(mx)n] + (im)(nx) + (in)(mx) + (ix)(mn) \in M.$$

(b) By identity (3.24) we have

$$(im^2)x = -2[i(mx)]m + (ix)m^2 + 2(im)(xm) \in M.$$

Applying (a), (b), and identity (3.22), we shall now prove points (c) and (d):

$$\begin{aligned}(((im^2)n)x)x &= -(((im^2)x)x)n - (im^2)((nx)x) \\ &\quad + ((im^2)n)x^2 + 2((im^2)x)(nx) \in M,\end{aligned}$$

$$(y(m^2n))x = -2((ym)n)m)x + ((yn)m^2)x + 2((ym)(mn))x \in M.$$

(e) By identity (3.24) we obtain

$$\begin{aligned}(((m^2n^2)y)x)x &= -2(((m^2(ny))n)x)x + (((m^2y)n^2)x)x \\ &\quad + 2(((m^2n)(ny))x)x.\end{aligned}$$

The first summand on the right side lies in M by (c), the second and third lie in M by (d) and the fact that M^3 is an ideal in I.

(f) By (d) and identity (3.22) we have

$$((M^3x)x)x \subseteq (M^3x)x^2 + M^3x^3 \subseteq M.$$

This proves the lemma.

PROOF OF THEOREM 12. Let M not be an ideal in J. Then there exists an element $x \in J$ such that $xM \not\subseteq M$. In this case C_x/M is a nonzero ideal in I/M. We shall show that $(C_x/M)^{(3)} = (0)$, that is, that the ideal C_x/M is solvable of index $\leqq 3$. Since by Lemma 19 $C_x^2 \subseteq M + xM$, it suffices for this to prove that $(xM)^2(xM)^2 \subseteq M$.

From congruence (23) it follows that each element from $(xM)^2$ is representable in the form of a linear combination of elements of the form $(m^2x)x$ and elements from M. By Lemma 17 $m^2x \in M$, $n^2x \in M$ for any elements $m, n \in M$, so that applying (23) once again we obtain

$$((m^2x)x)((n^2x)x) \equiv \tfrac{1}{2}(((m^2x)(n^2x))x)x \qquad (\text{mod}\, M).$$

Hence it follows that each element from $(xM)^2(xM)^2$ is representable in the form of a linear combination of elements from M and elements of the form $(((m^2x)(n^2x))x)x$. We show that the latter lies in M. By (3.24)

$$\begin{aligned}(((m^2x)(n^2x))x)x &= -\tfrac{1}{2}(((m^2n^2)x^2)x)x + (((m^2(n^2x))x)x)x \\ &\quad + \tfrac{1}{2}(((m^2x^2)n^2)x)x.\end{aligned}$$

The first summand belongs to M by Lemma 20(e), the second by Lemma 20(f), and the third by Lemma 20(d).

Thus we have proved that $(C_x/M)^{(3)} = (0)$. However, by Lemma 4.4 the quotient algebra I/M does not contain nonzero solvable ideals. Consequently

$C_x/M = (0)$ and $xM \subseteq C_x \subseteq M$. But this contradicts the choice of the element x.

This proves the theorem.

A radical $\mathscr{R}$ in a class of algebras $\mathfrak{K}$ is called *hypernilpotent* if all nilpotent algebras from $\mathfrak{K}$ are $\mathscr{R}$-radical.

COROLLARY 1. Any hypernilpotent radical $\mathscr{R}$ in the class of Jordan algebras satisfies condition (H).

PROOF. We suppose that in an $\mathscr{R}$-semisimple Jordan algebra J an ideal I is not $\mathscr{R}$-semisimple. Then $\mathscr{R}(I) \neq (0)$, and by Theorem 12 $\mathscr{R}(I)$ is an ideal in J. But then it is contained in $\mathscr{R}(J)$, and this is a zero ideal. This is a contradiction, and the corollary is proved.

Nikitin recently proved the assertion of Corollary 1 for an arbitrary radical $\mathscr{R}$ in the class of Jordan algebras [161].

COROLLARY 2. In the class of Jordan algebras the radicals $\mathscr{L}$, $\mathfrak{N}$, $\mathscr{M}$, $\mathscr{A}$, $\mathscr{J}$ are hereditary.

PROOF. All the enumerated radicals are hypernilpotent, and by Corollary 1 satisfy condition (H). By Theorem 8.3 it remains to prove for these radicals condition (G), that is, that an ideal of a radical algebra is radical. For the radicals $\mathscr{L}$, $\mathfrak{N}$, $\mathscr{J}$ this fact is trivial. Condition (G) for the Andrunakievich radical is proved the same as the corollary to Lemma 8.13. We shall prove condition (G) for the McCrimmon radical.

Let I be an ideal in J, $\mathscr{M}(J) = J$, and $\mathscr{M}(I) \neq I$. By Theorem 12 $\mathscr{M}(I)$ is an ideal in J. Factoring out $\mathscr{M}(I)$, it is possible to assume that $\mathscr{M}(J) = J$, $\mathscr{M}(I) = (0)$. We shall now prove that $I \cap \mathscr{M}(J) = (0)$, from which it will follow that $I = (0)$. Let $z \in \mathscr{M}_1(J)$ be some absolute zero divisor. For an arbitrary element $i \in J$ the element zU_i is an absolute zero divisor in I. In view of the $\mathscr{M}$-semisimplicity of I, we have $zU_i = 0$ for all $i \in I$. Then $\{izj\} = \frac{1}{2}(zU_{i+j} - zU_i - zU_j) = 0$ for any $i, j \in I$, and consequently,

$$\{I\mathscr{M}_1(J)I\} = (0).$$

If the ideal $I \cap \mathscr{M}_1(J)$ was different from zero, then by the identity $(ab)c = \frac{1}{2}\{abc\} + \frac{1}{2}\{bac\}$ it would be nilpotent of index 3, which would contradict the $\mathscr{M}$-semisimplicity of I. Consequently, $I \cap \mathscr{M}_1(J) = (0)$. This fact serves as a basis for the subsequent induction.

We assume that $I \cap \mathscr{M}_\beta(J) = (0)$ for all ordinal numbers $\beta < \alpha$. Then if α is a limit ordinal, it is obvious that $I \cap \mathscr{M}_\alpha(J) = (0)$. If α is not a limit ordinal, then $\mathscr{M}_\alpha(J)$ is the Φ-module generated by all possible elements $z \in J$ such that $JU_z \subseteq \mathscr{M}_{\alpha-1}(J)$. Let z be some element of that sort. We consider zU_i, where i is some element from I. By Macdonald's identity and the induction assumption

$$JU(zU_i) = JU_iU_zU_i \subseteq I \cap \mathcal{M}_{\alpha-1}(J) = (0).$$

But I does not contain absolute zero divisors, and consequently, $zU_i = 0$ for all $i \in I$. Now analogous to the proof of the inductive premise we obtain that $I \cap \mathcal{M}_\alpha(J) = (0)$.

This proves the corollary.

LITERATURE

McCrimmon [114, 117, 124, 125, 128, 134], Nikitin [161], Osborn [163, 165], Skosyrskiy [194], Slin'ko [208, 209, 211], Erickson and Montgomery [284].

CHAPTER **15**

Structure Theory of Jordan Algebras with Minimal Condition

The development of the structure theory for Jordan algebras basically went the same as the development of the structure theories for associative and alternative algebras. However, there were also exceptions in this development. After the finite-dimensional Jordan algebras were studied by Albert, it remained misunderstood for a long time which subsets in Jordan algebras are analogous to one-sided ideals. This became clear in 1965 after the introduction by Topping of the concept of a quadratic ideal. In 1966 Jacobson considered nondegenerate Jordan algebras with minimal condition for quadratic ideals, and proved a theorem for them analogous to the theorem of Wedderburn–Artin on semisimple Artinian associative algebras. However, its classification of the simple algebras was incomplete. The division algebras and simple Jordan algebras of capacity 2 were not described. Osborn later described the algebras of capacity 2. The question on the description of Jordan division algebras is at the present time unsolved.

Since the nondegenerate Jordan algebras are precisely the $\mathscr{M}$-semisimple algebras, there arises the question whether the radical $\mathscr{M}(J)$ is nilpotent in a Jordan algebra J with minimal condition for quadratic ideals. The answer to this question proved to be yes. In 1972 Slin'ko proved that the radical $\mathscr{J}(J)$ (and all the more $\mathscr{M}(J)$) is nilpotent in the case when the algebra J is special. Zel'manov recently removed from this theorem the restriction to special algebras.

Morgan proved earlier that in a Jordan algebra with minimal condition for quadratic ideals, any nilpotent ideal is finite-dimensional.

Thus, as in the associative case, the structure of Jordan algebras with minimal condition turns out to be completely determined up to division algebras.

1. QUADRATIC IDEALS

In this section are considered Jordan algebras over an associative–commutative ring Φ containing an element $\frac{1}{2}$.

Let J be a Jordan algebra. A submodule Q in J is called a *quadratic ideal* if $JU_q \subseteq Q$ for all $q \in Q$. It is clear that any ideal in J is a quadratic ideal, but the converse is not true. If $J = A^{(+)}$ for some associative algebra A, then the quadratic ideals of the algebra J are those Φ-submodules Q in A such that $qAq \subseteq Q$ for all $q \in Q$. In particular, all one-sided ideals of the algebra A are quadratic ideals in $A^{(+)}$.

Let Q be a quadratic ideal in J and $q_1, q_2 \in Q$. Then for any $a \in J$ we have $aU_{q_1,q_2} = \frac{1}{2}(aU_{q_1+q_2} - aU_{q_1} - aU_{q_2}) \in Q$. This means that $JU_{q_1,q_2} \subseteq Q$. If there is an identity element in the algebra, then Q is a subalgebra since $q_1q_2 = \{q_1 1 q_2\} \in Q$. In the general case this is not so. However, by the relation $(q_1q_2)q_3 = \frac{1}{2}\{q_1q_2q_3\} + \frac{1}{2}\{q_2q_1q_3\}$, it is possible to assert that $Q^3 \subseteq Q$.

Let Q be a quadratic ideal of the algebra J and $a \in J$. Then for any $q \in Q$ by Macdonald's identity

$$JU(qU_a) = JU_aU_qU_a \subseteq QU_a.$$

This means that the set QU_a is a quadratic ideal. In particular, JU_a is a quadratic ideal. It is called the *principal quadratic ideal generated by the element a.* By linearized identity (3.47) we have

$$\{axa\}\{aya\} = \{a\{xa^2y\}a\}.$$

Hence it is possible to conclude that a principal quadratic ideal is always a subalgebra.

We note that if z is an absolute zero divisor in J, then the submodule $\Phi \cdot z$ is a quadratic ideal.

Another natural way of obtaining quadratic ideals is contained in the following lemma.

LEMMA 1. Let J be a Jordan algebra, I be an ideal in J, and K be an ideal in I. Then

(a) KI is an ideal in I;
(b) KI is a quadratic ideal in J;
(c) the chain $(K, I)_1 \supseteq (K, I)_2 \supseteq \cdots$ consists of ideals of the algebra I and quadratic ideals of the algebra J.

PROOF. Assertion (a) is obvious. For the proof of (b) one uses the identity

$$(xy, z, t) + (xt, z, y) + (yt, z, x) = 0.$$

We have

$$\begin{aligned}\{(KI)J(KI)\} &\subseteq (J(KI))(KI) + J(KI)^2 \subseteq KI + (J, KI, KI)\\ &\subseteq KI + (K, KI, JI) + (I, KI, KJ) \subseteq KI,\end{aligned}$$

that is, (b) is proved. Assertion (c) follows from (a) and (b).
This proves the lemma.

We now proceed to the study of properties of quadratic ideals.

LEMMA 2. If the quadratic ideal Q of the algebra J is a subalgebra, then an element $q \in Q$ is quasi-regular in J if and only if it is quasi-regular in Q. In addition, $\mathscr{J}(J) \cap Q \subseteq \mathscr{J}(Q)$.

PROOF. Let q be quasi-regular with quasi-inverse element p. It is necessary to prove that $p \in Q$. In view of Theorem 14.7, we have

$$\begin{aligned}0 &= p + q - pq = p + q + q^2 - (p + q)q\\ &= p + q + q^2 - (pq)q = p + q + q^2 - pU_q,\end{aligned}$$

whence $p = pU_q - q^2 - q \in Q$. With this the first assertion of the lemma is proved. The second assertion follows from the first in an obvious manner. This proves the lemma.

For principal quadratic ideals generated by idempotents, which play an important role in the structure theory, it is possible to assert more.

THEOREM 1 (*McCrimmon*). The equality

$$\mathscr{J}(JU_e) = JU_e \cap \mathscr{J}(J) = \mathscr{J}(J)U_e$$

is valid for any idempotent e of a Jordan algebra J.

PROOF. The latter equality is obvious. We shall prove the first one. We note first that containment in one direction was proved by us in the previous lemma. Let $z \in \mathcal{J}(JU_e)$. Then by Theorem 14.11 z is quasi-regular in any homotope $(JU_e)^{(xU_e)}$ of the algebra J. By the shift principle we obtain the element zU_e is quasi-regular in any homotope $J^{(x)}$. But since $z = zU_e$, by Theorem 14.11 $z \in \mathcal{J}(J)$. This gives us the containment $\mathcal{J}(JU_e) \subseteq JU_e \cap \mathcal{J}(J)$. Thus the reverse containment is also proved, and together with it the theorem.

A convenient sufficient test for the presence of an idempotent is

LEMMA 3. Let Q be a nonzero quadratic ideal of a Jordan algebra J, $a \in Q$, and $QU_a = Q$. Then Q contains an idempotent.

PROOF. Let $c \in Q$ be such that $cU_a = a$. Then by identity (3.47) $a^2 = a^2U_cU_a \in Q$. For $q \in Q$ we denote by $\bar{U}_q$ the restriction of the operator U_q on Q. Then by Macdonald's identity $U_a = U_aU_cU_a$, whence it follows that $\bar{U}_c\bar{U}_a = E$. Therefore, for $f = a^2U_c$ we have

$$f^3 = fU_f = a^2U_cU_cU_aU_aU_c = a^2U_c = f.$$

But then $f^4 = f^2$, and we have two possibilities: (1) f^2 is an idempotent, (2) $f^2 = 0$. In the second case $f = f^3 = 0$, whence $0 = U_f = U_cU_aU_aU_c$, which gives us $\bar{U}_a\bar{U}_c = 0$. But then $(\bar{U}_c\bar{U}_a)^2 = 0$. This contradiction shows that the second case is impossible, and proves the lemma.

THEOREM 2. A Jordan algebra J does not have proper quadratic ideals if and only if J is a division algebra or $J = \Phi \cdot x$ and $x^2 = 0$.

PROOF. If J is a division algebra, then for any $0 \neq a \in J$ the operator U_a is invertible, and therefore there are no proper quadratic ideals in the algebra J.

Suppose there are no proper quadratic ideals in J. If there is an absolute zero divisor z in J, then $\Phi \cdot z$ is a quadratic ideal and $J = \Phi \cdot z$. If the algebra J is nondegenerate, then $J = JU_a$ for any $a \neq 0$. Let $b \in J$ be such that $bU_a = a$. Then $U_a = U_aU_bU_a$, and this means that $U_bU_a = E$. Hence it follows that the mapping U_aU_b is idempotent, and since it is surjective, it is the identity map. Consequently, the mapping U_a is invertible for each $a \neq 0$.

We note that by Lemma 3 there is an idempotent e in the algebra J. Then $(ex - x)U_e = 0$ for any $x \in J$. By the invertibility of U_e we have $ex = x$, that is, e is an identity element in J. Since the identity element is in the image of each operator U_a for $a \neq 0$, then by Lemma 14.5 each nonzero element in J is invertible, and J is a division algebra.

In the light of this theorem, the analogy between quadratic ideals of Jordan algebras and one-sided ideals of associative algebras becomes more clear.

Exercises

In all of the exercises J is a Jordan Φ-algebra.

1. For any $a \in J$ the set $\Phi \cdot a + JU_a$ is a quadratic ideal in J.

2. If I is an ideal in J and Q is a quadratic ideal in J, then $I + Q$ is a quadratic ideal in J.

3. Prove that in the algebra $\Phi_n^{(+)}$ the singly generated submodules $\Phi \cdot e_{ij}$ are quadratic ideals. Prove that the sum of such quadratic ideals may not be a quadratic ideal.

4. (*Zel'manov*) We call the set $\xi(a) = \{x \in J \mid ax = (a, J, x) = 0\}$ the *annihilator of the element* $a \in J$, and the set $\xi(A) = \cap_{a \in A} \xi(a)$ the *annihilator of the subset* $A \subseteq J$. Prove that $\xi(A)$ is a quadratic ideal in J.

In addition, let $a, b \in J$, $A \subseteq J$. Then

(a) $a \in \xi(b) \Leftrightarrow b \in \xi(a)$;
(b) $\xi(\xi(\xi(A))) = \xi(A)$;
(c) if A is an ideal in J, then $\xi(A)$ is also an ideal;
(d) $\xi(\{bab\}) \supseteq \xi(b)$;
(e) if $ab = ab^2 = 0$, then $b^2 \in \xi(a)$.

5. (*Jacobson*) If Q is a minimal quadratic ideal in J, then one of the following possibilities takes place:

(1) $Q = \Phi \cdot z$, where z is an absolute zero divisor;
(2) $Q = JU_q$ for each nonzero $q \in Q$ and $Q^2 = (0)$;
(3) $Q = JU_e$ for some idempotent $e \in Q$ and Q is a division algebra.

6. Let the elements $a, b \in J$ be such that $a^2 = 0 = b^2$ and $2ab = 1$. Then there is an idempotent $e \neq 0, 1$ in J.

7. (*Jacobson*) Let J be a nondegenerate Jordan algebra, with an identity element 1, which contains a minimal quadratic ideal. Then either J is a division algebra or it contains an idempotent $e \neq 0, 1$. *Hint.* Use Exercises 5 and 6.

2. RADICAL IDEALS OF JORDAN ALGEBRAS WITH MINIMAL CONDITION

In this section all Jordan algebras are considered over an arbitrary field F of characteristic $\neq 2$.

LEMMA 4 (*McCrimmon*). Let I be a quasi-regular ideal of a Jordan algebra J which satisfies the minimal condition for principal quadratic ideals contained in I. Then $I \subseteq \mathcal{M}(J)$.

PROOF. By the heredity of the McCrimmon radical, it suffices to prove that $I = \mathcal{M}(I)$. We assume the contrary, $I \neq \mathcal{M}(I)$. Then since by Theorem 14.12 $\mathcal{M}(I)$ is an ideal in J, factoring out this ideal we can assume that $\mathcal{M}(I) = (0)$, that is, I is a nondegenerate algebra.

We consider a minimal principal quadratic ideal $Q = JU_a$ contained in I. Then for any $0 \neq q \in Q$, in view of the $\mathcal{M}$-semisimplicity of the algebra I we have $JU_q \neq (0)$, and consequently, $JU_q = Q$. Thus each nonzero element $q \in Q$ is von Neumann regular. But by Theorem 14.9 there cannot be such elements in a quasi-regular ideal I. This contradiction proves the lemma.

LEMMA 5 (*Morgan*). Let I be a solvable ideal of a Jordan algebra J with minimal condition for quadratic ideals contained in I. Then the ideal I is nilpotent and finite-dimensional.

PROOF. By Lemma 4.4 for some natural number n

$$I = I^{[0]} \supset I^{[1]} \supset \cdots \supset I^{[n]} = (0),$$

where $I^{[i+1]} = (I^{[i]})^3$. Since for any k the set $I^{[k]}$ is an ideal in J, it suffices to prove that any ideal L such that $L^3 = (0)$ is finite-dimensional. Actually, then all quotients $I^{[k-1]}/I^{[k]}$ will be finite-dimensional, and therefore the ideal I itself will be finite-dimensional. But then by Theorem 4.2 it will also be nilpotent.

We consider the subspace L^2. Since any element of L^2 is an absolute zero divisor, this subspace is finite-dimensional, that is, $L^2 = F \cdot x_1 + \cdots + F \cdot x_p$. We now consider the subspace $L_1 = L^2 + L^2 J$ and prove its finite-dimensionality. Let $a, b \in J$, $x \in L^2$. Then by identity (3.22)

$$2(xa)(xb) = -(ab)x^2 + 2[(xa)b]x + a(x^2 b) = 0,$$

and consequently, $(xa)(xb) = 0$. We now note that identity (3.23) can be written in the form

$$(x, yt, z) = (x, y, z)t + (x, t, z)y.$$

By this identity, for any $x \in L^2$, $a, b \in J$,

$$bU_{xa} = 2[b(xa)](xa) = 2(b, xa, xa) = 2(b, x, xa)a + 2(b, a, xa)x = 0.$$

We have established that each element of the set xJ is an absolute zero divisor, and consequently, $\dim_F xJ < \infty$. In particular, the sets $x_i J$ are finite-dimensional. Since

$$L_1 = F \cdot x_1 + \cdots + F \cdot x_p + x_1 J + \cdots + x_p J,$$

then $\dim_F L_1 < \infty$.

By identity (3.22) the subspace L_1 is an ideal in J. Since in the quotient algebra J/L_1 the ideal L/L_1 squares to zero, then $\dim_F L/L_1 < \infty$, whence also $\dim_F L < \infty$.

This proves the lemma.

LEMMA 6. Let I be an ideal of a Jordan algebra J. Then

(a) $\mathfrak{Z}(I) = \{z \in J \mid zI^2 = (zI)I = (0)\}$ is an ideal in J;
(b) if $I^2 = I$, then $\mathrm{Ann}(I)$ is an ideal in J.

PROOF. (a) follows from identities (3.23) and (3.24). (b) follows from (a). This proves the lemma.

LEMMA 7 (*Slin'ko*). Let L be a locally nilpotent ideal of a Jordan algebra J, and I be a minimal ideal in J. If J satisfies the minimal condition for quadratic ideals contained in I, then $I \subseteq \mathfrak{Z}(L)$.

PROOF. In view of Lemma 1(c), the chain

$$(I, L)_1 \supseteq (I, L)_2 \supseteq \cdots \supseteq (I, L)_k \supseteq \cdots$$

consists of quadratic ideals. Let k be a number such that $(I, L)_k = (I, L)_{k+1} = \cdots$. If $(I, L)_k = (0)$, then $I \cap \mathfrak{Z}(L) \neq (0)$, and by Lemma 6 $I \subseteq \mathfrak{Z}(L)$.

Let $(I, L)_k \neq (0)$. Then the set M of quadratic ideals Q such that (1) $Q \subseteq I$, (2) Q is an ideal in L, and (3) $QL = Q$, is nonempty. Let P be a minimal element in M. We take an arbitrary element $p \in P$. The set $pR(L)$ is a nonzero ideal in L, and therefore by Lemma 1 the set $P' = pR(L)L$ is a quadratic ideal in J, and an ideal of the algebra L contained in I. By the local nilpotency of the ideal L and Theorem 4.1, $p \notin P'$, and consequently, $P' \subset P$. Since properties (1) and (2) are satisfied for the ideal P' and for all the terms of the chain

$$P' \supseteq (P', L)_1 \supseteq \cdots \supseteq (P', L)_l \supseteq \cdots,$$

this chain can only stabilize at zero, that is, $(P', L)_l = (0)$ for some l. But in this case we again have $I \cap \mathfrak{Z}(L) \neq (0)$, and consequently, $I \subseteq \mathfrak{Z}(L)$.

This proves the lemma.

LEMMA 8. (*Slin'ko*). Let L be a locally nilpotent ideal of a Jordan algebra J which satisfies the minimal condition for quadratic ideals contained in L. Then the ideal L is nilpotent and finite-dimensional.

PROOF. By Lemma 5 it suffices to prove the solvability of the ideal L. We suppose that it is not solvable. Then the chain

$$L \supseteq L^{[1]} \supseteq \cdots \supseteq L^{[n]} \supseteq \cdots$$

of ideals of the algebra J stabilizes at a nonzero term:

$$L^{[k]} = L^{[k+1]} = \cdots = L_1 \neq (0).$$

Since $L_1^3 = L_1$, then also $L_1^2 = L_1$. By Lemmas 6 and 7 we conclude that $\operatorname{Ann}(L_1)$ is a nonzero ideal in J. In the quotient algebra $\bar{J} = J/\operatorname{Ann}(L_1)$ the ideal L_1 has a nonzero image $\bar{L}_1$. Let $\bar{I}$ be a minimal ideal of the algebra $\bar{J}$ contained in $\bar{L}_1$, and let I be its complete inverse image in J. Since I strictly contains $\operatorname{Ann}(L_1)$, we have $IL_1 \neq (0)$. We shall now prove that $IL_1^3 = (0)$, and by the same token, with this arrive at a contradiction to $L_1 = L_1^3$. In fact, by Lemma 7

$$(IL_1)L_1 \subseteq \operatorname{Ann} L_1, \qquad IL_1^2 \subseteq \operatorname{Ann} L_1,$$

and therefore

$$((IL_1)L_1)L_1 = (IL_1^2)L_1 = (0).$$

But now by identity (3.23) we have

$$IL_1^3 \subseteq ((IL_1)L_1)L_1 + (IL_1^2)L_1 = (0).$$

This proves the lemma.

For special algebras we can already prove the basic result of this section.

COROLLARY (*Slin'ko*). Let I be a quasi-regular ideal of a special Jordan algebra J which satisfies the minimal condition for quadratic ideals contained in I. Then the ideal I is nilpotent and finite-dimensional.

PROOF. By Lemma 4 $I \subseteq \mathscr{M}(J)$. However, since J is special, by Theorem 4.4 $\mathscr{M}(J) \subseteq \mathscr{L}(J)$, and our assertion follows from Lemma 8. This proves the corollary.

In order to get rid of the restriction of special algebras, it is necessary to still carry out some arguments for which the idea belongs to Zel'manov.

LEMMA 9. Let A be a finitely generated associative algebra over a field F, and B be an ideal of A with finite co-dimension. Then the algebra B is also finitely generated.

PROOF. Let $a_1, \ldots, a_k$ be generating elements of the algebra A. Without loss of generality it is possible to assume that A has an identity element 1. Let $\{u_1, \ldots, u_n\}$ be a basis for the algebra A modulo B such that $u_1 = 1$, and let V be the subspace spanned by the elements of this basis. We have the relations

$$a_i = \sum \alpha_i^{(j)} u_j + b_i, \qquad u_r u_s = \sum \beta_{rs}^{(i)} u_i + b_{rs}, \tag{1}$$

where $\alpha_i^{(j)}$, $\beta_{rs}^{(i)} \in F$, b_i, $b_{rs} \in B$. We consider the subalgebra C generated by elements of the form $u_i b_j u_l$ and $u_i b_{rs} u_l$. By relation (1) C is an ideal in A since $a_i C \subseteq C$ and $C a_i \subseteq C$. By induction on the degree of a polynomial $f(x_1, \ldots, x_k)$, it is easy to show that whatever this polynomial

$$f(a_1, \ldots, a_k) = \sum \gamma_p u_p + c,$$

where $\gamma_p \in F$, $c \in C$. If $f(a_1, \ldots, a_k) = b \in B$, then $\sum \gamma_p u_p \in B \cap V$. Thus

$$B = (B \cap V) + C,$$

and the algebra B is finitely generated since the ideal C is finitely generated and the subspace $B \cap V$ is finite-dimensional. This proves the lemma.

LEMMA 10. Let w, z be absolute zero divisors of a Jordan algebra J. Then all elements of the set $JU_{w,z}$ are absolute zero divisors.

PROOF. We note that $U_{w,z} = \frac{1}{2}U_{w+z}$, so that it suffices to prove that for any $a \in J$ the element aU_{w+z} is an absolute zero divisor. We denote $w + z$ by v. We shall show that for all $a, b \in J$ the equality

$$\{v\{avb\}v\} = 0 \tag{2}$$

holds. In fact, $\{v\{ava\}v\} = 2\{z\{a(z + w)a\}w\} = 2\{z\{aza\}w\} + 2\{z\{awa\}w\} = 2\{\{zaz\}aw\} + 2\{za\{waw\}\} = 0$. Relation (2) now follows from the obtained relation by linearization in a. We note, too, that by the result of Shirshov–Macdonald the identity

$$\{\{cac\}b\{cac\}\} = 2\{\{c\{acb\}c\}ac\} - \{cb\{c\{aca\}c\}\} \tag{3}$$

is valid in all Jordan algebras, since it is true in all special algebras. Substituting in (3) the element v in place of c, we obtain by (2) that

$$\{\{vav\}b\{vav\}\} = 0,$$

and this means that aU_v is an absolute zero divisor in J. This proves the lemma.

Let A be an associative algebra over a field F of characteristic $\neq 2$. We denote by $A^{((+))}$ the algebra for which the vector space coincides with the vector space of the algebra A and the multiplication is given by the formula $a \circ b = ab + ba$. It is easy to see that the algebra $A^{((+))}$ is Jordan. Any subalgebra J of this algebra is isomorphically imbeddable in the algebra $\tilde{A}^{(+)}$, where $\tilde{A}$ is the associative algebra obtained from the algebra A by introduction of the new multiplication $a \cdot b = 2ab$. Therefore J is special. We shall need these arguments in the proof of the following lemma.

LEMMA 11 (*Zel'manov*). Let the Jordan algebra J be generated by a finite collection of absolute zero divisors and each principal quadratic ideal of J be finite-dimensional. Then J is nilpotent.

PROOF. In the multiplication algebra $R(J)$ of the algebra J we consider the ideal $W = \{w \in R(J) \mid \dim_F Jw < \infty\}$. Since $JU_{a,b} \subseteq JU_{a+b} + JU_a + JU_b$, then $\dim_F JU_{a,b} < \infty$, and $U_{a,b} \in W$ for any $a, b \in J$. In view of the fact that $U_{a,b} = R_a \circ R_b - R_{ab}$, the mapping $\varphi: a \to R_a + W$ is a homomorphism of the algebra J into $(R(J)/W)^{((+))}$. Thus J^φ is a special Jordan algebra generated by a finite number of absolute zero divisors. By Theorem 14.3 the algebra $R(J)/W$ is nilpotent and finite-dimensional. Since the algebra $R(J)$ is finitely generated, then W is also a finitely generated algebra by Lemma 9. Let $a_1, \ldots, a_n$ be generators of the algebra J, $v_1, \ldots, v_s$ be a basis of the algebra $R(J)$ modulo W, and $w_1, \ldots, w_t$ be generators of the algebra W. Then $R(J) = W + Fv_1 + \cdots + Fv_s$, and

$$J^2 = \sum_{i=1}^{n} a_i R(J) = \sum_{i,j} F \cdot a_i v_j + JW = \sum_{i,j} F \cdot a_i v_j + \sum_{q=1}^{t} Jw_q.$$

Hence, it follows that the square of the algebra J is finite-dimensional. By the corollary to Theorem 4.3 and Theorem 14.2, the ideal J^2 is nilpotent. Consequently J is a solvable finitely generated algebra, which by Zhevlakov's Theorem 4.2 is nilpotent.

This proves the lemma.

THEOREM 3 (*Slin'ko–Zel'manov*). Let I be a quasi-regular ideal of a Jordan algebra J which satisfies the minimal condition for quadratic ideals contained in I. Then the ideal I is nilpotent and finite-dimensional.

PROOF. We carry out the proof by contradiction. By Theorem 14.12 the locally nilpotent radical $\mathscr{L}(I)$ of the ideal I is an ideal in J. By Lemma 8 it is nilpotent and finite-dimensional. Factoring it out, without loss of generality it is possible to assume that $\mathscr{L}(I) = (0)$. On the other hand, by Lemma 4 $I \subseteq \mathscr{M}(J)$. We shall show that $I_1 = I \cap \mathscr{Z}(J)$ is a nonzero ideal. By the heredity of the McCrimmon radical $I = \mathscr{M}(I)$, and the algebra I has an absolute zero divisor z. Either $JU_z = (0)$ and z is an absolute zero divisor in J, or $z_1 = aU_z \neq 0$ for some $a \in J$. But then $JU_{z_1} \subseteq IU_z = (0)$, and z_1 is a nonzero element from I_1. In both cases $I_1 \neq (0)$.

Let x be an arbitrary element from I_1. Then $x = z_1 + \cdots + z_n$, where the z_i are absolute zero divisors in J. We consider the principal quadratic ideal $JU_x = \sum_{i<j} JU_{z_i,z_j}$. By the minimal condition and Lemma 10, each of the subspaces JU_{z_i,z_j} is finite-dimensional. Consequently, the principal quadratic ideal JU_x is finite-dimensional, and so all the more the quadratic ideal I_1U_x of the algebra I_1 is finite-dimensional. We can now conclude by

Lemma 11 that the ideal I_1 is locally nilpotent. This contradicts that $\mathscr{L}(I) = (0)$.

This proves the theorem.

Exercises

1. (*Slin'ko*) Let J be a Jordan algebra over a ring $\Phi \ni \frac{1}{2}$, I be an ideal of J, and $I \subseteq \mathscr{B}(J)$. Prove that if J satisfies the maximal condition for quadratic ideals contained in I, then the ideal I is nilpotent and finitely generated as a Φ-module. *Hint*. Reason analogous to the proof of Lemma 5.

2. (*Zel'manov*) Let J be a Jordan algebra over a ring $\Phi \ni \frac{1}{2}$ and I be a nil-ideal of J. Prove that if J satisfies the maximal condition for quadratic ideals contained in I, then I is a nilpotent ideal which is finitely generated as a Φ-module. *Hint*. In view of the previous exercise, it suffices to show that $\mathfrak{N}(I) \neq (0)$ implies $I \cap \mathscr{B}(J) \neq (0)$. By means of Exercise 4 of Section 1 prove that $\mathscr{M}(J) \cap I \neq (0)$. Next, by means of Lemma 11 show that $\mathscr{L}(J) \cap I \neq (0)$. The final step of the proof is also deduced by means of Exercise 4 of Section 1.

3. Let J be a finitely generated Jordan algebra over a field F of charactertistic $\neq 2$, and let B be an ideal of J with finite co-dimension. Prove that B is a finitely generated algebra. *Hint*. Use Lemma 4.5.

3. SEMISIMPLE JORDAN ALGEBRAS WITH MINIMAL CONDITION

Throughout this section all Jordan algebras are considered over a field F of characteristic $\neq 2$. Our goal is to determine the structure of a Jordan algebra satisfying the minimal condition in the case when the algebra is semisimple, that is, does not contain nilpotent (or equivalently, quasi-regular) ideals. The fact that a semisimple Jordan algebra with minimal condition contains "sufficiently many" idempotents will play a key role here. We shall prove this fact in the following lemma.

LEMMA 12. Let Q be a quadratic ideal of a Jordan algebra J with the minimal condition for quadratic ideals. Then either each element in Q is nilpotent, or Q contains an idempotent.

PROOF. Let x be a nonnilpotent element in Q. We consider the chain of quadratic ideals

$$QU_x \supseteq QU_{x^2} \supseteq \cdots \supseteq QU_{x^n} \supseteq \cdots.$$

By the minimal condition $QU(x^k) = QU(x^{k+1}) = \cdots$ for some number k. The quadratic ideal $QU(x^k)$ is different from zero since $x^{2k+1} = xU(x^k) \in$

$QU(x^k)$, and in addition $QU(x^k)U(x^{2k+1}) = QU(x^{3k+1}) = QU(x^k)$. By Lemma 3 there is an idempotent in the quadratic ideal $QU(x^k)$, which proves the lemma.

We now consider a Jordan algebra J with idempotent e. If necessary we formally adjoin to the algebra J an identity element 1. Then the elements e and $1 - e$ satisfy the relations

$$e + (1 - e) = 1, \qquad e(1 - e) = 0. \tag{4}$$

Therefore, $U_e + 2U_{e,1-e} + U_{1-e} = U_{e+(1-e)} = E$ is the identity mapping of the algebra $J^{\#}$, and for any element $x \in J$

$$x = xU_e + 2xU_{e,1-e} + xU_{1-e}. \tag{5}$$

We introduce the notation $x_1 = xU_e$, $x_{1/2} = 2xU_{e,1-e}$, $x_0 = xU_{1-e}$. The elements x_1, $x_{1/2}$, x_0 are in the algebra J. We also define the subspaces $J_1 = JU_e$, $J_{1/2} = JU_{e,1-e}$, $J_0 = JU_{1-e}$. It is clear that by (5) $J = J_1 + J_{1/2} + J_0$.

If there will appear several idempotents in the course of a proof, we shall write $x_i(e)$ and $J_i(e)$ instead of x_i and J_i.

We now note that

$$U_e = R_e(2R_e - E), \qquad U_{e,1-e} = R_e(R_e - E), \qquad U_{1-e} = (R_e - E)(2R_e - E),$$

and that by identity (3.25)

$$R_e(R_e - E)(2R_e - E) = 0. \tag{6}$$

Therefore

$$U_eR_e = U_e, \qquad U_{e,1-e}R_e = \tfrac{1}{2}U_{e,1-e}, \qquad U_{1-e}R_e = 0.$$

Hence it follows that $x_iR_e = ix_i$ for $i = 0, \frac{1}{2}, 1$. It is now easy to see that the sum of the subspaces J_i is direct:

$$J = J_1 \oplus J_{1/2} \oplus J_0. \tag{7}$$

This decomposition into a sum of subspaces is called the *Peirce decomposition*, and the subspaces J_i are called the *Peirce components*.

THEOREM 4 (*Albert*). Let J be a Jordan algebra with idempotent e. Then J decomposes into a direct sum or Peirce components

$$J = J_1 \oplus J_{1/2} \oplus J_0,$$

where $J_i = \{x \in J \mid xe = ix\}$ for $i = 0, \frac{1}{2}, 1$. Also $J_1 = JU_e$, $J_{1/2} = JU_{e,1-e}$, $J_0 = JU_{1-e}$ (the identity element here is from the algebra $J^{\#}$). The components J_0 and J_1 are quadratic ideals in J. The multiplication table for the Peirce decomposition is

$$\begin{array}{lll} J_1^2 \subseteq J_1, & J_1 J_0 = (0), & J_0^2 \subseteq J_0, \\ J_0 J_{1/2} \subseteq J_{1/2}, & J_1 J_{1/2} \subseteq J_{1/2}, & J_{1/2}^2 \subset J_0 + J_1. \end{array} \tag{8}$$

PROOF. It only remains to eastablish the rules of multiplication for the Peirce components. By identity (3.22) for any $x, y \in J$

$$(xy)e = (xy)e^2 = -2(xe)(ye) + 2((xe)y)e + x(ye).$$

If $x \in J_i$, $y \in J_j$, this relation implies

$$(2i - 1)(xy)e = j(2i - 1)xy. \tag{9}$$

For $i = 1$ we obtain that $J_1 J_j \subseteq J_j$, and for $i = 0$ we have $J_0 J_j \subseteq J_j$. Hence all relations from (8) follow except the second and the last. But $J_1 J_0 = J_0 J_1 \subseteq J_1 \cap J_0 = (0)$, and therefore it remains to prove the last relation. By (3.22) for any $x, y \in J_{1/2}$

$$((xy)e)e = -((xe)e)y - x((ye)e) + (xy)e + 2(xe)(ye) = (xy)e.$$

This means that $(xy)_1 + \frac{1}{4}(xy)_{1/2} = (xy)_1 + \frac{1}{2}(xy)_{1/2}$. Consequently, $(xy)_{1/2} = 0$, and $xy = (xy)_1 + (xy)_0$. Thus, the last relation from equality (8) is also proved.

THEOREM 5 (*McCrimmon*). If J is a Jordan algebra with idempotent e, then

$$\mathcal{J}(J_i) = J_i \cap \mathcal{J}(J) \qquad \text{for} \quad i = 0, 1.$$

PROOF. Since $J_1(e) = JU_e$, in the case $i = 1$ the assertion follows from Theorem 1. If the algebra J has an identity element, then the second assertion is also true since $J_0(e) = J_1(1 - e)$.

In order to prove the assertion for $i = 0$ in the general case, we consider the algebra $J^{\#}$ obtained by the formal adjoining of an identity element 1 to J. In view of the fact that J_0 is an ideal in $J_0^{\#}$, we have

$$\mathcal{J}(J_0) = J_0 \cap \mathcal{J}(J_0^{\#}) = J_0 \cap (J_0^{\#} \cap \mathcal{J}(J^{\#})) = J_0 \cap \mathcal{J}(J).$$

This proves the theorem.

We recall that idempotents e and f are called *orthogonal* if $ef = 0$. An idempotent e of an algebra J is called *principal* if there are no idempotents in J orthogonal to e. This is equivalent to the component $J_0(e)$ not containing idempotents.

LEMMA 13. Any nonnilpotent ideal H of a Jordan algebra J with minimal condition for quadratic ideals contains a principal idempotent e for which $H_0(e)$ is a nil-algebra.

PROOF. By Lemma 12 there is an idempotent in the ideal H, and we denote it by e_1. We consider the component $H_0(e_1)$ which, as is not difficult to see, is a quadratic ideal in J. If all the elements in $H_0(e_1)$ are nilpotent, then the idempotent e_1 is principal and everything is proved. If there is a nonnilpotent element in $H_0(e_1)$, then there is also an idempotent e_2 in $H_0(e_1)$. We now consider the component $H_0(e_1 + e_2)$. Since $1 - e_1 - e_2 \in J_0^{\#}(e_1)$, then $H_0(e_1 + e_2) = HU_{1-e_1-e_2} \subseteq H_0(e_1)$. This containment is strict, since $e_2 \notin H_0(e_1 + e_2)$. We perform the same procedures with $H_0(e_1 + e_2)$ as with $H_0(e_1)$, etc.

Since the chain of quadratic ideals

$$H_0(e_1) \supset H_0(e_1 + e_2) \supset \cdots$$

cannot be infinite, it follows from this that there is a principal idempotent e in H for which $H_0(e)$ is a nil-algebra. This proves the lemma.

THEOREM 6 (*Jacobson*). A semisimple Jordan algebra J with minimal condition for quadratic ideals has an identity element and decomposes into a finite direct sum of simple Jordan algebras, which each have an identity element and satisfy the minimal condition.

PROOF. Let H be a minimal ideal of the algebra J and e be a principal idempotent in H, which exists by Lemma 13. By that same lemma $H_0(e)$ is a nil-algebra, and by Theorem 5

$$\mathscr{J}(H) \supseteq H_0(e).$$

In view of the semisimplicity of the algebra J and the heredity of the quasi-regular radical, this means that $H_0(e) = (0)$.

We consider an arbitrary element $h_{1/2} = \{eh(1-e)\}$ from $H_{1/2}(e)$. Since the subalgebra generated by e and h is special, computing in an enveloping algebra we have

$$\begin{aligned} h_{1/2}^2 &= \tfrac{1}{4}[eh(1-e)eh(1-e) + (1-e)he(1-e)he \\ &\quad + (1-e)he^2h(1-e) + eh(1-e)^2he] \\ &= \tfrac{1}{4}eh(1-e)he = \tfrac{1}{4}\{e\{h(1-e)h\}e\}. \end{aligned}$$

Therefore for any element $x \in J$

$$xU(h_{1/2}^2) = \tfrac{1}{16}xU_eU_hU_{1-e}U_hU_e \subseteq H_0U_hU_e = (0),$$

that is, $h_{1/2}^2$ is an absolute zero divisor in J. By Theorem 14.2 we conclude that $h_{1/2}^2 = 0$ for any $h \in H$, and consequently $[H_{1/2}(e)]^2 = (0)$. It now follows from relations (8) that $H_{1/2}(e)$ is a trivial ideal in H, and therefore it equals zero.

It is now easy to observe that $J_{1/2}(e) = H_{1/2}(e)$, and therefore $H = H_1(e) = J_1(e)$ and $J_{1/2}(e) = (0)$. The idempotent e is an identity element in H, and the algebra J decomposes into a direct sum of ideals

$$J = H \oplus K,$$

where $K = J_0(e)$. Each ideal of the algebra H is an ideal in J, and therefore H is a simple algebra. It is obvious that H and K satisfy the minimal condition for quadratic ideals.

We denote H by H_1 and K by K_1. We consider, further, a minimal ideal H_2 of the algebra K_1. The same arguments as above give us a decomposition

$$J = H_1 \oplus H_2 \oplus K_2,$$

etc. Since the chain of ideals $K_1 \supset K_2 \supset \cdots$ cannot descend infinitely, for some number s we shall have $K_s = (0)$, that is,

$$J = H_1 \oplus H_2 \oplus \cdots \oplus H_s.$$

This proves the theorem.

The structure of simple Jordan algebras with minimal condition is described by

THEOREM 7 (*Jacobson–Osborn*). Let J be a simple Jordan algebra with minimal condition for quadratic ideals. Then J is one of the following:

(1) J is a division algebra;

(2) J is an algebra of a nondegenerate symmetric bilinear form defined on a vector space V over some extension K of the base field such that $\dim_K V > 1$;

(3) $J = H(D_n, S_A)$ where $n \geqq 2$, and (D, j) is either $\Delta \oplus \Delta^0$, where Δ is an associative division algebra, and j is the involution transposing the components; or D is an associative division algebra and j is an involution on D; or D is a split quaternion algebra over some extension of the base field and j is the standard involution; or D is a Cayley–Dickson algebra over some extension of the base field, j is the standard involution, and $n = 3$.

The proof of this theorem is very cumbersome. It is well set forth in the book of Jacobson [47], where we also recommend the reader to read it.

Finite-dimensional simple Jordan algebras over an algebraically closed field are especially simply constructed. We shall describe them in the exercises for this section.

Exercises

1. Let J be a Jordan algebra with an identity element $1 = \sum_{i=1}^{n} e_i$, which decomposes into a sum of pairwise orthogonal idempotents e_i. Then J decomposes into a direct sum of subspaces $J = \bigoplus_{1 \leqq i \leqq j \leqq n} J_{ij}$, where

$$J_{ii} = \{x \in J \mid xe_i = x\},$$
$$J_{ij} = \{x \in J \mid xe_i = xe_j = \tfrac{1}{2}x\}.$$

In addition, the components J_{ij} are connected by the relations

$$J_{ii}^2 \subseteq J_{ii}, \qquad J_{ij}J_{ii} \subseteq J_{ij}, \qquad J_{ij}^2 \subseteq J_{ii} + J_{jj},$$
$$J_{ii}J_{jj} = J_{ii}J_{jk} = J_{ij}J_{kl} = (0), \qquad J_{ij}J_{jk} \subseteq J_{ik},$$

where the indices i, j, k, l are distinct. This decomposition is called the *Peirce decomposition* with respect to the system of idempotents $E = \{e_1, \ldots, e_n\}$.

2. Let the indices i, j, k, l be distinct, and x_{pq}, y_{rs}, z_{tl} be arbitrary elements from the components J_{pq}, J_{rs}, J_{tl}, respectively. Then

(a) $(x_{ij}^2 e_i)x_{ij} = (x_{ij}^2 e_j)x_{ij}$;
(b) $(x_{jk}y_{ij})z_{ij} + (x_{jk}z_{ij})y_{ij} = x_{jk}(y_{ij}z_{ij})$;
(c) $(x_{ij}, y_{jk}, z_{kl}) = 0$;
(d) $(x_{ij}, y_{jk}, z_{ki})e_i = 0$.

3. Let $H(D_n, S_A)$ be a Jordan matrix algebra. Prove that if $n \geqq 4$, then D is associative. *Hint*. Use Exercise 2(c).

An idempotent e of a Jordan algebra J over a field F is called *primitive* if it does not decompose into the sum of two orthogonal idempotents, and it is called *absolutely primitive* if each element of the Peirce component $J_1(e)$ has the form $\alpha \cdot e + z$, where $\alpha \in F$ and z is a nilpotent element.

4. An absolutely primitive idempotent is primitive.

Now let J be a finite-dimensional Jordan algebra over a field F of characteristic $\neq 2$.

5. Prove that any idempotent $e \in J$ (and in particular the identity element 1) decomposes into a sum of pairwise orthogonal primitive idempotents.

6. Prove that for a primitive idempotent e the minimal annihilating polynomial of each element of the Peirce component $J_1(e)$ is the power of an irreducible polynomial over F.

7. If F is algebraically closed, then each primitive idempotent is absolutely primitive.

Henceforth we shall assume the field F is algebraically closed and the algebra J is semisimple.

8. The identity element of the algebra J_n decomposes into a sum of pairwise orthogonal idempotents $1 = \sum_{i=1}^n e_i$ such that $J_{ii} = F \cdot e_i$ for all i. *Hint.* Use Exercise 2 from Section 14.2, and also Theorems 3 and 5.

9. Let $x, y \in J_{ij}$, $i \neq j$. Then

$$xy = \alpha(e_i + e_j), \qquad \alpha \in F.$$

Hint. Using Exercise 2(a) prove this assertion in the case $y = x$, and note that this is sufficient.

10. Let $x \in J_{ij}$. Then either x is invertible in $J_1(e_i + e_j) = J_{ii} + J_{ij} + J_{jj}$, or $xJ_{ij} = (0)$. In the latter case x is an absolute zero divisor in J.

11. If $J_{ij} \neq (0)$, then there exists $u_{ij} \in J_{ij}$ such that

$$u_{ij}^2 = e_i + e_j.$$

12. For any $x_{jk} \in J_{jk}$, it follows from the fact $u_{ij}x_{jk} = 0$ that $x_{jk} = 0$.

13. Let $s_{ij} = \dim J_{ij}$. Prove that if $s_{ij} > 0$ and $s_{jk} > 0$, then $s_{ij} = s_{jk} = s_{ik}$.

Now let J be a simple algebra.

14. The dimensions of all the components J_{ij} for $i \neq j$ are the same.

15. If $n = 1$, then $J = F \cdot 1$. If $n = 2$, then $J = F \cdot 1 + V$ is the algebra of a nondegenerate symmetric bilinear form. *Hint.* In the case $n = 2$, it is possible to take as V the subspace $F(e_1 - e_2) + J_{12}$.

We subsequently assume that $n \geqq 3$.

16. Let $x_{ij} \in J_{ij}$. By Exercise 9 $x_{ij}^2 = f_{ij}(x_{ij})(e_i + e_j)$ where $f_{ij}(x_{ij}) \in F$. Prove that f_{ij} is a strictly nondegenerate quadratic form on J_{ij}.

17. $f_{ik}(2x_{ij}y_{jk}) = f_{ij}(x_{ij})f_{jk}(y_{jk})$.

18. There exist $n(n-1)/2$ elements $e_{ij} \in J_{ij}$, $i < j$, such that $e_{ij}^2 = e_i + e_j$, $2e_{ij}e_{jk} = e_{ik}$.

19. Prove that the form f_{12} admits the composition

$$f_{12}(x_{12})f_{12}(y_{12}) = f_{12}(8(x_{12}e_{23})(y_{12}e_{13})).$$

Hint. Apply Exercise 17 repeatedly.

20. Let $p_{ij} = 1 - e_i - e_j + e_{ij}$. Then the mapping $U_{(ij)} = U_{p_{ij}}$ is an automorphism of the algebra J of order 2 with the following properties:

(a) $U_{(ij)}$ interchanges the places of the idempotents e_i and e_j, translates the subspaces J_{ii} and J_{jj} one into the other, and translates the subspace J_{ij} into itself;

(b) $U_{(ij)}$ is the identity map on any J_{kl} for $\{i, j\} \cap \{k, l\} = \varnothing$;

(c) $U_{(ij)}$ interchanges the places of the subspaces J_{ik} and J_{jk} for $k \neq i, j$, and coincides with $2R_{e_{ij}}$ on these subspaces.

21. Let Σ be the group of automorphisms of the algebra J generated by the automorphisms $U_{(ij)}$. Then we have a unique isomorphism $\pi \to U_\pi$ of the symmetric group S_n onto Σ such that $(ij) \to U_{(ij)}$. We have $e_i U_\pi = e_{i\pi}$ for all $i = 1, \ldots, n$ and $\pi \in S_n$. If π and π' are such that $i\pi = i\pi'$ and $j\pi = j\pi'$ (i and j can be identical), then π and π' act the same on J_{ij}. *Hint.* For the proof of the existence of a homomorphism from S_n to Σ is used the fact that the homomorphism θ of the free $(n-1)$-generated group with generators $x_2, x_3, \ldots, x_n$ onto S_n such that $\theta(x_i) = (1i)$ has kernel the normal subgroup generated by the elements

$$x_j^2, \quad (x_j x_k)^3, \quad (x_j x_k x_j x_l)^2,$$

where j, k, l are distinct.

22. Let D be the Peirce subspace J_{12}. We define a multiplication on this subspace by the formula

$$x \cdot y = 2xU_{(23)} yU_{(13)}.$$

Prove that D is a composition algebra with respect to this multiplication, e_{12} is its identity element, and $d \to \bar{d} = dU_{(12)}$ is its involution. *Hint.* Use Exercises 19 and 20.

23. For each pair of indices i, j we define a mapping $x \mapsto x_{ij}$ of the algebra D into J_{ij}. If $i \neq j$, let π be a permutation such that $1\pi = i$, $2\pi = j$, and if $i = j$ then π is a permutation such that $1\pi = i$. We set

$$x_{ij} = xU_\pi, \qquad i \neq j; \qquad x_{ii} = (xe_{12})e_1 U_\pi.$$

Prove that $\bar{x}_{ii} = x_{ii}$ and $\bar{x}_{ij} = x_{ji}$.

24. Prove that

$$2x_{ij} y_{jk} = (x \cdot y)_{ik}.$$

Hint. Note that by the definition of the operation of multiplication in D we have $(x \cdot y)_{12} = 2x_{13} y_{32}$. Apply to both sides of this equality the automorphism U_π, where $1\pi = i$, $2\pi = k$, $3\pi = j$.

25. Prove the relations

$$x_{ii} y_{ij} = ((x + \bar{x}) \cdot y)_{ij},$$
$$(x_{ij} y_{ji}) e_i = (x \cdot y)_{ii},$$
$$x_{ii}^2 = (x + \bar{x})_{ii}^2.$$

Hint. Prove these relations for $i = 1$, $j = 2$, and afterwards apply a suitable automorphism U_π.

26. Prove the theorem:

THEOREM (*Albert*). Any simple finite-dimensional Jordan algebra over an algebraically closed field F of characteristic not equal 2 is isomorphic to one of the following algebras:

(1) $F \cdot 1$, the base field;

(2) $F \cdot 1 + V$, the algebra of a symmetric nondegenerate bilinear form on a vector space V;

(3) the Jordan matrix algebra $H(D_n)$, where $n \geqq 3$ and D is a composition algebra which is associative for $n > 3$.

LITERATURE

Albert [2, 5], Jacobson [47], Zel'manov [82], McCrimmon [117, 124], Morgan [155], Osborn [162], Slin'ko [209, 212, 213], Topping [228].

CHAPTER **16**

Right Alternative Algebras

Along with alternative algebras, there are often encountered in mathematics algebras with weaker conditions of associativity. The most important of these conditions is power-associativity, that is, every subalgebra generated by a single element is associative. The study of this broad class of algebras (in particular, it contains all anticommutative algebras) meets with insurmountable obstacles. However, some of its subclasses, which also contain the alternative algebras, have been adequately studied. Right alternative algebras, which arose from the study of a certain class of projective planes, have been studied the most intensively. In this chapter we shall briefly set forth the basic facts which are known up to the present time about right alternative algebras.

1. ALGEBRAS WITHOUT NILPOTENT ELEMENTS

An algebra is called *right alternative* if it satisfies the identities

$$(xy)y = xy^2, \tag{1}$$

$$((xy)z)y = x((yz)y). \tag{2}$$

We recall that (1) is called the *right alternative identity*, and (2) is the *right Moufang identity*. By Proposition 1.4 the linearizations of these identities

$$(xy)z + (xz)y = x(y \circ z), \tag{1'}$$

$$((xy)z)t + ((xt)z)y = x((yz)t + (tz)y), \tag{2'}$$

are also valid in every right alternative algebra.

If the algebra is without elements of additive order 2, then the right Moufang identity is a consequence of the right alternative identity. In fact, we have $2(yz)y = (z \circ y) \circ y - z \circ y^2$, whence by (1′) and (1)

$$\begin{aligned}
2x[(yz)y] &= x[(z \circ y) \circ y] - x(z \circ y^2) \\
&= [x(z \circ y)]y + (xy)(z \circ y) - (xz)y^2 - (xy^2)z \\
&= [(xz)y]y + [(xy)z]y + [(xy)z]y + [(xy)y]z - (xz)y^2 - (xy^2)z \\
&= 2[(xy)z]y.
\end{aligned}$$

We note that the right Moufang identity is equivalent to the identity

$$(x, yz, y) = (x, z, y)y. \tag{3}$$

Actually, we have

$$\begin{aligned}
0 &= [(xy)z]y - x[(yz)y] = (x, y, z)y + (x, yz, y) \\
&= (x, yz, y) - (x, z, y)y.
\end{aligned}$$

THEOREM 1. Every right alternative algebra is power-associative.

PROOF. It suffices to prove that for arbitrary nonassociative words $u_i = u_i(x)$, where $i = 1, 2, 3$, and an arbitrary element a of a right alternative algebra A, it is true $(\bar{u}_1, \bar{u}_2, \bar{u}_3) = 0$ where $\bar{u}_i = u_i(a)$. We shall carry out the proof by induction on the total length $d(u_1) + d(u_2) + d(u_3)$. By the fact that in any algebra there is the equality

$$(xy, z, t) - (x, yz, t) + (x, y, zt) = x(y, z, t) + (x, y, z)t,$$

we can assume that $u_3(x) = x$. Furthermore, by the induction assumption one can assume that $u_2 = x \cdot u_2'$. It is now clear that in view of (3),

$$(\bar{u}_1, \bar{u}_2, \bar{u}_3) = (\bar{u}_1, a \cdot \bar{u}_2', a) = 0.$$

This proves the theorem.

With a view toward a compact presentation of calculations, we shall make use of notation which was introduced into use by Smiley. We denote by c' the operator of right multiplication R_c. We then set $c^d = (cd)' - c'd'$ and

$c_d = (cd)' - d'c'$. Identities (1), (1′), (2), and (2′) imply the relations

$$c'c' = (c^2)', \qquad (cd + dc)' = c'd' + d'c',$$
$$c^c = 0, \qquad c^d + d^c = 0,$$
$$c'd'c' = [(cd)c]', \qquad c'd's' + s'd'c' = [(cd)s + (sd)c]'.$$

THEOREM 2 (*Mikheev*). In every right alternative algebra the (Mikheev) identity

$$(x, x, y)^4 = 0$$

is valid.

PROOF. Let a, b, c be arbitrary elements of a right alternative algebra A. We set $q = [a, b]$ and $p = (a, a, b)$. We shall prove the relations

$$p = -aa^b, \tag{4}$$
$$a^b a_b = 0, \qquad a^b a_c + a^c a_b = 0, \tag{5}$$
$$a_b a^b = -(q, a, b)', \tag{6}$$
$$a_b a' a^b = -(qa, a, b)'. \tag{7}$$

Relation (4) is obvious. Moreover,

$$\begin{aligned} a^b a_b &= (ab)'(ab)' - (ab)'b'a' - a'b'(ab)' + a'b'b'a' \\ &= [(ab)(ab)]' - [((ab)b)a + (ab)(ab)]' + [(ab^2)a]' = 0. \end{aligned}$$

The second part of (5) is linearization of the first. We shall now prove (7):

$$\begin{aligned} a_b a' a^b &= (ab)'a'(ab)' - (ab)'a'a'b' - b'a'a'(ab)' + b'a'a'a'b' \\ &= \{[(ab)a](ab) - [(ab)a^2]b - (ba^2)(ab) + (ba^3)b\}' \\ &= -(qa, a, b)'. \end{aligned}$$

Relation (6) is proved analogously. Now by (4), (5), (6), and (7)

$$p(q, a, b) = -aa^b(q, a, b)' = aa^b a_b a^b = 0, \tag{8}$$
$$p(qa, a, b) = -aa^b(qa, a, b)' = aa^b a_b a' a^b = 0. \tag{9}$$

In addition, we have the relations

$$p' = a^{ba} - a'a^b = a_{ba} - a_b a'. \tag{10}$$

We shall only prove the first:

$$\begin{aligned} p' &= -(a, b, a)' = [a(ba) - (ab)a]' = [a(ba)]' - a'b'a' \\ &= [a(ba)]' + a'a'b' - a'(ba)' - a'(ab)' = a^{ba} - a'a^b. \end{aligned}$$

Applying consecutively (10), (5), (10) again, (6), (7), (8), and (9), we have

$$\begin{aligned} a^b p'p'p' &= a^b(a_{ba} - a_b a')(a^{ba} - a'a^b)(a_{ba} - a_b a') \\ &= a^b a_{ba}(a^{ba} - a'a^b)(a_{ba} - a_b a') = -a^b a_{ba} a^{ba} a_b a' - a^b a_{ba} a' a^b a_{ba} \\ &= -a^{ba} a_b a^b a_{ba} a' + a^{ba} a_b a' a^b a_{ba} = -p' a_b a^b a_{ba} a' + p' a_b a' a^b a_{ba} \\ &= -p' a_b a^b p' a' + p' a_b a' a^b p' = p'(q, a, b)' p' a' - p'(qa, a, b)' p' \\ &= \{[p(q, a, b)]p\}' a' - \{[p(qa, a, b)]p\}' = 0. \end{aligned}$$

Now $p^4 = -aa^b p'p'p' = 0$. This proves the theorem.

COROLLARY 1 (*Kleinfeld*). Every right alternative algebra without nilpotent elements is alternative.

A much stronger assertion, due to Albert [3, 7], is valid for finite-dimensional algebras. He proved that every finite-dimensional right alternative algebra over a field of characteristic $\neq 2$, which does not contain nil-ideals, is alternative. The proof is technically complicated, and we shall not produce it here.

COROLLARY 2 (*Skornyakov*). Every right alternative division algebra is alternative.

Skornyakov proved by means of this result that the realization of little Desargues' theorem on two lines of a projective plane implies its projective realization in that plane.

The question on the structure of simple right alternative algebras is at present not clear to its conclusion. Thedy proved that under certain additional restrictions, simple right alternative algebras are alternative [232, 233]. For example, this is so if the simple algebra A contains an idempotent $e \neq 0, 1$ for which $(e, e, A) = (0)$. There was the conjecture that simple right alternative algebras which are not alternative do not exist. However, Mikheev [152] recently disproved this conjecture by constructing an appropriate example. It is interesting that the algebra in Mikheev's example satisfies the identity $x^3 = 0$, that is, is a nil-algebra of bounded index. The hope, nevertheless, remains that right alternative algebras which do not contain nil-ideals are alternative. Some basis for such a conjecture is the above-mentioned result of Albert.

We now study in somewhat more detail alternative (and consequently, also right alternative) algebras without nilpotent elements.

LEMMA 1. In every alternative algebra without nilpotent elements

$$(ab)c = 0 \Leftrightarrow a(bc) = 0. \tag{11}$$

PROOF. If $(ab)c = 0$, then using the middle Moufang identity we have

$$\begin{aligned}[a(bc)]^3 &= a(bc \cdot (a \cdot bc \cdot a) \cdot bc) = a(bc \cdot (ab \cdot ca) \cdot bc) \\ &= a((bc \cdot ab)(ca \cdot bc)) = a((bc \cdot ab)(c \cdot ab \cdot c)) = 0,\end{aligned}$$

whence $a(bc) = 0$. The reverse implication is proved analogously.

Algebras satisfying condition (11) are called *associative modulo zero*. They were introduced into consideration by Ryabukhin. It is easy to see that if in an associative modulo zero algebra a product of some n elements equals zero for one arrangement of parentheses, then it also equals zero for any other arrangement.

LEMMA 2. In every associative modulo zero algebra without nilpotent elements

$$ab = 0 \Leftrightarrow ba = 0. \tag{12}$$

PROOF. $ab = 0 \Rightarrow b(ab) = 0 \Rightarrow (ba)b = 0 \Rightarrow [(ba)b]a = 0 \Rightarrow (ba)^2 = 0 \Rightarrow ba = 0$.

LEMMA 3. Let $v = v(x_1, \ldots, x_n)$ be a multilinear nonassociative word. If $v(a_1, a_2, \ldots, a_n) = 0$ for elements $a_1, a_2, \ldots, a_n$ of an associative modulo zero algebra A without nilpotent elements, then also $v((a_1), a_2, \ldots, a_n) = (0)$, where (a_1) is the ideal generated in A by the element a_1.

PROOF. Let $v = x_1x_2$ and a be an arbitrary element of A. By (11) and (12) $a(a_1a_2) = 0 \Rightarrow (aa_1)a_2 = 0$ and $(a_2a_1)a = 0 \Rightarrow a_2(a_1a) = 0 \Rightarrow (a_1a)a_2 = 0$. Repeating this argument we obtain $(a_1) \cdot a_2 = (0)$.

We carry out an induction on the length of the word v. Let $d(v) = n > 2$. We consider one of the cases that arises here: $v = (v_1v_2)v_3$. If x_1 appears in v_3, then by the induction assumption there is nothing to prove. Let x_1 appear in v_1 (the argument for v_2 is analogous). The induction assumption allows us to assume that $d(v_2) = d(v_3) = 1$. Let $v_2 = x_{n-1}$, $v_3 = x_n$. Then

$$0 = v(a_1, \ldots, a_n) = (v_1(a_1, \ldots, a_{n-2})a_{n-1})a_n = v_1(a_1, \ldots, a_{n-2})(a_{n-1}a_n).$$

By the induction assumption,

$$\begin{aligned}(0) &= v_1((a_1), a_2, \ldots, a_{n-2})(a_{n-1}a_n) = (v_1((a_1), a_2, \ldots, a_{n-2})a_{n-1})a_n \\ &= v((a_1), a_2, \ldots, a_n),\end{aligned}$$

which is what was to be proved.

LEMMA 4. If $v(a_1, \ldots, a_n) = 0$ for some multilinear nonassociative word $v(x_1, \ldots, x_n)$ and elements $a_1, \ldots, a_n$ of an associative modulo zero algebra A without nilpotent elements, then for any nonassociative word $u(x_1, \ldots, x_n)$ of the same type as v it is also true $u(a_1, \ldots, a_n) = 0$.

PROOF. $\bar{u} = u(a_1, \ldots, a_n)$ is in the ideal (a_i) for any $i = 1, 2, \ldots, n$. Therefore $0 = v(\bar{u}, \bar{u}, \ldots, \bar{u}) = \bar{u}^n$ by Lemma 3, whence also follows the assertion.

THEOREM 3 (*Ryabukhin*). An algebra A is a subdirect sum of algebras without divisors of zero if and only if A is an associative modulo zero algebra without nilpotent elements.

PROOF. Let us assume that A is associative modulo zero and does not have nilpotent elements. Then the multiplicative subgroupoid generated by a nonzero element $a \in A$ does not contain zero, and by Zorn's lemma it is contained in some maximal subgroupoid which does not contain zero. We denote this groupoid by G_a. We shall prove that $I_a = A \backslash G_a$ is an ideal in A. We show that the set I_a is closed with respect to subtraction. Let $x \in I_a$. Then the subgroupoid $\langle G_a, x \rangle$ contains 0, and consequently there exists a nonassociative word $v(x_1, \ldots, x_n)$ and elements $h_1, h_2, \ldots, h_n \in \langle G_a, x \rangle$, not all belonging to G_a, such that $v(h_1, \ldots, h_n) = 0$. Hence by Lemma 4 it follows that $x^i g = 0$ for some $i \geq 1$ and $g \in G_a$. By that same lemma $x^i g = 0$ implies $(xg)^i = 0$, whence we obtain $xg = 0$. If y is another element from I_a, then there can be found for it a $g' \in G_a$ such that $yg' = 0$. By Lemma 2 we have $gxg' = gyg' = 0$, whence $g(x - y)g' = 0$. But this means that $x - y \notin G_a$, that is, $x - y \in I_a$.

Moreover, from Lemma 3 it follows that $x \in I_a$ implies $(x) \subseteq I_a$, so that I_a is an ideal in A.

As is easy to see, the algebra A/I_a does not contain divisors of zero and $a \notin I_a$. It is obvious that the algebra A is a subdirect sum of the algebras A/I_a, which proves the theorem in one direction. The proof in the other direction is clear.

COROLLARY (*L'vov*). Every alternative algebra without nilpotent elements is a subdirect sum of algebras without divisors of zero.

Andrunakievich and Ryabukhin proved this theorem earlier for associative algebras.

2. NIL-ALGEBRAS

We consider the free right alternative algebra RA$[X]$ from the set of free generators $X = \{x_1, x_2, \ldots\}$. We shall call an element of the algebra RA$[X]$

a *right alternative j-polynomial* if it is expressible from elements of the set X by means of addition, multiplication by elements from Φ, squaring, and the "quadratic multiplication" $xU_y = (yx)y$. We denote by $j_{\mathrm{RA}}[X]$ the set of all right alternative j-polynomials. If π is the canonical homomorphism of the algebra $\mathrm{RA}[X]$ onto the free associative algebra $\mathrm{Ass}[X]$, then it is obvious $\pi(j_{\mathrm{RA}}[X]) = j[X]$.

LEMMA 5. Let $f = f(x_1, \ldots, x_n) \in j_{\mathrm{RA}}[X]$. Then

$$R_{f(x_1, \ldots, x_n)} = f^{\pi}(R_{x_1}, \ldots, R_{x_n}).$$

The proof repeats, verbatim, the proof of Lemma 5.12, where only the right alternative and right Moufang identities are used.

Let A be a right alternative algebra and $M = \{m_i\}$ be a subset in A. We denote by $j_{\mathrm{RA}}[M]$ the set of elements of the form $f(m_1, \ldots, m_k)$ where $f \in j_{\mathrm{RA}}[X]$. We shall call elements of the set $j_{\mathrm{RA}}[M]$ j-polynomials from elements of the set M.

LEMMA 6. In the right alternative Φ-algebra A let all j-polynomials from elements of some finite set $M = \{a_1, a_2, \ldots, a_k\}$ be nilpotent, and in addition let their indices of nilpotency be bounded overall. Then the subalgebra M^* of the algebra of right multiplications which is generated by the operators $R_{a_1}, R_{a_2}, \ldots, R_{a_k}$ is nilpotent.

PROOF. Let $f \in j_{\mathrm{RA}}[X]$ and $f^m(a_1, \ldots, a_k) = 0$. Then by Lemma 5

$$(f^{\pi})^m(R_{a_1}, \ldots, R_{a_k}) = R_{f^m(a_1, \ldots, a_k)} = 0.$$

Thus all j-polynomials from $R_{a_1}, \ldots, R_{a_k}$ are nilpotent with an overall bound on the indices of nilpotency, whence by the Corollary to Theorem 5.3 follows the nilpotency of the subalgebra M^*. This proves the lemma.

We shall say that a subset M of an algebra A is *right nilpotent* if for some number N all r_1-words from elements of the set M with length N (and consequently, also greater length) equal zero. An algebra is called *locally r_1-nilpotent* if any finite subset of it is right nilpotent.

THEOREM 4 (*Shirshov*). Every right alternative nil-algebra of bounded index is locally r_1-nilpotent.

The proof consists of applying Lemma 6.

This theorem has a series of important corollaries.

COROLLARY 1. Every finite-dimensional right alternative nil-algebra A over an arbitrary field Φ is right nilpotent.

PROOF. Let $a \in A$ and n be the index of nilpotency of the element a. The chain of subspaces

$$A \supset AR_a \supset AR_{a^2} \supset \cdots \supset AR_{a^n} = (0)$$

has length no greater than the dimension of the algebra A, and consequently, the indices of nilpotency of all the elements are bounded overall by the number $1 + \dim_\Phi A$. By Theorem 4 the algebra A is locally r_1-nilpotent. In particular, its basis is right nilpotent. But this means that the algebra itself is also right nilpotent.

COROLLARY 2. Every simple finite-dimensional right alternative algebra over a field is alternative.

PROOF. By Corollary 1 a simple finite-dimensional right alternative algebra cannot be a nil-algebra. Consequently, it does not contain nil-ideals, and by the theorem of Albert which we mentioned in Section 1 it is alternative.

It is natural to ask whether or not every finite-dimensional right alternative nil-algebra is also nilpotent? It turns out that they are not. The following example of a five-dimensional right nilpotent but not nilpotent algebra belongs to Dorofeev [58]. Its basis is $\{a, b, c, d, e\}$, and the multiplication is given by the table (zero products of basis vectors are omitted)

$$ab = -ba = ae = -ea = db = -bd = -c, \qquad ac = d, \qquad bc = e.$$

The assertion of Corollary 1 can be strengthened. As shown by Shestakov, every right alternative Φ-algebra which is finitely generated as a Φ-module by nilpotent generating elements is right nilpotent.

The property of local r_1-nilpotency was studied by Mikheev. He proved that this is a radical property in the sense of Amitsur–Kurosh, and that it is different from the usual local right nilpotency, which is not a radical property [151].

Exercises

1. (*Shirshov*) We define an operation $\langle\,\rangle$ on nonassociative words of the free algebra $\Phi[X]$ which cancels the arrangement of parentheses possessed by a nonassociative word and distributes them anew in the standard fashion from the right. For example: $\langle x_3((x_1x_4)x_2)\rangle = ((x_3x_1)x_4)x_2$. We also linearly extend the operation $\langle\,\rangle$ to polynomials. Let T_{RA} be the T-ideal of the variety of right alternative algebras. Prove that $\langle T_{\mathrm{RA}}\rangle = (0)$, and by the same token show that the operation $\langle\,\rangle$ is reasonably defined for the free right alternative algebra $\mathrm{RA}[X] = \Phi[X]/T_{\mathrm{RA}}$. Prove that for any j-polynomial $f \in j_{\mathrm{RA}}[X]$ we have $f = \langle f \rangle$.

2. (*Mikheev*) Prove that for any natural numbers n and t there exists a natural number $M = M(n, t)$ such that for any M elements $x_{i_1}, x_{i_2}, \ldots, x_{i_M} \in \{x_1, \ldots, x_n\} \subseteq X$ in the right multiplication algebra of the free right alternative algebra $\mathrm{RA}[X]$ the following equality holds:

$$R_{x_{i_1}} R_{x_{i_2}} \cdots R_{x_{i_M}} = \sum_k R_{j_k(x)} R_{x_{l_1}} \cdots R_{x_{l_k}},$$

where $j_k(x)$ is a j-polynomial of degree $\geq t$.

3. (*Mikheev*) If an ideal I and the quotient algebra A/I of a right alternative algebra A are locally r_1-nilpotent, then A is also locally r_1-nilpotent. Prove this assertion and deduce from this the existence of a locally r_1-nilpotent radical in the class of right alternative algebras. *Hint.* Use Exercises 1 and 2.

4. (*Skosyrskiy*) Prove that a right alternative algebra A is locally r_1-nilpotent if and only if the algebra $A^{(+)}$ is locally nilpotent.

5. Prove that in every right alternative algebra are valid the relations

$$w(x \odot y)' = (w \odot x)y' + (w \odot y)x' - \{xwy\},$$
$$wx'y' + w(xy)' = 2(w \odot x)y' + 2(w \odot y)x' + 2w \odot (xy) - 4(w \odot y) \odot x,$$

where u' is the Smiley notation for the operator of right multiplication R_u.

6. Let A be a right alternative algebra over a ring Φ containing $\frac{1}{2}$, and let I_m be the right ideal of A generated by the set $(A^{(+)})^m$, where the power is understood with respect to the operation of multiplication in the algebra $A^{(+)}$. Prove that for any elements $a_1, a_2, a_3, a_4 \in A$

$$I_m a_1' a_2' a_3' a_4' \subseteq I_{m+1}.$$

Hint. Use the relations of Exercise 5.

7. (*Skosyrskiy*) A right alternative algebra A over a ring Φ containing $\frac{1}{2}$ is right nilpotent if and only if the associated Jordan algebra $A^{(+)}$ is nilpotent. *Hint.* Use the previous exercise.

8. Prove that if I is an ideal of a right alternative algebra A, then AI is also an ideal in A.

9. (*Slin'ko*) Prove that a left and right nilpotent right alternative algebra is nilpotent.

10. Prove that for each natural number k there exists a natural number $h(n, k)$ such that in the right multiplication algebra of the free right alternative algebra $\mathrm{RA}[x_1, \ldots, x_n]$ any word of degree $h(n, k)$ is representable in the form of a linear combination of words each of which contains the operator $R_{u(x)}$, where $u(x)$ is a monomial of degree $\geq k$. *Hint.* See the proof of Lemma 4.2.

11. (*Slin'ko*) Every left nilpotent right alternative algebra with a finite number of generators is nilpotent. *Hint*. Use Exercises 8–10.

LITERATURE

Albert [3, 7], Andrunakievich and Ryabukhin [19], Dorofeev [58], Kleinfeld [91, 98, 99], McCrimmon [138], Mikheev [147–152], Pchelintsev [174], Ryabukhin [187], Skornyakov [191, 192], Skosyrskiy [194], Slin'ko [207], Smiley [218], Thedy [232–234], Hentzel [245], Humm [255], Shestakov [267], Shirshov [276, 279].

Bibliography[†]

1. Albert, A. A., Quadratic forms permitting composition, *Ann. of Math.* **43** (1942), 161–177.
2. Albert, A. A., A structure theory for Jordan algebras, *Ann. of Math.* **48** (1947), 546–567.
3. Albert, A. A., On right alternative algebras, *Ann. of Math.* **50** (1949), 318–328.
4. Albert, A. A., Almost alternative algebras, *Portugal. Math.* **8** (1949), 23–36.
5. Albert, A. A., A theory of power-associative commutative algebras, *Trans. Amer. Math. Soc.* (1950), 503–527.
6. Albert, A. A., On simple alternative rings, *Canad. J. Math.* **4** (1952), 129–135.
7. Albert, A. A., The structure of right alternative algebras, *Ann. of Math.* **59** (1954), no. 3, 408–417.
8.* Albert, A. A., "Studies in Modern Algebra" (Studies in Mathematics, Vol. 2). Math. Assoc. Amer.; distributor, Prentice-Hall, Englewood Cliffs, New Jersey, 1963.
9. Albert, A. A., and Paige, L. J., On a homomorphism property of certain Jordan algebras, *Trans. Amer. Math. Soc.* **93** (1959), 20–29.
10. Allen, H. P., Jordan algebras and Lie algebras of type D_4, *J. Algebra* **5** (1967), no. 2, 250–265.
11. Allen, H. P., and Ferrar, J. C., Jordan algebras and exceptional subalgebras of the exceptional algebra E_6, *Pacific J. Math.* **32** (1970), no. 2, 283–297.
12. Amitsur, S., A general theory of radicals. I, II, and III, *Amer. J. Math.* **74** (1952), 774–786; **76** (1954), 100–125; 126–136.

[†] The order of the entries in this list of literature is, as in the Russian text, Cyrillic alphabetical order (Translator).

13. Anderson, T., The Levitzki radical in varieties of algebras, *Math. Ann.* **194** (1971), no. 1, 27–34.
14. Anderson, T., On the Levitzki radical, *Canad. Math. Bull.* **17** (1974), no. 1, 5–11.
15. Anderson, T., Divinsky, N., and Sulinski, A., Hereditary radicals in associative and alternative rings, *Canad. J. Math.* **17** (1965), no. 4, 594–603.
16. Andrunakievich, V. A., Antisimple and strongly idempotent rings (Russian), *Izv. Akad. Nauk SSSR Ser. Mat.* **21** (1957), 125–144.
17. Andrunakievich, V. A., Radicals of associative rings. I (Russian), *Mat. Sb.* **44** (1958), no. 2, 179–212.
18. Andrunakievich, V. A., Radicals of associative rings. II (Russian), *Mat. Sb.* **55** (1961), no. 3, 329–346.
19. Andrunakievich, V. A., and Ryabukhin, Yu. M., Rings without nilpotent elements and completely simple ideals (Russian), *Dokl. Akad. Nauk SSSR* **180** (1968), no. 1, 9–11.
20. Babich, A. M., On the Levitzki radical (Russian), *Dokl. Akad. Nauk SSSR* **126** (1959), no. 2, 242–243.
21. Belkin, V. P., On varieties of right alternative algebras (Russian), *Algebra i Logika* **15** (1976), no. 5, 491–508.
22. Block, R. E., Determination of $A^{(+)}$ for the simple flexible algebras, *Proc. Nat. Acad. Sci. USA* **61** (1968), 394–397.
23. Block, R. E., A unification of the theories of Jordan and alternative algebras, *Amer. J. Math.* **94** (1972), 389–412.
24. Bruck, R. H., and Kleinfeld, E., The structure of alternative division rings, *Proc. Amer. Math. Soc.* **2** (1951), no. 6, 878–890.
25. Brown, R. B., On generalized Cayley–Dickson algebras, *Pacific J. Math.* **20** (1967), 415–422.
26.* Braun, H., and Koecher, M., "Jordan-Algebren." Berlin, Göttingen, Heidelberg; Springer-Verlag, 1966.
27. Britten, D. J., On prime Jordan rings $H(R)$ with chain condition, *J. Algebra* **27** (1973), no. 2, 414–421.
28. Britten, D. J., On semiprime Jordan rings $H(R)$ with ACC, *Proc. Amer. Math. Soc.* **45** (1974), no. 2, 175–178.
29. Britten, D. J., Prime Goldie-like Jordan matrix rings and the common multiple property, *Comm. Algebra* **3** (1975), no. 4, 365–389.
30. Britten, D. J., On prime Goldie-like quadratic Jordan matrix algebras, *Canad. Math. Bull.* **20** (1977), no. 1, 39–45.
31. Gaynov, A. T., Monocomposition algebras (Russian), *Sibirsk. Mat. Z.* **10** (1969), no. 1, 3–30.
32. Gaynov, A. T., Some classes of monocomposition algebras (Russian), *Dokl. Akad. Nauk SSSR* **201** (1971), no. 1, 19–21.
33. Gaynov, A. T., Subalgebras of non-degenerate commutative KM-algebras (Russian), *Algebra i Logika* **15** (1976), no. 4, 371–383.
34. Hein, W., A connection between Lie algebras of type F_4 and Cayley algebras, *Indag. Math.* **38** (1976), no. 5, 419–425.
35. Glennie, C. M., Some identities valid in special Jordan algebras but not valid in all Jordan algebras, *Pacific J. Math.* **16** (1966), no. 1, 47–59.
36. Glennie, C. M., Identities in Jordan algebras, *in* "Computational Problems in Abstract Algebra" (*Proc. Conf., Oxford, 1967*), pp. 307–313. Pergamon, Oxford, 1970.
37. Golod, E. S., On nil-algebras and finitely-approximable groups (Russian), *Izv. Akad. Nauk SSSR Ser. Mat.* **28** (1964), 273–276.
38. Gordon, S. R., On the automorphism group of a semisimple Jordan algebra of characteristic zero, *Bull. Amer. Math. Soc.* **75** (1969), 499–504.

39. Gordon, S. R., The components of the automorphism group of a Jordan algebra, *Trans. Amer. Math. Soc.* **153** (1971), 1–52.
40. Gordon, S. R., An integral basis theorem for Jordan algebras, *J. Algebra* **24** (1973), no. 2, 258–282.
41. Gordon, S. R., On the structure group of a split semisimple Jordan algebra, I and II, *Comm. Algebra* **5** (1977), no. 10, 1009–1024, 1025–1056.
42. Gordon, S. R., Associators in simple algebras, *Pacific J. Math.* **51** (1974), no. 1, 131–141.
43. Jacobson, N., A Kronecker factorization theorem for Cayley algebras and exceptional simple Jordan algebras, *Amer. J. Math.* **76** (1954), 447–452.
44. Jacobson, N., Structure of alternative and Jordan bimodules, *Osaka Math. J.* **6** (1954), 1–71.
45. Jacobson, N., Composition algebras and thier automorphisms, *Rend. Circ. Mat. Palermo* **7** (1958), 55–80.
46.* Jacobson, N., "Structure of Rings" (Russian). M., IL, 1961.
47.* Jacobson, N., "Structure and Representations of Jordan Algebras" (Amer. Math. Soc. Colloq. Publ., Vol. 39). Providence, Rhode Island, 1968.
48. Jacobson, N., Structure groups and Lie algebras of symmetric elements of associative algebras with involution, *Adv. Math.* **20** (1976), no. 2, 106–150.
49. Jacobson, N., and McCrimmon, K., Quadratic Jordan algebras of quadratic forms with base points, *J. Indian Math. Soc.* **35** (1971), 1–45.
50.* Divinsky, N., "Rings and Radicals." Toronto, Ontario, Canada, 1965.
51. Divinsky, N., Krempa, J., and Sulinski, A., Strong radical properties of alternative and associative rings, *J. Algebra* **17** (1971), no. 3, 369–388.
52. Divinsky, N., and Sulinski, A. Kurosh radicals of rings with operators, *Canad. J. Math.* **17** (1965), 278–280.
53. Dickson, L. E., On quaternions and their generalizations and the history of eight square theorem. *Ann. of Math.* **20** (1919), 151–171.
54.* Dnestrovskaya Text, "Unsolved Problems in the Theory of Rings and Modules" (Russian). Novosibirsk, 1976.
55. Dorofeev, G. V., An example of a solvable but not nilpotent alternative ring (Russian), *Uspehi Mat. Nauk* **15** (1960), no. 3, 147–150.
56. Dorofeev, G. V., Alternative rings with three generators (Russian), *Sibirsk. Mat. Z.* **4** (1963), no. 5, 1029–1048.
57. Dorofeev, G. V., One example in the theory of alternative rings (Russian), *Sibirsk. Mat. Z.* **4** (1963), no. 5, 1049–1052.
58. Dorofeev, G. V., On nilpotency of right alternative rings (Russian), *Algebra i Logika* **9** (1970), no. 3, 302–305.
59. Dorofeev, G. V., On the locally nilpotent radical of nonassociative rings (Russian), *Algebra i Logika* **10** (1971), no. 4, 355–364.
60. Dorofeev, G. V., Centers of nonassociative rings (Russian), *Algebra i Logika* **12** (1973), no. 5, 530–549.
61. Dorofeev, G. V., On the varieties of generalized standard and generalized accessible algebras (Russian), *Algebra i Logika* **15** (1976), no. 2, 143–167.
62. Dorofeev, G. V., A unification of varieties of algebras (Russian), *Algebra i Logika* **15** (1976), no. 3, 267–291.
63. Dorofeev, G. V., On some properties of a unification of varieties of algebras (Russian), *Algebra i Logika* **16** (1977), no. 1, 24–39.
64. Dorofeev, G. V., and Pchelintsev, S. V., On the varieties of standard and accessible algebras (Russian), *Sibirsk. Mat. Z.* **18** (1977), no. 5, 995–1001.
65. Dorfmeister, J., and Koecher, M., Relative Invarianten und nichtassociative Algebren, *Math. Ann.* **228** (1977), no. 1, 147–186.

66. Zhevlakov, K. A., Solvability of alternative nil-rings (Russian), *Sibirsk. Mat. Z.* **3** (1962), no. 3, 368–377.
67. Zhevlakov, K. A., On radical ideals of alternative rings (Russian), *Algebra i Logika* **4** (1965), no. 4, 87–102.
68. Zhevlakov, K. A., Alternative Artinian rings (Russian), *Algebra i Logika* **5** (1966), no. 3, 11–36; **6** (1967), no. 4, 113–117.
69. Zhevlakov, K. A., Solvability and nilpotency of Jordan rings (Russian), *Algebra i Logika* **5** (1966), no. 3, 37–58.
70. Zhevlakov, K. A., Remarks on simple alternative rings (Russian), *Algebra i Logika* **6** (1967), no. 2, 21–33.
71. Zhevlakov, K. A., Nil-ideals of an alternative ring satisfying the maximal condition (Russian), *Algebra i Logika* **6** (1967), no. 4, 19–26.
72. Zhevlakov, K. A., On the radicals of Kleinfeld and Smiley in alternative rings (Russian), *Algebra i Logika* **8** (1969), no. 2, 176–180.
73. Zhevlakov, K. A., The coincidence of the radicals of Smiley and Kleinfeld in alternative rings (Russian), *Algebra i Logika* **8** (1969), no. 3, 309–319.
74. Zhevlakov, K. A., On radicals and von Neumann ideals (Russian), *Algebra i Logika* **8** (1969), no. 4, 425–439.
75. Zhevlakov, K. A., Quasi-regular ideals in finitely-generated alternative rings (Russian), *Algebra i Logika* **11** (1972), no. 2, 140–161.
76. Zhevlakov, K. A., The radical and representations of alternative rings (Russian), *Algebra i Logika* **11** (1972), no. 2, 162–173.
77. Zhevlakov, K. A., Remarks on locally nilpotent rings with cut-off conditions (Russian), *Mat. Zametki* **12** (1972), no. 2, 121–126.
78. Zhevlakov, K. A., Nilpotency of ideals in alternative rings with minimal condition (Russian), *Trudy Moskov. Mat. Obsc.* **29** (1973), 133–146.
79.* Zhevlakov, K. A., Alternative rings and algebras (Russian), *in* "Mathematical Encyclopedia," Vol. 1, pp. 237–240. M., 1977.
80. Zhevlakov, K. A., and Shestakov, I. P., On local finiteness in the sense of Shirshov (Russian), *Algebra i Logika* **12** (1973), no. 1, 41–73.
81. Zhelyabin, V. N., A theorem on the splitting of the radical for alternative algebras over a Hensel ring (Russian), Theses reports 14th All-Union algebraic conf., part 2, Novosibirsk, 1977, p. 28.
82. Zel'manov, E. I., Jordan algebras with a finiteness condition (Russian), Theses reports 14th All-Union algebraic conf., part 2, Novosibirsk, 1977, p. 31.
83. Jordan, P., von Neumann, J., and Wigner, E., On an algebraic generalization of the quantum mechanical formalism, *Ann. of Math.* **36** (1934), 29–64.
84. Iordanescu, R., and Popovici, I., Sur les representations des algebres de Jordan et leur interpretation geometrique, *Rev. Roumaine Math. Pures Appl.* **13** (1968), no. 3, 399–416.
85. Iordanescu, R., and Popovici, I., On representations of special Jordan algebras, *Rev. Roumaine Math. Pures Appl.* **13** (1968), no. 8, 1089–1100.
86. Camburn, M. E., Local Jordan algebras, *Trans. Amer. Math. Soc.* **202** (1975), 41–50.
87. Kaplansky, I., Semisimple alternative rings, *Portugal. Math.* 10 (1951), 37–50.
88. Kaplansky, I., Infinite-dimensional quadratic forms permitting composition, *Proc. Amer. Math. Soc.* **4** (1953), 956–960.
89. Koecher, M., Imbedding of Jordan algebras into Lie algebras. I and II, *Amer. J. Math.* **89** (1967), no. 3, 787–816; **90** (1968), no. 2, 476–510.
90. Kleinfeld, E., An extension of the theorem on alternative division rings, *Proc. Amer. Math. Soc.* **3** (1952), no. 3, 348–351.
91. Kleinfeld, E., Right alternative rings, *Proc. Amer. Math. Soc.* **4** (1953), 939–944.
92. Kleinfeld, E., Simple alternative rings, *Ann. of Math.* **58** (1953), no. 3, 545–547.

93. Kleinfeld, E., On simple alternative rings, *Amer. J. Math.* **75** (1953), no. 1, 98–104.
94. Kleinfeld, E., Generalization of a theorem on simple alternative rings, *Portugal. Math.* **14** (1955), no. 3–4, 91–94.
95. Kleinfeld, E., Primitive rings and semisimplicity, *Amer. J. Math.* **77** (1955), 725–730.
96. Kleinfeld, E., Standard and accessible rings, *Canad. J. Math.* **8** (1956), 335–340.
97. Kleinfeld, E., Alternative nil rings, *Ann. of Math.* **66** (1957), 395–399.
98.* Kleinfeld, E., On alternative and right alternative algebras, Report of a Conference on Linear Algebras, *Nat. Acad. Sci.–Nat. Res. Council Publ.* **502** (1957), 20–23.
99. Kleinfeld, E., On right alternative rings without proper right ideals, *Pacific J. Math.* **31** (1969), 87–102.
100. Kleinfeld, E., Kleinfeld, M., and Kosier, F., A generalization of commutative and alternative rings, *Canad. J. Math.* **22** (1970), 348–362.
101.* Cohn, P., "Universal Algebra" (Russian). M., Mir, 1968.
102. Krempa, J., Lower radical properties for alternative rings, *Bull. Acad. Polon. Sci. Ser. Sci. Math. Astronom. Phys.* **23** (1975), no. 2, 139–142.
103. Kuz'min, E. N., On the Nagata–Higman theorem (Russian), Trudy, dedicated to the 60th year of the academician L. Iliev, Sofia, 1975, pp. 101–107.
104. Kurosh, A. G., Problems in the theory of rings connected with the Burnside problem on periodic groups (Russian), *Izv. Akad. Nauk SSSR Ser. Mat.* **5** (1941), 233–340.
105. Kurosh, A. G., Radicals of rings and algebras (Russian), *Mat. Sb.* **33** (1953), 13–26.
106. Linnik, Yu. V., A generalization of the theorem of Frobenius and establishment of its connection with the theorem of Hurwitz on composition of quadratic forms (Russian), *Izv. Akad. Nauk SSSR Ser. Mat.* (1938), no. 1, 41–52.
107. Loos, O., Existence and conjugacy of Cartan subalgebras of Jordan algebras, *Proc. Amer. Math. Soc.* **50** (1975), 40–44.
108. Loustau, J. A., Radical extensions of Jordan rings, *J. Algebra* **30** (1974), no. 1–3, 1–11.
109. Loustau, J. A., The structure of algebraic Jordan algebras without nonzero nilpotent elements, *Comm. Algebra* **4** (1976), 1045–1070.
110. L'vov, I. V., On varieties of associative rings. I (Russian), *Algebra i Logika* **12** (1973), no. 3, 269–297.
111. L'vov, I. V., On some finitely-based nonassociative rings (Russian), *Algebra i Logika* **14** (1974), no. 1, 15–27.
112. Lyu Shao-syue, On decomposition of infinite algebras (Russian), *Mat. Sb.* **42** (1957), 327–352.
113. Macdonald, I. G., Jordan algebras with three generators, *Proc. London Math. Soc.* **10** (1960), 395–408.
114. McCrimmon, K., Jordan algebras of degree 1, *Bull. Amer. Math. Soc.* **70** (1964), no. 5, 702.
115. McCrimmon, K., Norms and noncommutative Jordan algebras, *Pacific J. Math.* **15** (1965), no. 3, 925–956.
116. McCrimmon, K., Structure and representations of noncommutative Jordan algebras, *Trans. Amer. Math. Soc.* **121** (1966), no. 1, 187–199.
117. McCrimmon, K., A general theory of Jordan rings, *Proc. Nat. Acad. Sci. USA* **56** (1966), no. 4, 1072–1079.
118. McCrimmon, K., Bimodules for composition algebras, *Proc. Amer. Math. Soc.* **17** (1966), 480–486.
119. McCrimmon, K., A proof of Schafer's conjecture for infinite-dimensional forms admitting composition, *J. Algebra* **5** (1967), no. 1, 72–83.
120. McCrimmon, K., Generically algebraic algebras, *Trans. Amer. Math. Soc.* **127** (1967), 527–551.
121. McCrimmon, K., Macdonald's theorem with inverses, *Pacific J. Math.* **21** (1967), no. 2, 315–325.

122. McCrimmon, K., Jordan algebras with interconnected idempotents, *Proc. Amer. Math. Soc.* **19** (1968), 1327–1336.
123. McCrimmon, K., The Freudenthal–Springer–Tits constructions of exceptional Jordan algebras, *Trans. Amer. Math. Soc.* **139** (1969), 495–510.
124. McCrimmon, K., The radical of a Jordan algebra, *Proc. Nat. Acad. Sci. USA* **62** (1969), 671–678.
125. McCrimmon, K., On Herstein's theorems relating Jordan and associative algebras, *J. Algebra* **13** (1969), 382–392.
126. McCrimmon, K., The Freudenthal–Springer–Tits constructions revisited, *Trans. Amer. Math. Soc.* **148** (1970), 293–314.
127. McCrimmon, K., Koecher's principle for quadratic Jordan algebras, *Proc. Amer. Math. Soc.* **28** (1971), 39–43.
128. McCrimmon, K., A characterization of the radical of a Jordan algebra, *J. Algebra* **18** (1971), 103–111.
129. McCrimmon, K., Speciality of quadratic Jordan algebras, *Pacific J. Math.* **36** (1971), no. 3, 761–773.
130. McCrimmon, K., Homotopes of alternative algebras, *Math. Ann.* **131** (1971), no. 4, 253–262.
131. McCrimmon, K., Noncommutative Jordan rings, *Trans. Amer. Math. Soc.* **158** (1971), no. 1, 1–33.
132. McCrimmon, K., A characterization of the Jacobson–Smiley radical, *J. Algebra* **18** (1971), 565–573.
133. McCrimmon, K., Noncommutative Jordan division algebras, *Trans. Amer. Math. Soc.* **163** (1972), 215–224.
134. McCrimmon, K., Quadratic Jordan algebras whose elements are all regular or nilpotent, *Proc. Amer. Math. Soc.* **45** (1974), no. 1, 1–27.
135.* McCrimmon, K., Quadratic methods in nonassociative algebras, *Proc. Internat. Congr. Math., Vancouver, 1974*, pp. 325–330.
136. McCrimmon, K., Mal'cev's theorem for alternative algebras, *J. Algebra* **28** (1974), no. 3, 484–495.
137. McCrimmon, K., Absolute zero divisors and local nilpotence in alternative algebras, *Proc. Amer. Math. Soc.* **47** (1975), 293–299.
138. McCrimmon, K., Finite-dimensional left Moufang algebras, *Math. Ann.* **224** (1976), no. 2, 179–187.
139. McCrimmon, K., Malcev's theorem for Jordan algebras, *Comm. Algebra* **5** (1977), no. 9, 937–968.
140. McCrimmon, K., Speciality and reflexivity of quadratic Jordan algebras, *Comm. Algebra* **5** (1977), no. 9, 903–936.
141. McCrimmon, K., Axioms for inversion in Jordan algebras, *J. Algebra* **47** (1977), no. 1, 201–222.
142.* Mal'cev, A. I., "Selected Works," Vol. 1 (Russian), Classical Algebra. M., Nauka, 1976.
143. Markovichev, A. S., Nil-systems and the radical in alternative Artinian rings (Russian), *Mat. Zametki* **11** (1972), no. 3, 299–306.
144. Markovichev, A. S., On rings of type (γ, δ) (Russian), Theses reports 14th All-Union algebraic conf., part 2, Novosibirsk, 1977, pp. 55–56.
145. Medvedev, Yu. A., Local finiteness of periodic subloops of an alternative PI-ring (Russian), *Mat. Sb.* **103** (1977), no. 6, 309–315.
146. Medvedev, Yu. A., Finite baseability of varieties with a two-term identity (Russian), Theses reports 14th All-Union algebraic conf., part 2, Novosibirsk, 1977, pp. 59–60.
147. Mikheev, I. M., On an identity in right alternative rings (Russian), *Algebra i Logika* **8** (1969), no. 3, 357–366.

148. Mikheev, I. M., The locally right nilpotent radical in the class of right alternative rings (Russian), *Algebra i Logika* **11** (1972), no. 2, 174–185.
149. Mikheev, I. M., The theorem of Wedderburn on the splitting of the radical for $(-1, 1)$ rings (Russian), *Algebra i Logika* **12** (1974), no. 3, 298–304.
150. Mikheev, I. M., On prime right alternative rings (Russian), *Algebra i Logika* **14** (1975), no. 1, 56–60.
151. Mikheev, I. M., On right nilpotency in right alternative rings (Russian), *Sibirsk. Mat. Z.* **17** (1976), no. 1, 225–227.
152. Mikheev, I. M., On simple right alternative rings (Russian), Theses reports 14th All-Union algebraic conf., part 2, Novosibirsk, 1977, p. 61.
153. Montgomery, S., Chain conditions on symmetric elements, *Proc. Amer. Math. Soc.* **46** (1974), no. 3, 325–331.
154. Montgomery, S., Rings of quotients for a class of special Jordan algebras, *J. Algebra* **31** (1974), no. 1, 154–165.
155. Morgan, D. L., Jordan algebras with minimum condition, *Trans. Amer. Math. Soc.* **155** (1971), 161–173.
156. Moufang, R., Zur Struktur von Alternativ Korpern, *Math. Ann.* **110** (1935), 416–430.
157. Nagata, M., On the nilpotency of nil-algebras, *J. Math. Soc. Japan* **4** (1952), 296–301.
158. Niewieczerzal, D., and Terlikowska, B., A note on alternative semiprime rings, *Bull. Acad. Polon. Sci. Ser. Sci. Math. Astronom. Phys.* **20** (1972), no. 4, 265–268.
159. Nikitin, A. A., On hyper-nilpotent radicals of $(-1, 1)$ rings (Russian), *Algebra i Logika* **12** (1973), no. 3, 305–311.
160. Nikitin, A. A., Almost alternative algebras (Russian), *Algebra i Logika* **13** (1974), no. 5, 501–533.
161. Nikitin, A. A., On ideal heredity of radicals in Jordan rings (Russian), Theses reports 14th All-Union algebraic conf., part 2, Novosibirsk, 1977, pp. 61–62.
162. Osborn, J. M., Jordan algebras of capacity two, *Proc. Nat. Acad. Sci. USA* **57** (1967), 582–588.
163. Osborn, J. M., Jordan and associative rings with nilpotent and invertible elements, *J. Algebra* **15** (1970), no. 3, 301–308.
164.* Osborn, J. M., Varieties of algebras, *Adv. Math.* **8** (1972), 163–369.
165. Osborn, J. M., Representations and radicals of Jordan algebras, *Scripta Math.* **29** (1973), no. 3–4, 297–329.
166. Oslowski, B. J., A note on alternative antisimple rings with finiteness condition, *Bull. Acad. Polon. Sci. Ser. Sci. Math. Astronom. Phys.* **23** (1975), no. 12, 1241–1245.
167. Patil, K. B., and Racine, M. L., Central polynomials for Jordan algebras, II, *J. Algebra* **41** (1976), no. 1, 238–241.
168. Petersson, H. P., Borel subalgebras of alternative and Jordan algebras, *J. Algebra* **16** (1970), no. 4, 541–560.
169. Polin, S. V., On identities of finite algebras (Russian), *Sibirsk. Mat. Z.* **17** (1976), no. 6, 1356–1366.
170.* Procesi, C., "Rings with Polynomial Identities." Dekker, New York, 1973.
171. Pchelintsev, S. V., The free $(-1, 1)$ algebra with two generators (Russian), *Algebra i Logika* **13** (1974), no. 4, 425–449.
172. Pchelintsev, S. V., Nilpotency of the associator ideal of a free finitely-generated $(-1, 1)$ ring (Russian), *Algebra i Logika* **14** (1975), no. 5, 543–571.
173. Pchelintsev, S. V., The defining identities of one variety of right alternative algebras (Russian), *Mat. Zametki* **20** (1976), no. 2, 161–176.
174. Pchelintsev, S. V., On the locally nilpotent radical in some classes of right alternative rings (Russian), *Sibirsk. Mat. Z.* **17** (1976), no. 2, 340–360.

175. Razmyslov, Yu. P., Identities with a trace of complete matrix algebras over a field of characteristic 0 (Russian), *Izv. Akad. NaukSSSR Ser. Mat.* **38** (1974), no. 4, 723–756.

176.* Racine, M. L., "The Arithmetics of Quadratic Jordan Algebras" (Mem. Amer. Math. Soc., No. 136). Providence, Rhode Island, 1973.

177. Racine, M. L., On maximal subalgebras, *J. Algrbra* **30** (1974), no. 1–3, 155–180.

178. Racine, M. L., Central polynomials for Jordan algebras. I, *J. Algebra* **41** (1976), no. 1, 224–237.

179. Racine, M. L., Maximal subalgebras of exceptional Jordan algebras, *J. Algebra* **46** (1977), no. 1, 12–21.

180. Racine, M. L., Point spaces in exceptional quadratic Jordan algebras, *J. Algebra* **46** (1977), no. 1, 22–36.

181. Robbins, D. P., Jordan elements in a free alternative algebra, *J. Algebra* **19** (1971), no. 3, 354–374.

182. Roomel'di, R. E., The lower nil-radical of $(-1, 1)$ rings (Russian), *Algebra i Logika* **12** (1973), no. 3, 323–332.

183. Roomel'di, R. E., Nilpotency of ideals in $(-1, 1)$ rings with minimal condition (Russian), *Algebra i Logika* **12** (1973), no. 3, 333–348.

184. Roomel'di, R. E., Solvability of $(-1, 1)$ nil-rings (Russian), *Algebra i Logika* **12** (1973), no. 4, 478–489.

185. Roomel'di, R. E., Centers of a free $(-1, 1)$ ring (Russian), *Sibirsk. Mat. Z.* **18** (1977), no. 4, 861–876.

186. Ryabukhin, Yu. M., On one class of locally nilpotent rings (Russian), *Algebra i Logika* **7** (1968), no. 5, 100–108.

187. Ryabukhin, Yu. M., Algebras without nilpotent elements. I and II (Russian), *Algebra i Logika* **8** (1969), no. 2, 181–214; 215–240.

188. Ryabukhin, Yu. M., Incomparable nil-radicals and non-special hyper-nilpotent radicals (Russian), *Algebra i Logika* **14** (1975), no. 1, 86–99.

189. Ryabukhin, Yu. M., On countably generated locally $\mathfrak{M}$-algebras (Russian), *Izv. Akad. Nauk SSSR* **40** (1976), no. 6, 1203–1223.

190. Skornyakov, L. A., Alternative division rings (Russian), *Ukrain. Mat. Z.* **2** (1950), no. 1, 70–85.

191. Skornyakov, L. A., Right alternative division rings (Russian), *Izv. Akad. Nauk SSSR Ser. Mat.* **15** (1951), 177–184.

192.* Skornyakov, L. A., Projective planes (Russian), *Uspehi Mat. Nauk* **4** (1951), no. 6, 112–154.

193.* Skornyakov, L. A., Alternative rings, *Rend. Mat.* **24** (1965), no. 3–4, 360–372.

194. Skosyrskiy, V. G., On nilpotency in Jordan and right alternative algebras (Russian), Theses reports 14th All-Union algebraic conf., part 2, Novosibirsk, 1977, pp. 68–69.

195. Scribner, D. R., Lie admissible, nodal, noncommutative Jordan algebras, *Trans. Amer. Math. Soc.* **154** (1971), 105–111.

196. Scribner, D. R., Infinite nodal noncommutative Jordan algebras, differentially simple algebras, *Trans. Amer. Math. Soc.* **156** (1971), 381–389.

197. Slater, M., Nucleus and center in alternative rings, *J. Algebra* **7** (1967), no. 3, 372–388.

198. Slater, M., Ideals in semiprime alternative rings, *J. Algebra* **8** (1968), no. 1, 60–76.

199. Slater, M., The open case for simple alternative rings, *Proc. Amer. Math. Soc.* **19** (1968), 712–715.

200. Slater, M., Alternative rings with D.C.C., I, *J. Algebra* **11** (1969), 102–110.

201. Slater, M., The socle of an alternative ring, *J. Algebra* **14** (1970), 443–463.

202. Slater, M., Alternative rings with D.C.C., II, *J. Algebra* **14** (1970), 464–484.

203. Slater, M., Prime alternative rings. I and II, *J. Algebra* **15** (1970), 229–243; 244–251.

204. Slater, M., Alternative rings with D.C.C., III, *J. Algebra* **18** (1971), 179–200.
205. Slater, M., Prime alternative rings. III, *J. Algebra* **21** (1972), no. 3, 394–409.
206. Slater, M., Free alternative rings, *Notices Amer. Math. Soc.* **21** (1974), no. 5, A–480.
207. Slin'ko, A. M., On the equivalence of certain nilpotencies in right alternative rings (Russian), *Algebra i Logika* **9** (1970), no. 3, 342–348.
208. Slin'ko, A. M., On radicals of Jordan rings (Russian), *Algebra i Logika* **11** (1972), no. 2, 206–215.
209. Slin'ko, A. M., On the Jacobson radical and absolute zero divisors in special Jordan algebras (Russian), *Algebra i Logika* **11** (1972), no. 6, 711–723.
210. Slin'ko, A. M., A remark on radicals and derivations of rings (Russian), *Sibirsk. Mat. Z.* **13** (1972), no. 6, 1395–1397.
211. Slin'ko, A. M., Radicals of Jordan rings that are attached to alternative ones (Russian), *Mat. Zametki* **16** (1974), no. 1, 135–140.
212. Slin'ko, A. M., Jordan algebras without nilpotent elements which satisfy finiteness conditions (Russian), *Mat. Issled.* **38** (1976), Kishinev, "Shtiintsa," 170–176.
213. Slin'ko, A. M., The lower nil-radical in Jordan algebras with maximal condition (Russian), *Algebra i Logika* **16** (1977), no. 1, 98–100.
214. Slin'ko, A. M., and Shestakov, I. P., Right representations of algebras (Russian), *Algebra i Logika* **13** (1974), no. 5, 544–587.
215. Smiley, M. F., The radical of an alternative ring, *Ann. of Math.* **49** (1948), no. 3, 702–709.
216.* Smiley, M. F., Some questions concerning alternative rings, *Bull. Amer. Math. Soc.* **57** (1951), 36–43.
217. Smiley, M. F., Kleinfeld's proof of the Bruck–Kleinfeld–Skornyakov theorem, *Math. Ann.* **134** (1957), no. 1, 53–57.
218. Smiley, M. F., Jordan homomorphisms and right alternative rings, *Proc. Amer. Math. Soc.* **8** (1957), 668–671.
219. Smith, B. D., A standard Jordan polynomial, *Comm. Algebra* **5** (1977), no. 2, 207–218.
220. Smith, K. C., Noncommutative Jordan algebras of capacity two, *Trans. Amer. Math. Soc.* **158** (1971), 151–159.
221. Smith, K. C., Extending Jordan ideals and Jordan homomorphisms of symmetric elements in a ring with involution, *Canad. J. Math.* **24** (1972), no. 1, 50–59.
222. Taft, E. J., Invariant Wedderburn factors, *Illinois J. Math.* **1** (1957), 565–573.
223. Taft, E. J., The Whitehead first lemma for alternative algebras, *Proc. Amer. Math. Soc.* **8** (1957), 950–956.
224. Taft, E. J., Invariant splitting in Jordan and alternative algebras, *Pacific J. Math.* **15** (1965), no. 4, 1421–1427.
225. Taft, E. J., On the Whitehead first lemma for Jordan algebras, *Math. Z.* **107** (1968), no. 2, 83–86.
226. Tits, J., Une classe d'algebres de Lie en relation avec les algèbres de Jordan, *Nederl. Akad. Wetensch. Proc. Ser. A* **65**; *Indag. Math.* **24** (1962), 530–535.
227. Tits, J., Algèbres alternatives, algèbres de Jordan et algèbres de Lie exceptionnelles, I, Construction, *Proc. Koninkl. Nederl. Akad. Wetensch.* **A69** (1966), no. 2, 223–237.
228.* Topping, D. M., "Jordan Algebras of Self-adjoint Operators" (Mem. Amer. Math. Soc., No. 53). Providence, Rhode Island, 1965.
229. Thedy, A., Mutationen und polarisierte Fundamentalformul, *Math. Ann.* **177** (1968), 235–246.
230. Thedy, A., Zum Wedderburnschen Zerlegungsatz, *Math. Z.* **113** (1970), 173–195.
231. Thedy, A., On rings with completely alternative commutators, *Amer. J. Math.* **93** (1971), 42–51.
232. Thedy, A., Right alternative rings, *J. Algebra* **37** (1975), no. 1, 1–43.

233. Thedy, A., Right alternative rings with Peirce decomposition, *J. Algebra* **37** (1975), no. 1, 44–63.
234. Thedy, A., Right alternative rings with minimal condition, *Math. Z.* **155** (1977), no. 3, 277–286.
235.* Faulkner, J. R., "Octonion Planes Defined by Quadratic Jordan Algebras" (Mem. Amer. Math. Soc., No. 104). Providence, Rhode Island, 1970.
236. Faulkner, J. R., Orbits of the automorphism group of the exceptional Jordan algebra, *Trans. Amer. Math. Soc.* **151** (1970), 433–441.
237. Foster, D. M., On Cartan subalgebras of alternative algebras, *Trans. Amer. Math. Soc.* **162** (1971), 225–238.
238. Foster, D. M., Generalizations of nilpotence and solvability in universal classes of algebras, *J. Algebra* **26** (1973), 536–555.
239.* Freudenthal, H., "Oktaven, Ausnahmegruppen and Oktaven-geometrie." Utrecht, 1951 (Russian transl. *J. "Matematika"* **1** (1957), no. 1, 117–154).
240. Helwig, K. H., Über Mutationen von Jordan-Algebren, *Math. Z.* **103** (1968), no. 1, 1–7.
241. Helwig, K. H., Halbeinfache reele Jordan-Algebren, *Math. Z.* **109** (1969), no. 1, 1–28.
242. Helwig, K. H., Involutionen von Jordan-Algebren, *Manuscripta Math.* **1** (1969), no. 3, 211–229.
243. Helwig, K. H., Jordan-Algebren und symmetrische Räume. I, *Math. Z.* **115** (1970), 315–349.
244. Helwig, K. H., and Hirzebruch, U., Uber reguläre Jordan-Algebren und eine Aquivalenz-Relation von M. Koecher, *Proc. Koninkl. Nederl. Akad. Wetensch.* **A71** (1968), no. 5, 460–465; *Indag. Math.* **30** (1968), no. 5, 460–465.
245. Hentzel, I. R., Right alternative rings with idempotent, *J. Algebra* **17** (1971), no. 3, 303–309.
246. Hentzel, I. R., Nil semi-simple $(-1, 1)$ rings, *J. Algebra* **22** (1972), no. 3, 442–450.
247. Hentzel, I. R., The characterization of $(-1, 1)$ rings, *J. Algebra* **30** (1974), no. 1–3, 236–258.
248. Hentzel, I. R., and Cattaneo, G. M., Simple (γ, δ)-algebras are associative, *J. Algebra* **47** (1977), no. 1, 52–76.
249. Hentzel, I. R., and Slater, M., On the Andrunakievich lemma for alternative rings, *J. Algebra* **27** (1973), 243–256.
250. Herstein, I. N., Sugli anneli semplici alternative, *Rend. Mat. Appl.* **23** (1964), 9–13.
251. Herstein, I. H., Anneli alternativi ed algebre di compozisione, *Rend. Math. Appl.* **23** (1964), 364–393.
252.* Herstein, I. N., "Noncommutative Rings" (Russian). M., Mir, 1972.
253. Higman, G., On a conjecture of Nagata, *Proc. Cambridge Philos. Soc.* **52** (1956), 1–4.
254. Holgate, P., Jordan algebras arising in population genetics, *Proc. Edinburgh Math. Soc.* **15** (1967), no. 4, 291–294.
255. Humm, M., On a class of right alternative rings without nilpotent elements, *J. Algebra* **5** (1967), no. 2, 164–175.
256. Humm, M., and Kleinfeld, E., On free alternative rings, *J. Combin. Theory* **2** (1967), no. 2, 140–144.
257. Tsai, C. I., The Levitzki radical in Jordan rings, *Proc. Amer. Math. Soc.* **24** (1970), 119–123.
258. Zorn, M., Theorie der alternativen Rings, *Abh. Math. Sem. Univ. Hamburg* **8** (1930), 123–147.
259. Zorn, M., Alternative rings and related questions. I. Existence of the radical, *Ann. of Math.* **42** (1941), 676–686.
260. Schafer, R. D., On forms of degree n permitting composition, *J. Math. Mech.* **12** (1963), 777–792.

261.* Schafer, R. D., "An Introduction to Nonassociative Algebras." Academic Press, New York, 1966.

262. Schafer, R. D., Generalized standard algebras, *J. Algebra* **12** (1969), no. 3, 376–417.

263. Schafer, R. D., Forms permitting composition, *Adv. Math.* **4** (1970), 127–148.

264. Schafer, R. D., A coordinatization theorem for commutative power-associative algebras, *Scripta Math.* **29** (1973), no. 3–4, 437–442.

265. Shelipov, A. N., Some properties of the nucleus of an alternative ring (Russian), *Mat. Issled.* **28** (1973), Kishinev, "Shtiintsa", 183–187.

266. Shestakov, I. P., Finite-dimensional algebras with a nil-basis (Russian), *Algebra i Logika* **10** (1971), no. 1, 87–99.

267. Shestakov, I. P., On some classes of noncommutative Jordan rings (Russian), *Algebra i Logika* **10** (1971), no. 4, 407–448.

268. Shestakov, I. P., Generalized standard rings (Russian), *Algebra i Logika* **13** (1974), no. 1, 88–103.

269. Shestakov, I. P., Radicals and nilpotent elements of free alternative algebras (Russian), *Algebra i Logika* **14** (1975), no. 3, 354–365.

270. Shestakov, I. P., Centers of alternative algebras (Russian), *Algebra i Logika* **15** (1976), no. 3, 343–362.

271. Shestakov, I. P., Absolute zero divisors and radicals of finitely-generated alternative algebras (Russian), *Algebra i Logika* **15** (1976), no. 5, 585–602.

272. Shestakov, I. P., On a problem of Shirshov's (Russian), *Algebra i Logika* **16** (1977), no. 2, 227–246.

273. Shestakov, I. P., Zhevlakov's radical and representations of alternative algebras (Russian), Theses reports 14th All-Union algebraic conf., part 2, Novosibirsk, 1977, pp. 83–84.

274. Shirshov, A. I., On special *J*-rings (Russian), *Mat. Sb.* **38** (1956), no. 2, 149–166.

275. Shirshov, A. I., Some theorems on imbedding for rings (Russian), *Mat. Sb.* **40** (82) (1956), 65–72.

276. Shirshov, A. I., On some nonassociative nil-rings and algebraic algebras (Russian), *Mat. Sb.* **41** (83) (1957), 381–394.

277. Shirshov, A. I., On rings with polynomial identities (Russian), *Mat. Sb.* **43** (1957), no. 2, 277–283.

278. Shirshov, A. I., On free Lie rings (Russian), *Mat. Sb.* **45** (1958), no. 2, 113–122.

279.* Shirshov, A. I., Some questions in the theory of rings that are nearly associative (Russian), *Uspehi Mat. Nauk* **13** (1958), no. 6, 3–20.

280. Springer, T. A., "Jordan Algebras and Algebraic Groups" (Ergebnisse Math. 75). Springer-Verlag, Berlin, 1973.

281. Strade, H., Nodal nichtkommutative Jordan-Algebren und Lie-Algebren bei charakteristic $p > 2$, *J. Algebra* **21** (1972), no. 3, 353–377.

282. Eilenberg, S., Extensions of algebras, *Ann. Polon. Math.* **21** (1948), 125–134.

283. Erickson, T. S., Martindale, W. S., and Osborn, J. M., Prime nonassociative algebras, *Pacific J. Math.* **60** (1975), no. 1, 49–63.

284. Erickson, T. S., and Montgomery, S., The prime radical in special Jordan rings, *Trans. Amer. Math. Soc.* **156** (1971), 155–164.

List of Notations

$V[X]$	Set of nonassociative words from elements of the set X, 2
$d(v)$	Length of the word v, 2
$\Phi[X]$	Free algebra over the ring Φ from set of generators X, 3
$T(A)$	Ideal of identities of the algebra A, 4
$T(\mathfrak{M})$	Ideal of identities of the variety $\mathfrak{M}$, 4
$I(A)$	4
$\Phi_{\mathfrak{M}}[X]$	Free algebra of the variety $\mathfrak{M}$ from the set of generators X, 5
$\mathrm{Ass}[X]$	Free associative algebra, 5
$\mathrm{Alt}[X]$	Free alternative algebra, 5
$\mathrm{RA}[X]$	Free right alternative algebra, 5
$\mathrm{J}[X]$	Free Jordan algebra, 5
$\Delta(y)$	9
$\Delta_i^k(y)$	9
$f_i^{(k)}(x_1, \ldots, x_n; x_j)$	Partial linearization of the polynomial f in x_i of degree k, 11
$f\Delta$	11
L_i^k	14
$A^{\#}$	Algebra obtained by the formal adjoining of an identity element to the algebra A, 17
Δ'	18
R_a, (L_a)	Operator of right (left) multiplication by the element a, 25, 66
(A, α)	Algebra obtained by means of the Cayley–Dickson process, 28
$K(\mu)$	30
$Q(\mu,\beta)$	Algebra of generalized quaternions, 30
$C(\mu,\beta,\gamma)$	Cayley–Dickson algebra, 31

$\mathrm{Ann}_r(X)$	Right annihilator of the set X, 168
$\mathrm{Ann}_l(X)$	Left annihilator of the set X, 168
$\mathrm{Ann}(X)$	Annihilator of the set X, 168
$D(A)$	Associator ideal of the algebra A, 170
$U(A)$	Associative nucleus of the algebra A, 171
$\mathscr{A}$	Andrunakievich radical or antisimple radical, 173
$\mathscr{A}(A)$	Andrunakievich radical of the algebra A, 174
$(I:A)$	177
$\check{I}$	Largest ideal contained in the set I, 177
$S^{-1}R$	185
$\mathscr{K}(A)$	Kleinfeld radical of the algebra A, 202
$\mathscr{S}(A)$	Smiley radical of the algebra A, 207
$\mathscr{J}(A)$	Quasi-regular radical of the algebra A, 210, 310
$r_1^{\mathrm{reg}}(X)$	211
$T(V)$	Tensor algebra of the module V, 218
$T(\varphi)$	218
$\mathrm{Ker}_\rho(A)$, $K_\rho(A)$	219
$I_{\mathfrak{M}}[A]$	219
$\mathscr{R}_{\mathfrak{M}}(A)$	Universal algebra for right $\mathfrak{M}$-representations of the algebra A, 220
$R^*_{\mathfrak{M}}(A)$	Universal algebra of right multiplications for the algebra A in the variety $\mathfrak{M}$, 224
$[m,a,b]$	228
$\xi(M)$	Centralizer of the module M, 237
$(X,Y)_k$	243
$r_b(\mathfrak{M})$	Base rank of the variety $\mathfrak{M}$, 258
$S^n_{n_1,\ldots,n_k}$	259
$V^n_{n_1,\ldots,n_k}$	259
$K[X]$	272
ZN_{Alt}	283
U_{Alt}	283
$N_f^{(1)}(A)$	285
$\mathscr{Z}(J)$	297
$\mathscr{M}(J)$	McCrimmon radical of the algebra J, 298
$x \cdot_a y$	305
$J^{(a)}$	305
$\mathrm{PQI}(J)$	312
$V_{y,z}$	312
$R_x^{(a)}$, $U_x^{(a)}$, $V_{x,y}^{(a)}$	312
$T_{x,y}$	312
$j_{\mathrm{RA}}[X]$	Set of right alternative j-polynomials, 348

Subject Index

C

D

E

F

H

I

J

K

L

M

N

O